Outstanding Contributions to Logic

Volume 21

Outstanding Contributions to Logic puts focus on important advances in modern logical research. Each volume is devoted to a major contribution by an eminent logician. The series will cover contributions to logic broadly conceived, including philosophical and mathematical logic, logic in computer science, and the application of logic in linguistics, economics, psychology, and other specialized areas of study.

A typical volume of Outstanding Contributions to Logic contains:

- A short scientific autobiography by the logician to whom the volume is devoted
- The volume editor's introduction. This is a survey that puts the logician's contributions in context, discusses its importance and shows how it connects with related work by other scholars
- The main part of the book will consist of a series of chapters by different scholars that analyze, develop or constructively criticize the logician's work
- Response to the comments, by the logician to whom the volume is devoted
- A bibliography of the logician's publications Outstanding Contributions to Logic is published by Springer as part of the Studia Logica Library.

This book series, is also a sister series to Trends in Logic and Logic in Asia: Studia Logica Library. All books are published simultaneously in print and online. This book series is indexed in SCOPUS.

Proposals for new volumes are welcome. They should be sent to the editor-in-chief sven-ove.hansson@abe.kth.se

More information about this series at http://www.springer.com/series/10033

Ofer Arieli · Anna Zamansky
Editors

Arnon Avron on Semantics and Proof Theory of Non-Classical Logics

Springer

Editors
Ofer Arieli
School of Computer Science
Academic College of Tel Aviv-Yafo
Tel Aviv, Israel

Anna Zamansky
Information Systems Department
University of Haifa
Haifa, Israel

ISSN 2211-2758 ISSN 2211-2766 (electronic)
Outstanding Contributions to Logic
ISBN 978-3-030-71257-0 ISBN 978-3-030-71258-7 (eBook)
https://doi.org/10.1007/978-3-030-71258-7

This Springer imprint is published by the registered company Springer Nature Switzerland AG
The registered company address is: Gewerbestrasse 11, 6330 Cham, Switzerland

Preface

It is our honor and great pleasure to introduce this volume of *Outstanding Contributions to Logic*, honoring Arnon Avron's work on semantics and proof theory of non-classical logics.

Arnon Avron is a faculty member at the School of Computer Science, Tel Aviv University, since 1988. His research interests are very broad, spanning over proof theory, automated reasoning, non-classical logics, foundations of mathematics, and applications of logic in computer science and artificial intelligence. His foundational and pioneering contributions have been widely acknowledged and adopted by the scientific community. This was reflected in an international workshop celebrating his 60th birthday, held on November 2012 in Tel Aviv University, and followed by a special issue of *The Journal of Logic and Computation* (Volume 26, Number 1), published on February 2016.

This volume is another appreciation of Arnon Avron's seminal work over the years. It contains contributions of worldwide leading experts in semantic and proof-theoretical aspects of computer science logic. We are grateful to the authors for their positive response to our invitations as well as their cooperation in preparing inspiring papers in rather limited timeframes. Each submission has gone through a single-blinded peer-refereeing process by at least two reviewers. It is our pleasant duty to cordially thank all those who have acted as reviewers of the manuscripts submitted to this volume:

Leila Amgoud	Marcello D'Agostino	Edwin Mares
Michał Baczyński	J. Michael Dunn	Hitoshi Omori
Paolo Baldi	Melvin Fitting	Graham Priest
Libor Běhounek	Andrea Iacona	Ricardo Oscar Rodriguez
Katalin Bimbó	Rosalie Iemhoff	Yaroslav Shramko
Carlos Caleiro	Norihiro Kamide	Heinrich Wansing
Federico Cerutti	Paolo Maffezioli	Yoni Zohar
Liron Cohen	Sérgio Marcelino	

We hope that this book will be useful for scholars who are interested in the foundations of non-classical logics. This is an outcome of an initiative by Heinrich Wansing, who kindly invited us to be this volume's editors. We would like to thank the series editor, Sven-Ove Hansson, for his valuable assistance during the preparation of this book. We also acknowledge with gratitude the financial support by the Israel Science Foundation (grant numbers 817/15 and 550/19).

Tel Aviv, Israel
Haifa, Israel
July 2020

Ofer Arieli
Anna Zamansky

An Uncertain Road to Certainty

Early Years: 1952–1970

The first error connected with my life was the date of my birth. I was planned to be born on the day of Lenin's Russian revolution that was supposed to be the beginning of an era of freedom and justice to all men and women all over the world. Unfortunately, I had disappointed my parents even before the revolution itself did so, and came to the world 3 days later, on November 10, 1952. I have been the youngest child in the family, having two sisters, 10 years and 5 years, respectively, older than me. In addition to being a boy, the first thing my parents noted about me after my birth was that unlike my sisters, I was a redhead—and being a redhead remained (and still is) one of my main characteristics, even though I lost most of my hair many years ago.

My parents have by far been the most dominant and influential persons in my life. My father worked at the small harbor of Tel Aviv (and has always been the leader of the workers there). This is why I grew up in the area of Tel Aviv that is still called "the harbor workers' neighborhood" (although there is no active harbor in Tel Aviv anymore, and all the harbor's workers that were living in that neighborhood are long dead). My mother was a housewife until I was about 9 years old. Then she learned librarianship, and not long after that she became the legendary librarian of one of the most prestigious (at least at that time) high schools in Tel Aviv. Objectively (and certainly from today's point of view), we were poor (at least until my elder sister left home, and my mother started to work and earn money too). However, we did not feel so, and my parents always found the money for what they considered as important. This did not include pocket money for us, but did include everything connected with culture, and especially books. They have never saved money on *that*, and so we had a huge and versatile collection of books. (Many of these books are still in my own private library.) Therefore, although neither of my parents had formal education beyond few years in high school in the countries in which they were born (my father at Belarus; my mother at Poland), their informal education was greater than most people I have met in my life. They were also very ideological, and my basic values and beliefs are still those that were installed in me by their education. (In the daily newsletter we were reading at home, immediately below its title, these values were summarized every day as follows: Zionism; socialism; comradeship among nations.) Another crucial feature that I think I got from my parents is the urge to fight for my values and for what I believe is right. This short scientific autobiography does not deal with those aspects and events in my life that are directly connected with those values and that urge. Still, it is worth noting that I think that they have indirectly influenced also my mathematical career and research, especially in seeking (and even willing to fight for) absoluteness: absolute certainty and absolute rigor.

My parents were very proud members of the working class. Nevertheless, they strongly did not want that their children will also be workers like them. Accordingly, when it came to us, learning and studying were by far their top priority. Not being good pupils was simply not an option for us. Luckily for my parents, we were all able to be *very* good pupils, and in my case—even an excellent one, the best in class in most theoretical topics. (In gymnastics, craft, etc. I was horrible...) Strangely, at the first half of my 8 years in primary school, I was not the best in class in mathematics. I liked the humanities much more, and my big dream as a child was to be a writer when I grow up. These tendencies still did not change at the last 2 years of the primary school, in which I unquestionably became the best in class in mathematics too.

The big change in my attitude to mathematics came in my first year (out of four) in high school. We had at that year a very good teacher, and with him we started to learn Euclidean geometry, with its theorems, proofs, and constructions. I simply fell in love with geometry then (and I still have a great interest in it). I was enchanted by the realization that interesting geometrical facts can actually be *proved* from some obviously true, simple axioms, and I found very great pleasure in solving difficult geometrical problems. Thus, I devoted the whole of the Passover vacation of that year (almost 3 weeks) to a problem that was given to us as a challenge by our teacher: to show that a triangle in which two angle bisectors are equal is necessarily isosceles. I still view my success in solving it at that time as one of the greatest mathematical achievements in my life. (The fact that my big competitor in class failed to solve that problem gave me extra satisfaction, of course. That competitor too is now a professor in Tel Aviv University, but in economics.) The love of geometry made me interested also in other branches of mathematics. So during high school I started to read books in mathematics whose content was well beyond what we were learning in class. I was not able to understand at that time everything I read, but I understood enough to become even more fascinated with mathematics. It became my favorite subject, and so I decided quite early that it would be what I would study at the university.

University and Army Years: 1970–1978

The title of this section might be confusing, because almost all my years since 1970 (when I finished high school and started to learn mathematics at Tel Aviv university) can be described as "university years". But what I mean by my "university years" is the 5 years in which I was only a student of mathematics, without any teaching tasks.

Here I should explain, first of all, that most Israelis do not go to university or college at the age of 18. They go to the army for 3 years instead (if they are male; female serve less.) However, I was accepted by the army to a special program called "academic reserve". In other words, I was a student with call-up deferred until finishing B.Sc., and in my case (since I got it with distinction)—even until finishing M.Sc. That meant that I spent in a boot camp a great part of the summer vacation between my first and second years as a student. Otherwise I was living the usual life of an Israeli male student at the years before and after the 1973 war. (This includes being called from time to time for a short period of army service.)

As a student, I discovered rather quickly that I am unable to follow the teachers in class. When I was still trying to understand what is written on one side of the blackboard, the teacher has already been writing on the other side. So I usually gave up going to lectures, and instead learned from books, and from lecture notes taken by students I knew. I was very successful at that, and so I got the highest possible grade in almost all the courses I took. However, already at that stage there began to be some gaps between the knowledge I showed at the examinations, and the feeling of "cheating" that I really felt about some of the proofs we were learning. A few years later I understood that all the proofs that I had found suspicious either include implicit applications of the axiom of choice, or introduce sets by using impredicative definitions. I should emphasize that nobody has told me at that time that there might be something problematic about such proofs. I also knew then nothing about the historical debates concerning them. I simply felt uneasy about them, but said (at that time) nothing about it to anybody. The result was that I began to be more and more interested in logic and foundations. However, at my first 2 years as a student of mathematics I knew nobody I could ask about these topics. Luckily for me, this state changed at my third (and last) year as a B.Sc. student. At that year a professional logician joined the department of mathematics of Tel Aviv University for the first time: Yoram Hirshfeld. So at that year I had the opportunity to take courses and seminars by Yoram on mathematical logic, set theory, computability, and model theory.

Yoram was a good teacher, and he was also open to talk about things. His courses and our private discussions made it clear to me that logic and foundations are going to be my mathematical subjects. The unavoidable conclusion was that I should do my M.Sc. thesis at the Hebrew University at Jerusalem, whose department of mathematics had then (as I learned from Yoram) one of the strongest group of logicians in the world. It included Michael Rabin, Azriel Levy, Saharon Shelah, and Haim Gaifman. (M. Megidor came a few years later.) In my first year as an M.Sc. student, I took courses and had conversations with all of them, and saw that of them Gaifman was the most philosophically inclined, and is the one to whom I was closer in spirit and interest. So in my second year at Jerusalem I did my M.Sc. thesis under him. The subject of that thesis (which has never been published) was progressions of arithmetical theories. It was strongly connected with Feferman's famous work on this subject, and this was the first time I heard Feferman's name and studied two of his classical papers.

Although I was an M.Sc. student in Jerusalem, I still spent most of my time at that period in Tel Aviv. (In Jerusalem I spent at most 2 days, including one night, each week.) I had good reasons for that: studying mathematics and working on my M.Sc. thesis occupied only a part of my time at that two crucial years in my life. Thus, I was working as a teaching assistant (who checks assignments of students) in both Tel Aviv and Jerusalem; I gave a lot of private lessons, and I was very involved in political activity (as a leftist, of course). However, what was most important of all at those 2 years were the time and energy which I devoted to what is in the center of life of most young men at their early 20s. In my case, this type of activity had ended already before I finished my M.Sc. thesis: At the beginning of June 1975 I married my wife, Tsipi, whom I met for the first time about 2 years before. I was 22.5 years old then, and she was (still is...) 2 years younger than me. In the 5 months that followed our marriage I finished (with great hurry) my M.Sc. thesis; we bought (with the help of our parents, of course) an apartment in Petah Tiqva (a town near Tel Aviv), and we moved there a month before I started my full 3 years of military service.

Despite having an M.Sc. in mathematics when I joined it, my abilities and knowledge in mathematics were not used by the army. This was in sharp contrast to what happened with most of those who were in the "academic reserve" with me. The reason was almost certainly my political activity at my university days, together with the political background of my parents. As a result, I was just waiting for my service to end, hating almost every moment of it (even though I was not a combat soldier either). However, I did find enough free time at that period to expand my knowledge in logic, and in particular to learn that branch of it that I had never learned at the university: Proof theory. In addition, I started also to study Philosophy on my own, since I reached at that time the conclusion (which is now even more valid than it was then) that adequate philosophy is the most important thing for humanity in the crazy times in which we live.

Two very important events in my life took place at my last year at the army. My father died at the beginning of that year. About 8 months later my first child, my son Haim (which is named after him, and is now a Professor of Mathematics at Tel Aviv University himself) was born. This happened exactly 1 month before my return to civil life. I was exactly 26 years old then.

Ph.D. Student 1978–1984

If you wonder why it took me 6 years to finish my Ph.D., the answer is that only before the end of those 6 years the head of the school of mathematics called me to tell me that if I do not submit my thesis within 4 months, I would not be able to work as an instructor in the school anymore. So I had no choice, but to sit and write down all the research I did over those years (which was sufficient for two Ph.D. theses)—and then to arrange for it to be typed. (There was still no LaTeX then, and I had to change four different typists before the work was done!) Anyway, I managed to do it in time.

But had nobody given me a strict deadline, I would probably have never finished my Ph.D... The trouble was that in addition to being a Ph.D. student, I was occupied with many other tasks. I was a father of two small children, for whom I was responsible in most of the afternoons (my daughter Noa was born at 1982); I was teaching in several places (in addition to the university), and I gave many private lessons too. I had no choice: we moved at that period to Tel Aviv, to the vicinity of the university, and our apartment there cost twice as much as our apartment in Petah Tiqva. So we had to take several loans, in addition to the mortgage on our new apartment. Another problem was that it was much easier and tempting for me to make progress in my research than to write down what I had already done in a form which is suitable for publication (a boring task).

The name of my thesis has been: "The semantics and proof theory of relevance logics and nontrivial theories containing contradictions". As this name suggests, its area was relevance logics, and more generally: paraconsistent logics. How did I arrive at this subject? Well, after the army I returned to Prof. Gaifman, and he agreed to serve again as my supervisor. (But since he was at Jerusalem University, while I myself was already strongly connected at that time with the School of Mathematics of Tel Aviv University, I needed to have a supervisor from Tel Aviv too, and Yoram Hirshfeld agreed to be the one.) I told Gaifman that I am interested in Philosophy, and I would be happy if my thesis in mathematics will be connected with philosophical problems. He, in turn, suggested two topics for my thesis. One of them was a concrete mathematical problem which was open then, and is connected with Gödel's incompleteness theorems: to extend to the first-order level Solovey's theorem about the completeness of the propositional provability modal logic **GL** for its arithmetical interpretation. The other subject was completely different in nature: to provide good explanations and model(s) for the fact that in both mathematics and physics, there have been useful inconsistent theories, even though such theories are, in principle, logically trivial (from a classical, and even intuitionistic, point of view). At the end, my thesis was devoted to the second topic.

The truth is that at the beginning, the first subject suggested by Gaifman was more appealing to me. Therefore, it was the one to which my main efforts then were devoted to. I even had some plan how to attack the problem. Its first step was to find a cut-free Gentzen-type system for **GL**. (Already then I was very fond of Gentzen-type systems, a topic I learned by myself at my years in the army, according to the advice given to me by Gaifman.) The next step was to show cut-elimination for QGL, the natural extension of that system to the first-order level. Then I hoped to show directly that the arithmetical interpretation obeys the same reduction steps as QGL. I succeeded (or so I thought) to implement a part of this plan. First, I completed the first two steps. This involved a complicated proof by triple induction of cut-elimination for QGL. I succeeded also to show the arithmetical validity of some of its reduction steps. However, I was completely stuck with others. Then a worse thing happened: I did the horrible mistake of checking again and again the correctness of my proof of cut-elimination, to make sure that I missed no possible case. And sure enough, I did discover a case in which something was going wrong. I tried to overcome the problem, but could not. Then came what I thought to be the ultimate disaster: by pursuing the small subcase in which my proof went wrong, I finally arrived at an example of a sequent which is derivable in QGL, but has there no cut-free proof. Not only the proof was wrong—so was the theorem itself!

Luckily, by that time I had also some ideas concerning the other subject suggested to me by Gaifman. So I decided to leave for a while the first subject, and to turn to the other one. And since I began to make a real progress in it, I had no motivation to go back to the difficult problem I had left unsolved. However, there was to be one more chapter in this story. Three years later, I looked at a new issue of the journal of symbolic logic, and found there ... a paper which presents "my" QGL, including exactly my faulty proof of cut-elimination for it! I was shocked. Until that moment in my life (and I was 30 years old then) I did not believe that it is possible that a respectable journal like JSL may publish a paper with wrong theorems and mistaken proofs! I told Yoram the story, and he advised me to submit to JSL a paper about this. So I took from my drawer the stuff that lied there 3 years, turned it into a paper, and submitted that paper to JSL. It was very quickly accepted, and then was one of my two first published papers [2].[1] (The other one was published at the same year, 1984, and at the same journal, but it already was in relevance logic.) That paper actually included some good results and proofs. Thus, it includes correct (but semantic) proofs of cut-elimination for Gentzen-type versions of the provability logics **GL** and **Grz**. (It turned out that the one for **GL** had been known before, but that for **Grz** was new.) It also contained some nice applications of the arithmetical fixed point theorem. However, what turned out to be its most significant contribution was the simplest (and least appreciated by me): the demonstration that QGL does not admit cut-elimination. Several years later Sergei Artemov told me that reading my paper was a turning point in his research on the subject, since this has been the first *negative* result concerning it. Therefore, it gave the first hint that Solovey's results in the propositional case fail in the first-order one. When I told him the above story he said that I would have saved him a lot of time and efforts had I published it before...

I devoted above a relatively big portion of my academic biography to the story of QGL. The reason is that I believe that one can learn a lot from it. I personally certainly did. Nevertheless, what unfortunately I could not internalize is what some might take as its main practical moral: do not check your proofs—publish them quickly instead. Checking is at best a waste of time, and you might even lose papers because of such a dangerous activity!

While still working mainly on provability logics, I started to think also on the other problem. Again my starting point was the use of Gentzen-type systems. I introduced three such systems, inducing three different logics, and proved (using Gentzen's syntactic method) cut-elimination for them. All my systems were based on the idea of weakening the structural rule of weakening. My real first achievement was proving (weak)[2] completeness of one of them relative to a certain three-valued logic which I thought had never been investigated before. I was wrong, of course. A

[1] From now on the numbers in square brackets refer to the articles in my list of publications.

[2] At that time, I was not aware yet of the importance of consequence relations, and the difference between strong and weak completeness.

few months later I discovered, to my dismay, that it was introduced and axiomatized by Sobociński in the year I was born. Unfortunately, this happened to me again and again during my academic career. For almost every new interesting idea of mine, it turned out that someone, somewhere, has thought about it before...

In parallel to my independent thinking about the problem of inconsistent theories, I started also to read the relevant literature. The only pointer that Gaifman was able to give me was the classical paper of Anderson and Belnap on Entailment. Soon I discovered that meanwhile they had published a big book, **Entailment**. (Volume 1; the second volume appears almost 20 years later.) Almost everything in this book was completely new to me. Therefore, I studied it extensively. and I was frustrated to find there all "my" three logics. For one of them, which was $\mathbf{R}_{\overset{\sim}{\rightarrow}}$ (the implication-negation fragment of the famous relevant logic **R**), the book presented even $G\mathbf{R}_{\overset{\sim}{\rightarrow}}$—which was identical to one of "my" Gentzen-type systems. To the other two logics it presents only equivalent Hilbert-type systems. In particular, there was there a Hilbert-type counterpart of the logic I preferred most (and later called $\mathbf{RMI}_{\overset{\sim}{\rightarrow}}$). Like $G\mathbf{R}_{\overset{\sim}{\rightarrow}}$, my Gentzen-type system for it was obtained from the classical one by deleting the weakening rule, but unlike in $G\mathbf{R}_{\overset{\sim}{\rightarrow}}$ the two sides of a sequent consist in it of *sets* of formulas rather than multisets or sequences. (I found that, and I still do, much more natural.) In addition, I had a rather natural idea of possible semantics for $\mathbf{RMI}_{\overset{\sim}{\rightarrow}}$, based on using "relevance domains". I also had a nice conjecture that a particular instance of the general semantics, in the form of a simple infinite-valued matrix, is characteristic for $\mathbf{RMI}_{\overset{\sim}{\rightarrow}}$. I was, therefore, particularly disappointed at first to find this logic too in **Entailment**, even though very little information was given there about it. Luckily, Gaifman saw things differently. He told me that it is actually *good* for me that "my" logic had been introduced by others before. He also told me that if I succeeded to prove my conjecture, I would be able to publish that in the JSL. So I made an effort, and did manage to prove it. Then, exactly as Gaifman had predicted, this result, together with related ones, was accepted to the JSL [1]. It was my very first published paper.

By that time I had completely abandoned provability logics, and concentrated on relevance logics and paraconsistent logics. In the following years, I developed my own theory of relevance, and I collected more and more results. However, during my time as a Ph.D. student I published only a very small part of them. The main body of my thesis was finally published in three parts [12–14] only several years later. The main difference between my approach there and that of the main relevantists' school was that my systems were purely relevant, and any attempt to add, e.g., an extensional conjunction $\wedge$ to them (for which $\varphi \wedge \psi \rightarrow \varphi$, $\varphi \wedge \psi \rightarrow \psi$, and the adjunction rule are all valid) means losing the variable-sharing property. It seems to me that this was one of the main reasons that this big work of mine was almost totally ignored, despite my effort in [21] to get the community's attention to it. However, the time and efforts I devoted to my thesis had their merit. First, during those years I became a real expert about Logic and logics. Second, my thesis contains one (almost) new idea that did become rather well-known and popular: the use of *hypersequents*.[3] This was due to the fact that I could not find a cut-free Gentzen-type system for my main logic, **RMI**. My supervisor, Gaifman, asked me if this really matters to me. (Being more a model-theorist, this did not look so important to *him*.) My answer was positive, since already then my view was that a useful logic should have both an analytic proof system and an effective semantics. **RMI** did have the latter, but I needed to introduce hypersequents in order to be able to provide a decent proof system for it. I proved cut-elimination for my hypersequential formulation of **RMI** using an extremely complicated syntactic proof. This proof was later published in [14]. However, I had published before that a somewhat easier (but still very complex) versions of the calculus and of my cut-elimination proof for it in the case of the simpler, and better known, semi-relevant system **RM**. A few years later I found cut-free hypersequential calculus for Gödel–Dummett logic **G** (also known as **LC**), in which I introduced the "communication" rule. The latter, and the use of hypersequents in general, have become since then the basis of the proof theory of all fuzzy logics. As a result, my name started to be known among logicians. Well, I always took it as ironical that what was of a secondary importance in my thesis, done only because of my insistence, is the only part of my thesis that has given me some fame...

Post-Doc 1984–1988

Tel Aviv 1984–1986

I submitted my thesis at the end of 1984, but it was finally approved (with distinction) near the end of 1985. Meanwhile, I was forced to think at last about the problem: "What do I want to do when I grow up?". My wish was, of course, to get a position in mathematics in one of the few universities in Israel. (There were only five then.) Unfortunately, there was then very little hope for that to happen. There were at that time almost no jobs in mathematics at the Israeli universities, especially for logicians, and certainly not to someone like me, who had done his thesis on what was then a particularly esoteric subject. So how did I become a professor at Tel Aviv University nevertheless? The simple answer is that I have been lucky.

A few years before I finished my Ph.D., our school of mathematics had decided to establish a new department: computer science. (Already this decision was a part of my luck.) In its first years it was very small, and I even have not heard about it.

[3] As usual, it turned out that similar structures had been used before. Thus, the referee of my first paper on hypersequents, [6], pointed out that Pottinger had used what I called "hypersequents" in a small abstract concerning modal logics. Years later I was shown that Mints too had used a similar structure in his proof systems for modal logics. Strangely, although I discussed with Mints my use of hypersequents for **G** when I was on sabbatical at Stanford, he has never mentioned to me this fact! Anyway, it was certainly me who first developed the general theory of hypersequents, applied it for various families of logics, and made it known. (Even the name "hypersequent" is due to me.)

However, the wish to make it bigger and important led to my first big luck. In 1981, B. Trakhtenbrot immigrated from the USSR to Israel, and joined the new department of computer science in our school. He was then a very famous logician and computer scientist, but I myself had not heard about him before he came, and knew very little about him also after that. Therefore, it did not even occur to me to make contact with him. Instead, he made contact with me, since he was looking for people to work with, and someone had told him that I am one of the few people in the whole school who have some knowledge in logic. My thesis was not ready yet then, and I was used to work in isolation from others, and on what was of interest for me. So at first I viewed my connections with Trakhtenbrot (that he practically forced on me) as a headache—especially that he was very demanding, and assigned me tasks that were of no interest for me at that time. The first such task was to read and then explain to him Martin-Löf's type theory and its purported relations to computer science, as described in his famous 1982 paper: "Constructive mathematics and computer programming". (Funny, but the fact is that my acquaintance with computer science started with this paper...) I did not want to do that, but I had no choice. So I did it the best way I could, and Trakhtenbrot was very impressed. He decided that I have a great potential—and in few years I found out how crucial for my future was this good impression I made on whom I viewed then as an old, imperious person! There is no question that I owe him my career. More than that: after finishing my thesis, he practically became my mentor for many years, and once I began to really appreciate the knowledge, insights, and vision of this great scientist, I also learned from him a lot.

Stupid as I was, after submitting my thesis it did not take me too long to realize that turning to computer science was my only chance and hope. So I started to try to learn it. As a part of this attempt to become a computer scientist, I gave an advance course on automated theorem proving (which I practically learned and taught in parallel). At the end of that course I was looking for problems to include at the final home examination of the course. By chance, a friend gave me at about the same time the following problem as a challenge: Given two points in the plane, construct by means of a compass alone the corresponding midpoint. I solved it easily using a bottom-up search. Then it occurred to me that asking how to attack it using a computer would be a good problem for the home examination. However, before giving it to my students, I wanted to check whether I can do it myself. This was also an opportunity to do some programming for the first time after 15 years. (I did have to write some programs in FORTRAN at my first years at the university.) So I learned PROLOG, which was very fashionable at that time, because of the Japanese fifth-generation computer systems project. Then I wrote a program in PROLOG that could indeed solve the above problem, but was able also to find many other points that can be constructed by means of a compass alone, given two points in the plane. An inspection of the output of my program led me to a conjecture about what points it can produce. So I tried to prove that conjecture, but instead I proved that it is wrong. (However, the computer itself gave in to my result only after outputting several thousands points that confirmed my wrong conjecture...) Next I noticed that my proof of what my program can do can be turned without too many difficulties into a simple proof of a *strengthening* of the classical Mohr–Mascheroni theorem about constructions by means of a compass alone (which I had heard about by that time). So I wrote my proof down, and submitted it to the Journal of Geometry, where it was published [5]. However, the referee has some reservation about my proof, since in addition to proper intersection points, it allowed also the use of intersection points of two *tangent* circles. (This was something that both Mohr and Mascheroni avoided.) Nevertheless, I was able to remedy this problematic feature of my proof with the help of a construction given to me by the program, and this was again published at the Journal of Geometry [11]. I was, and still am, very proud of these two short papers. I fulfilled in them a big dream that I had had when I was in high school: to prove an interesting new theorem in Euclidean Geometry that anybody can understand. In addition, this experience showed me for the first time (but not the last) how fruitful for my research can teaching a course be.[4]

While I was devoting most of my working time at that period to standard academic activity (new research; turning parts of my thesis into papers; teaching; learning new subjects), my main goal then was to find a post-doc position in a good place abroad—something that was a necessary condition for getting a permanent position at a university in Israel. This was a rather frustrating task, involving getting one negative answer after another, even from places that had shown at first some interest. However, at a certain point my luck (together with the great help of Trakhtenbrot), found a place for me that proved to be really great: the Department of Computer Science of Edinburgh University. That department had already been one of the best in the world, with giants like Robin Milner, Gordon Plotkin, and Rod Burstall. Luckily for me, at that time they decided to find a new internal research institute called LFCS (Laboratory for Foundations of Computer Science). Moreover, one of the first two big projects that were planned for the new LFCS was the construction of the first *Logical Framework* (LF): a general computerized system for implementing a variety of logical systems of all sorts. Therefore, the department was looking for post-docs who might be able to contribute to this big project, and Trakhtenbrot convinced them that I am a good candidate. At that academic year (1985–86), I also won the Rothschild Fellowship for the following year (which I failed to get the year before, because my thesis was not approved yet). This meant that I would not have financial problems during my 2 years as a post-doc at Edinburgh. At last, things fell in place for me!

Edinburgh 1986–1988

I truly fell in love in Edinburgh during the 2 years I spent there. I believe that it is the most beautiful town in the world. I also like its atmosphere. (Especially in the summer; the winter is somewhat problematic, because you almost never see the sun.) Therefore, I am always happy to return to Edinburgh for a visit. However, my

[4] The full story, together with an analysis, can be found in [24].

first months there were not easy at all. I was older than all the other postdoctoral researchers there, and while almost all of them were not even married, I had two children: a boy at the age of 8 and a daughter at the age of 4.5. Neither of my children knew a word in English when we came to Edinburgh, so taking care of them was a major (and uneasy) task for me at those first months. In addition, it took me time to find myself in this new environment. I was unable at first to follow what the people there were talking about; I have difficulties in communicating with them, and I was still desperate to understand what is computer science.[5] Luckily for me, the atmosphere in the LFCS was rather relaxed and tolerant, unlike the pressure I heard about from friends that went to the USA. This fact was of great help for me. There was also a *person* who helped me a lot at those difficult days. His name was Furio Honsell. He was an Italian who started his post-doc in Edinburgh a few months before me.[6] Like me, Furio belonged to the LF group, and at the beginning he was the only one in the department with whom I was able to communicate. Thanks to him, I started to understand the ideas that underlie the planed design of the LF. On the other hand, he has learned from me too. In fact, at a certain point I realized that although logic has a central place in the research made at the LFCS, most (if not all) of the people there had a rather limited knowledge and narrow view about it. So I wrote for Furio by hand some pages of remarks about logics and logical systems that I thought every logician know (or should know). Furio became very enthusiastic about those remarks. He started to spread their content, and the ideas presented there had great effect on the theoretical development of the LF, as well as its practical use for implementing logical systems. Furio then encouraged me to turn my notes into a paper. I did so several months later. The resulting paper, "Simple Consequence Relations", was first published in the form of a technical report of the LFCS about a year after I came to Edinburgh. Already in this form it became rather popular, and one of my most successful papers.[7]

My notes to Furio were one reason for the change in my status in the LFCS from the sort of an outsider that I was in my first months there, to a respected member, whose knowledge in logic was much appreciated. The other reason was due again to a great piece of luck. A few months after I had joined LFCS, another member of the LF group, Bob Harper,[8] returned from a visit in Paris, and brought with him a preprint of a new big paper of J.Y. Girard. Bob told us that it was what everyone was talking about at that time at the University of Paris, although he himself could not

[5] I frequently say, as a half joke, that I finally understood what is computer science only when I became an editor of "Theoretical Computer Science", because from that point something belongs to computer science if I decide so...

[6] Later Furio became the Rector of the University of Udine, and then he served many years as the Mayor of Udine. He still divides his time between research and politics.

[7] The paper was finally published 4 years later in "Information and Computation" [15]. The truth is that this was not the most appropriate place for it. I chose it only because it was crucial for me at that time to have papers in respectable journals of *computer science*.

[8] Bob came to Edinburgh a year before me, and by the time I came he has already been a well-established member of the LFCS. At the time of writing, he is a well-known professor of computer science at Carnegie Mellon University.

really understand what is written there. The name of the paper was "Linear Logic". Naturally, everyone in the LFCS wanted to understand what it is about—and I was the first one to succeed. The reason was simple: I soon realized that except for the notations, linear logic was a very close relative of relevance logics. Accordingly, there was nothing mysterious for *me* at Girard's paper, and so I volunteered to read it and to lecture about its content. The three lectures I gave on it made great impression on everyone, including Gordon, Robin, and Rod. Therefore, they were the turning point of my time in Edinburgh—and this was only the beginning of the great positive effect that Girard's linear logic had on my career. Thus, I could rather easily adapt to linear logic some of the easier things that I had done in my thesis. The result was submitted to TCS, not long after Girard's original paper had been published there. It was quickly accepted and published there too ([7]). This was my very first paper in a computer science journal, and also the second published paper on linear logic. (The first was Girard's paper itself, of course.) Since there was a huge interest in linear logic at that time, this paper of mine really helped to make my name known. Ironically, its most important contribution for many was a table I included in it of translations from Girard's notations to those which had been used in the relevance logic literature. That table was called by some relevantists "the Rosetta stone" that let them understand the very fast growing literature on linear logic, a subject that was very hot at the end of the 80s and at the 90s. Consequently, my name became known also in the community of philosophical logic.

I have always found as ironical that the two papers that gave me some fame at the first stages of my academic career were papers that I myself did not appreciate much, since they did not have real mathematical depth, and their popularity was mainly due to ignorance of people about issues that I believed should have been well known. But I was not complaining, of course. Thanks to these two papers, most of my time in Edinburgh was rather nice and fruitful. I had other works there, some of them more important in the long run than the two mentioned above (like [16] and [17]). I even had there my very first joint paper with other people: the second big paper on the LF, written by Furio, Ian Mason, and me [22].[9] This was significant, since until then I had worked in complete isolation. (Even with Gaifman, my supervisor, I met very rarely, and there were years in which we did not meet at all, since he was abroad.) Even more important was the fact that I met and talked with many scholars. First, the LFCS has a very international atmosphere. There were people there from Sweden, Denmark, Poland, Italy, Germany, China, Japan, India, England, Canada, USA, Australia—and even Scotland! Some of them became good friends of mine. Second, during my 2 years at Edinburgh I took part in several scientific meetings, and visited several countries. This was almost a new experience for me. Before coming to Edinburgh I had participated just in three international meetings, and the first of them (which was also my very first trip abroad after high school) took part only when I was 31 years old. (It was the big 7th congress on logic, methodology, and philosophy of science in Salzburg, 1983.) Things completely changed during my

[9] [9] was an earlier, shorter version of it, while R. Polack's contribution was added later.

2 years at Edinburgh, and only there I started to be (and to feel as) a part of an international scientific community.

In my second year at beautiful Scotland, there was only one thing that prevented me from fully enjoying my stay there: the worry about my future, and the great uncertainty whether I would find a position in Israel. This uncertainty remained in force most of the year. But at the end fortune again smiled at me. The computer science department of Tel Aviv University has only six faculty members when I left to Edinburgh, but it had 14 (including me) when I returned. My great luck was that the university decided exactly at that time to really expand it, and the year in which I applied was the main one in which this decision was implemented. I was one of five new faculty members that joined it at the end of 1988—the biggest expansion in the history of our debarment until present, and practically the last time in which someone whose Ph.D. had not been in computer science was given a tenure-track position in it. Even so, I would not have been one of the fortunate five had not Trakhtenbrot strongly fought for me. Without him, even the good letters that were sent on my behalf by Gordon, Girard, and others would not have helped.

Climbing the Academic Ladder: 1988–1999

As my story so far showed, it might be necessary to be lucky in order to get a position in a place like the School of Mathematics at Tel Aviv University. (I am not saying, of course, that it can be *only* a matter of luck.) However, from the point in which one gets the *chance* to have a place there, by being given a tenure-track position, luck has nothing to do with the rest, i.e., actually getting tenure and then being promoted. It completely depends on what one does. I knew that, and I worked hard during the period of 10–11 years that followed my return in order to successfully pass through all the stages of an academic career in Israel, until I became a full professor at 1999. This time I was not lucky at all, since in my case the process consists of no less than four different stages, with all the agony and the complicated process that each such stage involves. How come? Well, like all those who joined the computer science department at those years, I started as a lecturer. (This position practically does not exist any more in mathematics or computer science; now every new comer immediately starts as a senior lecturer.) After 1 year, I was promoted to a senior lecturer.[10] In 1992 I got the tenure, in 1995 I became an associate professor, and since 1999 I am a full professor.[11]

[10] This was an initiative of Trakhtenbrot, and it was a mistake: it was better to wait and get this promotion together with the tenure.

[11] Nowadays, there are just two stages: getting tenure together with a promotion to associate professor, and becoming a full professor.

Getting Tenure: 1988–1992

My main worry after my return was still to secure my future by getting tenure. It was not at all guaranteed, and I even had to work for that harder than others, because my being a real computer scientist was somewhat suspicious. (On the other hand, when it came to promotions, it helped my cause that we were then still a part of the school of mathematics, and so were judged by its standards.) At the first 4 years, my main research activity was devoted mainly to continuing, finishing, and writing down works that I started before returning. However, I was trying to look at new subjects as well. One of them was concurrency, because this was a subject in which Trakhtenbrot was very interested at that time. Just one paper directly came out of this [28], but learning it gave me ideas how one might use calculi of hypersequents for modeling parallel computations or processes. I have never managed to seriously pursue those ideas myself, but recently this was successfully done by others.[12] A particularly fruitful source of ideas and research was my teaching courses in computer science, since this involved a lot of thinking on the topics of the courses. Thus, [23] was a direct product of teaching automated reasoning, while [19] and its full version [26] were inspired by my course on databases.[13] The reason that it was me who taught databases at those years was that I had heard in Edinburgh that its theory has connections with logic, and there was no real expert then on it in our department. (Now there are two.) At the first year after my return, I both learned the theory of databases and taught it. After that I started also to look at interesting research topics that I had found in the material on databases that I was teaching. The abovementioned work with Yoram was the first result. However, the really great outcome of my teaching databases for 4 years came several years later. It will be described in the sequel.[14]

Another academic activity that I continued, of course, at that time was presenting my work in meetings abroad, and getting to know more people there. The trips that I made in those 4 years were very important for my career. However, there was among them one that proved to be *particularly* important: On April 1991, I returned to Edinburgh for the first time (out of many) in order to participate in a conference there, and my wife Tsipi joined me on that trip. Exactly 9 months later my third and youngest child, Uri, was born (about 10 years after his sister and 13 years after his brother). This has by far been the most productive academic trip I have ever made!

[12] F. Aschieri, A. Ciabattoni, and F. A. Genco; A. Beckman and N. Preining.

[13] It should be noted that this was the first, and so far the only, time I had a paper with one of my three mentors; I have never had a joint paper with either Gaifman or Trakhtenbrot.

[14] I should admit that since I left to my course assistant the responsibility for the practical project that the students of my course had to do, I have never used a database system myself. Well, I have also taught my children how to ride a bicycle, even though I cannot do it myself...

At Stanford: 1992–1993

As I wrote at the beginning, I chose mathematical logic as my area because of my deep interest in foundations of mathematics and in philosophy of mathematics. Unfortunately, for almost 17 years after finishing my M.Sc. thesis, I did not have time to do research on these related subjects. However, I did find time during those years to think and to read about them. Thus, already when I was a Ph.D. student I learned about platonism, formalism, logicism, and intuitionism. Surprisingly for me, my own views about the foundations and nature of mathematics did not fit to any of these major schools. Then, at a certain point in my conversations on foundations with Prof. Jonathan Stavi, he told me that views like mine are known as *predicativism*, and suggested that I read Feferman's papers about it. That is how I have come to realize that I am a predicativist.[15] Needless to say, after that I returned to study Feferman's papers. Since my M.Sc. thesis had been based on his first two main papers, it was as if I am simply continuing to the next ones. Anyway, Sol Feferman became my academic hero, so I strongly wished to have the opportunity to work with him. Our first meeting took place in the congress at Salzburg in 1983. I simply came to him and introduced myself as a student of Gaifman, and told him that I am very interested in his work. Our conversation was rather short. Therefore, I was surprised that he remembered it when in 1990 I wrote him, and asked to spend some weeks of that summer in Stanford. He agreed, and arranged the financial side of my 6 weeks visit. That visit was the beginning of my connections with Sol. It turned out to be also an opportunity to renew my connections with the LF group: There was a week in which Gordon, Furio, Ian, and me were all there. (Feferman too was at that time interested in logical frameworks, so some of our conversations were on that subject.)

At 1992, I knew that I am about to get tenure at last, and so I decided that it is a good time to take sabbatical abroad. Stanford was, of course, the place I wanted, but at first it seemed impossible to go there, because of the strong recession that USA was experiencing at that time. Luckily, Feferman was then the chair of Stanford's Department of Mathematics, and at the last moment, when I was about to go elsewhere, he found a possibility to support my visit. So again I went abroad with my family (which now included also a baby). This time it was for just 1 year (1992–3), and to California instead of Scotland. Another difference was that not a long time after we came to Stanford, I officially got my tenure. Therefore, I was free from worrying about my future for the first time in my life (at the age of 40). This fact allowed me to start devoting a part of my research to foundations of mathematics. My meetings with Feferman during that year helped me to put my views on that matter in context, and to understand the related research that had been made up to that point. As usual with me, no joint paper came out of our discussions. Still, I was able to surprise Feferman by showing that a certain significant improvement of his own suggested logical framework, which he had believed to be impossible, is possible

[15] Laura Crosilla once asked me how I became a predicativist. My answer was that it seems that I was born one, and that I only can tell when and how I *discovered* that I am a predicativist.

after all. This improvement was strongly connected with my first foundational idea and subject of research: the use of **AL** (ancestral logic, also known as transitive closure logic) as the basic logic that underlies absolutely certain mathematics. (For me the latter is identical with predicative mathematics.) My study of **AL** actually started at that year in Stanford, even though my first paper on this logic [56] was published only 10 years later.

There were two other important (from my scientific point of view) developments that took place at that year in Stanford. One was my first meeting with Mike Dunn, with whom I only had had some correspondence before that. Mike invited me to visit Indiana University at Bloomington, and to give a lecture there. This has been a very nice visit, and it included several very useful discussions with Mike and with other people there.[16] The other development was again a meeting. This time it was with a young fellow I had not heard about before. His name was Richard Zach, and now he is a famous Professor of Philosophy at the University of Calgary. At that time, he was still only a research student in TU Wien. He came to a short visit at Stanford near the end of my year there, and during that visit he initiated a meeting with me. I could not guess it then, but that meeting was the beginning of my close relations and friendship with the great group of logic in TU Wien and with many people that have spent time there over the years. (Including, of course, some which are still there.)

Becoming a Full Professor: 1993–1999

The next stage in my academic career came relatively quickly. Two years after returning from Stanford I became an associate professor. There was a price to pay for that, though. Not long later, between 1996 and 1998, I had to serve as the chair of the CS department. The reason was that there were very few professors (either full or associate) in our department at that time, and each of the others had already done this job. Therefore, I had no choice but to agree. Being the chair demanded a great part of my time, and was a very hard test for my nerves. Nevertheless, at the end, both I and my department somehow survived (not without great difficulties) those 2 years. Happily, this was the last time I had to take on myself such a big administrative task.

As for my research activity, a part of it continued to be devoted to subjects I had worked on before, like substructural logics, including relevance logics and linear logic ([35, 40]). However, the main subject I was working on at that period was bilattices, and their use for uncertainty reasoning. Originally, bilattices had been introduced by Ginsberg, but I learned about them from Mel Fitting at a conference in Varna in 1990. After that I read Fitting's papers on this subject, and became interested in it. Therefore, I soon began to be engaged with research on it myself. Most of this

[16] One side effect of that visit was that I started to use the name "Dunn–Belnap logic" for the famous four-valued logic that everyone, including myself, had called "Belnap's four-valued logic" until then. Slowly but surely, this more correct name has been adopted by others too. I am really glad that I have helped in giving my friend Mike the credit he deserves here!

research was done after returning from Stanford. One of its main achievements, with which I was particularly pleased, was to be the first to prove a conjecture of Fitting about the structure of interlaced bilattices ([33]).[17] However, this result was a purely mathematical one. More important from the practical point of view was to investigate the applications of bilattices for logic and reasoning. I carried this investigation together with my first Ph.D. student, Ofer Arieli. Our fruitful cooperation led to some papers on logics for uncertain reasoning that are based on using what we call logical bilattices [29, 32, 38, 39]. Those papers became rather popular.

The logic induced by bilattices is paraconsistent. Moreover, inconsistencies in our knowledge is one of the major concerns of the vast area of uncertain reasoning, which may be classified as my area at that time. Accordingly, I naturally returned to make research on paraconsistent logics. A good opportunity to become updated about the state of the research on paraconsistency was the first congress on it that took part at Ghent in 1997. It was a big event, with many invited speakers, but I was not one of them then. Still, I was happy to present there my work with Ofer in a contributed talk. That congress was rather fruitful for me: I got to know most people who were working on the subject, and with many of them I have been keeping close connections ever since. (And not long after that congress, the community started to recognize me as one of the main experts in the field of paraconsistency.)

Another meeting that turned out to be very fruitful for my research was the Tableaux conference that took place in Pont a Mousson at the same year. It was the first Tableaux meeting (out of many) in which I took part, and I gave there a tutorial on the proof theory of propositional modal logics. But again the most important outcome for me were the new personal connections I made there. By far, the most important among them was the acquaintance I made with Beata Konikowska from the Polish academy. It was the beginning of my first (and so far the only) long-term cooperation, leading to several joint papers, with a researcher other than my past or present students.

In addition to the developments in my research, there was also a crucial development at those years in my other academic activity: teaching. After returning from Stanford I became one of the two main teachers of the very first course which is given by us since then to our first-year students: discrete mathematics. This course has in our department two parts: an introduction to set theory and general mathematical concepts, and standard subjects in combinatorics and graph theory. When I started, I knew almost nothing about the second part. However, I was the department's expert on the first, and I was very enthusiastic about this chance to shape the course according to my views about what every student of mathematics or computer science should know and understand. Accordingly, I turned the first part of the course into an advance introduction to the language of mathematics and its logic. I put a

[17] Unfortunately, there was also a very unpleasant affair connected with this nice result. One of those who independently (and using a different method) proved Fitting's conjecture after me, a guy named Pinko, refused to recognize my priority. More than that, he blamed *me* of stealing his result—even though my paper on the subject had already been published by the time he submitted his. What Pinko wrote me and tried to do next was simply unbelievable. But something I would better skip the details here.

particular emphasis on the correct use of formal expressions. Thus, I introduced the use of λ-notation and its associated rules into the material of the course, and used it myself consistently. (The other main teacher did not like it at first, but then learned to appreciate the advantages of using λ.) This was a small revolution. Indeed, the course we developed was different from any other one taught in Israel, or presented in textbooks on discrete mathematics. Therefore, I realized at a certain point that I should write my own book on discrete mathematics. It took me several years to complete it. It is a book in Hebrew that only our students can get. However, I do hope to find one day the time and energy needed in order to translate it into English and publish it. Anyway, writing this book and teaching this course involved a lot of thinking. As will be described in the next section that thinking had very significant consequences for my research as well.

In addition to my book on discrete mathematics, I wrote in those years another book in Hebrew. It was a small, popular book on Gödel incompleteness theorem and the problem of the foundations of mathematics. The book was based on 13 short lectures I gave on this subject in the Israeli radio, in the framework of what is called "broadcast university". In contrast to my other book, this one *was* published in the broadcast university's series of books, and it has been rather successful. My book and a similar one by David Harel on the foundations of computer science are still the only two books (out of hundreds) in this series that are about mathematical subjects. I have been urged many times by colleagues to translate it into English, and I hope that one day I will.

Full Professor 1999–

My becoming a full professor was a very important event for my department. By this it has reached the minimal number of full professors which is needed in order to be able to become an independent school. So in the following year we left the school of mathematics, and became the school of computer science. (Ironically, I myself was not happy with this move...) As for myself, since I have never wanted to be the head of anything, my being a full professor was for me the height of my professional career. I was at last free from any worry about promotions, etc., and could do research on whatever I like. Accordingly, since then I have been having peaceful academic life, with no important turn points. I have even been avoiding any trip abroad that is more than 18 days long. Therefore, during my sabbaticals I remained in Tel Aviv. On the other hand, there was a very important change in my status outside the academy: at the end of 2010 I became a grandfather. Since then, my two elder children gave me more grandchildren, and at the time of writing the set of my grandchildren includes three grandsons and two granddaughters. I hope and believe that one day my youngest son will add new members to this exclusive set.

One remarkable event in my academic life, that did take place at the period described in this last section, was the workshop "Logic: Between Semantics and Proof Theory", which was held at Tel Aviv University on November 2012, on the occasion of my 60th birthday. I was really moved and very grateful to see so many friends coming from all over the world to take part in this celebration. Later most of them (as well as others) also contributed papers to the proceedings of that workshop, which was published in February 2016 in the form of a huge special issue in my tribute of the Journal of Logic and Computation. The workshop itself was organized by my previous Ph.D. student Anna Zamansky, with the help of Ofer Arieli, and the Ph.D. students I had at that time.

Talking about students, I have not had too many. Still, I was very lucky with those that I did have. All of them were exceptionally good (from any point of view). I should add here that I am proud about the fact that all the students who have finished their Ph.D. under my supervision have found positions in the Israeli academy: Ofer Arieli is a Professor at the Academic College of Tel Aviv, Anna Zamansky is an Associate Professor at Haifa University, Ori Lahav is at Tel Aviv University, Liron Cohen at Beer-Sheva University, and Yoni Zohar is joining Bar-Ilan University next year. (I am sure that they will all become full professors in the future.)

So far about honor. In the rest of this section, I describe my major ideas and directions of research during the years that passed since I became a full professor.

Non-deterministic Matrices

At the abovementioned congress on paraconsistency at Ghent, I met Diderik Batens for the first time, and heard from him about his adaptive logics. I wished to know more about this approach to paraconsistency, and so suggested adaptive logics to a new M.Sc. student of mine, Iddo Lev, as the topic of his M.Sc. thesis. While studying together Batens' papers I discovered a very interesting idea hidden in one of them. It was connected with the semantics that Batens gave to the basic ordinary (i.e., not adaptive) logic **CLuN**, on which adaptive logics are based. **CLuN** is obtained from positive classical logic by adding to it the axiom of excluded middle, and it is easy to see that a corresponding cut-free Gentzen-type system is obtained from the classical one by deleting the left introduction rule for negation. Batens' semantics for that system looks at first strange to me, and in trying to understand better what is going on there, I realized that its presentation can be simplified and better understood if it is put in the form of what I immediately called "two-valued non-deterministic matrix (Nmatrix)". (Here being in a computer science department was rather helpful!)

The notion of an Nmatrix is a generalization of the usual (algebraic/truth-functional) notion of a logical many-valued matrix, in which the "truth-tables" that correspond to the connectives may be non-deterministic. Once this idea occurred to me, I immediately saw its great potential.[18] So after Iddo submitted his M.Sc. thesis and became my Ph.D. student, we devoted our joint research to it. We started with

[18] As in the case of hypersequents, it turned out that I had not been the first to have this idea. Ori Lahav found at a certain point that both Schütte and Girard applied certain three-valued Nmatrices in their books on proof theory. Much before Ori's discovery, J. Marcos sent me a paper of Crawford and Etherington, which uses another three-valued Nmatrix. Years later I discovered by accident that

generalizing the case of **CLuN**, by showing that every logic which is obtained from classical logic by deleting some of its Gentzen-type logical rules has a characteristic two-valued Nmatrix. Then we generalized this too, by introducing the notion of a canonical Gentzen-type system, and showing that such a system admits cut-elimination iff it is not trivial; iff it has a characteristic two-valued Nmatrix; and iff it satisfies a certain simple, easy to check, coherence criterion. ([52,61]. The implications of these results to the "tonk" problem are described in [92].)

After completing the study of two-valued Nmatrices, the turn came of multiple-valued Nmatrices. The first related main result was due to Iddo: He proved that the compactness theorem applied for every logic that has a characteristic finite Nmatrix.[19] Unfortunately, after that Iddo decided to switch into the area that really interested him: AI. So although he had already made a nice progress, he left it all, and went to Stanford in order to do there his Ph.D. in AI and NLP. So I continued without him (sometimes with the help of Beata Konikowska). Thus, following meetings and discussions with W. Carnielli and J. Marcos, I found that the use of Nmatrices is particularly efficient for the study of their big family of paraconsistent logics called LFIs (logics of formal inconsistency). This study revealed again the big advantage of the semantic framework of Nmatrices: the modularity it allows in developing effective semantics as well as cut-free Gentzen-type systems, for families that contain thousands of logics.

The framework of Nmatrices has been one of my main research topics as a full professor. Many of my papers at this period, either alone, or with others, are devoted to various directions of its applications. Some examples of such directions are first-order languages and beyond (together with Anna Zamansky); constructive logics and non-deterministic Kripke frames (together with Ori Lahav); fuzzy logics (together with Yoni Zohar); proof theory (together with Beata and Anna); and knowledge bases (together with Beata, J. Ben-Naim, and Y. Dvir). [97] is a survey of most of the results in this area at its first 10 years.

---

the Russian logician Y. V. Ivlev practically introduced Nmatrices not long before me. He called them "quasi-matrices" and applied certain special such "quasi-matrices" in modal logics. (This discovery was rather embarrassing for me, because I saw that Ivlev had a talk about it in a conference at Torun at 1998, in which I took part too. The name of the talk was "Quasi-matrix logic as a para-consistent logic for dubitable information", and I do not even remember whether I attended it or not. Even if I did, I certainly did not understand then what the speaker was saying—which is usually what happens to me in talks... Strangely, nobody has ever told me about Ivlev's work. Not even Ivlev himself!) Nevertheless, I did reach the idea independently, and as in the case of hypersequents, I was the one who turned it into a new subject of its own, with many diverse applications.

[19] Later, I found out that this generalizes a similar theorem of Shoesmith and Smiley for ordinary matrices.

The Book on Paraconsistent Logics

In 2010, Ofer had a sabbatical, and he had the idea to use it in order to write a first extensive book on paraconsistent logics. (There was no such a book then.[20]) He suggested to Anna and me to join, and we both liked the idea. We thought then that it would take us about a year. However, the project turned out to involve much more work than we had anticipated at its beginning. At a certain point, we realized that our initial plan for the book (which includes topics like non-monotonic inference mechanisms, first-order systems, and several more) was too ambitious. In order to be sure that we finish the project one day, we decided to restrict it to the propositional level, within it to ordinary (monotonic) logics, and among them only to what we call *effective* logics. For the latter, we made a list of criteria that a logic should satisfy in order to count as such. However, even with this restricted scope, we soon saw that a lot of research is needed in order to write a book of the type we want.

- First, we had to provide exact definitions for many fuzzy notions that had been used in the literature on paraconsistent logics. This even includes the very notion of a "paraconsistent" logic itself. Our precise definitions naturally led, in turn, to the need for precise propositions about the various defined notions and about the relations among them (together with precise proofs of those propositions).
- Second, we discovered that many of the logics that we had thought should unquestionable be dealt with in our book were not meeting yet all our criteria for effectivity. So we had to fill in many serious gaps that existed in the knowledge about those logics, while we were writing our book.

As a result of all these circumstance, our goal of writing a book on paraconsistent logics developed into a massive research program, which led to many new ideas and results, as well as to papers that described them. The work took us, therefore, more than 8 years, and the book was published only in 2018. On the other hand, I (at least) am very pleased with the final outcome—I see it as the climax and ultimate conclusion of my 40 years of research on paraconsistent logics of all sorts.

Safety Relations and Predicative Mathematics

In my course on discrete mathematics, I was putting a lot of emphasis on teaching students how to correctly manipulate formal expressions, like abstract set terms and λ-terms. With the former I had the problem that not every such term can be taken as denoting a set. ($\{x \mid x \notin x\}$ is a case in point.) So I developed (and taught) a system of syntactic rules, closely connected to the axioms of the formal set theory **ZF**, for writing legal abstract set terms. At a certain point, I noticed some similarity between those rules and Ullman's rules for writing safe (i.e., domain independent)

[20] In contrast, by the time we finished writing our book there were at least two. Neither of them has the broad scope of our book, though.

queries in database theory. To exploit this similarity, I had to generalize the domain independence *property* of formulas to a *relation* between a formula and its set of free variables. Following Ullman, I called such a relation "*safety relation*". Using one particular such relation, I was able to provide a rather convenient formalization of **ZF** in a language which allows the use of abstract set terms. My formulation was based on purely syntactical principles (most of them taken from database theory), and has very natural axioms. I believed that my system may be useful for MKM (Mathematical Knowledge Management). So I published it [58], and then generalized it to a general syntactic framework for formalizing set theories [77].

The next step in this line of research came when I observed that there is actually a property of formulas (due to Gödel) called *absoluteness*, which is very important in the meta-theory of set theories on one hand, and on the other it is really close (semantically and syntactically) to the property of domain independence (d.i.) which is used in database theory. Moreover, my notion of a safety relation unifies these two properties in a rather nice way: a formula is d.i. if it is safe with respect to its whole set of free variables; a formula is absolute if it is safe with respect to the empty set of free variables. This observation, together with the recognition (already due to Poincaré and Weyl) that absoluteness is the key idea and notion in the predicativist program, made it possible for me to contribute at last to the research on predicative mathematics, and even to develop my own version of predicativity. At first, there was one obstacle, though, to do so in a fully satisfactory way: As long as I confined myself to the use of a first-order language (augmented with variable binding term operators), there was essentially just one way to introduce the natural numbers as a set into my framework: by using brute force. However, I saw that this problem could be solved easily and naturally if ancestral logic and its language are used as the underlying logic and language. I was, of course, happy to introduce both into my framework. Combining that with the basic principles of safety, I developed a syntactically defined predicative set theory which I am calling **PZF**. (See [91].)

Predicative mathematics has been (and still is) one of my major research topics in recent years. One direction of this research is developing classical analysis within **PZF** and related systems. This was one of the main subjects investigated by Liron Cohen at her Ph.D. thesis. (See [127, 128].) Understanding other approaches to predicativity and comparing them to mine is another important current direction of research. Thus, my latest (so far) rather big paper [139] is doing that to Weyl's original system in his classical book "**Das Kontinuum**" from 1918, while my Ph.D. student Nissan Levy and I are at present investigating the relations of my systems with those which have been studied in Friedman-Simpson's program of Reverse Mathematics.

The part of my research that is connected with the use of safety relation has one more branch, whose ultimate goal is to develop a general, unified theory of constructions and computations. It started with yet another observation about safety relations, which this time I made when I was teaching an advance course devoted to Gödel's incompleteness theorems: that the same principles that underlie d.i. in databases, and absoluteness in set theory, can be used to characterize decidability of formulas in computability theory and formal number theory. Some steps toward the aims of this branch have been made in [78] and [130].[21]

A Proof from THE BOOK

I would like to end my story with a very short paper of mine (together with my old friend and colleague Nachum Dershowitz) that I admit to be particularly proud of. Like my papers on geometric constructions with a compass, it is not a paper on logical matters, and like those papers (and many other works of mine), it has grown out from a course I was teaching: the course on discrete mathematics (again).

As I said above, our course in discrete mathematics includes a chapter on graph theory. The choice what to include in it was made by Prof. Michael (Miki) Tarsi, one of our experts in combinatorics, who was the other principal teacher of the course. Despite the very short time that we were able to allocate to this topic, Miki wanted to teach in our course at least one nontrivial nice result in graph theory, and he chose for that Cayley's formula for the number of trees. The only proof that he knew (and so also I, who have learned the subject from him) was the one which is based on Prüfer's code. Accordingly, Cayley's formula has been taught since then by all of us using basically that proof. However, every year when I was reaching the subject, I gave some thought to it, trying to get deeper understanding of the theorem and its proof, and better ways to present it to the students. At a certain point, I decided that it would be easier to derive Cayley's formula from another formula (which I learned later that Cayley had known too), for which using a Prúfer's code is somewhat easier and clearer. After few years it occurred to me that the easier formula can further be generalized. So I devoted some thinking to the whole subject—and at a certain point I was surprised to realize that I have found a proof of Cayley's formula which is totally different from the one I had known. I also discovered a sequence related to the topic whose limit was the famous number e .

With all my past experience, I could not believe that the proof and limit that I had found were new. Therefore, I sent them first to the experts in combinatorics in Israel that I knew. None of them had been acquainted with either, but I was told by Noga Alon that there are many known proofs of Cayley's formula. So I started to look at the literature, and indeed saw many proofs. In fact, in the famous **Proofs from THE BOOK** (of M. Aigner, G. M. Ziegler) alone I found four.[22] However, none of the proofs I read could be viewed as identical to mine (even though one of the proofs in **Proofs from THE BOOK** was based on a similar approach). Meanwhile, Nachum found a significant simplification of one important step in my proof. With

[21] Like Iddo Lev, Shahar Lev (no family connections) is a former brilliant M.Sc. student and then Ph.D. student of mine, who did not finish his thesis. In his case, the reason was that he got tired of the academy, and left it for challenges at the industry, which he finds as more exciting.

[22] This is a book that presents selected particularly nice proofs from all branches of mathematics—proofs of the type that its authors believe should belong to "THE BOOK", in which, according to Erdös, God keeps the most beautiful mathematical proofs.

that improvement, the proof became the simplest one that Nachum and I know. Therefore, we decided to dare and submit a short notice (about half a page long) to the American Mathematical Monthly, containing our proof and the limit I had found. Its very pretentious title was *Cayley's Formula: A Page from* The Book, and under this name it was accepted and published [121]. Our choice of title was vindicated when a new 6th edition of **Proofs from THE BOOK** was published. In that edition our proof replaces one of the previous four, and it is described there as a "marvelous proof".

And this, I believe, is a marvelous place to end this autobiography.

Arnon Avron

Tel Aviv, Israel
June 2020

Contents

1 Introduction: Non-classical Logics— Between Semantics and Proof Theory (In Relation to Arnon Avron's Work) 1
Ofer Arieli and Anna Zamansky
1.1 Motivation and Scope 1
1.2 Paraconsistency and Nondeterminism 2
1.3 Relevance Logics 4
1.4 Bilattice-Valued Logics 5
1.5 Modal Logics 6
1.6 Other Forms of Non-Classical Reasoning 7
1.7 Conclusion 8
References 8

2 Interpretations of Weak Positive Modal Logics 13
Katalin Bimbó
2.1 Introduction 13
2.2 Groups of Positive Modal Logics 15
2.3 Semantics for the Kernel Logic with Disjunction 19
2.4 Interpreting the Modalities 26
2.5 Semantics for Logics with Conjunction 29
2.6 Further Additions to the Kernel Logics 33
2.7 Conclusions 36
References 36

3 On Axioms and Rexpansions 39
Carlos Caleiro and Sérgio Marcelino
3.1 Introduction 39
3.2 Preliminaries 40
3.3 Adding Axioms 43
3.3.1 A General Construction 43
3.3.2 A Better (Less General) Construction 46
3.4 Worked Examples 49

3.5 Analytic Multiple-Conclusion Calculi 57
3.6 Concluding Remarks 67
References 67

4 Credal Calculi, Evidence, and Consistency 71
Walter Carnielli and Juliana Bueno-Soler
4.1 How Should Logic, Probability, and Their Generalizations Be Related? 71
4.2 Paraconsistency, in the Guise of LFIs 74
4.3 Formal Consistency, Possibilistic Measures, and Knowledge Representation 77
4.3.1 Possibility and Necessity Measures over **Cie** 77
4.3.2 The Principle of Minimum Specificity (**PMS**) and Paraconsistency 82
4.4 Contradictory Evidence via Multi-source Reasoning 85
4.5 Conclusion and Further Challenges 87
References 88

5 Theorems of Alternatives for Substructural Logics 91
Almudena Colacito, Nikolaos Galatos, and George Metcalfe
5.1 Introduction 91
5.2 Preliminaries 93
5.3 Theorems of Alternatives 96
5.4 Interpolation 100
5.5 Density 102
References 104

6 Degree-Preserving Gödel Logics with an Involution: Intermediate Logics and (Ideal) Paraconsistency 107
Marcelo E. Coniglio, Francesc Esteva, Joan Gispert, and Lluis Godo
6.1 Introduction 107
6.2 Preliminaries 112
6.2.1 Truth-Preserving Gödel Logics 112
6.2.2 Degree-Preserving Gödel Logics with Involution 115
6.3 Logics Defined by Matrices Over $[0, 1]_{G\sim}$ by Means of Order Filters 117
6.4 Logics Between $G^{\leq}_{n\sim}$ and CPL 123
6.4.1 The Intermediate Logics of $G^{\leq}_{n\sim}$ for $n > 4$ 124
6.4.2 Example: the Case $n = 5$ 127
6.5 Ideal and Saturated Paraconsistent Extensions of $G^{\leq}_{n\sim}$ 129
6.6 Saturated Paraconsistency and Finite-Valued Łukasiewicz Logics 132
6.7 A Final Remark: Relationship to Logics of Formal Inconsistency 135
6.8 Conclusions 136
References 137

7 R-Mingle is Nice, and so is Arnon Avron 141
J. Michael Dunn
7.1 Introduction 141
7.2 Background. The Systems R of Relevant Implication and E of Entailment 143
7.3 The Creation of R-Mingle 145
7.4 Ohnishi and Matsumoto's System $\mathfrak{S}$ of "Strict Implication" 147
7.5 Various Fragments of R-Mingle—What a Nightmare! 152
7.6 Some of Arnon Avron's Contributions to the Study of R-Mingle 155
7.6.1 Arnon Avron and Hypersequents 155
7.6.2 Arnon Avron's Characterization of Semi-relevant Logics 156
7.6.3 Arnon Avron's Characterization of Relevance Logics 157
7.7 A *Consumer Report* Style Checklist 157
References 163

8 The Strict/Tolerant Idea and Bilattices 167
Melvin Fitting
8.1 Introduction 167
8.2 ST, Classical Logic, and One New Example 168
8.3 ST and $\mathcal{FOUR}$ 170
8.4 Bilattices 172
8.5 Consistent, Anticonsistent, Exact 175
8.6 Logical Bilattices 176
8.7 A Family of Strict/Tolerant Logics 177
8.8 Bilattice Representation Theorems 181
8.9 Logical De Morgan Algebras 184
8.10 Generating Strict/Tolerant Examples 186
8.11 And More? 189
References 190

9 What Is Negation in a System 2020? 193
Dov M. Gabbay
9.1 Negation in Deductive (Monotonic or Non-monotonic) Systems with Cut 193
9.2 Calculus of Failure 209
9.3 Conclusion and Future Research 219
References 220

10 Relevance Domains and the Philosophy of Science 223
Edwin Mares
10.1 Introduction 223
10.2 Relevant Disjunction Structures 225

10.3 Relevance Domains 228
10.4 A Dappled World? 230
10.5 The Problems of Inconsistency 234
10.6 RDS and the Limits of Paraconsistency 237
10.7 A Syntactic Turn 239
10.8 Towards a Theory of RDS Revision 244
10.9 Summing Up 245
References 246

11 Consequence Relations with Real Truth Values 249
Daniele Mundici
11.1 Introduction 249
11.2 Syntax and Semantics of Sentential Logics 252
11.3 Dropping the "Consequentia Mirabilis" Axiom 254
11.4 The Differential Semantics of Łukasiewicz Infinite-Valued Logic 258
11.5 Concluding Remarks 262
References 263

12 Geometric Rules in Infinitary Logic 265
Sara Negri
12.1 Introduction 265
12.2 Syntax 269
12.3 Geometric Implications 270
12.3.1 Examples of Geometric Axioms and Rules 273
12.4 Structural Properties 274
12.4.1 Cut Admissibility 278
12.5 An Intuitionistic Infinitary Calculus 282
12.5.1 A Proof of the Infinitary Barr Theorem 289
12.6 Concluding Remarks and Further Work 290
References 291

13 Connexive Variants of Modal Logics Over FDE 295
Sergei Odintsov, Daniel Skurt, and Heinrich Wansing
13.1 Introduction 296
13.2 Semantics for Connexive FDE-Based Modal Logics 300
13.2.1 Semantics for KFDE 300
13.2.2 Connexive Extensions of KFDE 301
13.2.3 Semantics for cMBL and scMBL 302
13.3 Tableau Calculi for Connexive FDE-Based Modal Logics 303
13.3.1 Tableau Calculi for cBK^{-}, cKN4, $scBK^{-}$, and scKN4 303
13.3.2 Tableau Calculi for cMBL and scMBL 304
13.4 Soundness and Completeness 305
13.4.1 cBK^{-}, cKN4, $scBK^{-}$, and scKN4 305
13.4.2 cMBL and scMBL 307

13.5 Some Properties of cBK^-, cKN4, $scBK^-$, scKN4, cMBL, and scMBL 308
13.6 On Algebraizability of the Connexive Logics cBK^-, cKN4, and cMBL 309
13.7 Conclusion 316
References 316

14 Comments on the Papers 319
Arnon Avron
14.1 Bimbo 319
14.2 Caleiro, Marcelino 320
14.3 Carnielli and Bueno-Soler 321
14.4 Colacito, Galatos, and Metcalfe 322
14.5 Coniglio, Esteva, Gispert, and Godo 323
14.6 Dunn 324
14.7 Fitting 325
14.8 Gabbay 325
14.9 Mares 326
14.10 Mundici 327
14.11 Negri 328
14.12 Odintsov, Skurt, and Wansing 329
References 329

List of Publication 331

Editors and Contributors

About the Editors

Ofer Arieli is a Professor of Computer Science at the School of Computer Science, the Academic College of Tel Aviv. He received his B.Sc. in Mathematics and Computer Science from the Hebrew University in Jerusalem, and M.Sc. and Ph.D. in Computer Science from Tel Aviv University. Afterward (2000–2001), he was a Postdoc Researcher at the Department of Computer Science, the University of Leuven, Belgium. His main research interests are related to the applications of non-classical logics in Artificial Intelligence and to reasoning with incomplete and inconsistent information.

Anna Zamansky is an Associate Professor at the Information Systems Department, University of Haifa. She received her B.A. and M.Sc. in Computer Science from the Technion, Israeli Institute of Technology and her Ph.D. in Computer Science from Tel Aviv University. Afterward (2010–2012), she was a Marie Curie Postdoctoral Researcher in the Computational Logic Group at the Technical University of Vienna, Austria. Her research interests include applied logic and reasoning with inconsistent information.

Both Ofer Arieli and Anna Zamansky were Ph.D. students of Arnon Avron, and have continued collaborating with him ever since, co-authoring a book on the theory of paraconsistent logics (College Publications, 2018).

Contributors

Ofer Arieli School of Computer Science, Tel Aviv Academic College, Tel Aviv, Israel

Arnon Avron School of Computer Science, Tel Aviv University, Tel Aviv, Israel

Katalin Bimbó Department of Philosophy, University of Alberta, Edmonton, AB, Canada

Juliana Bueno-Soler School of Technology, University of Campinas - Unicamp, Limeira, SP, Brazil

Carlos Caleiro SQIG - Instituto de Telecomunicações, Dep. Matemática - Instituto Superior Técnico, Universidade de Lisboa, Lisbon, Portugal

Walter Carnielli Centre for Logic, Epistemology and the History of Science and Department of Philosophy, University of Campinas - Unicamp, Campinas, SP, Brazil

Almudena Colacito Mathematical Institute, University of Bern, Bern, Switzerland

Marcelo E. Coniglio Dept. of Philosophy - IFCH and Centre for Logic, Epistemology and the History of Science, University of Campinas, Campinas, Brazil

J. Michael Dunn Luddy School of Informatics, Computing and Engineering, and Department of Philosophy, Indiana University, Bloomington, IN, USA

Francesc Esteva Artificial Intelligence Research Institute (IIIA) - CSIC, Barcelona, Spain

Melvin Fitting The Graduate School and University Center, City University of New York, New York, NY, USA

Dov M. Gabbay University of Luxembourg, Luxembourg, Luxembourg;
King's College London, London, England

Nikolaos Galatos Department of Mathematics, University of Denver, Denver, CO, USA

Joan Gispert Departament de Matemàtiques i Informàtica, Universitat de Barcelona, Barcelona, Spain

Lluis Godo Artificial Intelligence Research Institute (IIIA) - CSIC, Barcelona, Spain

Sérgio Marcelino SQIG - Instituto de Telecomunicações, Dep. Matemática - Instituto Superior Técnico, Universidade de Lisboa, Lisbon, Portugal

Edwin Mares School Philosophy, Victoria University, Wellington, New Zealand

George Metcalfe Mathematical Institute, University of Bern, Bern, Switzerland

Daniele Mundici Department of Mathematics and Computer Science "Ulisse Dini", University of Florence, Florence, Italy

Sara Negri Department of Mathematics, University of Genoa, Genoa, Italy

Sergei Odintsov Sobolev Institute of Mathematics, Novosibirsk, Russia

Daniel Skurt Department of Philosophy I, Ruhr University Bochum, Bochum, Germany

Heinrich Wansing Department of Philosophy I, Ruhr University Bochum, Bochum, Germany

Anna Zamansky Department of Information Systems, University of Haifa, Haifa, Israel

Chapter 1
Introduction: Non-classical Logics—Between Semantics and Proof Theory (In Relation to Arnon Avron's Work)

Ofer Arieli and Anna Zamansky

Abstract We recall some of the better known approaches to non-classical logics, with an emphasis on the contributions of Arnon Avron to the subject and in relation to the papers in this volume.

1.1 Motivation and Scope

Classical logic is by all means the most extensively studied and applied logic in Mathematics, Philosophy, Engineering, Computer Science, Economy, and other areas. Yet, its original motivation was to capture *mathematical* reasoning, which is monotonic in nature, and does not aim at handling incomplete, imprecise, and/or inconsistent information. Indeed, a major shortcoming of classical logic is that any conclusion whatsoever may be inferred from a classically inconsistent set of premises, thus a single contradiction "pollutes" the whole set of premises. Moreover, situations involving inductive definitions, reasoning over time, representations of norms and obligations, fuzzy concepts, and so forth are not always well captured by "pure" classical logic.

Nowadays, there is an increasing quest for alternative, non-classical formalisms, stimulated by various practical considerations. Reasoning about time, resources, or programs; reasoning with uncertainty or inconsistency; commonsense reasoning—all these gave rise to a plethora of different formalisms: temporal, constructive, substructural, nonmonotonic, paraconsistent, and many more. Many of them have become active fields of research with numerous applications.

O. Arieli (✉)
School of Computer Science, Tel-Aviv Academic College, Tel-Aviv, Israel
e-mail: oarieli@mta.ac.il

A. Zamansky
Department of Information Systems, University of Haifa, Haifa, Israel

O. Arieli and A. Zamansky (eds.), *Arnon Avron on Semantics and Proof Theory of Non-Classical Logics*, Outstanding Contributions to Logic 21,
https://doi.org/10.1007/978-3-030-71258-7_1

In Avron (1999), Arnon Avron indicates that "there is no limit to the number of logics that logicians (and non-logicians) can produce", and identifies three "ingredients" that a "natural" logics should have:

- "natural" primitives, which intuitively correspond to concepts informally used outside the realm of formal logic, such as implication, negation, conjunction, and necessity;
- a simple, illuminating semantics; and
- a "nice" proof system making it easy to find proofs in the corresponding logic.

Indeed, working on the intersection between semantics and proof theory, and investigating families of intuitively motivated (non-classical) formalisms that have "nice" and meaningful proof-theoretical and semantical characterizations, is a main theme in Avron's seminal contributions, to which this volume of the OCL series is devoted.

Practically, it is of course impossible to cover even a small fragment of the non-classical disciplines that have been introduced over the years. (Some introductory books on the subject are, e.g., Bell et al. 2001; Priest 2012; Schechter 2005; van Benthem et al. 2009.) We chose to concentrate here on some fields and approaches in which Avron has made significant contributions at different times of his research career, such as paraconsistent logics, substructural logics, or logics that are based on many-valued semantics, either algebraic or non-deterministic.[1] In the following sections, we recall the subareas of the disciplines that are relevant to the contributions in this volume, and briefly summarize the topics discussed in the related chapters.

1.2 Paraconsistency and Nondeterminism

One of the key principles of classical logic is that of explosion, "ex contradictione sequitur quodlibet", allowing the inference of any proposition from a single pair of contradicting statements. It has been repeatedly attacked on various philosophical grounds, as well as because of practical reasons: in its presence every inconsistent theory or knowledge-base is totally trivial. Paraconsistent logics are alternatives to classical logic which do not have this drawback.

While the roots of paraconsistent thinking may be traced back already to Aristotelian logic, it is commonly agreed that the foundations of paraconsistent reasoning in modern times where laid at the beginning of the twentieth century by the Russian logician (Vasilev, 1993) and the Polish philosopher Łukasiewicz (Lukasiewicz and Wedin 1971). Other paraconsistent systems were later introduced independently by the pioneering works of Jaśkowski (1948), Nelson (1959), Anderson and Belnap

[1] Some other contributions of Avron, which are not related to the theme of this volume, are not covered here. This includes his research on the foundations of mathematics, especially predicative mathematics (e.g., Avron 2008a, 2010; Avron and Cohen 2016), logical frameworks (e.g., Avron et al. 1992; Avron 2008b), as well as purely mathematical results (like Avron 1990a; Avron and Dershowitz 2016).

(1962), and da Costa (1974). In recent years, paraconsistent reasoning is a very active research topic with many applications. Some collections of papers on this subject and further references can be found in Batens et al. (2000), Béziau et al. (2007, 2015), and Carnielli et al. (2001).

Nowadays, Avron is one of the most influential and leading figures in the study of paraconsistent logics, having a variety of works on different aspects and approaches to inconsistent information. In a number of papers (e.g., Avron 2002; Arieli et al. 2011), he has given syntactic and semantic characterizations of what should be regarded a "negation operator", and has defined in a clear and precise way what properties are expected from a logic for reasoning with inconsistency (called "ideal" in Arieli et al. 2011). These and other issues are presented in a comprehensive textbook on paraconsistent logics that Avron co-authored (see Avron et al. 2018).

Definitions of negation operators and the study of their properties and characterizations in different contexts have been a subject for long-standing debates and works in the last decades. We recall, for instance, the collection of papers on this subject in Gabbay and Wansing (1999). The paper of Dov Gabbay in this volume, titled "*What is negation in a system 2020?*", is concerned with these very issues. In the paper, Gabbay recalls his 1986 publication on the concept of negation (see Gabbay 1986) and expands it according to the developments and findings in recent years.

Desirable properties of paraconsistent logics, and in particular the notions of their maximality, are reexamined for degree-preserving Gödel logics in the paper of Marcelo Coniglio, Francesc Esteva, Joan Gispert, and Lluis Godo, titled "*Degree-preserving Gödel logics with an involution: intermediate logics and (ideal) paraconsistency*". The authors introduce the notion of saturated paraconsistency (which relaxes the condition of ideal paraconsistency by not requiring maximality with respect to classical propositional logic), and fully characterize the saturated paraconsistent logics between the degree-preserving finite-valued extensions of Gödel's fuzzy logic with an involutive negation and classical logic. They also identify a large family of saturated paraconsistent logics in the family of intermediate logics for degree-preserving finite-valued Łukasiewicz logics.

One of the most prominent approaches to paraconsistent reasoning, originally developed by da Costa's Brazilian School, encompasses a large family of paraconsistent logics known now as *Logics of Formal Inconsistency* (LFIs). These logics are based on the idea of internalizing the notion of (in)consistency at the object language level. The efforts of a very active group of Brazilian logicians on this family of logics are summarized in Carnielli et al. (2007) as well as a in more recent book (Carnielli and Coniglio 2016). The paper "*Credal calculi, evidence, and consistency*", by Walter Carnielli and Juliana Bueno-Soler studies credal calculi, which are possibility and necessity measures, based on LFIs. These can be used as belief and plausibility measures supporting artificial reasoning that not only automatically practices suspension of judgment, but also respects the beliefs of agents, even when they are contradictory (and so acting as a belief revision procedure).

A significant contribution to the study and understanding of LFIs was obtained by Avron's ideas on generalizing the notion of a multi-valued matrix. In (Avron and Lev, 2005), he introduced a natural generalization of the class of standard multi-valued matrices, called *non-deterministic matrices (Nmatrices)*, which (among others) provide in a modular way simple, useful, and finite semantics for LFIs (as well as for many important logics lacking finite semantics that is based on ordinary, deterministic matrices). This also guides a systematic process in defining analytic Gentzen-type proof systems for LFIs (see Avron et al. 2013, 2015). The paper "*On axioms and rexpansions*" by Carlos Caleiro and Sérgio Marcelino is directly related to the line of research on non-deterministic semantics. In particular, the authors study the general problem of strengthening the logic of a given (partial) (non-deterministic) matrix with a set of axioms, using the idea of rexpansion, a notion introduced by Avron and Zohar (2019).

1.3 Relevance Logics

Relevance logics, introduced by Anderson and Belnap in (Anderson and Belnap, 1962), aim to capture the common view that in valid inferences the assumptions should be *relevant* to the conclusion. As such, relevant logics are non-explosive and so they may be viewed as a kind of paraconsistent logics. We refer the readers to Anderson and Belnap (1975), Read (1988), Anderson et al. (1992), Dunn and Restall (2002), Mares (2004), Bimbó (2006), Avron (2014b), Avron et al. (2018) for some extensive books and surveys on the subject. Avron has contributed to the study on relevance logics throughout his academic career. Examples of recent contributions are Avron (2014a, 2014b, 2016). His most important footprints in this study are his investigations of the semi-relevant logic RM, and the introduction of the relevant logics RMI (Avron 1990b), RMI_m (or $RMI_{\stackrel{\neg}{\rightarrow}}$) (Avron 1984b), and $SRMI_m$ (Avron 1997).

- The logic RMI is a relevant version of the logic RM (see below), in which the implication $\to$ and the additive conjunction $\wedge$ both have the variable sharing property. Its semantics is based on the idea that propositions may be divided into "domains of relevance", with a "relevance relation" R defined on the collection of these domains. Classical logic is valid within each domain, while the propositions $\varphi \to \psi$ and $\varphi \wedge \psi$ are necessarily false if the domains of φ and ψ are not related by R.
- The logic RMI_m is the purely intensional (or "multiplicative") fragment of RMI, and has particularly nice properties. In particular, it has the Scroggs property (Anderson and Belnap 1975; Avron 2016), an ordinary cut-free Gentzen-type formulation, and is sound and weakly complete with respect to a very simple infinite matrix called $\mathscr{A}_\omega$.
- The logic $SRMI_m$ is the extension of RMI_m with its admissible rule $\varphi \otimes \psi / \varphi$, where $\otimes$ is the multiplicative "conjunction". Avron showed that unlike RMI_m, the logic $SRMI_m$ is *strongly* sound and complete with respect to $\mathscr{A}_\omega$, and he provided for it too a corresponding cut-free Gentzen-type formulation.

In the more general context of the study of substructural logics, it is worth mentioning the influential work of Avron (1988), where he has pinpointed the relation between relevance logic and linear logic.

In this volume, relevant reasoning is considered from several perspectives, some of them are related to the contributions of Avron to the subject.

- In "*R-mingle has nice properties, and so does Arnon Avron*", Michael Dunn provides a nice overview on the relevant logic RM, introduced by him and McCall.[2] As noted in Dunn and Restall (2002), RM is "*by far the best understood of the Anderson-Belnap style systems*". Indeed, RM has sound and complete Hilbert- and Gentzen-type proof systems and a clear semantics in terms of Sugihara matrices. Moreover, RM has some other desirable characteristics, such as being decidable, paraconsistent, and it satisfies the Scroggs' property. In his paper, Dunn describes the history of RM, including the system of Ohnishi and Matsumoto as well as his own experience with a problem concerning entailment (the logic E) as an intersection of a pair of logics that led to the invention of RM. In the process, a series of results about RM are mentioned, also in relation to the extensive study of RM by Avron.
- In his paper, "*Relevance domains and the philosophy of science*", Edwin Mares applies a variant of Avron's logic RMI to model what the philosopher of science Nancy Cartwright has called the "dappled world". In this world, scientific theories represent restricted aspects and regions of the universe. Mares characterizes such theories by Avron's algebraic structures that are used for giving semantics to RMI, and shows, among others, how the paraconsistent nature of RMI can be used for dealing with inconsistencies within and between the scientific theories.
- The paper of Almudena Colacito, Nikolaos Galatos, and George Metcalfe, titled "*Theorems of alternatives for substructural logics*", consists of a generalization of an early result by Avron about the logic RM. It shows that Avron's theorem belongs to a family of results that may be understood as "theorems of alternatives" for substructural logics. It is shown that a variety of logics admit such a theorem, and the relation with interpolation and density is discussed.

1.4 Bilattice-Valued Logics

In (Belnap 1977a, 1977b), Belnap introduced a framework for collecting and processing information coming from different sources. His formalism is based on Dunn's four-valued algebraic structure (Dunn 1966), in which the elements are simultaneously arranged in two lattice orders. This structure may be viewed as a particular case of Ginsberg's *bilattices* (Ginsberg 1988), which have been shown particularly useful for providing fixpoint semantics to logic programs (Fitting 1991, 1993, 2002), and

[2] Sadly, Prof. J. Michael Dunn passed away prior to witnessing the publication of this volume. We shall cherish memories of him as a great logician, and as the title of his manuscript suggests, a very kind and humble man.

for fuzzy (Cornelis et al. 2007) and paraconsistent (Arieli and Avron 1996, 1998, 2000a) reasoning (see also the survey in Fitting 2006).

In (Arieli and Avron, 1998), Avron has shown that the Dunn–Belnap four-valued logic is a characteristic logic among bilattice-based logics, and related these logics to nonmonotonic and preferential reasoning (Arieli and Avron 2000b). He also made an important contribution to the algebraic study of bilattices by investigating the structure of a family of bilattices, called interlaced bilattices (see Avron 1996b).

This volume contains two contributions that are related to bilattice-based reasoning. One, by Melvin Fitting (titled "*The strict/tolerant idea and bilattices*"), presents a general theory of strict/tolerant versus classical counterparts for non-distributive (as well as distributive) De Morgan logics. The algorithm for constructing the strict/tolerant logic makes use of bilattice products, which provide interlaced logical bilattices with negation and conflation. In process, Fitting gives an overview of the essentials of the bilattice theory.

In the other contribution, "*Connexive variants of modal logics over FDE*", Sergei Odintsov, Daniel Skurt, and Heinrich Wansing relate connexive logics (Wansing 2014), modalities, and bilattice-valued semantics, through a series of (paraconsistent and decidable) logics, to which they provide sound and complete tableau calculi. To some of the systems, algebraizability in the sense of Blok and Pigozzi is also established.

1.5 Modal Logics

The incorporation in the language of modal operators is a well-established and common method for non-classical reasoning that has many successful applications (see, e.g., Bull and Segerberg 2001 and Chellas 1980 for some introductory manuscripts to the subject). The main contributions of Avron to this area are threefold:

- Avron's first contribution was his paper in (Avron, 1984a). Among other things, in this paper, he introduced sequent calculi for the modal provability logics GL and Grz, and proved cut-elimination for both of them.[3] He further showed that contrary to what was believed and even published before, the natural first-order extension of the sequent calculus for GL does *not* admit cut-elimination. This was the first negative result in this area.
- The second contribution of Avron has been due to his generalization of Gentzen's sequents, known as "*hypersequents*" (Avron 1987, 1996a). A hypersequent is a finite set (or sequence) of sequents, which may be regarded as their disjunction. The original motivation for introducing these structures was to provide cut-free Gentzen-type systems for some relevance logics. However, Avron soon discovered that it is useful for other families of logics too. Thus, beginning with his paper in (Avron, 1991), hypersequents provide a major framework for the proof theory

[3] In the case of Grz, he was the first to do so.

of fuzzy logics (see Metcalfe et al. 2009). In the context of modal logics, the incorporation of hypersequents allowed him to provide a cut-free calculus for *S5* (to which an ordinary cut-free sequent calculus is not known, see Avron 1996a; Avron and Lahav 2018).

- Avron's third contribution to this area is related to his work on paraconsistent reasoning, and it is concerned with establishing its connections to two famous modal logics, *B* and *S5* (see Avron and Zamansky 2016 and Avron et al. 2018, Chap. 9). In particular, it was shown that the minimal paraconsistent logic which satisfies the replacement property (i.e., equivalence of two formulas implies their congruence) is equivalent to the well-known Brouwerian modal logic *B*. Interestingly, *B* is a very robust paraconsistent logic, in the sense that almost any axiom which has been used in the context of LFIs (see Sect. 1.2) is either already a theorem of *B*, or its addition to it leads to a logic which is no longer paraconsistent. There is only one exceptional axiom, the addition of which leads to another famous modal logic: *S5* (the modal logic which is induced by the class of Kripke frames in which the accessibility relation is an equivalence relation).

Modal logics are discussed in this volume with respect to different frameworks. We have already mentioned in the previous section the paper of Odintsov, Skurt, and Wansing that studies various connexive modal logics. Modal notions are also central in the paper of Carnielli and Bueno-Soler, which is mentioned in Sect. 1.2. In another paper on the subject, titled "*Interpretations of weak positive modal logics*", Katalin Bimbó examines relational semantics for two positive (negation-free) modal logics: one contains conjunction but not disjunction, and the other contains disjunction but not conjunction. Both of these logics have implication, fusion, and fission, and they make room for the development of sequent calculi in which the two basic modal connectives may be introduced independently (but can be defined from each other in the presence of a suitable negation). The two logics are inspired by related works in the context of relevance and linear logics.

1.6 Other Forms of Non-Classical Reasoning

This volume contains some chapters that are related to further applications of non-classical logics, which are not covered in the previous sections. They are summarized below.

In his paper "*Consequence relations with real truth-values*", Daniele Mundici draws inspiration from Avron's paper in (Avron, 2015), where he investigates a general notion of implication that does not assume the availability of any proof system and thus does not depend on the notion of a "use" of a formula in a given proof system. This notion typically occurs in relevance logics, suggesting a generalized semi-implication, which leads to a weak form of the classical-intuitionistic deduction theorem. Mundici builds on these ideas using a similar approach in the context of a [0,1]-valued Łukasiewicz logic, also revising the Bolzano–Tarski paradigm of "semantic consequence". It is shown that the Łukasiewicz axiom guarantees the continuity and the piecewise linearity of the implication operation, a desirable fault-tolerance property of any real-valued logic.

Avron has also contributed to the mechanization of mathematics. One of the main tools he has suggested for this (e.g., in Avron 2003; Avron and Cohen 2016) is the use of ancestral logic (an extension of first-order logic with an operation for transitive closure) in order to overcome the insufficiency of first-order logic in dealing with some notions and constructions in mathematics. Together with Liron Cohen, Avron has considered and applied two versions of ancestral logic: classical and intuitionistic. In her paper *"Geometric rules in infinitary logic"*, Sara Negri takes a different approach. Instead of using transitive closure, she concentrates on the theories which are based on the very large and central class of what are called geometric axioms. On the other hand, she allows the use in the language and in proofs of infinitary disjunctions. Again, her system has two versions: classical and intuitionistic. As an application, Negri presents a simple proof, in which the axioms of choice is not used, of the infinitary Barr's theorem. This theorem connects classical derivability of geometric implications with their intuitionistic derivability.

1.7 Conclusion

The content of this volume is very diverse, representing the state of the art of the logical study on reasoning with non-classical logics in different contexts and for different purposes. As we have already noted previously, it is not possible to have a complete coverage of the area in one volume. In fact, this volume does not even pretend to provide an exhaustive reference of all the contributions of Avron to the subject. Having said this, we believe that the chapters of this book, written by worldwide leading experts in the area, cover many of the active research subjects in contemporary (non-classical) logic, and faithfully reflect the diversity and mathematical depth of Arnon Avron's work.

References

Anderson, A. R., & Belnap, N. D. (1962). The pure calculus of entailment. *Journal of Symbolic Logic*, *27*, 19–52.

Anderson, A. R., & Belnap, N. D. (1975). *Entailment: The Logic of Relevance and Necessity* (Vol. I). Princeton: Princeton University Press.

Anderson, A. R., Belnap, N. D., & Dunn, J. M. (1992). *Entailment: The Logic of Relevance and Necessity* (Vol. II). Princeton: Princeton University Press.

Arieli, O., & Avron, A. (1996). Reasoning with logical bilattices. *Journal of Logic, Language, and Information*, *5*(1), 25–63.

Arieli, O., & Avron, A. (1998). The value of the four values. *Artificial Intelligence*, *102*(1), 97–141.

Arieli, O., & Avron, A. (2000a). Bilattices and paraconsistency. In D. Batens, C. Mortensen, G. Priest, & J. Van Bendegem (Eds.), *Frontiers of Paraconsistent Logic* (Vol. 8, pp. 11–27). Studies in Logic and Computation. Research Studies Press.

Arieli, O., & Avron, A. (2000b). General patterns for nonmonotonic reasoning: from basic entailments to plausible relations. *Logic Journal of the IGPL*, *8*(2), 119–148.

Arieli, O., Avron, A., & Zamansky, A. (2011). Ideal paraconsistent logics. *Studia Logica*, *99*(1–3), 31–60.

Avron, A. (1984a). On modal systems having arithmetical interpretations. *Journal of Symbolic Logic*, *49*, 935–942.

Avron, A. (1984b). Relevant entailment - semantics and formal systems. *Journal of Symbolic Logic*, *49*, 334–342.

Avron, A. (1987). A constructive analysis of RM. *Journal of Symbolic Logic*, *52*, 939–951.

Avron, A. (1988). The semantics and proof theory of linear logic. *Theoretical Computer Science*, *57*, 161–184.

Avron, A. (1990a). On strict strong constructibility with a compass alone. *Journal of Geometry*, *38*, 12–15.

Avron, A. (1990b). Relevance and paraconsistency - A new approach. *Journal of Symbolic Logic*, *55*, 707–732.

Avron, A. (1991). Hypersequents, logical consequence and intermediate logics for concurrency. *Annals of Mathematics and Artificial Intelligence*, *4*, 225–248.

Avron, A. (1996a). The method of hypersequents in proof theory of propositional non-classical logics. In *Logic: Foundations to Applications* (pp. 1–32). Oxford Science Publications.

Avron, A. (1996b). The structure of interlaced bilattices. *Journal of Mathematical Structures in Computer Science*, *6*, 287–299.

Avron, A. (1997). Multiplicative conjunction as an extensional conjunction. *Logic Journal of the IGPL*, *5*, 181–208.

Avron, A. (1999). On the proof theory of natural many-valued logics. *Collegium Logicum (Annals of the Kurt Gödel Society)*, *3*, 51–59.

Avron, A. (2002). On negation, completeness and consistency. In *Handbook of Philosophical Logic* (Vol. 9, pp. 287–319). Kluwer.

Avron, A. (2003). Transitive closure and the mechanization of mathematics. In *Thirty Five Years of Automating Mathematics* (pp. 149–171). Springer.

Avron, A. (2008a). Constructibility and decidability versus domain independence and absoluteness. *Theoretical Computer Science*, *394*, 144–158.

Avron, A. (2008b). A framework for formalizing set theories based on the use of static set terms. In A. Avron, N. Dershowitz, & A. Rabinovich (Eds.), *Pillars of Computer Science* (Vol. 4800, pp. 87–106). LNCS. Springer.

Avron, A. (2010). A new approach to predicative set theory. In R. Schindler (Ed.), *Ways of Proof Theory* (pp. 31–63). Onto series in mathematical logic. Onto verlag.

Avron, A. (2014a). The classical constraint on relevance. *Logica Universalis*, *8*, 1–15.

Avron, A. (2014b). What is relevance logic? *Annals of Pure and Applied Logic*, *165*, 26–48.

Avron, A. (2015). Semi-implication: A chapter in universal logic. In *The Road to Universal Logic* (pp. 59–72). Springer.

Avron, A. (2016). RM and its nice properties. In K. Bimbó (Ed.), *J. Michael Dunn on Information Based Logics* (Vol. 8, pp. 15–43). Outstanding Contributions to Logic. Springer.

Avron, A., Arieli, O., & Zamansky, A. (2018). *Theory of Effective Propositional Paraconsistent Logics* (Vol. 75), Studies in Logic. College Publications.

Avron, A., & Cohen, L. (2016). Formalizing scientifically applicable mathematics in a definitional framework. *Journal of Formalized Reasoning*, *9*(1), 53–70.

Avron, A., & Dershowitz, N. (2016). Cayley's formula: A page from the book. *American Mathematical Monthly*, *123*(7), 699–700.

Avron, A., Honsell, F. A., Mason, I. A., & Pollack, R. (1992). Using typed lambda calculus to implement formal systems on a machine. *Journal of Automated Reasoning*, *9*, 309–354.

Avron, A., Konikowska, B., & Zamansky, A. (2013). Cut-free sequent calculi for C-systems with generalized finite-valued semantics. *Journal of Logic and Computation*, *23*(3), 517–540.
Avron, A., Konikowska, B., & Zamansky, A. (2015). Efficient reasoning with inconsistent information using C-systems. *Information Science*, *296*, 219–236.
Avron, A., & Lahav, O. (2018). A simple cut-free system for a paraconsistent logic equivalent to S5. In *Advances in Modal Logic* (Vol. 12, pp. 29–42). College Publications.
Avron, A., & Lev, I. (2005). Non-deterministic multi-valued structures. *Journal of Logic and Computation*, *15*, 241–261.
Avron, A., & Zamansky, A. (2016). A paraconsistent view on B and S5. In *Advances in Modal Logic* (Vol. 11, pp. 21–37). College Publications.
Avron, A., & Zohar, Y. (2019). Rexpansions of nondeterministic matrices and their applications in nonclassical logics. *The Review of Symbolic Logic*, *12*(1), 173–200.
Batens, D., Mortensen, C., Priest, G., & Van Bendegem, J. (Eds.). (2000). In *Frontiers of Paraconsistent Logic, Proceedings of the first World Congress on Paraconsistency* (Vol. 8). Studies in Logic and Computation. Research Studies Press.
Bell, J., DeVidi, D., & Solomon, G. (2001). *Logical Options: An Introduction to Classical and Alternative Logics*. Broadview Press.
Belnap, N. D. (1977a). How a computer should think. In G. Ryle (Ed.), *Contemporary Aspects of Philosophy* (pp. 30–56). Oriel Press.
Belnap, N. D. (1977b). A useful four-valued logic. In J. M. Dunn, & G. Epstein (Eds.), *Modern Uses of Multiple-Valued Logics* (pp. 7–37). Reidel Publishing Company.
Béziau, J. Y., Carnielli, W., & Gabbay, D. (Eds.). (2007). *New Directions in Paraconsistent Logic: Proceedings of the 3rd World Congress of Paraconsistent Logic* (Vol. 9). Studies in Logic. College Publications.
Béziau, J. Y., Chakraborty, M., & Dutta, S. (Eds.). (2015). *New Directions in Paraconsistent Logic: Proceedings of the 5th World Congress of Paraconsistent Logic* (Vol. 152). Lecture Notes in Mathematics and Statistics. Springer.
Bimbó, K. (2006). Relevance logics. In D. Jacquette (Ed.), *Philosophy of Logic* (pp. 723–789). Elsevier.
Bull, R. A., & Segerberg, K. (2001). Basic modal logic. In D. Gabbay, & F. Guenther (Eds.), *Handbook of Philosophical Logic* (2nd ed., Vol. 3, pp. 1–81). Kluwer.
Carnielli, W. A., & Coniglio, M. E. (2016). *Paraconsistent Logic: Consistency, Contradiction and Negation*. Number 40 in Logic, Epistemology, and the Unity of Science. Springer.
Carnielli, W. A., Coniglio, M. E., & D'Ottaviano, I. (Eds.) (2001). *Paraconsistency: The Logical Way to the Inconsistent – Proceedings of the second World Congress on Paraconsistency*. Number 228 in Lecture Notes in Pure and Applied Mathematics. Marcel Dekker.
Carnielli, W. A., Coniglio, M. E., & Marcos, J. (2007). Logics of formal inconsistency. In D. M. Gabbay & F. Guenthner (Eds.), *Handbook of Philosophical Logic* (2nd ed., Vol. 14, pp. 1–95). Springer.
Chellas, B. F. (1980). *Modal Logic – An introduction*. Cambridge University Press.
Cornelis, C., Arieli, O., Deschrijver, G., & Kerre, E. (2007). Uncertainty modeling by bilattice-based squares and triangles. *IEEE Transactions on Fuzzy Systems*, *15*(2), 161–175.
da Costa, N. C. A. (1974). On the theory of inconsistent formal systems. *Notre Dame Journal of Formal Logic*, *15*, 497–510.
Dunn, J. M. (1966). *The Algebra of Intensional Logics*. PhD thesis, University of Pittsburgh, Ann Arbor (UMI).
Dunn, J. M., & Restall, G. (2002). Relevance logic. In D. Gabbay & F. Guenther (Eds.), *Handbook of Philosophical Logic* (2nd ed., Vol. 6, pp. 1–136). Kluwer.
Fitting, M. (1991). Bilattices and the semantics of logic programming. *Journal of Logic Programming*, *11*(2), 91–116.
Fitting, M. (1993). The family of stable models. *Journal of Logic Programming*, *17*, 197–225.
Fitting, M. (2002). Fixpoint semantics for logic programming a survey. *Theoretical Computer Science*, *278*(1–2), 25–51.

Fitting, M. (2006). Bilattices are nice things. In T. Bolander, V. Hendricks & S. A. Pedersen (Eds.), *Self Reference* (Vol. 178, pp. 53–77). CSLI Lecture Notes. CLSI Publications.
Gabbay, D. (1986). What is negation in a system? In F. R. Drake & J. K. Truss (Eds.), *Logic Colloquium '86* (pp. 95–112). North Holland: Elsevier Science Publishers.
Gabbay, D., & Wansing, H. (Eds.). (1999). *What is Negation?* Applied Logic Series (Vol. 13). Springer.
Ginsberg, M. (1988). Multi-valued logics: A uniform approach to reasoning in AI. *Computer Intelligence*, *4*, 256–316.
Jaśkowski, S. (1999). A propositional calculus dor inconsistent deductive systems. *Logic, Language and Philosophy*, *7*, 35–56. Translation of the original paper in *Studia Societatis Scienctiarun Torunesis*, Sectio A, *I*(5), 57–77 (1948).
Lukasiewicz, J. & Wedin, V. (1971). On the principle of contradiction in aristotle. *The Review of Metaphysics*, pp. 485–509.
Mares, E. D. (2004). *Relevant Logic*. Cambridge University Press.
Metcalfe, G., Olivetti, N., & Gabbay, D. (2009). *Proof Theory for Fuzzy Logics*. Springer.
Nelson, D. (1959). Negation and separation of concepts in constructive systems. In A. Heyting (Ed.), *Constructivity in Mathematics* (pp. 208–225). North Holland.
Priest, G. (2012). *An Introduction to Non-Classical Logic*. Cambridge University Press.
Read, S. (1988). *Relevant Logic*. Basil Blackwell.
Schechter, E. (2005). *Classical and Nonclassical Logics: An Introduction to the Mathematics of Propositions*. Princeton University Press.
van Benthem, J., Heinzmann, M., Rebuschi, G., & Visser, H. (2009). *The Age of Alternative Logics* (2nd ed.). Springer.
Vasilev, N. A. (1993). Imaginary (non-aristotelian) logic. *Axiomathes*, *4*(3), 353–355.
Wansing, H. (2014). Connexive logic. Stanford Encyclopedia of Philosophy. https://plato.stanford.edu/entries/logic-connexive/.

Ofer Arieli is a Professor of Computer Science at the School of Computer Science, the Academic College of Tel-Aviv. He received his B.Sc. in Mathematics and Computer Science from the Hebrew University in Jerusalem, and M.Sc. and Ph.D. in Computer Science from Tel-Aviv University. Afterward (2000-2001), he was a Postdoc Researcher at the Department of Computer Science, the University of Leuven, Belgium. His main research interests are related to the applications of non-classical logics in Artificial Intelligence and to reasoning with incomplete and inconsistent information.

Anna Zamansky is an Associate Professor at the Information Systems Department, University of Haifa. She received her B.A. and M.Sc. in Computer Science from the Technion, Israeli Institute of Technology and her Ph.D. in Computer Science from Tel-Aviv University. Afterward (2010-2012), she was a Marie Curie Postdoctoral Researcher in the Computational Logic Group at the Technical University of Vienna, Austria. Her research interests include applied logic and reasoning with inconsistent information.

Chapter 2
Interpretations of Weak Positive Modal Logics

Katalin Bimbó

Abstract This paper investigates *set-theoretical semantics* for logics that contain unary connectives, which can be viewed as modalities. Indeed, some of the logics we consider are closely related to linear logic. We use insights from the relational semantics of relevance logics together with a new version of the *squeeze lemma* in our semantics for logics with disjunction (but no conjunction). The ideal-based semantics, which takes co-theories to be situations, *dualizes* the theory-based semantics for logics with conjunction (but no disjunction).

Keywords Relational semantics · Three-termed relation · Modal logic · Relevance logic · Sequent calculus · Semi-lattice

2.1 Introduction

Modal concepts—especially *necessity* and to a lesser extent *possibility*—intrigued thinkers for many centuries, as attested by Cresswell et al. (2016). The modern approach to modal logics that contain unary connectives as formal counterparts of necessity and possibility originates in the work of C. I. Lewis. He preferred a particular modal logic, nonetheless, defined several logics of strict implication (a connective, which combines necessity and material conditional). Gödel (1986) introduced a provability interpretation for modal connectives, and with the set-theoretical semantics blossoming, a variety of informal meanings were attached to unary connectives from the 1950s onward. On the other hand, investigations into proof systems shifted from an exclusively axiomatic approach to more constrained calculuses. Different viewpoints do not always lead to a preference for one logic. For example, S5 is arguably the simplest normal modal logic from a semantical point of view, because in its models all worlds can be linked through the total relation on worlds. And, of course, S5

K. Bimbó (✉)
Department of Philosophy, University of Alberta, Edmonton, AB T6G 2E7, Canada
e-mail: bimbo@ualberta.ca
URL: http://www.ualberta.ca/~bimbo

O. Arieli and A. Zamansky (eds.), *Arnon Avron on Semantics and Proof Theory of Non-Classical Logics*, Outstanding Contributions to Logic 21,
https://doi.org/10.1007/978-3-030-71258-7_2

has a rather uncomplicated axiomatic formulation too. However, there is no elegant sequent calculus as an extension of LK from Gentzen (1964) that formalizes S5. However, a more general class of proof systems, the so-called hypersequent calculuses, can be used to formalize S5 too. Hypersequent calculuses were introduced by Arnon Avron (1987), and they proved to be a versatile framework with wide-ranging applications. See Avron's articles Avron (1991, 1996).

The normal modal system S4 turned out to be an all-around pleasant normal modal logic. It has natural connections to intuitionistic logic and topology; it can be modeled in set-theoretical semantics on quasi-ordered frames; it has a sequent calculus formalization. Normal modal logics are often axiomatized in a language with necessity ($\Box$) where possibility ($\Diamond$) is merely an afterthought (or a defined connective). Accordingly, one might completely omit $\Diamond$, and then two straightforward sequent calculus rules can introduce $\Box$, which happens to be S4's necessity. These rules were added to relevance logics too, in particular, Meyer (1966) added them to the sequent calculus formulation of lattice-R (the non-distributive version of the logic of relevant implication). Kripke (1963) suggested sequent calculus rules that allow the proof of the equivalence of $\neg\Box\neg\mathcal{A}$ and $\Diamond\mathcal{A}$. Thereby, he made room for the development of sequent calculuses in which $\Box$ and $\Diamond$ (or similar connectives) may be introduced independently, but they can be proved to be definable from each other in the presence of a suitable negation.

The logics that we consider below have further motivations. Girard (1987) introduced linear logic that uses two punctuation symbols for a pair of unary connectives. Avron (1988) noted that the "of-course" connective (i.e., $!$) is similar to $\Box$ and the "why-not" connective (i.e., $?$) is like $\Diamond$. The analogy cannot go all the way, because linear logic is not the same logic as S4, for instance, conjunction and disjunction cannot be proved to distribute over each other due to the absence of unrestricted weakening and contraction rules.[1] The modalities of linear logic, which are also called exponentials, are essential to linear logic. MALL (linear logic without modalities) is a somewhat uninteresting logic, and it is, for example, easily shown to be decidable. The introduction of the exponentials allows for a modeling of the use of resources. Contingent facts must be handled with tight control, whereas information that is necessary can be always assumed, and information that is possible may be concluded at any time.

In the logics we consider, we will omit some connectives, but these will not be the modalities. First of all, negation is completely excluded from our logics; hence, the adjective "positive" in the title. It is an interesting question that was investigated by Dunn (1995) which axioms would suffice in the context of a negation-free logic with distributive conjunction and disjunction to force $\Box$ and $\Diamond$ to have an interpretation through a shared binary relation. However, it seems to us that those axioms would require the inclusion of the usual structural rules into a sequent calculus; hence, we

[1] Semantical considerations may suggest that $!$ and $\Diamond$, on one hand, plus $?$ and $\Box$, on the other hand, are better pairings (cf. Bimbó 2007; Bimbó and Dunn 2008). However, both the proof-theoretical and the semantic analogies are mere analogies, which we underline by not using a notation (like $\Box$ or $\Diamond$) that have deeply ingrained connotations in modal logic.

do not aim to focus on logics in which modalities share part of their interpretation. Second, we want to omit one of conjunction or disjunction too. Scrutinizing fragments of a logic help us understand the whole logic. It may appear that omitting two connectives will unreasonably diminish the expressive power of our logics. However, it is not so; one of the logics that we consider is equivalent to full linear logic in a smaller vocabulary. (See Kopylov 2001 for a reduction of full linear logic.) This is a strong motivation to investigate this group of logics.

We start the next section by introducing some positive modal logics through *sequent calculuses*. We will select certain rules as immutable, and permit some variations of the other rules. Section 2.3 defines a *dual semantics* for logics that contain disjunction but no conjunction. The inspiration here comes from the idea that it depends on our viewpoint whether a semi-lattice is a meet semi-lattice or a join semi-lattice. Then, in Sect. 2.5, we proceed to defining a semantics for those logics that have only conjunction. In Sect. 2.6, we consider the semantic impact of some further variations in the logics. Finally, we add some concluding remarks in Sect. 2.7.

2.2 Groups of Positive Modal Logics

Sequent calculuses provide an elegant and controlled way to define logics. We want to limit our consideration to logics that share certain connectives except that in some of the logics there is a conjunction connective but there is no disjunction, whereas in the other logics there is a disjunction connective but no conjunction.

We already mentioned that investigating fragments of logics adds to our understanding of a logic. Also, some of the logics are of independent interest, because they were introduced without considerations for the other logics in their group. For instance, conjunction, which is in many ways the easiest to deal with among the connectives, has been added to pure implicational relevance logics. Beyond these reasons to consider some of these logics, we think that there is a theoretical interest in investigating the semantics for our two groups of logics; furthermore, it is fruitful to do this by placing them side by side.

Further, we will allow some variance in the choice of (pairs of) structural rules. The structural rules impact the properties of all the connectives except the two lattice connectives (conjunction and disjunction), which we introduce through rules that are independent of the structural connective (i.e., ;). The changes in the meaning of the affected connectives will be reflected algebraically and in the matching semantical conditions. (We give at once all the connective and structural rules from which we will select certain rules for our logics.)

Definition 2.2.1 (*Languages*) The languages are defined over a denumerable set of *propositional variables*, which we denote by $\mathbb{P}$. The *signatures* in the two groups of logics are $\langle t^0, \circ^2, \to^2, +^2, \wedge^2, \rhd^1, \lhd^1\rangle$ and $\langle t^0, \circ^2, \to^2, +^2, \vee^2, \rhd^1, \lhd^1\rangle$, where a superscript indicates the number of arguments for a connective. The *set of formulas* is generated by the following (context-free) grammar, where $\mathbb{P}$ may be rewritten as any of its elements, and $\curlywedge$ is either $\wedge$ or $\vee$—depending on the signature.

$$\mathcal{A} := \mathbb{P} \mid t \mid (\mathcal{A} \circ \mathcal{A}) \mid (\mathcal{A} \to \mathcal{A}) \mid (\mathcal{A} + \mathcal{A}) \mid (\mathcal{A} \curlywedge \mathcal{A}) \mid \rhd\mathcal{A} \mid \lhd\mathcal{A}$$

Remark 2.2.1 We use the symbols $\rhd$ and $\lhd$ for the unary connectives that we consider *modalities*. If we would have the same background logic as in normal modal logics (i.e., if we would have two-valued propositional logic as a basis), then $\rhd$ could be viewed as $\Box$ and $\lhd$ as $\Diamond$ (using usual notation from alethic modal logics) due to the rules that introduce these connectives. However, the logics we start with lack any sort of negation and some of the usual structural rules may be omitted too. To guard against misleading connotations, we use the symbols $\rhd$ and $\lhd$ for the unary connectives.[2] Should it be required, we shall call $\rhd$ a *solid* and $\lhd$ a *fluid* modality. The intuition behind the labels is that $\rhd A$ can be introduced from A in the antecedent, while $\lhd A$ can be obtained from A in the succedent. That is, $\rhd A$ is a very firm premise, and $\lhd A$ is a flaccid conclusion.

We will use *finite multisets* of formulas in the sequent calculuses below. (We will drop the adjective "finite" from now on, because we nowhere use infinite multisets.) A multiset comprises finitely many *tokens* that are of finitely many *types*. A usual notation for a multiset employs brackets and commas. For instance, $[\mathcal{A}, \mathcal{A}, \mathcal{B}, \mathcal{B}, \mathcal{B}, \mathcal{D}, \mathcal{E}]$ is a multiset with exactly *seven tokens* that fall into *at most four types*. In a multiset, the order of listing the elements does not matter, but the number of occurrences of an element makes a difference, that is, multisets are halfway between sets and lists. We use $\Gamma, \Theta, \ldots$ as variables for multisets of formulas (including the empty multiset). As usual in sequent calculuses, we will omit the delimiters, and we will replace commas with semicolons, because of a link to intensional connectives.[3] $\mathcal{A}; \Gamma$ is the same multiset as $\Gamma; \mathcal{A}$, that is, the multiset union of Γ and $[\mathcal{A}]$.

Definition 2.2.2 (*Sequents*) A *sequent* is an ordered pair of multisets of formulas. Instead of $\langle \Gamma, \Theta \rangle$, we write the sequent as $\Gamma \Vdash \Theta$.

All the rules that we may include in any of the logics we consider are collected together in the next definition. $\Gamma^{\rhd}$ indicates that each element of Γ starts with $\rhd$; similarly, for a Θ with $\lhd$.

Remark 2.2.2 None of the rules below are new with us. The rule and the axiom for t originated in relevance logic (cf. Dunn 1973, 1986). The rules for $\wedge$ and $\vee$ are independent of ; , because only one of A and B occurs in the one-premise rules, and Γ and Θ are shared in the two-premise rules. The rules $\circ$, $\to$, and $+$ are the usual rules for fusion, implication, and fission in the context of relevance logic (see, e.g., Dunn 1986 and Bimbó and Dunn 2015). In the linear logic framework, these are the rules

---

[2] The notation used here is the same notation that we used in Bimbó (2017).

[3] Semicolons were introduced in sequent calculuses for positive relevance logics, where it is essential to have multiple structural connectives. See, for example, Dunn (1973, 1986).

for $\otimes$, $\multimap$ and $\invamp$. These three connectives are intensional connectives—informally speaking—hence, their rules are structure dependent. We already discussed connections between S4 and the introduction rules for the modalities. We may mention though that weaker (than S4) modal logics have been formalized with more restricted rules.

Definition 2.2.3 The list of axioms and rules is as follows.

Axioms: $\mathcal{A} \Vdash \mathcal{A}$ (I) $\quad \Vdash t$ $(\Vdash t)$

Operational rules:

$$\frac{\Gamma; \mathcal{A} \Vdash \Theta}{\Gamma; \mathcal{A} \wedge \mathcal{B} \Vdash \Theta}\,(\wedge\Vdash_1) \qquad \frac{\Gamma; \mathcal{B} \Vdash \Theta}{\Gamma; \mathcal{A} \wedge \mathcal{B} \Vdash \Theta}\,(\wedge\Vdash_2) \qquad \frac{\Gamma \Vdash \mathcal{A}; \Theta \quad \Gamma \Vdash \mathcal{B}; \Theta}{\Gamma \Vdash \mathcal{A} \wedge \mathcal{B}; \Theta}\,(\Vdash\wedge)$$

$$\frac{\Gamma; \mathcal{A} \Vdash \Theta \quad \Gamma; \mathcal{B} \Vdash \Theta}{\Gamma; \mathcal{A} \vee \mathcal{B} \Vdash \Theta}\,(\vee\Vdash) \qquad \frac{\Gamma \Vdash \mathcal{A}; \Theta}{\Gamma \Vdash \mathcal{A} \vee \mathcal{B}; \Theta}\,(\Vdash\vee_1) \qquad \frac{\Gamma \Vdash \mathcal{B}; \Theta}{\Gamma \Vdash \mathcal{A} \vee \mathcal{B}; \Theta}\,(\Vdash\vee_2)$$

$$\frac{\Gamma \Vdash \Theta}{\Gamma; t \Vdash \Theta}\,(t\Vdash) \qquad \frac{\Gamma; \mathcal{A}; \mathcal{B} \Vdash \Theta}{\Gamma; \mathcal{A} \circ \mathcal{B} \Vdash \Theta}\,(\circ\Vdash) \qquad \frac{\Gamma \Vdash \mathcal{A}; \Theta \quad \Lambda \Vdash \mathcal{B}; \Psi}{\Gamma; \Lambda \Vdash \mathcal{A} \circ \mathcal{B}; \Theta; \Psi}\,(\Vdash\circ)$$

$$\frac{\Gamma \Vdash \mathcal{A}; \Theta \quad \Lambda; \mathcal{B} \Vdash \Psi}{\Gamma; \Lambda; \mathcal{A} \to \mathcal{B} \Vdash \Theta; \Psi}\,(\to\Vdash) \qquad \frac{\Gamma; \mathcal{A} \Vdash \mathcal{B}; \Theta}{\Gamma \Vdash \mathcal{A} \to \mathcal{B}; \Theta}\,(\Vdash\to)$$

$$\frac{\Gamma; \mathcal{A} \Vdash \Theta \quad \Lambda; \mathcal{B} \Vdash \Psi}{\Gamma; \Lambda; \mathcal{A} + \mathcal{B} \Vdash \Theta; \Psi}\,(+\Vdash) \qquad \frac{\Gamma \Vdash \mathcal{A}; \mathcal{B}; \Theta}{\Gamma \Vdash \mathcal{A} + \mathcal{B}; \Theta}\,(\Vdash+)$$

$$\frac{\Gamma; \mathcal{A} \Vdash \Theta}{\Gamma; \rhd\mathcal{A} \Vdash \Theta}\,(\rhd\Vdash) \qquad \frac{\Gamma^{\rhd} \Vdash \mathcal{A}; \Theta^{\lhd}}{\Gamma^{\rhd} \Vdash \rhd\mathcal{A}; \Theta^{\lhd}}\,(\Vdash\rhd)$$

$$\frac{\Gamma^{\rhd}; \mathcal{A} \Vdash \Theta^{\lhd}}{\Gamma^{\rhd}; \lhd\mathcal{A} \Vdash \Theta^{\lhd}}\,(\lhd\Vdash) \qquad \frac{\Gamma \Vdash \mathcal{A}; \Theta}{\Gamma \Vdash \lhd\mathcal{A}; \Theta}\,(\Vdash\lhd)$$

Structural rules:

$$\frac{\Gamma; \mathcal{A}; \mathcal{A} \Vdash \Theta}{\Gamma; \mathcal{A} \Vdash \Theta}\,(W\Vdash) \qquad \frac{\Gamma \Vdash \mathcal{A}; \mathcal{A}; \Theta}{\Gamma \Vdash \mathcal{A}; \Theta}\,(\Vdash W)$$

$$\frac{\Gamma; \rhd\mathcal{A}; \rhd\mathcal{A} \Vdash \Theta}{\Gamma; \rhd\mathcal{A} \Vdash \Theta}\,(\rhd W\Vdash) \qquad \frac{\Gamma \Vdash \lhd\mathcal{A}; \lhd\mathcal{A}; \Theta}{\Gamma \Vdash \lhd\mathcal{A}; \Theta}\,(\Vdash\lhd W)$$

$$\frac{\Gamma \Vdash \Theta}{\Gamma; \rhd\mathcal{A} \Vdash \Theta}\,(\rhd K\Vdash) \qquad \frac{\Gamma \Vdash \Theta}{\Gamma \Vdash \lhd\mathcal{A}; \Theta}\,(\Vdash\lhd K)$$

$$\frac{\Gamma \Vdash \Theta}{\Gamma; \mathcal{A} \Vdash \Theta}\,(K\Vdash) \qquad \frac{\Gamma \Vdash \Theta}{\Gamma \Vdash \mathcal{A}; \Theta}\,(\Vdash K)$$

In all the sequent calculuses we consider, *proofs* are defined standardly as a tree with all the leaves being axioms and other nodes justified by applications of rules (see, e.g., Bimbó 2015b for a precise definition and sample proofs). A sequent $\Gamma \Vdash \Theta$ is *provable* if there is a proof rooted in the sequent $\Gamma \Vdash \Theta$. $\mathcal{A}$ is a *theorem* when $\boldsymbol{t} \Vdash \mathcal{A}$ (or equivalently, $\Vdash \mathcal{A}$) is provable.

The *cut rule* takes the following form, and it is known to be admissible for the calculuses that do or do not include some of the pairs of the structural rules.[4]

$$\frac{\Gamma \Vdash \Theta; \mathcal{A} \qquad \mathcal{A}; \Lambda \Vdash \Psi}{\Gamma; \Lambda \Vdash \Theta; \Psi}\ \text{cut}$$

Now that we have reasonable syntactic specifications (namely, sequent calculuses for the logics we want to consider, we will algebraize them. For the latter, we use the equivalence relation that holds between $\mathcal{A}$ and $\mathcal{B}$ when $\mathcal{A} \Vdash \mathcal{B}$ and $\mathcal{B} \Vdash \mathcal{A}$ are both provable.

We start with two logics—one with disjunction, the other with conjunction—that we call *kernel* logics. These logics form the modality-free core of our logics, and isolating them allows us to interpret them first, which is quite complicated in itself.

Definition 2.2.4 (*Kernel logics*) The logic $\mathfrak{C}^{\wedge}$ is defined by the axioms, and the rules for $\wedge$, $\boldsymbol{t}$, $\circ$, $\rightarrow$, and $+$, and the logic $\mathfrak{C}^{\vee}$ is defined by the axioms, and the rules for $\vee$, $\boldsymbol{t}$, $\circ$, $\rightarrow$, and $+$.

Lemma 2.2.5 *The Lindenbaum algebra of* $\mathfrak{C}^{\wedge}$ *is* $\mathfrak{A}^{\wedge}$ *and the Lindenbaum algebra of* $\mathfrak{C}^{\vee}$ *is* $\mathfrak{A}^{\vee}$. *The algebras are* $\mathfrak{A}^{\wedge} = \langle A; \wedge, \boldsymbol{t}, \circ, \rightarrow, + \rangle$, *where* (a1)–(a4) *hold, and* $\mathfrak{A}^{\vee} = \langle A; \vee, \boldsymbol{t}, \circ, \rightarrow, + \rangle$, *where* (a4)–(a7) *hold.*

(a1) $\langle A; \wedge \rangle$ *is a meet semi-lattice (an msl, for short);*
(a2) $\langle A; \boldsymbol{t}, \circ, \rightarrow \rangle$ *is a residuated Abelian monoid;*
(a3) $+$ *is an msl-ordered Abelian semi-group operation;*
(a4) $a \circ (b + c) \leq (a \circ b) + c$ *hemi-distributivity of* $\circ$ *over* $+$;
(a5) $\langle A; \vee \rangle$ *is a join semi-lattice (a jsl, for short);*
(a6) $\langle A; \boldsymbol{t}, \circ, \rightarrow \rangle$ *is a jsl-ordered residuated Abelian monoid;*
(a7) $+$ *is a monotone Abelian semi-group operation.*

Proof The proof proceeds along well-known lines; we provide two sample steps (and leave the rest of the details to the reader).

1. Let us assume that $\mathcal{A} \Vdash \mathcal{B}$ as well as $\mathcal{B} \Vdash \mathcal{A}$ are provable. We wish to show that $\mathcal{A} \vee \mathcal{C} \Vdash \mathcal{B} \vee \mathcal{C}$ and $\mathcal{B} \vee \mathcal{C} \Vdash \mathcal{A} \vee \mathcal{C}$ are provable too. The following proof segments suffice:

$$\dfrac{(\Vdash\vee_1)\ \dfrac{\overset{\vdots}{\mathcal{A} \Vdash \mathcal{B}}}{\mathcal{A} \Vdash \mathcal{B} \vee \mathcal{C}} \qquad \dfrac{\mathcal{C} \Vdash \mathcal{C}}{\mathcal{C} \Vdash \mathcal{B} \vee \mathcal{C}}\ (\Vdash\vee_2)}{\mathcal{A} \vee \mathcal{C} \Vdash \mathcal{B} \vee \mathcal{C}}\ (\vee\Vdash) \qquad \dfrac{(\Vdash\vee_1)\ \dfrac{\overset{\vdots}{\mathcal{B} \Vdash \mathcal{A}}}{\mathcal{B} \Vdash \mathcal{A} \vee \mathcal{C}} \qquad \dfrac{\mathcal{C} \Vdash \mathcal{C}}{\mathcal{C} \Vdash \mathcal{A} \vee \mathcal{C}}\ (\Vdash\vee_2)}{\mathcal{B} \vee \mathcal{C} \Vdash \mathcal{A} \vee \mathcal{C}}\ (\vee\Vdash)$$

[4] For example, $\langle (\triangleright K \Vdash), (\Vdash \triangleleft K) \rangle$ is a pair that may be included or omitted. The cut theorem for the calculuses here is Theorem 15 in Bimbó (2017). The proof of the cut theorem in Bimbó (2015a, §3) is triple-inductive proof for the intensional part of our calculuses with modalized structural rules. See also Chap. 7 in Bimbó (2015b).

Proving the equivalence of $\mathcal{C} \vee \mathcal{A}$ and $\mathcal{C} \vee \mathcal{B}$ is similar.

2. Again, let us suppose that $\mathcal{A}$ and $\mathcal{B}$ are equivalent. We will show that so are $\mathcal{C} \circ \mathcal{A}$ and $\mathcal{C} \circ \mathcal{B}$. Consider the following proofs (with the parts yielding the assumption omitted):

$$\dfrac{\dfrac{\mathcal{C} \Vdash \mathcal{C} \quad \mathcal{A} \Vdash \mathcal{B}}{\mathcal{C}; \mathcal{A} \Vdash \mathcal{C} \circ \mathcal{B}}\,(\Vdash\circ)}{\mathcal{C} \circ \mathcal{A} \Vdash \mathcal{C} \circ \mathcal{B}}\,(\circ\Vdash) \qquad \dfrac{\dfrac{\mathcal{C} \Vdash \mathcal{C} \quad \mathcal{B} \Vdash \mathcal{A}}{\mathcal{C}; \mathcal{B} \Vdash \mathcal{C} \circ \mathcal{A}}\,(\Vdash\circ)}{\mathcal{C} \circ \mathcal{B} \Vdash \mathcal{C} \circ \mathcal{A}}\,(\circ\Vdash)$$

The equivalence of $\mathcal{A} \circ \mathcal{C}$ and $\mathcal{B} \circ \mathcal{C}$ may be proved similarly. ∴

Remark 2.2.3 The labels we used in the lemma are fairly standard; nevertheless, we fix their meaning with (tacitly) universally quantified axioms. Both $\mathfrak{A}^{\wedge}$ and $\mathfrak{A}^{\vee}$ are definable by finitely many equations, but we utilize some inequations and even quasi-inequations for the sake of familiarity. ((a4) is an inequation itself, and we do not repeat it.)

(a1) $a \wedge a = a,\ a \wedge b = b \wedge a,\ a \wedge (b \wedge c) = (a \wedge b) \wedge c;\ a \wedge b = a$ iff $a \leq b$;
(a2) $t \circ a = a,\ a \circ b = b \circ a,\ a \circ (b \circ c) = (a \circ b) \circ c,\ a \circ b \leq c$ iff $a \leq b \to c$;
(a3) $a + b = b + a,\ a + (b + c) = (a + b) + c,\ a + (b \wedge c) = (a + b) \wedge (a + c)$;
(a5) $a \vee a = a,\ a \vee b = b \vee a,\ a \vee (b \vee c) = (a \vee b) \vee c;\ a \vee b = b$ iff $a \leq b$;
(a6) $t \circ a = a,\ a \circ b = b \circ a,\ a \circ (b \circ c) = (a \circ b) \circ c,\ a \circ b \leq c$ iff $a \leq b \to c$;
(a7) $a + b = b + a,\ a + (b + c) = (a + b) + c,\ a + b \leq a + (b \vee c)$.

2.3 Semantics for the Kernel Logic with Disjunction

The disjunction connective is often thought to be more problematic than conjunction is. There are many reasons for this—from the natural language equivalents of $\wedge$ and $\vee$ to the natural deduction rules for these connectives. However, $\wedge$ and $\vee$ share many similarities, especially when they are viewed abstractly as lattice operations. A lattice turned upside down, so to speak, is a lattice, and some of its features (such as being modular or distributive) are unaffected. Similarly, a meet semi-lattice flipped over is a join semi-lattice, and vice versa.

Semantically speaking, sets of filters give a representation for conjunction through intersection, and the presence of conjunction guarantees that filters exist. Dually, sets of ideals allow disjunction to be represented by intersection. Most often though, when both connectives are in a logic, moreover, they distribute over each other, a subset of filters is used with union standing in for disjunction. In general, filters are prevailing in logicians' thinking about semantics, because they can be viewed as *theories*, that is, deductively closed sets of formulas. (Of course, equivalence classes of formulas and the formulas themselves are different kinds of entities, but a 1–1 correspondence justifies the view purported in the previous sentence.) In this paper, we develop the view that takes the *duality* of conjunction and disjunction seriously when only one of them is in a logic. We will provide semantics for logics with *disjunction* in terms of *ideals*.[5]

Definition 2.3.1 A *frame* for $\mathfrak{C}^{\vee}$ is $\mathfrak{F} = \langle U, I, \sqsubseteq, R, R_{+} \rangle$, where the elements of the quintuple satisfy conditions (f0)–(f10).[6]

(f0) $U \neq \emptyset, \quad I \subseteq U, \quad R \subseteq U^3, \quad R_{+} \subseteq U^3, \quad \sqsubseteq \subseteq U^2, \quad I \neq \emptyset, \quad \uparrow I = I,$
(f1) $\forall\alpha\, \alpha \sqsubseteq \alpha, \quad \forall\alpha\, \forall\beta\, \forall\gamma\, ((\alpha \sqsubseteq \beta \land \beta \sqsubseteq \gamma) \Rightarrow \alpha \sqsubseteq \gamma),$
(f2) $\forall\alpha\, \forall\beta\, \forall\gamma\, \forall\alpha'\, \forall\beta'\, \forall\gamma'\, ((\alpha \sqsubseteq \alpha' \land \beta \sqsubseteq \beta' \land \gamma' \sqsubseteq \gamma \land \overline{R}\alpha\beta\gamma) \Rightarrow \overline{R}\alpha'\beta'\gamma'),$
(f3) $\forall\alpha\, \exists\beta\, (\overline{R}\beta\alpha\alpha \land \beta \notin I),$
(f4) $\forall\alpha\, \forall\beta\, \forall\gamma\, ((\overline{R}\alpha\beta\gamma \land \alpha \notin I) \Rightarrow \gamma \sqsubseteq \beta),$
(f5) $\forall\alpha\, \forall\beta\, \forall\gamma\, (\overline{R}\alpha\beta\gamma \Rightarrow \overline{R}\beta\alpha\gamma),$
(f6) $\forall\alpha\, \forall\beta\, \forall\gamma\, \forall\delta\, (\overline{R}(\alpha\beta)\gamma\delta \Leftrightarrow \overline{R}\alpha(\beta\gamma)\delta),$
(f7) $\forall\alpha\, \forall\beta\, \forall\gamma\, \forall\alpha'\, \forall\beta'\, \forall\gamma'\, ((\alpha' \sqsubseteq \alpha \land \beta' \sqsubseteq \beta \land \gamma \sqsubseteq \gamma' \land R_{+}\alpha\beta\gamma) \Rightarrow R_{+}\alpha'\beta'\gamma'),$
(f8) $\forall\alpha\, \forall\beta\, \forall\gamma\, (R_{+}\alpha\beta\gamma \Rightarrow R_{+}\beta\alpha\gamma),$
(f9) $\forall\alpha\, \forall\beta\, \forall\gamma\, \forall\varepsilon\, (\exists\delta\, (R_{+}\alpha\delta\varepsilon \land R_{+}\beta\gamma\delta) \Leftrightarrow \exists\delta\, (R_{+}\alpha\beta\delta \land R_{+}\delta\gamma\varepsilon)),$
(f10) $\forall\alpha\, \forall\gamma\, \forall\delta\, \forall\varepsilon\, \forall\vartheta\, ((R_{+}\varepsilon\gamma\delta \land \overline{R}\alpha\vartheta\delta) \Rightarrow \exists\beta\, (\overline{R}\alpha\beta\varepsilon \land R_{+}\beta\gamma\vartheta)).$

Remark 2.3.1 U is a set of situations. I is a distinguished subset of U, which is non-empty and is upward closed with respect to the pre-order ($\sqsubseteq$) on U. ($\uparrow I$ denotes the cone generated by I.) R is related to both $\to$ and $\circ$, and hence we do not attach a subscript. R and $\overline{R}$ are each other's complements, that is, $\overline{R} = U^3 - R$. We write the arguments of R and $\overline{R}$ following the relation symbol, (usually) without parentheses or commas (as customary in the literature). Although R is the relation associated to $\circ$ and $\to$, it is clearer if we express some of the properties of R in terms of $\overline{R}$. $\overline{R}(\alpha\beta)\gamma\delta$ is shorthand for $\exists\zeta\, (\overline{R}\alpha\beta\zeta \land \overline{R}\zeta\gamma\delta)$, and similarly, $\overline{R}\alpha(\beta\gamma)\delta$ stands for $\exists\zeta\, (\overline{R}\beta\gamma\zeta \land \overline{R}\alpha\zeta\delta)$. For the next definition, we note that $\mathcal{C}$ is the set of cones on U, that is, $\mathcal{C} = \{\, W \subseteq U : \forall\alpha\, \forall\beta\, ((\alpha \in W \land \alpha \sqsubseteq \beta) \Rightarrow \beta \in W) \,\}$.

We have already explained informally a couple of the frame conditions. Some of the lengthy stipulations, for example (f2) and (f7), may be simply rephrased as $\overline{R}$ is monotone increasing in its first two argument places and monotone decreasing in its third, and the other way around for R_{+}. Or to put it even more informally and concisely, $\overline{R}\uparrow\uparrow\downarrow$ and $R_{+}\downarrow\downarrow\uparrow$ characterize the tonicity of the relations with respect to $\sqsubseteq$. To give another example, (f5) and (f8) express the commutativity of $\circ$ and $+$, respectively. These operations may be thought of as "living" in the third argument place with their arguments in the first two.

Definition 2.3.2 A *model* for $\mathfrak{C}^{\vee}$ is $\mathfrak{M} = \langle \mathfrak{F}, v \rangle$, where $\mathfrak{F}$ is a frame (as in Definition 2.3.1), and v is a valuation function of type $v : \mathbb{P} \longrightarrow \mathcal{C}$. v gives rise to a satisfiability relation via (m0)–(m5).[7]

---

[5] The duality between filters and ideals, which is rooted in the duality of $\land$ and $\lor$, has a storied past in logic (see Halmos 1962, p. 22 and Dunn 2019, p. 28). Our semantics are also motivated by semantics for relevance logics in a broad sense of relevance (see Avron 1984, 2014; Bimbó and Dunn 2009).

[6] Lowercase Greek letters range over elements of U, and we will use logical symbols in the meta-language which may be thought—for the sake of simplicity—to be those in two-valued logic.

[7] As a rule, we only indicate the situation that makes a formula true in a model in the $\vDash$ notation, and we omit mentioning $\mathfrak{F}$ or v—to enhance readability.

(m0) $\alpha \vDash p$ iff $\alpha \in v(p)$;
(m1) $\alpha \vDash \boldsymbol{t}$ iff $\exists \iota \in I\, \iota \sqsubseteq \alpha$;
(m2) $\alpha \vDash \mathcal{A} \vee \mathcal{B}$ iff $\alpha \vDash \mathcal{A}$ and $\alpha \vDash \mathcal{B}$;
(m3) $\alpha \vDash \mathcal{A} \circ \mathcal{B}$ iff $\forall \beta\, \forall \gamma\, (R\beta\gamma\alpha \vee \beta \vDash \mathcal{A} \vee \gamma \vDash \mathcal{B})$;
(m4) $\alpha \vDash \mathcal{A} \to \mathcal{B}$ iff $\exists \beta\, \exists \gamma\, (\overline{R}\alpha\beta\gamma \wedge \beta \nvDash \mathcal{A} \wedge \gamma \vDash \mathcal{B})$;
(m5) $\alpha \vDash \mathcal{A} + \mathcal{B}$ iff $\exists \beta\, \exists \gamma\, (R_{+}\beta\gamma\alpha \wedge \beta \vDash \mathcal{A} \wedge \gamma \vDash \mathcal{B})$.

Lemma 2.3.3 (Heredity lemma) *For all formulas* $\mathcal{A}$, $\{\alpha : \alpha \vDash \mathcal{A}\} \in \mathfrak{C}$.

Proof The lemma expresses that all formulas are interpreted by cones of situations. The proof is fairly standard. The cases for p and $\boldsymbol{t}$ follow from Definition 2.3.2. The case for $\mathcal{A} \vee \mathcal{B}$ is immediate, because the set of cones $\mathfrak{C}$ is closed under intersection. The last three cases follow from (m3) and (m4) using (f2), and from (m5) using (f7). (We omit further details.) ∴

Remark 2.3.2 The idea that propositions are upward closed sets of situations originated in Kripke (1965) in the modeling of intuitionistic logic, where situations are not maximally consistent. In relevance logic, additionally, the situations may not be negation consistent, and hence the Meyer–Routley semantics utilizes upward closure too.

A model $\mathfrak{M}$ comprises a set of propositions, that is, cones of situations, some of which correspond to formulas. We will denote the proposition of $\mathcal{A}$ by $\|\mathcal{A}\|$.

Theorem 2.3.4 (Soundness) *If* $\Vdash\!\mathcal{A}$ *is provable in* $\mathfrak{C}^{\vee}$, *then* $\|\mathcal{A}\| \subseteq I$ *in any* $\mathfrak{M}$ *that is a model in the sense of Definition 2.3.2.*

Proof We detail two steps in the proof, and leave the rest of the proof—which is straightforward—to the reader.
1. First, we show that $\|\boldsymbol{t} \circ \mathcal{A}\| = \|\mathcal{A}\|$. Let us assume that $\alpha \in \|\mathcal{A}\|$, as well as, $\overline{R}\beta\gamma\alpha$ and $\beta \notin \|\boldsymbol{t}\|$. By (f4), it follows that $\alpha \sqsubseteq \gamma$, and hence by Lemma 2.3.3, $\gamma \in \|\mathcal{A}\|$. We can write this as $\forall \beta\, \forall \gamma\, (R\beta\gamma\alpha \vee \beta \vDash \boldsymbol{t} \vee \gamma \vDash \mathcal{A})$, which means that $\alpha \in \|\boldsymbol{t} \circ \mathcal{A}\|$. For the other direction, let us assume that $\alpha \notin \|\mathcal{A}\|$. This can be merged with (f3) into $\exists \beta\, (\overline{R}\beta\alpha\alpha \wedge \beta \nvDash \boldsymbol{t} \wedge \alpha \nvDash \mathcal{A})$. The latter means that $\alpha \nvDash \boldsymbol{t} \circ \mathcal{A}$.
2. The $\to$ operation on propositions is, indeed, the residual of $\circ$. For one direction, we show that if $\gamma \notin \|\mathcal{A} \circ \mathcal{B}\|$ and $\|\mathcal{B} \to \mathcal{C}\| \subseteq \|\mathcal{A}\|$, then $\gamma \notin \|\mathcal{C}\|$. From the former, we get that $\exists \alpha\, \exists \beta\, (\overline{R}\alpha\beta\gamma \wedge \alpha \nvDash \mathcal{A} \wedge \beta \nvDash \mathcal{B})$. Then, $\alpha \nvDash \mathcal{B} \to \mathcal{C}$ is immediate. The latter means that $\forall \delta\, \forall \zeta\, ((\overline{R}\alpha\delta\zeta \wedge \delta \nvDash \mathcal{B}) \Rightarrow \zeta \nvDash \mathcal{C})$. Obviously, $\gamma \nvDash \mathcal{C}$ follows, as we wanted to show.

For the other direction, we start with the assumptions that $\alpha \vDash \mathcal{B} \to \mathcal{C}$ and $\|\mathcal{C}\| \subseteq \|\mathcal{A} \circ \mathcal{B}\|$. The first can be spelled out as $\exists \beta\, \exists \gamma\, (\overline{R}\alpha\beta\gamma \wedge \beta \nvDash \mathcal{B} \wedge \gamma \vDash \mathcal{C})$, and then $\gamma \vDash \mathcal{A} \circ \mathcal{B}$ is immediate. By (m3), $\forall \delta\, \forall \zeta\, ((\overline{R}\delta\zeta\gamma \wedge \delta \nvDash \mathcal{A}) \Rightarrow \zeta \vDash \mathcal{B})$. Since we already have $\overline{R}\alpha\beta\gamma$ and $\beta \nvDash \mathcal{B}$, we obtain $\alpha \vDash \mathcal{A}$. ∴

Remark 2.3.3 An impetus in this paper is that we want to take seriously the duality between conjunction and disjunction to the extent that we want to replace theories with co-theories in the canonical model. When $\vee$ is in the language of a logic, then it is natural to consider the weakest formula implied by a pair of formulas and then take all the formulas that could imply that. This is just like considering the strongest formula that implies a pair of formulas in the presence of $\wedge$, and afterward gathering all the formulas that are implied by that formula. (Technically speaking, if there is no conjunction, but there is a disjunction, then theories are informationally too weak.) This has an impact on the shape of the whole semantics—as the frame conditions already suggest. The book Bimbó and Dunn (2008, Ch. 4) used prime cones in the semantics of logics that algebraize into a join semi-lattice, whereas Bimbó and Dunn (2005) proposed a range of semantics for Kleene logic and action logic. Here we build models for logics with disjunction exclusively from co-theories, or equivalently, ideals. (See also Bimbó 2007, 2009, where duality is explored for relevance logics with both $\wedge$ and $\vee$.)

Definition 2.3.5 The *canonical frame* for $\mathfrak{C}^{\vee}$ is $\mathfrak{F} = \langle \mathfrak{I}, I, \subseteq, R, R_{+} \rangle$, where the elements are specified by (c0)–(c3).

(c0) $\mathfrak{I}$ is the set of non-empty ideals on $\mathfrak{A}^{\vee}$; $\subseteq$ is set inclusion;
(c1) $\iota = \{ a \in A : a \leq [\, \boldsymbol{t}\,] \}$ and $I = \{ \delta \in \mathfrak{I} : \iota \subseteq \delta \}$;
(c2) $\overline{R}\alpha\beta\gamma$ iff $\forall a\, \forall b\, (a \circ b \in \gamma \Rightarrow (a \in \alpha \vee b \in \beta))$;
(c3) $R_{+}\alpha\beta\gamma$ iff $\forall a\, \forall b\, ((a \in \alpha \wedge b \in \beta) \Rightarrow a + b \in \gamma)$.

We will obtain completeness in two stages—just like we built up a model in two steps.

Lemma 2.3.6 *The canonical frame (from Definition 2.3.5) is* in the class of frames *(from Definition 2.3.1).*

Proof **1.** We take it that it is obvious that I is a non-empty cone of ideals, and $\subseteq$ is a pre-order (indeed, a weak partial order) on the set of ideals. R and R_{+} are also appropriately defined with respect to their type. Hence, (f0) and (f1) hold. We leave the verification of conditions (f2), (f3), and (f7) to the reader.
2. Now, we show that (f4) holds. Let us assume that $\overline{R}\alpha\beta\gamma$ and $\alpha \notin I$. First of all, we note that by the definition of I, for any δ, $[\, \boldsymbol{t}\,] \in \delta$ if and only if $\delta \in I$. $\gamma \neq \emptyset$, and hence, for some c, $c \in \gamma$. However, $[\, \boldsymbol{t}\,] \circ c = c$, and if $[\, \boldsymbol{t}\,] \notin \alpha$, then by the definition of $\overline{R}$, $c \in \beta$. This means that $\gamma \subseteq \beta$.
3. To show that (f5) is true on the canonical frame, we will assume $\overline{R}\alpha\beta\gamma$ and $b \circ a \in \gamma$. $\circ$ is commutative in $\mathfrak{A}^{\vee}$, which means that $a \circ b \in \gamma$. By the definition of $\overline{R}$, it follows that $a \in \alpha$ or $b \in \beta$. Since, a and b are arbitrary in our assumption, we have that $\forall a\, \forall b\, (b \circ a \in \gamma \Rightarrow (b \in \beta \vee a \in \alpha))$, that is, $\overline{R}\beta\alpha\gamma$, which is what we wanted to prove.
4. For (f6), we assume $\overline{R}(\alpha\beta)\gamma\delta$. By eliminating the abbreviation, we have that $\overline{R}\alpha\beta\zeta$ and $\overline{R}\zeta\gamma\delta$. We need an ideal for β and γ to bear $\overline{R}$ to. We define $\vartheta = (\{ b \circ c : \exists a\, (a \notin \alpha \wedge a \circ b \circ c \in \delta) \}]$, that is, we collect those $b \circ c$'s together for which a fusion with an a that is not an element of α is an element of δ, and we close the set under join. From the definition of ϑ, $\overline{R}\alpha\vartheta\delta$ follows. If $b \circ c \in \vartheta$, then by the tonicity of $\circ$ and the definition of ϑ, we have that $a \circ b \circ c \in \delta$ for some a. However, $\overline{R}(\alpha\beta)\gamma\delta$ implies

that $a \in \alpha$ or $b \in \beta$ or $c \in \gamma$; hence, $b \in \beta$ or $c \in \gamma$, if $a \notin \alpha$. Showing that the condition holds in the other direction is similar.

5. Let us assume that $R_{+}\alpha\beta\gamma, a \in \alpha$, and $b \in \beta$. Then $a + b$ and $b + a$ are elements of γ. Obviously, $\forall b \forall a\, ((b \in \beta \wedge a \in \alpha) \Rightarrow b + a \in \gamma)$, which establishes $R_{+}\beta\alpha\gamma$; hence, (f8) holds.

6. To prove that (f9) is true, we assume that $R_{+}\alpha\delta\varepsilon$ and $R_{+}\beta\gamma\delta$. We take arbitrary elements of α, β and γ, namely, a, b, and c. We define $\vartheta = (\alpha + \beta]$, that is, $\vartheta = (\{e\colon \exists d \in \alpha\, \exists g \in \beta\; e \leq d + g\}]$. The definition of ϑ guarantees that $R_{+}\alpha\beta\vartheta$. If $e \leq (a_1 + b_1) \vee (a_2 + b_2)$, then $a_1 + b_1 + c \in \delta$ and $a_2 + b_2 + c \in \delta$. However, $a_1 \vee a_2 \in \alpha$ and $b_1 \vee b_2 \in \beta$, and $a_1 + b_1 + c \leq (a_1 \vee a_2) + (b_1 \vee b_2) + c$ and $a_2 + b_2 + c \leq (a_1 \vee a_2) + (b_1 \vee b_2) + c$; hence, $(a_1 + b_1 + c) \vee (a_2 + b_2 + c) \leq (a_1 \vee a_2) + (b_1 \vee b_2) + c$. It is immediate from $(a_1 \vee a_2) + (b_1 \vee b_2) + c \in \delta$ that $(a_1 + b_1 + c) \vee (a_2 + b_2 + c) \in \delta$. This completes showing that $R_{+}\vartheta\gamma\delta$, and therefore the frame condition guaranteeing the associativity of $+$ holds.

7. Lastly, we prove that (f10) holds. Let us assume that $R_{+}\varepsilon\gamma\delta$ and $\overline{R}\alpha\vartheta\delta$ hold. Let us consider for β the ideal $(\{b\colon \exists a\, (a \circ b \in \varepsilon \wedge a \notin \alpha)\}]$. The definition guarantees that $\overline{R}\alpha\beta\varepsilon$ holds. We have to show that $R_{+}\beta\gamma\vartheta$ also holds with our β. Let us assume that $b \in \beta$ and $c \in \gamma$. Then $a \circ b \in \varepsilon$ with some $a \notin \alpha$, hence, $(a \circ b) + c \in \delta$, because $R_{+}\varepsilon\gamma\delta$. However, by (a4), $a \circ (b + c) \leq (a \circ b) + c$, which means that $a \circ (b + c) \in \delta$ (because $\delta \in \mathfrak{I}$). But $a \notin \alpha$, and $\overline{R}\alpha\vartheta\delta$ is true, which means that by the definition of $\overline{R}$, $b + c \in \vartheta$. This is what is needed for $R_{+}\beta\gamma\vartheta$. ⁖

Before we proceed to the definition of the canonical model, we introduce a lemma, which may be viewed as a version of the squeeze lemma. The squeeze lemma was originally proved for relevance logics. (See part of Lemma 12 in Routley and Meyer 1973, and its generalization, Lemma 2.3.19 in Bimbó and Dunn 2008.) The idea is that a relation R may hold between theories α, β and the prime theory γ as $R\alpha\beta\gamma$. However, the situations in a model must be prime, and hence α and β have to be extended to prime theories α' and β' while maintaining R (i.e., $R\alpha'\beta'\gamma$ should hold). In the case when R is associated with fusion, α and β cannot be arbitrarily expanded (lest R cease to hold). Hence, α' and β' are "squeezed" into the space between α and β, and where R fails. The notion of primeness, which (in logic) most commonly used in connection to filters, can be adapted to co-theories and cones too—as the following definition shows. We use our version of the squeeze lemma in the proof of Lemma 2.3.12.

Definition 2.3.7 Let $\langle X, \sqsubseteq \rangle$ be a set with a pre-order. The sets of *cones*, *principal cones*, *downward directed cones*, and *prime cones* are defined by (1), (2), (3), and (4), respectively. ($Y \subseteq X$ everywhere.)

(1) Y is a *cone* iff $\forall x \forall y\, ((x \in Y \wedge x \sqsubseteq y) \Rightarrow y \in Y)$.
(2) Y is a *principal cone* iff $\exists y\, (y \in Y \wedge \forall x\, (x \in Y \Leftrightarrow y \sqsubseteq x))$.
(3) Y is a *downward directed cone* iff Y is a cone and
$\forall x \forall y\, ((x \in Y \wedge y \in Y) \Rightarrow \exists z (z \in Y \wedge z \sqsubseteq x \wedge z \sqsubseteq y))$.
(4) If X is a jsl, then Y is a *prime cone* iff $\forall x \forall y\, (x \vee y \in Y \Leftrightarrow (x \in Y \vee y \in Y))$.

We denote the set of cones on X by $\mathcal{C}(X)$ (or simply by $\mathcal{C}$—as before—when the underlying set X is clear from context). Similarly, $\mathcal{C}^{\curlyvee}$, $\mathcal{C}^{\simeq}$ and $\mathcal{C}_{\mathcal{P}}$ denote the sets of principal, downward directed and prime cones, respectively.

We used similar notation for sets of various kinds of cones starting with Bimbó and Dunn (2008). As $\mathcal{F}$ or $\mathcal{I}$ denote the set of *f*ilters or *i*deals, $\mathcal{C}$ denotes the set of *c*ones. The decorations on $\mathcal{C}$ intend to suggest that a principal cone is generated by a single element ($\curlyvee$), a downward directed cone includes a lesser element for each pair ($\simeq$) and $\mathcal{P}$ indicates *p*rimeness.

Definition 2.3.8 Let $\overline{R}$ be defined as in (c2) in Definition 2.3.5. Then Q is defined as $Qxy\overline{\gamma} \Leftrightarrow \overline{R}\overline{x}\,\overline{y}\gamma$, that is, $Qxy\overline{\gamma} \Leftrightarrow \forall a\,\forall b\,((a \in x \wedge b \in y) \Rightarrow a \circ b \in \overline{\gamma})$, where $\overline{x}, \overline{y}, \overline{\gamma}$ are the complements of x, y, and γ, and $\overline{x}, \overline{y}, \gamma \in \mathcal{I}$. Furthermore, we extend the latter definition to Q' by allowing $x, y \in \mathcal{C}$ (rather than $x, y \in \mathcal{C}_{\mathcal{P}}$).

Remark 2.3.4 We note that if $x \in \mathcal{I}$, then $\overline{x} \in \mathcal{C}_{\mathcal{P}}$ (in any jsl). Q and Q' are similar to the definition of the ternary relation in the Meyer–Routley semantics for relevance logics in Routley and Meyer (1973), when $\circ$ is in the language. In that situation, the canonical accessibility relation holds on the set of prime filters, and then it is relaxed so that filters may appear in the first and second argument places. Just as in the case of R in the Meyer–Routley semantics, Q and Q' are antitone (monotone decreasing) in their first two argument places and monotone (monotone increasing) in the third.

Lemma 2.3.9 (Squeeze lemma) *If $Qxy\overline{\gamma}$, with $x, y \in \mathcal{C}^{\simeq}$ and $\gamma \in \mathcal{I}$ on the canonical frame of the logic $\mathfrak{C}^{\vee}$, then* there are $x', y' \in \mathcal{C}_{\mathcal{P}}$ *$\mathfrak{st}$ $x \subseteq x'$, $y \subseteq y'$ and $Qx'y'\overline{\gamma}$.*

Proof We will maximize elements of pairs by Zorn's lemma. Let us define $E = \{\langle C_1, C_2\rangle : x \subseteq C_1 \wedge y \subseteq C_2 \wedge Q'C_1C_2\overline{\gamma} \wedge C_1, C_2 \in \mathcal{C}^{\simeq}\}$. Clearly, E is not empty, because $\langle x, y\rangle \in E$. The inclusion $\subseteq$ between pairs is point-wise, that is, $\langle C_1, C_2\rangle \subseteq \langle D_1, D_2\rangle$ when $C_1 \subseteq D_1$ and $C_2 \subseteq D_2$. Similarly, $\bigcup_{i\in I}\langle C_{1i}, C_{2i}\rangle$ stands for $\langle\bigcup_{i\in I} C_{1i}, \bigcup_{i\in I} C_{2i}\rangle$. Then, if I ($I \neq \emptyset$) is a linear order with $\leq$, then $\bigcup_{i\in I}\langle C_{1i}, C_{2i}\rangle \in E$ assuming that each $\langle C_{1i}, C_{2i}\rangle \in E$. By Zorn's lemma, there is a maximal element in E, let us say, $\langle D_1, D_2\rangle$.

A maximal element is a pair of prime cones, as we show next. We prove that D_1 is prime; D_2 can be shown to be prime by switching indices. Let $a_1 \vee a_2 \in D_1$. If $a_1, a_2 \notin D_1$, then there are $b_1, b_2 \in D_2$ $\mathfrak{st}$ $a_1 \circ b_1 \notin \overline{\gamma}$ and $a_2 \circ b_2 \notin \overline{\gamma}$. By construction, $D_2 \in \mathcal{C}^{\simeq}$, and hence there is some $b \in D_2$ with the property that $b \leq b_1$ and $b \leq b_2$. Then by the monotonicity of $\circ$, $a_1 \circ b \notin \overline{\gamma}$ and $a_2 \circ b \notin \overline{\gamma}$, hence $(a_1 \circ b) \vee (a_2 \circ b) \notin \overline{\gamma}$. However, $(a_1 \circ b) \vee (a_2 \circ b) = (a_1 \vee a_2) \circ b$, which leads to a contradiction. ·:·

The proof of the squeeze lemma would be slightly simpler if we could restrict our attention to one of the argument places of Q', provided that in the other argument place we already have a prime cone.

Corollary 2.3.10 *Let $\alpha, \beta, \gamma \in \mathcal{I}$ and $x, y \in \mathcal{C}^{\simeq}$ on the canonical frame of $\mathfrak{C}^{\vee}$. If $Q'\overline{\alpha}x\overline{\gamma}$, then there is an $x' \in \mathcal{C}_{\mathcal{P}}$ $\mathfrak{st}$ $x \subseteq x'$ and $Q\overline{\alpha}x'\overline{\gamma}$. Similarly, from $Q'y\overline{\beta}\overline{\gamma}$, $Qy'\overline{\beta}\overline{\gamma}$ follows, for some $y' \in \mathcal{C}_{\mathcal{P}}$ extending y.*

Definition 2.3.11 The *canonical model* is $\mathfrak{M} = \langle \mathfrak{F}, v \rangle$, when $\mathfrak{F}$ is the canonical frame (from Definition 2.3.5), and v is defined by (c4).

(c4) $v(p) = \{ \delta \in \mathfrak{I} \colon [p] \in \delta \}$ and $v(\boldsymbol{t}) = I$.

As we indicated in footnote 7, $\alpha \vDash \mathcal{A}$ may be read as "$\mathcal{A}$ is true in situation α," then Lemma 2.3.12 is a "Truth Lemma" saying that a situation makes a formula true precisely, when it contains the equivalence class of the formula.

Lemma 2.3.12 *The canonical valuation v extended to the relation $\vDash$ has the property that $\alpha \vDash \mathcal{A}$ iff $[\mathcal{A}] \in \alpha$.*

Proof **1.** The claim is obviously true for p by (c4), and for $\boldsymbol{t}$ by (c1) and (c4).
2. Let us consider $\mathcal{A} \vee \mathcal{B}$. $\alpha \vDash \mathcal{A} \vee \mathcal{B}$ iff $\alpha \vDash \mathcal{A}$ and $\alpha \vDash \mathcal{B}$. By inductive hypothesis, $[\mathcal{A}] \in \alpha$ and $[\mathcal{B}] \in \alpha$. But this holds exactly when $[\mathcal{A} \vee \mathcal{B}] \in \alpha$, because $\alpha \in \mathfrak{I}$.
3. Next, let our formula be $\mathcal{A} \circ \mathcal{B}$. For one direction, we start with $\gamma \nvDash \mathcal{A} \circ \mathcal{B}$. Then, $\exists\alpha\, \exists\beta\, (\overline{R}\alpha\beta\gamma \wedge \alpha \nvDash \mathcal{A} \wedge \beta \nvDash \mathcal{B})$. By inductive hypothesis, $[\mathcal{A}] \notin \alpha$ and $[\mathcal{B}] \notin \beta$. Then $[\mathcal{A} \circ \mathcal{B}] \notin \gamma$, by the definition of $\overline{R}$.

For the converse, we assume that $[\mathcal{A} \circ \mathcal{B}] \notin \gamma$. We note that all principal cones are downward directed, and hence $\uparrow[\mathcal{A}], \uparrow[\mathcal{B}] \in \mathfrak{C}^{\asymp}$. Then Q' holds between these principal cones and $\overline{\gamma}$ (where Q' is the relation in Definition 2.3.8). By Lemma 2.3.9, there are x' and y' s.t. $\uparrow[\mathcal{A}] \subseteq x'$, $\uparrow[\mathcal{B}] \subseteq y'$ and $Qx'y'\overline{\gamma}$. Since $x', y' \in \mathfrak{C}_{\mathfrak{P}}$, we know that $\overline{x'}, \overline{y'} \in \mathfrak{I}$. Then $\overline{R}\,\overline{x'}\,\overline{y'}\gamma$; moreover, $[\mathcal{A}] \notin \overline{x'}$ and $[\mathcal{B}] \notin \overline{y'}$, hence, by hypothesis, $\overline{x'} \nvDash \mathcal{A}$, and, similarly, $\overline{y'} \nvDash \mathcal{B}$. In sum, $\gamma \nvDash \mathcal{A} \circ \mathcal{B}$.
4. Finally, let us consider $\mathcal{A} \to \mathcal{B}$. First, we assume that $\alpha \vDash \mathcal{A} \to \mathcal{B}$, that is, $\exists\beta\, \exists\gamma\, (\overline{R}\alpha\beta\gamma \wedge \beta \nvDash \mathcal{A} \wedge \gamma \vDash \mathcal{B})$. By the hypothesis of induction, $[\mathcal{A}] \notin \beta$ and $[\mathcal{B}] \in \gamma$. From the latter, it follows that $[(\mathcal{A} \to \mathcal{B}) \circ \mathcal{A}] \in \gamma$, hence by the definition of $\overline{R}$, $[\mathcal{A} \to \mathcal{B}] \in \alpha$, as we wanted to show.

Now, let us assume that $[\mathcal{A} \to \mathcal{B}] \in \alpha$. The relation Q' introduced in Definition 2.3.8 holds between $\overline{\alpha}$ and two cones as $Q'\overline{\alpha}[[\mathcal{A}])\overline{([\mathcal{B}]]}$ (where $[[\mathcal{A}])$ is the principal cone generated by $[\mathcal{A}]$ and $\overline{([\mathcal{B}]]}$ is the prime cone, which is the complement of the principal ideal generated by $[\mathcal{B}]$). To show that Q' indeed holds, let us assume that for some $\mathcal{C}$, $[\mathcal{C}] \in \overline{\alpha}$ but $[\mathcal{C} \circ \mathcal{A}] \in ([\mathcal{B}]]$. By residuation, $[\mathcal{C}] \leq [\mathcal{A} \to \mathcal{B}]$, which means that $[\mathcal{C}] \in \alpha$. From the contradiction, we may conclude that $[\mathcal{C} \circ \mathcal{A}] \notin ([\mathcal{B}]]$. We note that $\overline{([\mathcal{B}]]} \in \mathfrak{C}_{\mathfrak{P}}$ and $[[\mathcal{A}]) \in \mathfrak{C}^{\asymp}$. By an application of Corollary 2.3.10, there is a y s.t. $Q\overline{\alpha}y\overline{([\mathcal{B}]]}$ where $y \in \mathfrak{C}_{\mathfrak{P}}$ and $[[\mathcal{A}]) \subseteq y$. This means that $\overline{R}\alpha\overline{y}([\mathcal{B}]]$. By the hypothesis of induction, $\overline{y} \nvDash \mathcal{A}$ and $([\mathcal{B}]] \vDash \mathcal{B}$. Since $\overline{R}$ holds, we have that $\alpha \vDash \mathcal{A} \to \mathcal{B}$, as we had to prove.
5. We turn to $+$ now. Let $\gamma \vDash \mathcal{A} + \mathcal{B}$. Then $\exists\alpha\, \exists\beta\, (R_{+}\alpha\beta\gamma \wedge \alpha \vDash \mathcal{A} \wedge \beta \vDash \mathcal{B})$. By inductive hypothesis, $[\mathcal{A}] \in \alpha$ and $[\mathcal{B}] \in \beta$, which yields $[\mathcal{A} + \mathcal{B}] \in \gamma$ together with the definition of R_{+}.

For the converse, let us assume that $[\mathcal{A} + \mathcal{B}] \in \gamma$. We form two principal ideals $([\mathcal{A}]]$ and $([\mathcal{B}]]$ that we call α and β, respectively. By the hypothesis of induction, $\alpha \vDash \mathcal{A}$ and $\beta \vDash \mathcal{B}$. By the monotonicity of $+$, $R_{+}\alpha\beta\gamma$ holds. Then $\exists\alpha\, \exists\beta\, (R_{+}\alpha\beta\gamma \wedge \alpha \vDash \mathcal{A} \wedge \beta \vDash \mathcal{B})$, that is, $\gamma \vDash \mathcal{A} + \mathcal{B}$, as we aimed to show. ∴

Lemma 2.3.13 (Separation) *If $t \not\Vdash \mathcal{A} \to \mathcal{B}$, then there is a β st $\beta \in \|\mathcal{B}\|$ but $\beta \notin \|\mathcal{A}\|$.*

Proof To start with, notice that $[\mathcal{A}] \notin ([\mathcal{B}]]$. By Lemma 2.3.12, $([\mathcal{B}]] \in \|\mathcal{B}\|$ and $([\mathcal{B}]] \notin \|\mathcal{A}\|$. Of course, $([\mathcal{B}]] \in \mathfrak{I}$, and hence it can be taken as the β in the claim. We could state the lemma as a catchphrase: a pair of non-equivalent formulas can be separated by an ideal. ∴

Having collected all the bits together, we have completeness.

Theorem 2.3.14 (Completeness) $\mathfrak{C}^{\vee}$ *is* complete *with respect to the class of its frames.*

2.4 Interpreting the Modalities

Now we consider the addition of the two modal operators to the algebras of our kernel logics. Then we proceed to defining adequate semantics for them.

Definition 2.4.1 An *algebra* $\mathfrak{A}^{\wedge}_{\mathrm{m}}$ has similarity type $\langle 2, 0, 2, 2, 2, 1, 1\rangle$, and $\mathfrak{A}^{\wedge}_{\mathrm{m}} = \langle A; \wedge, \boldsymbol{t}, \circ, \to, +, \rhd, \lhd\rangle$,where (a1)–(a4) (from Lemma 2.2.5) and (a8)–(a13) (below) hold.

(a8) $\rhd(a \wedge b) \leq \rhd a$,
(a9) $\lhd(a \wedge b) \leq \lhd a$,
(a10) $\rhd a \leq a$,
(a11) $\rhd a \leq \rhd\rhd a$,
(a12) $a \leq \lhd a$,
(a13) $\lhd\lhd a \leq \lhd a$.

An *algebra* $\mathfrak{A}^{\vee}_{\mathrm{m}}$ has similarity type $\langle 2, 0, 2, 2, 2, 1, 1\rangle$, and $\mathfrak{A}^{\vee}_{\mathrm{m}} = \langle A; \vee, \boldsymbol{t}, \circ, \to, +, \rhd, \lhd\rangle$, where (a4)–(a7) (from Lemma 2.2.5) and (a10)–(a15) hold.

(a14) $\rhd a \leq \rhd(a \vee b)$,
(a15) $\lhd a \leq \lhd(a \vee b)$.

Remark 2.4.1 The content of (a8)–(a9) (and (a14)–(a15), respectively) is that both modalities are monotone. The four other inequations that we will consider are reminiscent of inequations that hold of $\Box$ and $\Diamond$ in an S4 modal algebra. $\Box\mathcal{A} \to \mathcal{A}$ and $\Box\mathcal{A} \to \Box\Box\mathcal{A}$ are theorems of S4 (when $\to$ is material implication)—together with their duals $\mathcal{A} \to \Diamond\mathcal{A}$ and $\Diamond\Diamond\mathcal{A} \to \Diamond\mathcal{A}$. In the algebra of S4, these would turn into $\Box a \leq a$, $\Box a \leq \Box\Box a$, $a \leq \Diamond a$, and $\Diamond\Diamond a \leq \Diamond a$. However, $\leq$ in that context is the order relation of a Boolean algebra. If we add the connective rules for $\rhd$ and $\lhd$ to our kernel logics $\mathfrak{C}^{\wedge}$ and $\mathfrak{C}^{\vee}$, then the logics we get algebraize into $\mathfrak{A}^{\wedge}_{\mathrm{m}}$ and $\mathfrak{A}^{\vee}_{\mathrm{m}}$, respectively. (We leave the easy verification of this to the reader.) Hence, we denote these logics by $\mathfrak{C}^{\wedge}_{\mathrm{m}}$ and $\mathfrak{C}^{\vee}_{\mathrm{m}}$.

Definition 2.4.2 A *frame for* $\mathfrak{C}^{\vee}_{\mathrm{m}}$ is as in Definition 2.3.1 with two new binary relations $R_{\rhd}$ and $R_{\lhd}$ added, for which (f11)–(f17) hold.

(f11) $R_{\triangleright} \subseteq U^2, \quad R_{\triangleleft} \subseteq U^2$;
(f12) $\forall\alpha\,\forall\beta\,\forall\alpha'\,\forall\beta'((\alpha' \sqsubseteq \alpha \wedge \beta \sqsubseteq \beta' \wedge R_{\triangleright}\alpha\beta) \Rightarrow R_{\triangleright}\alpha'\beta')$;
(f13) $\forall\alpha\,\forall\beta\,\forall\alpha'\,\forall\beta'((\alpha' \sqsubseteq \alpha \wedge \beta \sqsubseteq \beta' \wedge R_{\triangleleft}\alpha\beta) \Rightarrow R_{\triangleleft}\alpha'\beta')$;
(f14) $\forall\alpha\, R_{\triangleright}\alpha\alpha$;
(f15) $\forall\alpha\,\forall\beta\,\forall\gamma\,((R_{\triangleright}\gamma\alpha \wedge R_{\triangleright}\alpha\beta) \Rightarrow R_{\triangleright}\gamma\beta)$;
(f16) $\forall\alpha\,\forall\beta\,(R_{\triangleleft}\alpha\beta \Rightarrow \alpha \sqsubseteq \beta)$;
(f17) $\forall\alpha\,\forall\beta\,(R_{\triangleleft}\alpha\beta \Rightarrow \exists\gamma\,(R_{\triangleleft}\alpha\gamma \wedge R_{\triangleleft}\gamma\beta))$.

Informally, $R_{\triangleright}$ and $R_{\triangleleft}$ may be thought of as accessibility relations on situations (like the accessibility relation in normal modal logics, e.g., in Kripke 1963, 1959). (f12) and (f13) give the tonicity for $R_{\triangleright}$ and $R_{\triangleleft}$, moreover the same tonicity: $R_{\triangleright}\downarrow\uparrow$ and $R_{\triangleleft}\downarrow\uparrow$. (f14) and (f15) require $R_{\triangleright}$ be reflexive and transitive. (f17) is the dual of (f15) and expresses density (like the frame condition corresponding to $\Box\Box\mathcal{A} \supset \Box\mathcal{A}$ in the case of normal modal logics). Lastly, (f16) means that $R_{\triangleleft}$ is a subrelation of $\sqsubseteq$.

Remark 2.4.2 We could have defined our logics by axiom systems without including the analogs of axioms (T) and (4) (or ($\Diamond$T) and ($\Diamond$4)). Then, (a10)–(a13) would not hold in the Lindenbaum algebra, and (f13)–(f16) could be omitted from among the frame conditions.

Next, Definition 2.3.2 is extended with clauses for $\triangleright$ and $\triangleleft$.

Definition 2.4.3 A *model* for $\mathfrak{C}^{\vee}_{\mathrm{m}}$ is a model on a frame for $\mathfrak{C}^{\vee}_{\mathrm{m}}$ with Definition 2.3.2 extended with (m6) and (m7).

(m6) $\alpha \vDash \triangleright\mathcal{A}$ iff $\exists\beta\,(R_{\triangleright}\beta\alpha \wedge \beta \vDash \mathcal{A})$;
(m7) $\alpha \vDash \triangleleft\mathcal{A}$ iff $\exists\beta\,(R_{\triangleleft}\beta\alpha \wedge \beta \vDash \mathcal{A})$.

The truth of the following lemma, which adds two cases to Lemma 2.3.3, is practically immediate from (f12) and (f13) together with (m6) and (m7). (We omit the details of the proof.)

Lemma 2.4.4 $\|\triangleright\mathcal{A}\| \in \mathcal{C}(U)$ *and* $\|\triangleleft\mathcal{A}\| \in \mathcal{C}(U)$.

Now we can extend Theorem 2.3.4.

Theorem 2.4.5 *The modal logic* $\mathfrak{C}^{\vee}_{\mathrm{m}}$ *is* sound *for the class of models from Definition 2.4.3.*

Proof We give three of the cases that pertain to $\triangleleft$. One of the $\triangleright$ cases is analogous, and the two others are easy (hence, omitted).

1. Let us assume that $\beta \in \|\triangleleft(\mathcal{A} \vee \mathcal{B})\|$. Then $\exists\alpha\,(R_{\triangleleft}\alpha\beta \wedge \alpha \in \|\mathcal{A} \vee \mathcal{B}\|)$. Using (m2) from Definition 2.3.2 and properties of the metalanguage quantifier, we get that $\exists\alpha\,(R_{\triangleleft}\alpha\beta \wedge \alpha \in \|\mathcal{A}\|)$ and $\exists\alpha\,(R_{\triangleleft}\alpha\beta \wedge \alpha \in \|\mathcal{B}\|)$. This means, by (m2) again, that $\beta \in \|\triangleleft\mathcal{A}\|$, and symmetrically, $\beta \in \|\triangleleft\mathcal{B}\|$.

2. We prove that $\|\triangleleft\mathcal{A}\| \subseteq \|\mathcal{A}\|$. Let $\beta \in \|\triangleleft\mathcal{A}\|$, that is, $\exists\alpha\,(R_{\triangleleft}\alpha\beta \wedge \alpha \in \|\mathcal{A}\|)$. By (f16), we know that $\alpha \sqsubseteq \beta$, and by Lemma 2.4.4, $\beta \in \|\mathcal{A}\|$.

3. To show that (a13) holds when the frame has property (f17), let us assume that $\beta \in \|\triangleleft\mathcal{A}\|$. By (m7), $\exists\alpha\,(R_{\triangleleft}\alpha\beta \wedge \alpha \in \|\mathcal{A}\|)$. However, (f17) then guarantees that $R_{\triangleleft}\alpha\gamma$ and $R_{\triangleleft}\gamma\beta$ hold for some γ. Hence, $\gamma \in \|\triangleleft\mathcal{A}\|$ and so $\beta \in \|\triangleleft\triangleleft\mathcal{A}\|$. ⁙

We add the definitions of two relations to the previously described canonical frame.

Definition 2.4.6 The *canonical frame* for $\mathfrak{C}_{\mathrm{m}}^{\vee}$ is as in Definition 2.3.5 with two binary relations, $R_{\triangleright}$ and $R_{\triangleleft}$, added that satisfy (c4) and (c5).

(c4) $R_{\triangleright}\alpha\beta$ iff $\forall a\,(a \in \alpha \Rightarrow \triangleright a \in \beta)$,
(c5) $R_{\triangleleft}\alpha\beta$ iff $\forall a\,(a \in \alpha \Rightarrow \triangleleft a \in \beta)$.

Lemma 2.4.7 *The canonical frame of* $\mathfrak{C}_{\mathrm{m}}^{\vee}$ *is a* frame for the logic *according to Definition 2.4.2.*

Proof As sample steps we prove two properties, and leave the rest to the reader.
1. For (f12), let us assume the antecedent of the universally quantified condition. And let us also assume that $a \in \alpha'$. Then, $a \in \alpha$, because $\alpha' \sqsubseteq \alpha$, hence $\triangleright a \in \beta$. But $\beta \sqsubseteq \beta'$, which means that $R_{\triangleright}\alpha'\beta'$, as we needed to show.
2. Since (a10) holds, (f14) is stipulated in a structure for $\mathfrak{C}_{\mathrm{m}}^{\vee}$. Let us assume that $\alpha \in \mathfrak{I}$, that is, α is a canonical situation. Since (a10) holds, $\triangleright a \in \alpha$ whenever a is an element of α. That is, $R_{\triangleright}\alpha\alpha$. ⁘

We also have to prove that the canonical valuation—the definition of which is unchanged—but which is extended with (m6) and (m7) leaves Lemma 2.3.12 true. (We do not repeat the lemma, which remains the same word by word; however, we add two steps to the proof.)

Proof We add two cases to the previous proof.
6. To continue the structural induction, let us assume that the formula is $\triangleleft\mathcal{A}$. If $\alpha \in \|\triangleleft\mathcal{A}\|$, then $\exists\beta\,(R_{\triangleleft}\beta\alpha \wedge \beta \in \|\mathcal{A}\|)$. By hypothesis, $[\mathcal{A}] \in \beta$, and then $[\triangleleft\mathcal{A}] \in \alpha$ due to the definition of $R_{\triangleleft}$. For the other direction, let us start with $[\triangleleft\mathcal{A}] \in \alpha$. The principal cone $([\mathcal{A}]]$ is an ideal, and $\forall[\mathcal{C}]\,([\mathcal{C}] \in ([\mathcal{A}]] \Rightarrow [\triangleleft\mathcal{C}] \in \alpha)$, because of the monotonicity of $\triangleleft$. By hypothesis, $([\mathcal{A}]] \in \|\mathcal{A}\|$, hence, $([\mathcal{A}]]$ is suitable for β in $\exists\beta\,(R_{\triangleleft}\beta\alpha \wedge \beta \in \|\mathcal{A}\|)$. That is, $\alpha \in \|\mathcal{A}\|$, as we had to show. ⁘
7. Let us assume that $\alpha \vDash \triangleright\mathcal{A}$, that is, $\exists\beta\,(R_{\triangleright}\beta\alpha \wedge \beta \vDash \mathcal{A})$. By inductive hypothesis, $[\mathcal{A}] \in \beta$, hence by the definition of $R_{\triangleright}$, $[\triangleright\mathcal{A}] \in \alpha$.

For the other direction, we start with the latter as our assumption. We take for β the principal ideal generated by $[\mathcal{A}]$, that is, $([\mathcal{A}]]$. If $[\mathcal{B}] \in \beta$, then $[\mathcal{B}] \leq [\mathcal{A}]$, and by the monotonicity of $\triangleright$, $[\triangleright\mathcal{B}] \leq [\triangleright\mathcal{A}]$. However, then $[\triangleright\mathcal{B}] \in \alpha$, because $\alpha \in \mathfrak{I}$. In other words, $\forall[\mathcal{B}]\,([\mathcal{B}] \in \beta \Rightarrow [\triangleright\mathcal{B}] \in \alpha)$, which means that $R_{\triangleright}\beta\alpha$. Since $[\mathcal{A}] \in \beta$, $\beta \vDash \mathcal{A}$. Then, $\exists\beta\,(R_{\triangleright}\beta\alpha \wedge \beta \vDash \mathcal{A})$, and so $\alpha \vDash \triangleright\mathcal{A}$, as we need. ⁘

As before, Lemma 2.4.7 (canonical frame lemma), Theorem 2.4.5 (soundness theorem), Lemma 2.3.13 (separation lemma), and the expanded version of Lemma 2.3.12 (truth lemma) imply completeness, which is the next theorem.

Theorem 2.4.8 (Completeness) $\mathfrak{C}_{\mathrm{m}}^{\vee}$ *is* complete *with respect to its class of frames from Definition 2.4.2.*

Now we give a semantics for $\mathfrak{C}_{\mathrm{m}}^{\wedge}$, and then we will return to further expansions in both groups of logics.

2.5 Semantics for Logics with Conjunction

We have been emphasizing the duality between $\wedge$ and $\vee$ that stretches into the relational semantics. However, the following semantics utilizes an idea in the modeling of $+$, the analog of which we did not use in the previous section. The following semantics is more "usual" than the semantics we introduced in the previous sections; hence, we will present it in a more concise way.

Definition 2.5.1 A *structure for* $\mathfrak{C}^{\wedge}_{\mathrm{m}}$ (from Definition 2.4.1) is $\mathfrak{F} = \langle U, \iota, \sqsubseteq, R, R_+, R_\triangleright, R_\triangleleft \rangle$, where (s0)–(s11) hold. (X, Y, and Z are metavariables that range over propositions in a model on $\mathfrak{F}$.)

(s0) $U \neq \emptyset,\ \ \iota \in U,\ \ \sqsubseteq\, \subseteq U^2,\ \ R \subseteq U^3,\ \ R_+ \subseteq U^3,\ \ R_\triangleright \subseteq U^2,\ \ R_\triangleleft \subseteq U^2$;
(s1) $\forall\alpha\, \alpha \sqsubseteq \alpha,\ \ \forall\alpha\, \forall\beta\, \forall\gamma\, ((\alpha \sqsubseteq \beta \wedge \beta \sqsubseteq \gamma) \Rightarrow \alpha \sqsubseteq \gamma)$;
(s2) $\forall\alpha\, \forall\beta\, \forall\gamma\, \forall\alpha'\, \forall\beta'\, \forall\gamma'\, ((\alpha' \sqsubseteq \alpha \wedge \beta' \sqsubseteq \beta \wedge \gamma \sqsubseteq \gamma' \wedge R\alpha\beta\gamma) \Rightarrow R\alpha'\beta'\gamma')$;
(s3) $\forall\alpha\, \forall\beta\, (R\iota\alpha\beta \Leftrightarrow \alpha \sqsubseteq \beta)$;
(s4) $\forall\alpha\, \forall\beta\, \forall\gamma\, \forall\delta\, (\exists\vartheta\, (R\vartheta\gamma\delta \wedge R\alpha\beta\vartheta) \Leftrightarrow \exists\vartheta\, (R\alpha\vartheta\delta \wedge R\beta\gamma\vartheta))$;
(s5) $\forall\alpha\, \forall\beta\, \forall\gamma\, \forall\delta\, \forall\vartheta\, ((R\alpha\beta\vartheta \wedge R\vartheta\gamma\delta) \Rightarrow \exists\vartheta\, (R\alpha\gamma\vartheta \wedge R\vartheta\beta\delta))$;
(s6) $\forall\alpha\, \forall\beta\, \forall\gamma\, \forall\alpha'\, \forall\beta'\, \forall\gamma'\, ((\alpha' \sqsubseteq \alpha \wedge \beta' \sqsubseteq \beta \wedge \gamma \sqsubseteq \gamma' \wedge R_+\alpha\beta\gamma) \Rightarrow R_+\alpha'\beta'\gamma')$;
(s7) $\forall\alpha\, \forall\beta\, \forall\gamma\, (R_+\alpha\beta\gamma \Rightarrow R_+\beta\alpha\gamma)$;
(s8) $\forall\alpha\, \forall\beta\, \forall\gamma\, \forall\varepsilon\, (\exists\delta\, (R_+\alpha\beta\delta \wedge R_+\delta\gamma\varepsilon) \Leftrightarrow \exists\vartheta\, (R_+\alpha\vartheta\varepsilon \wedge R_+\beta\gamma\vartheta))$;
(s9) $\forall\alpha\, \forall\beta\, \forall\gamma\, \forall\delta\, \forall\upsilon\, ((R_+\alpha\gamma\delta \wedge R_+\beta\upsilon\delta \wedge \alpha \in X \wedge \beta \in Y \wedge \gamma \in Z \wedge \upsilon \in Z) \Rightarrow$
$\exists\varphi\, \exists\psi\, \exists\varepsilon\, \exists\pi\, (\varphi \in X \wedge \psi \in Y \wedge \pi \in Z \wedge \varphi \sqsubseteq \varepsilon \wedge \psi \sqsubseteq \varepsilon \wedge R_+\varepsilon\pi\delta))$;
(s10) $\forall\alpha\, \forall\beta\, \forall\gamma\, \forall\delta\, \forall\varepsilon\, ((R\alpha\varepsilon\delta \wedge R_+\beta\gamma\varepsilon) \Rightarrow \exists\vartheta\, (R\alpha\beta\vartheta \wedge R_+\vartheta\gamma\delta))$;
(s11) $\forall\alpha\, \forall\beta\, \forall\alpha'\, \forall\beta'\, ((\alpha' \sqsubseteq \alpha \wedge \beta \sqsubseteq \beta' \wedge R_\triangleright) \Rightarrow \alpha\beta R_\triangleright \alpha'\beta')$;
(s12) $\forall\alpha\, \forall\beta\, \forall\alpha'\, \forall\beta'\, ((\alpha' \sqsubseteq \alpha \wedge \beta \sqsubseteq \beta' \wedge R_\triangleleft) \Rightarrow \alpha\beta R_\triangleleft \alpha'\beta')$;
(s13) $\forall\alpha\, \forall\beta\, (R_\triangleright\beta\alpha \Rightarrow \beta \sqsubseteq \alpha)$;
(s14) $\forall\alpha\, \forall\beta\, (R_\triangleright\beta\alpha \Rightarrow \exists\gamma\, (R_\triangleright\beta\gamma \wedge R_\triangleright\gamma\alpha))$;
(s15) $\forall\alpha\, R_\triangleleft\alpha\alpha$;
(s16) $\forall\alpha\, \forall\beta\, \forall\gamma\, ((R_\triangleleft\beta\alpha \wedge R_\triangleleft\gamma\beta) \Rightarrow R_\triangleleft\gamma\alpha)$.

Remark 2.5.1 U is a set of situations with an element ι, which is the logical situation. The relation $\sqsubseteq$ pre-orders the situations, whereas the R relations cater for the modeling of the connectives. The tonicities of the relations are $R\, \downarrow\downarrow\uparrow$, $R_+\, \downarrow\downarrow\uparrow$, $R_\triangleright\, \downarrow\uparrow$, $R_\triangleleft\, \downarrow\uparrow$. Some of these conditions state more than what is required for $\mathfrak{C}^{\wedge}_{\mathrm{m}}$. In particular, (s6), (s11), and (s12) stipulate anti-tonicity for argument places of the respective accessibility relations that are not utilized. These stronger-than-necessary conditions are forward looking, in the sense that they do not interfere with completeness (or soundness), and they point toward straightforward extensions of $\mathfrak{C}^{\wedge}_{\mathrm{m}}$.

(s4) and (s5) guarantee associativity and commutativity for $\circ$. (s7) and (s8) do the same for $+$. (s9) may require some further explanation. There is a clear parallel between the definitions of the modal operations in the Kripke-style semantics of normal modal logics and how they distribute over conjunction or disjunction. The distribution type of $+$ (i.e., $+\colon \wedge, \wedge \longrightarrow \wedge$) would lead along those lines to the definition $\gamma \vDash \mathcal{A} + \mathcal{B}$ iff $\forall\alpha\, \forall\beta\, (\overline{R}_+\alpha\beta\gamma \Rightarrow (\alpha \vDash \mathcal{A} \vee \beta \vDash \mathcal{B}))$. Definition 2.5.2, however, defines the satisfaction for a formula $\mathcal{A} + \mathcal{B}$ differently, by (n5). Therefore, to guarantee that $\|(\mathcal{A} + \mathcal{C}) \wedge (\mathcal{B} + \mathcal{C})\| \subseteq \|(\mathcal{A} \wedge \mathcal{B}) + \mathcal{C}\|$, we have to impose a condition on the structure (with reference to models). (s9) refers to X, Y, and Z, which are sets of situations. Propositions are cones that are selected in a model. In the canonical model, they are easily distinguishable, because they are principal cones of situations that are generated by a principal cone in the Lindenbaum algebra. (That is, the proposition $\|\mathcal{A}\|$ is $[[[\mathcal{A}]))$.)

Definition 2.5.2 A *model* for $\mathfrak{C}^{\wedge}_{\mathrm{m}}$ is $\mathfrak{M} = \langle \mathfrak{F}, v \rangle$, where $\mathfrak{F}$ is a structure as in Definition 2.5.1, and v is a valuation of type $v \colon \mathbb{P} \longrightarrow \mathcal{C}$. v gives rise to a satisfiability relation by (n0)–(n7).

(n0) $\alpha \vDash p$ iff $\alpha \in v(p)$;
(n1) $\alpha \vDash \boldsymbol{t}$ iff $\iota \sqsubseteq \alpha$;
(n2) $\alpha \vDash \mathcal{A} \wedge \mathcal{B}$ iff $\alpha \vDash \mathcal{A}$ and $\alpha \vDash \mathcal{B}$;
(n3) $\alpha \vDash \mathcal{A} \circ \mathcal{B}$ iff $\exists \beta\, \exists \gamma\, (R\beta\gamma\alpha \wedge \beta \vDash \mathcal{A} \wedge \gamma \vDash \mathcal{B})$;
(n4) $\alpha \vDash \mathcal{A} \to \mathcal{B}$ iff $\forall \beta\, \forall \gamma\, ((R\alpha\beta\gamma \wedge \beta \vDash \mathcal{A}) \Rightarrow \gamma \vDash \mathcal{B})$;
(n5) $\alpha \vDash \mathcal{A} + \mathcal{B}$ iff $\exists \beta\, \exists \gamma\, (R_{+}\beta\gamma\alpha \wedge \beta \vDash \mathcal{A} \wedge \gamma \vDash \mathcal{B})$;
(n6) $\alpha \vDash \rhd\mathcal{A}$ iff $\exists \beta\, (R_{\rhd}\beta\alpha \wedge \beta \vDash \mathcal{A})$;
(n7) $\alpha \vDash \lhd\mathcal{A}$ iff $\exists \beta\, (R_{\lhd}\beta\alpha \wedge \beta \vDash \mathcal{A})$.

Lemma 2.5.3 (Heredity lemma) *For all formulas* $\mathcal{A}$, $\|\mathcal{A}\| \in \mathcal{C}(U)$.

Proof The claim holds for $p \in \mathbb{P}$ by the specification of v. $\|\boldsymbol{t}\| = \uparrow\iota$ by (n1). $\mathcal{C}$ is closed under $\cap$, hence, $\|\mathcal{A} \wedge \mathcal{B}\| \in \mathcal{C}$ according to (n2). The cases for the intensional connectives—including + with its unconventional definition—are all similar to each other with the exception of the case for $\to$, which we detail here. Let us assume that $\alpha \vDash \mathcal{A} \to \mathcal{B}$ and $\alpha \sqsubseteq \varepsilon$. Were $\varepsilon \nvDash \mathcal{A} \to \mathcal{B}$, we would have that $R\varepsilon\beta\gamma$ and $\beta \vDash \mathcal{A}$ but $\gamma \nvDash \mathcal{B}$. $R\varepsilon\beta\gamma$ implies $R\alpha\beta\gamma$ by (s2), and together with $\beta \vDash \mathcal{A}$, we have that $\gamma \vDash \mathcal{B}$. The contradiction means that $\varepsilon \in \|\mathcal{A} \to \mathcal{B}\|$, as we aimed to demonstrate. ⁖

Theorem 2.5.4 (Soundness) *If* $\mathcal{A}$ *is a theorem of* $\mathfrak{C}^{\wedge}_{\mathrm{m}}$, *then* $\uparrow\iota \subseteq \|\mathcal{A}\|$ *in any model* $\mathfrak{M}$, *from Definition 2.5.2.*

Proof Parts of the proof are easy or similar to proofs of soundness for certain implicational logics.[8] (Hence, we leave most of the proof to the reader.)

1. First, we prove that $\|\boldsymbol{t} \circ \mathcal{A}\| = \|\mathcal{A}\|$. From left to right, we assume that $\alpha \in \|\mathcal{A}\|$. Having combined (s1) and (s3), we have that $R\iota\alpha\alpha$. By (n1) and (n3), it is immediate that $\alpha \in \|\boldsymbol{t} \circ \mathcal{A}\|$. For the other inclusion, let us suppose the latter. By (n3), $\exists \iota'\, \exists \beta\, (R\iota'\beta\alpha \wedge \iota' \in \|\boldsymbol{t}\| \wedge \beta \in \|\mathcal{A}\|)$. From (n1), we obtain that $\iota \sqsubseteq \iota'$, hence, by (s2), $R\iota\beta\alpha$. Then, $\beta \sqsubseteq \alpha$ by (s3), and $\|\mathcal{A}\| \in \mathcal{C}$ implies that $\alpha \in \|\mathcal{A}\|$.

2. Next, we prove that $\|\mathcal{A} + \mathcal{B}\| = \|\mathcal{B} + \mathcal{A}\|$. We start with $\gamma \vDash \mathcal{A} + \mathcal{B}$, that is, $\exists \alpha\, \exists \beta\, (R_{+}\alpha\beta\gamma \wedge \alpha \vDash \mathcal{A} \wedge \beta \vDash \mathcal{B})$. The frame condition (s7) gives that $R_{+}\beta\alpha\gamma$, hence it is obvious that $\exists \beta\, \exists \alpha\, (R_{+}\beta\alpha\gamma \wedge \beta \vDash \mathcal{B} \wedge \alpha \vDash \mathcal{A})$, that is, $\gamma \vDash \mathcal{B} + \mathcal{A}$.

3. Third, we show that + distributes over $\wedge$ into $\wedge$—despite its somewhat unconventional definition. Let $\delta \in \|(\mathcal{A} \wedge \mathcal{B}) + \mathcal{C}\|$. (n5) gives us that $\exists \alpha\, \exists \gamma\, (R_{+}\alpha\gamma\delta \wedge \alpha \vDash$

[8] See, for example, Dunn (1991, 1993), Bimbó (2007, 2014) and Bimbó and Dunn (2008).

$\mathcal{A} \wedge \mathcal{B} \wedge \gamma \vDash \mathcal{C})$. However, then $\alpha \vDash \mathcal{A}$ as well as $\alpha \vDash \mathcal{B}$. We may recuperate the two α's separately, and thereby we get both $\delta \vDash \mathcal{A} + \mathcal{C}$ and $\delta \vDash \mathcal{B} + \mathcal{C}$, i.e., $\delta \vDash (\mathcal{A} + \mathcal{C}) \wedge (\mathcal{B} + \mathcal{C})$.

For the other direction, we assume that $\delta \in \|(\mathcal{A} + \mathcal{C}) \wedge (\mathcal{B} + \mathcal{C})\|$. That is, $\delta \in \|\mathcal{A} + \mathcal{C}\|$ and $\delta \in \|\mathcal{B} + \mathcal{C}\|$. By (n5), $\exists\alpha\, \exists\gamma\, (R_+\alpha\gamma\delta \wedge \alpha \vDash \mathcal{A} \wedge \gamma \vDash \mathcal{C})$ and $\exists\beta\, \exists\nu\, (R_+\beta\nu\delta \wedge \beta \vDash \mathcal{B} \wedge \nu \vDash \mathcal{C})$. $\|\mathcal{D}\|$ is a proposition in the model, for any $\mathcal{D}$, and hence we can use (s9). Having detached the antecedent of the universally quantified conditional, we have for some φ, ψ, π and ε that $R_+\varepsilon\pi\delta$ as well as $\varphi \in \|\mathcal{A}\|$, $\psi \in \|\mathcal{B}\|$, $\varphi \sqsubseteq \varepsilon$, $\psi \sqsubseteq \varepsilon$, and $\pi \in \|\mathcal{C}\|$. Using Lemma 2.5.3, we have that $\varepsilon \in \|\mathcal{A}\|$ and also $\varepsilon \in \|\mathcal{B}\|$. Then $\delta \vDash \mathcal{A} + \mathcal{C}$ as well as $\delta \vDash \mathcal{B} + \mathcal{C}$, giving $\delta \in \|(\mathcal{A} + \mathcal{C}) \wedge (\mathcal{B} + \mathcal{C})\|$ as we need for the distribution type of $+$.

4. We show that $\rhd$ behaves as desired. Since $\rhd$ does not distribute over $\wedge$ (unlike $\Box$ does in normal modal logics), (n6) resembles the truth condition for $\Diamond$. First, we prove that $\|\rhd(\mathcal{A} \wedge \mathcal{B})\| \subseteq \|\rhd\mathcal{A}\|$. If $\beta \vDash \rhd(\mathcal{A} \wedge \mathcal{B})$, then $\exists\alpha\, (R_\rhd\alpha\beta \wedge \alpha \vDash \mathcal{A} \wedge \mathcal{B})$. From $\alpha \vDash \mathcal{A}$ and $\alpha \vDash \mathcal{B}$, we use the former, and in one step, we obtain $\beta \vDash \rhd\mathcal{A}$.

5. Let us assume that $\alpha \vDash \rhd\mathcal{A}$. Then $\exists\beta\, (R_\rhd\beta\alpha \wedge \beta \vDash \mathcal{A})$, and by (s13), $\beta \sqsubseteq \alpha$. Then, Lemma 2.5.3 guarantees that $\alpha \vDash \mathcal{A}$.

6. Lastly, we show that the density of the $R_\rhd$ relation suffices for $\|\rhd\mathcal{A}\| \subseteq \|\rhd\rhd\mathcal{A}\|$. From $\alpha \vDash \rhd\mathcal{A}$, we get $R_\rhd\beta\alpha$ where $\beta \vDash \mathcal{A}$, for some β. But by (s14), there is a γ s̑t $R_\rhd\beta\gamma$; hence, $\gamma \vDash \rhd\mathcal{A}$, but then this, with $R_\rhd\gamma\alpha$, yields $\alpha \vDash \rhd\rhd\mathcal{A}$. ⁘

Definition 2.5.5 The *canonical structure* for $\mathfrak{C}^\wedge_{\mathrm{m}}$ is $\mathfrak{F} = \langle \mathcal{F}, \iota, \subseteq, R, R_+, R_\rhd, R_\lhd \rangle$, where the components are given by (c0)–(c5).

(c0) $\mathcal{F}$ is the set of non-empty filters on $\mathfrak{A}^\wedge_{\mathrm{m}}$; $\subseteq$ is set inclusion;
(c1) $\iota = \{a \in A \colon [\,\boldsymbol{t}\,] \leq a\}$;
(c2) $R\alpha\beta\gamma$ iff $\forall a\, \forall b\, ((a \in \alpha \wedge b \in \beta) \Rightarrow a \circ b \in \gamma)$;
(c3) $R_+\alpha\beta\gamma$ iff $\forall a\, \forall b\, ((a \in \alpha \wedge b \in \beta) \Rightarrow a + b \in \gamma)$;
(c4) $R_\rhd\alpha\beta$ iff $\forall a\, (a \in \alpha \Rightarrow \rhd a \in \beta)$;
(c5) $R_\lhd\alpha\beta$ iff $\forall a\, (a \in \alpha \Rightarrow \lhd a \in \beta)$.

Remark 2.5.2 The defining clauses for the accessibility relations associated to the connectives $\circ$, $+$, $\rhd$, and $\lhd$ are lookalikes, which is in concordance with (n3), (n5)–(n7). The differences between $\rhd$ and $\lhd$ are confined to their properties that are expressed by the inequations (a10)–(a13).

Lemma 2.5.6 *The canonical structure of* $\mathfrak{C}^\wedge_{\mathrm{m}}$ *(from Definition 2.5.5) is a* structure for the logic *in the sense of Definition 2.5.2.*

Proof Most of the proof goes along lines that are similar to proofs from relevance logics. We focus on the frame conditions for the connectives $+$ and $\lhd$ (and we omit some other details). Notice that (s6) and (s12) hold because of the clauses (c3) and (c5).

1. To prove (s7), we start with $R_+\alpha\beta\gamma$ and the assumption that some arbitrary elements a and b, $a \in \alpha$ and $b \in \beta$. Since $a + b \in \gamma$ and $b + a = a + b$, $R_+\beta\alpha\gamma$ holds.

2. Next, we assume that the antecedent of the conditional in (s8) is the case. We define $\vartheta = \{d\colon \exists b \exists c\,(b \in \beta \wedge c \in \gamma \wedge b + c \leq d)\}$. Informally, $\vartheta = [\beta + \gamma)$. If $b \in \beta$ and $c \in \gamma$, then obviously, $b + c \in \vartheta$, which means that $R_{+}\beta\gamma\vartheta$. If $d \in \vartheta$, then there are $b \in \beta$ and $c \in \gamma$ st $b + c \leq d$ by the definition of ϑ. Let also $a \in \alpha$. By the initial assumption, $a + b \in \delta$, hence $(a + b) + c \in \varepsilon$. $(a + b) + c = a + (b + c)$, and so $a + (b + c) \leq a + d$ implies that $a + d \in \varepsilon$. This means that $R_{+}\alpha\vartheta\varepsilon$.
3. The next definition will specify the canonical valuation, however in a rather unsurprising fashion. Thus, we will use here already what we prove a bit later, namely, for any $\mathcal{A}$, $\|\mathcal{A}\| = \{\alpha \in \mathfrak{F}\colon [\mathcal{A}] \in \alpha\}$. In other words, $\|\mathcal{A}\| = [[[\mathcal{A}]))$, the set of canonical situations that make the formula $\mathcal{A}$ true constitute an upward closed set of situations with the least element being the principal cone generated by the equivalence class of $\mathcal{A}$.

Now, if $R_{+}\alpha\gamma\delta$ and $R_{+}\beta\nu\delta$, where $\alpha \in \|\mathcal{A}\|$, $\beta \in \|\mathcal{B}\|$, and $\gamma, \nu \in \|\mathcal{C}\|$, for some $\mathcal{A}$, $\mathcal{B}$, and $\mathcal{C}$, then there are φ and ψ generating the former two propositions. That is, $\|\mathcal{A}\| = [\varphi)$ and $\|\mathcal{B}\| = [\psi)$. Since $[\mathcal{A}] \in \varphi$ and $[\mathcal{B}] \in \psi$, and $[\mathcal{C}] \in \gamma, \nu$, both $[\mathcal{A} + \mathcal{C}]$ and $[\mathcal{B} + \mathcal{C}]$ are elements of δ. Of course, then $[(\mathcal{A} + \mathcal{C}) \wedge (\mathcal{B} + \mathcal{C})] \in \delta$ too, because $\delta \in \mathfrak{F}$. The least element in $[\varphi) \cap [\psi)$, let us say, ε satisfies $\mathcal{A} \wedge \mathcal{B}$, and the least element in $\|\mathcal{C}\|$, let us say, π satisfies $\mathcal{C}$. Also, $R_{+}\varepsilon\pi\delta$, because $[(\mathcal{A} + \mathcal{C}) \wedge (\mathcal{B} + \mathcal{C})] = [(\mathcal{A} \wedge \mathcal{B}) + \mathcal{C}]$.
4. To prove (s15), we take an $\alpha \in \mathfrak{F}$ and assume that $a \in \alpha$. This condition is linked to inequation (a12), which together with the assumption implies that $\triangleleft a \in \alpha$. Then, by (c5) from Definition 2.5.5, we have that $R_{\triangleleft}\alpha\alpha$.
5. Finally, let $R_{\triangleleft}\beta\alpha$ and $R_{\triangleleft}\gamma\beta$ hold. Toward the goal of showing that $R_{\triangleleft}\gamma\alpha$ obtains, let $c \in \gamma$. Then $\triangleleft a \in \beta$, and further $\triangleleft\triangleleft a \in \alpha$. But (a13) guarantees that $\triangleleft\triangleleft a \leq \triangleleft a$, hence $\triangleleft a \in \alpha$. ⁘

Definition 2.5.7 The *canonical model* $\mathfrak{M}$ is defined on the canonical structure $\mathfrak{F}$ by choosing v as in (c6).

(c6) $v(p) = \{\alpha \in \mathfrak{F}\colon [p] \in \alpha\}$ and $v(\boldsymbol{t}) = [\iota)$.

In words, an atomic formula is mapped into the set of filters that contain its equivalence class, and the interpretation of $\boldsymbol{t}$ is the set of situations that expand the logical situation ι.

Lemma 2.5.8 *The canonical valuation extended to* $\vDash$ *has the property that* $\alpha \in \|\mathcal{A}\|$ *iff* $[\mathcal{A}] \in \alpha$.

Proof The cases for p and $\boldsymbol{t}$ are self-evident from the above definition.
1. The following series of iff's suffices for the conjunction case. $\alpha \in \|\mathcal{A} \wedge \mathcal{B}\|$ iff $\alpha \in \|\mathcal{A}\|$ and $\alpha \in \|\mathcal{B}\|$ iff $[\mathcal{A}] \in \alpha$ and $[\mathcal{B}] \in \alpha$ iff $[\mathcal{A} \wedge \mathcal{B}] \in \alpha$.
2. Let $\gamma \in \|\mathcal{A} + \mathcal{B}\|$. By (c3), $\exists\alpha\,\exists\beta\,(R_{+}\alpha\beta\gamma \wedge \alpha \in \|\mathcal{A}\| \wedge \beta \in \|\mathcal{B}\|)$. By hypothesis, $[\mathcal{A}] \in \alpha$ and $[\mathcal{B}] \in \beta$. The canonical definition of R_{+} gives us that $[\mathcal{A}] + [\mathcal{B}] \in \gamma$.

For the other direction, let $[\mathcal{A} + \mathcal{B}] \in \gamma$. The filters $[[\mathcal{A}])$ and $[[\mathcal{B}])$ are elements of $\|\mathcal{A}\|$ and $\|\mathcal{B}\|$, respectively, by the hypothesis of the induction. Were $[\mathcal{C}] \geq [\mathcal{A}]$ and $[\mathcal{D}] \geq [\mathcal{B}]$, we would have $[\mathcal{A} + \mathcal{B}] \leq [\mathcal{C} + \mathcal{D}]$. That is, for any such $\mathcal{C}$ and $\mathcal{D}$, $[\mathcal{C} + \mathcal{D}] \in \gamma$. By (c3), this establishes that $R_{+}[[\mathcal{A}])[[\mathcal{B}])\gamma$, hence $\gamma \in \|\mathcal{A} + \mathcal{B}\|$.

In lieu of three more cases, we note that the reasoning in case **2** depended on the tonicity of the connective, the satisfiability condition, and the kind of definition of the canonical relation, which are "shared" between $\circ$, $+$, $\rhd$, and $\lhd$. For $\to$, we note that this connective is the residual of $\circ$. The claim for $\mathcal{A} \to \mathcal{B}$ follows from (n4) along well-known lines. (There is no need for a squeeze lemma here—unlike in Routley and Meyer (1973), for instance—because the canonical situations are not prime theories.) ∴

Lemma 2.5.9 *If* $t \not\Vdash \mathcal{A} \to \mathcal{B}$, *then there is an* α *st* $\alpha \in \|\mathcal{A}\|$ *but* $\alpha \notin \|\mathcal{B}\|$.

Proof The condition means that $[\mathcal{B}] \notin [[\mathcal{A}])$, but of course, $[\mathcal{A}] \in [[\mathcal{A}])$. That is, $[[\mathcal{A}])$ will do for our α. ∴

Corollary 2.5.10 *Each proposition in the canonical model is of the form* $[[[A]))$.

Remark 2.5.3 Principal cones are *disagreeable* and *pleasant* at the same time. They are rare (which is unfortunate). But they provide an easy way to avoid complicated objects in a model, and they ground the propositions in the canonical model, which allows for an elegant characterization of propositions.

2.6 Further Additions to the Kernel Logics

The kernel logics did not include any of the structural rules—whether modalized or not. It is reasonable to assume that the structural rules are added in pairs, which ensures that the resulting sequent calculuses continue to admit the cut rule. Accordingly, we specify ten (in)equations, which we think of as five pairs.

Definition 2.6.1 The following five pairs of (in)equations may be added to the algebras $\mathfrak{A}_{\mathrm{m}}^{\wedge}$ and $\mathfrak{A}_{\mathrm{m}}^{\vee}$:

(k1)	$a \circ \rhd b \leq a$,	(k2)	$a \leq \lhd b + a$,
(w1)	$\rhd a \leq \rhd a \circ \rhd a$,	(w2)	$\lhd a + \lhd a \leq \lhd a$,
(l1)	$\rhd a \circ \rhd b = \rhd(a \wedge b)$,	(l2)	$\lhd(a \vee b) = \lhd a + \lhd b$,
(K1)	$a \circ b \leq a$,	(K2)	$a \leq b + a$,
(W1)	$a \leq a \circ a$,	(W2)	$a + a \leq a$.

Remark 2.6.1 The labels are intended to be suggestive: the k's and w's hint at the combinators K and W, and through them to the thinning and contraction rules. The lowercase letters indicate the result of the addition of modalized rules, and the l's include the k's and w's. (Of course, including (l1) requires $\wedge$, and (l2) needs $\vee$, and hence from this pair only one can be added to $\mathfrak{A}_{\mathrm{m}}^{\wedge}$ or $\mathfrak{A}_{\mathrm{m}}^{\vee}$.)

If all the structural rules are included, then the rules for $\wedge$ and $\circ$, and the rules for $\vee$ and $+$ become equivalent; furthermore, $\wedge$ and $\vee$ distribute over each other. That is, we can no longer exclude one or the other connective, and we essentially have a positive S4-like modal logic (with a distributive lattice in the background).

The semantics could be simplified in the latter case, giving a much more familiar semantics.

We will focus on the effect of the modalized structural rules in this brief section. We indicate the addition of pairs of rules to a logic or the matching pairs of inequations to an algebra by the subscripts $_{\mathfrak{k}}$, $_{\mathfrak{w}}$, and $_{\mathfrak{l}}$, which replace the subscript $_{\mathfrak{m}}$.

First, we expand the semantics that we defined for $\mathfrak{C}^{\vee}_{\mathfrak{m}}$.

Definition 2.6.2 A *frame* for $\mathfrak{C}^{\vee}_{\mathfrak{m}}$ is extended by the following clauses when the inequations in Definition 2.6.1 hold in $\mathfrak{A}^{\vee}_{\mathfrak{m}}$:

(fk1) $\forall\alpha\,\forall\beta\,\forall\gamma\,\forall\delta\,((\overline{R}\alpha\beta\gamma \wedge \overline{R}_{\rhd}\delta\beta) \Rightarrow \gamma \sqsubseteq \alpha)$, $\|\mathcal{A}\| \neq \emptyset$, for any $\mathcal{A}$;
(fk2) $\forall\alpha\,\forall\beta\,\forall\gamma\,\forall\delta\,((R_{+}\alpha\beta\gamma \wedge R_{\lhd}\delta\alpha) \Rightarrow \beta \sqsubseteq \gamma)$;
(fw1) $\forall\alpha\,\forall\beta\,\forall\gamma\,\forall\delta\,((R_{\rhd}\delta\gamma \wedge \overline{R}\alpha\beta\gamma) \Rightarrow (R_{\rhd}\delta\alpha \vee R_{\rhd}\delta\beta))$;
(fw2) $\forall\gamma\,\forall\delta\,(R_{\lhd}\delta\gamma \Rightarrow \exists\varepsilon\,(R_{\lhd}\delta\varepsilon \wedge R_{+}\varepsilon\varepsilon\gamma))$;
(fl2) $\forall\alpha\,\forall\beta\,\forall\gamma\,\forall\varepsilon\,\forall\vartheta\,((R_{+}\alpha\beta\gamma \wedge R_{\lhd}\varepsilon\alpha \wedge R_{\lhd}\vartheta\beta) \Rightarrow \exists\delta\,(R_{\lhd}\delta\gamma \wedge \varepsilon \sqsubseteq \delta \wedge \vartheta \sqsubseteq \delta))$,
$\forall\alpha\,\forall\beta\,(R_{\lhd}\alpha\beta \Rightarrow \exists\gamma\,(R_{\lhd}\alpha\gamma \wedge R_{+}\gamma\gamma\beta))$.

Remark 2.6.2 We have not introduced any new connectives or changed the satisfiability conditions for the existing ones. This means that we can use Definition 2.4.3 for a model without modifying it.

The proof of the following theorem is straightforward, and hence we omit the details.

Theorem 2.6.3 (Soundness) *If $\mathcal{A}$ is a theorem of $\mathfrak{C}^{\vee}_{\mathfrak{k}}$ ($\mathfrak{C}^{\vee}_{\mathfrak{w}}$, $\mathfrak{C}^{\vee}_{\mathfrak{l}}$), then $\|\mathcal{A}\| \subseteq I$ in any model on a frame for $\mathfrak{C}^{\vee}_{\mathfrak{m}}$ with the respective frame conditions added.*

The definitions of the canonical frame and of the canonical model are unaltered from Definitions 2.4.6 and 2.4.3.

Lemma 2.6.4 *The canonical frame satisfies* (fx) *when the algebra of the logic satisfies the inequation* (x)*, where $x \in \{k1, k2, w1, w2, l2\}$.*

Proof We provide a couple of sample steps from the proof, and leave the rest to the reader.

1. Let us prove that (fk1) holds. Let $\overline{R}\alpha\beta\gamma$ and $\overline{R}_{\rhd}\delta\beta$ hold, and let $a \in \gamma$. We want to show that $a \in \alpha$. $\gamma \in \mathfrak{I}$ and $a \circ \rhd b \leq a$, hence $a \circ \rhd b \in \gamma$. From the definition of $R_{\rhd}$, we get that $\exists c\,(c \in \delta \wedge \rhd c \notin \beta)$. But b was arbitrary, that is, $a \circ \rhd c \in \gamma$ too. Then, the definition of $\overline{R}$ with $\rhd c \notin \beta$ implies that $a \in \alpha$. In sum, $\gamma \subseteq \alpha$.
2. To establish (fw2) on the canonical frame, let us assume that $R_{\lhd}\delta\gamma$. Let us take ε be $(\lhd\delta]$, that is, the ideal generated by the elements $\lhd a$ where $a \in \delta$. Obviously, $R_{\lhd}\delta\varepsilon$, by (c5) in Definition 2.4.6. If $b \in \varepsilon$, then there is a $\lhd a \in \varepsilon$ s̃t $b \leq \lhd a$ and $a \in \delta$. However, the latter and our original assumption imply that $\lhd a \in \gamma$. By (w2), $\lhd a + \lhd a \leq \lhd a$, hence $\lhd a + \lhd a \in \gamma$, and so $R_{+}\varepsilon\varepsilon\gamma$. ∴

Theorem 2.6.5 *The logics $\mathfrak{C}^{\vee}_{\mathfrak{k}}$ ($\mathfrak{C}^{\vee}_{\mathfrak{w}}$, $\mathfrak{C}^{\vee}_{\mathfrak{l}}$) are* complete *with respect to their classes of frames.*

Now we turn to the group of logics that contain conjunction.

Definition 2.6.6 A *structure* for $\mathfrak{C}^{\wedge}_{\mathfrak{m}}$ is extended by the following clauses when the inequations in Definition 2.6.1 hold in $\mathfrak{A}^{\wedge}_{\mathfrak{m}}$.

(sk1) $\forall\alpha\,\forall\beta\,\forall\gamma\,\forall\delta\,((R\alpha\beta\gamma \wedge R_{\rhd}\delta\beta) \Rightarrow \alpha \sqsubseteq \gamma)$, $\|\mathcal{A}\| \neq \emptyset$, for any $\mathcal{A}$;
(sk2) $\forall\gamma\,\forall\delta\,\exists\alpha\,(R_{\lhd}\delta\alpha \wedge R_{+}\alpha\gamma\gamma)$;
(sw1) $\forall\gamma\,\forall\delta\,(R_{\rhd}\delta\gamma \Rightarrow \exists\varepsilon\,(R_{\rhd}\delta\varepsilon \wedge R\varepsilon\varepsilon\gamma))$;
(sw2) $\forall\alpha\,\forall\beta\,\forall\gamma\,\forall\varepsilon\,\forall\vartheta\,((R_{+}\alpha\beta\gamma \wedge R_{\lhd}\varepsilon\alpha \wedge R_{\lhd}\vartheta\beta) \Rightarrow \exists\delta\,(R_{\lhd}\delta\gamma \wedge (\varepsilon \sqsubseteq \delta \vee \vartheta \sqsubseteq \delta)))$;
(sl1) $\forall\alpha\,\forall\beta\,\forall\gamma\,\forall\varepsilon\,\forall\vartheta\,((R\alpha\beta\gamma \wedge R_{\rhd}\varepsilon\alpha \wedge R_{\rhd}\vartheta\beta) \Rightarrow \exists\delta\,(R_{\rhd}\delta\gamma \wedge \varepsilon \sqsubseteq \delta \wedge \vartheta \sqsubseteq \delta))$,
$\forall\gamma\,\forall\delta\,(R_{\rhd}\delta\gamma \Rightarrow \exists\alpha\,(R_{\rhd}\delta\alpha \wedge R\alpha\alpha\gamma))$.

Once again, we state the soundness theorem and leave filling out the details of its proof to the reader. (The steps are easy, because the conditions on the structure are directly applicable.)

Theorem 2.6.7 (Soundness) *If $\mathcal{A}$ is a theorem of $\mathfrak{C}^{\wedge}_{\mathfrak{k}}$ ($\mathfrak{C}^{\wedge}_{\mathfrak{w}}$, $\mathfrak{C}^{\wedge}_{\mathfrak{l}}$), then $I \subseteq \|\mathcal{A}\|$ in any model on a frame for $\mathfrak{C}^{\wedge}_{\mathfrak{m}}$ with the respective conditions for each logic added.*

Once again, the definitions of the canonical structure and of the canonical model are unchanged from Definitions 2.5.5 and 2.5.7.

Lemma 2.6.8 *The canonical structure satisfies* (s*x*) *when the algebra of the logic satisfies the inequation* (*x*), *where $x \in \{k1, k2, w1, l1\}$.*

Proof We prove some of the cases in detail and leave the others to the reader. For variety, we deal with the conditions for the three inequations that were not considered in the proof for the ideal semantics.

1. We prove that (sk2) holds on the canonical frame. Let γ and δ be given. We set α to be the filter generated by $\{\lhd b\colon\, b \in \delta\}$. This definition guarantees that $R_{\lhd}\delta\alpha$ holds. If $c \in \alpha$, then there is a $\lhd b \in \alpha$ st $\lhd b \leq c$ and $b \in \delta$. For any $a \in \gamma, a \leq \lhd b + a$, which puts $\lhd b + a$ into γ. Since $\lhd b \leq c$, $\lhd b + a \leq c + a$. This establishes that $R_{+}\alpha\gamma\gamma$.

2. For (sw1), let us assume that $R_{\rhd}\delta\gamma$ is the case. We define ε as $[\rhd\delta)$, which right away gives us that $R_{\rhd}\delta\varepsilon$. If $\rhd d_1, \rhd d_2 \in \varepsilon$, then $d_1, d_2 \in \delta$. Hence, $d_1 \wedge d_2 \in \delta$ and $\rhd(d_1 \wedge d_2) \in \varepsilon$, and also $\rhd(d_1 \wedge d_2) \in \gamma$. The following chain of inequations shows that $\rhd d_1 \circ \rhd d_2 \in \gamma$. $\rhd(d_1 \wedge d_2) \leq \rhd(d_1 \wedge d_2) \circ \rhd(d_1 \wedge d_2) \leq \rhd d_1 \circ \rhd d_2$. Thus, $R\varepsilon\varepsilon\gamma$, as we intended to prove.

3. (sl1) has two parts. Let us assume the antecedent of the first clause. We consider $[\varepsilon, \vartheta)$ for δ. This clearly ensures that $\varepsilon \subseteq \delta$ and $\vartheta \subseteq \delta$. If $d \in \delta$, then $a \wedge b \leq d$ for some $a \in \varepsilon$ and $b \in \vartheta$. By the assumption, $\rhd a \in \alpha$ and $\rhd b \in \beta$ as well as $\rhd a \circ \rhd b \in \gamma$. However, $\rhd a \circ \rhd b \leq \rhd(a \wedge b) \leq \rhd d$, meaning $\rhd d \in \gamma$ and $R_{\rhd}\delta\gamma$.

Now let $R_{\rhd}\delta\gamma$ hold. We define α as $[\rhd\delta)$ and show that it is in the desired relationship with δ and γ. Well, $R_{\rhd}\delta\alpha$ follows by α's definition. If $a, b \in \alpha$, then there are $\rhd c \leq a$ and $\rhd d \leq b$ st $c, d \in \delta$ and $c \wedge d \in \delta$. Then $\rhd(c \wedge d) \leq a \wedge b$, and $\rhd(c \wedge b) \in \gamma$. By the monotonicity of $\circ$, $\rhd c \circ \rhd d \leq a \circ b$, but $\rhd(c \wedge d) \leq \rhd c \circ \rhd d$, which shows that $a \circ b \in \gamma$, and therefore $R\alpha\alpha\gamma$. ⁞

Theorem 2.6.9 *The logic $\mathfrak{C}^{\wedge}_{\mathfrak{k}}$ is* complete *with respect to its class of structures.*

Remark 2.6.3 The last theorem does not claim completeness for $\mathfrak{C}^{\wedge}_{\mathfrak{w}}$ or $\mathfrak{C}^{\wedge}_{\mathfrak{t}}$, and Lemma 2.6.8 does not include the (sw2) condition. Although (sw2) is sufficient for $\|\triangleleft\mathcal{A} + \triangleleft\mathcal{A}\| \subseteq \|\triangleleft\mathcal{A}\|$ to hold in a model, and perhaps the condition is even minimal in some sense, it seems not to be provable on the canonical frame.

We leave dealing with the non-modalized inequations for another occasion. Here, we only mention some sample conditions. In $\mathfrak{C}^{\wedge}$, conditions that guarantee that (K1) and (W1) hold would be the familiar $\forall\alpha\,\forall\beta\,\forall\gamma\,(R\alpha\beta\gamma \Rightarrow \alpha \sqsubseteq \gamma)$ and $\forall\alpha\,\forall\beta\,\forall\gamma\,(R\alpha\beta\gamma \Rightarrow \exists\delta\,(R\alpha\beta\delta \wedge R\delta\beta\gamma))$. In $\mathfrak{C}^{\vee}$, we could stipulate $\forall\alpha\,\forall\beta\,\overline{R}\alpha\beta\alpha$ together with requiring that for no $\mathcal{A}$, $\|\mathcal{A}\| = \mathfrak{I}$, for (K1). And either $\forall\alpha\,\overline{R}\alpha\alpha\alpha$ or $\forall\alpha\,\exists\beta\,\exists\gamma\,(\overline{R}\beta\gamma\alpha \wedge \beta \sqsubseteq \alpha \wedge \gamma \sqsubseteq \alpha)$ would guarantee that (W1) holds.

2.7 Conclusions

We investigated some ways of providing set-theoretical semantics for logics without negation but with various intensional operations. In the two groups of logics we delineated, we retained exactly one of the two lattice connectives, and argued for lifting their duality to a duality in the whole outlook of the semantics. For our semantics based on ideals, we proved a *new version of the primeness (squeeze) lemma* yielding prime cones rather than prime filters. We also entertained a new interpretation of fission. Weak positive modal logics, at least some of them, were introduced outside of a systematic overview of variations across logics, because they are of independent importance. Our anticipation is that the relational semantics will facilitate applications of some related techniques to obtain further results about weak positive modal logics.

Acknowledgements I am grateful to Anna Zamansky and Ofer Arieli, the editors of the volume, for inviting me to contribute a paper. I would like to thank the organizers of and the audiences at IsraLog 2017 (*The Third Israeli Workshop on Non-Classical Logics and their Applications*) and AAL 2018 (*2018 Annual Meeting of the Australasian Association for Logic*) where I presented talks related to this paper. I am also grateful to the *IU Logic Group* at Indiana University–Bloomington for letting me talk about the results in this paper in November 2019. The research reported in this paper was funded by an *Insight Grant* (SSHRC IG #435–2014–0127) awarded by the Social Sciences and Humanities Research Council of Canada.

References

Avron, A. (1984). Relevant entailment - semantics and formal systems. *Journal of Symbolic Logic, 49*(2), 334–342.
Avron, A. (1987). A constructive analysis of *RM*. *Journal of Symbolic Logic, 52*, 939–951.
Avron, A. (1988). The semantics and proof theory of linear logic. *Theoretical Computer Science, 57*, 161–184.
Avron, A. (1991). Using hypersequents in proof systems for non-classical logics. *Annals of Mathematics and Artificial Intelligence, 4*, 225–248.

Avron, A. (1996). The method of hypersequents in the proof theory of propositional non-classical logics. In W. Hodges, M. Hyland, C. Steinhorn, & J. Truss (Eds.), *Logic: From Foundations to Applications* (pp. 1–32). Oxford: Oxford University Press.

Avron, A. (2014). What is relevance logic? *Annals of Pure and Applied Logic, 165*, 26–48.

Bimbó, K. (2007). Relevance logics. In D. Jacquette (Ed.), *Philosophy of Logic, Handbook of the Philosophy of Science* (D. Gabbay, P. Thagard & J. Woods (Eds.)) (Vol. 5, pp. 723–789). Amsterdam: Elsevier (North-Holland). https://doi.org/10.1016/B978-044451541-4/50022-1.

Bimbó, K. (2009). Dual gaggle semantics for entailment. *Notre Dame Journal of Formal Logic, 50*(1), 23–41. https://doi.org/10.1215/00294527-2008-025.

Bimbó, K. (2015a). The decidability of the intensional fragment of classical linear logic. *Theoretical Computer Science, 597*, 1–17. https://doi.org/10.1016/j.tcs.2015.06.019.

Bimbó, K. (2015b). *Proof Theory: Sequent Calculi and Related Formalisms*. Boca Raton, FL: CRC Press. https://doi.org/10.1201/b17294

Bimbó, K. (2016). Some relevance logics from the point of view of relational semantics. *Logic Journal of the IGPL, 24*(3), 268–287. O. Arieli & A. Zamansky (Eds.) *Israeli Workshop on Non-classical Logics and their Applications* (IsraLog 2014). https://doi.org/10.1093/jigpal/jzw003.

Bimbó, K. (2017). On the decidability of certain semi-lattice based modal logics. In R. A. Schmidt & C. Nalon (Eds.), *Automated Reasoning with Analytic Tableaux and Related Methods. Proceedings of the 26th International Conference, TABLEAUX 2017, Brasília, Brazil, September 25–28, 2017, no. 10501* (pp. 44–61). Lecture Notes in Artificial Intelligence. Switzerland: Springer Nature. https://doi.org/10.1007/978-3-319-66902-1_3.

Bimbó, K., & Dunn, J. M. (2005). Relational semantics for Kleene logic and action logic. *Notre Dame Journal of Formal Logic, 46*, 461–490. https://doi.org/10.1305/ndjfl/1134397663.

Bimbó, K., & Dunn, J. M. (2008). *Generalized Galois Logics: Relational Semantics of Nonclassical Logical Calculi, CSLI Lecture Notes* (Vol. 188). Stanford, CA: CSLI Publications.

Bimbó, K., & Dunn, J. M. (2009). Symmetric generalized Galois logics. *Logica Universalis, 3*(1), 125–152. https://doi.org/10.1007/s11787-009-0004-3.

Bimbó, K., & Dunn, J. M. (2015). Modalities in lattice-R. (manuscript, 34 pages).

Cresswell, M., Mares, E., & Rini, A. (Eds.). (2016). *Logical Modalities from Aristotle to Carnap. The Story of Necessity*. Cambridge, UK: Cambridge University Press.

Dunn, J. M. (1966). The algebra of intensional logics. Ph.D. thesis, University of Pittsburgh, Ann Arbor (UMI). (Published as v. 2 in the *Logic PhDs* series by College Publications, London (UK), 2019.)

Dunn, J. M. (1973). A 'Gentzen system' for positive relevant implication, (abstract). *Journal of Symbolic Logic, 38*, 356–357.

Dunn, J. M. (1986). Relevance logic and entailment. In D. Gabbay & F. Guenthner (Eds.), *Handbook of Philosophical Logic* (1st ed., Vol. 3, pp. 117–224). Dordrecht: D. Reidel.

Dunn, J. M. (1991). Gaggle theory: An abstraction of Galois connections and residuation, with applications to negation, implication, and various logical operators. In J. van Eijck (Ed.), *Logics in AI: European Workshop JELIA '90* (Vol. 478, pp. 31–51)., Lecture Notes in Computer Science Berlin: Springer.

Dunn, J. M. (1993). Partial gaggles applied to logics with restricted structural rules. In K. Došen & P. Schroeder-Heister (Eds.), *Substructural Logics* (Vol. 2, pp. 63–108)., Studies in Logic and Computation Clarendon, Oxford, UK.

Dunn, J. M. (1995). Positive modal logic. *Studia Logica, 55*, 301–317. https://doi.org/10.1007/BF01061239.

Gentzen, G. (1964). Investigations into logical deduction. *American Philosophical Quarterly, 1*, 288–306.

Girard, J. Y. (1987). Linear logic. *Theoretical Computer Science, 50*, 1–102.

Gödel, K. (1986). Eine Interpretation des intuitionistischen Aussagenkalküls. In S. Feferman (Ed.), *Collected Works* (Vol. I, pp. 300–303). New York, NY and Oxford, UK: Oxford University Press and Clarendon Press.

Halmos, P. R. (1962). *Algebraic Logic*. New York, NY: Chelsea Publishing Company.

Kopylov, A. P. (2001). Decidability of linear affine logic. In A. R. Meyer (Ed.), *Special issue: LICS 1995, Information and Computation* (Vol. 164, pp. 173–198). IEEE.
Kripke, S. A. (1959). A completeness theorem in modal logic. *Journal of Symbolic Logic, 24*, 1–14.
Kripke, S. A. (1963). Semantical considerations on modal logic. *Acta Philosophica Fennica, 16*, 83–94.
Kripke, S. A. (1965). Semantical analysis of intuitionistic logic I. In J. N. Crossley & M. A. E. Dummett (Eds.), *Formal Systems and Recursive Functions. Proceedings of the Eighth Logic Colloquium* (pp. 92–130). Amsterdam: North-Holland.
Meyer, R. K. (1966). Topics in modal and many-valued logic. Ph.D. thesis, University of Pittsburgh, Ann Arbor (UMI)
Routley, R., & Meyer, R. K. (1973). The semantics of entailment. In H. Leblanc (Ed.), *Truth, Syntax and Modality. Proceedings of the Temple University Conference on Alternative Semantics* (pp. 199–243). Amsterdam: North-Holland.

Katalin Bimbó is a Professor of Philosophy at the University of Alberta in Edmonton, Canada. Her research focuses on relevance, substructural, linear, and combinatory logics. She is an author of numerous articles, book chapters, and three books. Some of her latest results concern the decidability of certain non-classical logics, including the decidability of implicational ticket entailment that is a product of her joint work with J. Michael Dunn.

Chapter 3
On Axioms and Rexpansions

Carlos Caleiro and Sérgio Marcelino

Abstract We study the general problem of strengthening the logic of a given (partial) (non-deterministic) matrix with a set of axioms, using the idea of rexpansion. We obtain two characterization methods: a very general but not very effective one, and then an effective method which only applies under certain restrictions on the given semantics and the shape of the axioms. We show that this second method covers a myriad of examples in the literature. Finally, we illustrate how to obtain analytic multiple-conclusion calculi for the resulting logics.

3.1 Introduction

The work reported in this paper has three underlying aims.

First, and foremost, on a higher level reading, this paper is an acclamation of the modularization power enabled by *non-deterministic matrices (Nmatrices)*, as proposed and developed by Arnon Avron, along with his coauthors and students over the past 15 years (Avron and Lev 2005; Avron 2005a, b, 2007; Avron et al. 2007; Avron and Zamansky 2011; Avron et al. 2012, 2013; Avron and Zohar 2019), and used by many others (Marcelino and Caleiro 2017; Caleiro and Marcelino 2019; Ciabattoni et al. 2014; Baaz et al. 2013; Caleiro et al. 2019; Marcelino and Caleiro 2019; Coniglio and Golzio 2019) when seeking for a clear semantic rendering of logics resulting from strengthening a given base logic.

This research was funded by FCT/MCTES through national funds and when applicable co-funded EU funds under the project UIDB/50008/2020. Work done under the scope of the CT4L initiative of SQIG at Instituto de Telecomunicações. The authors are indebted to the anonymous referees for their valuable feedback, which helped improve a previous version of this paper.

C. Caleiro (✉) · S. Marcelino
SQIG - Instituto de Telecomunicações, Dep. Matemática - Instituto Superior Técnico, Universidade de Lisboa, Lisbon, Portugal
e-mail: ccal@math.tecnico.ulisboa.pt; caleiro@gmail.com
e-mail: smarcel@math.tecnico.ulisboa.pt

O. Arieli and A. Zamansky (eds.), *Arnon Avron on Semantics and Proof Theory of Non-Classical Logics*, Outstanding Contributions to Logic 21,
https://doi.org/10.1007/978-3-030-71258-7_3

Secondly, in the technical developments we propose, this paper can be seen as an application of the ideas behind *rexpansions* (Avron and Zohar 2019) of Nmatrices, in the form of a generalization of the systematic method put forth in Ciabattoni et al. (2014) for obtaining modularly a suitable semantics for a given logic strengthened with additional axioms (and new unary connectives). Expectedly, the method may yield in general a *partial non-deterministic matrix (PNmatrix)* (Baaz et al. 2013), partiality being a feature that adds to the conciseness of Nmatrices but which is known to contend with *analyticity*.

Last but not least, this paper is an opportunity for putting into practice the techniques developed in Marcelino and Caleiro (2019); Caleiro and Marcelino (2019) for obtaining an analytic multiple-conclusion calculus for the logic defined by any finite PNmatrix (under a reasonable expressiveness proviso). This is in contrast with comparable results for sequent-like calculi (Baaz et al. 2013; Ciabattoni et al. 2014), for which partiality seems to devoid them of a usable (even if generalized) subformula property capable of guaranteeing analyticity (and elimination of non-analytic cuts).

The paper is organized as follows. In Sect. 3.2, we recall (or suitably adapt) the necessary notions about logics, their syntax and semantics. Section 3.3 presents two methods for using rexpansions in order to obtain semantic characterizations of the strengthening with additional (schema) axioms Ax of the logic of a given PNmatrix $\mathbb{M}$. The first method, presented in Sect. 3.3.1, is completely general but unfortunately produces an infinite PNmatrix even when a finite one would be available. In order to overcome this drawback, in Sect. 3.3.2, we present another more economic method, generalizing (Ciabattoni et al. 2014), which, under suitable requirements, always provides a finite PNmatrix when starting from finite $\mathbb{M}$ and Ax. Section 3.4 is devoted to illustrating the application of the method of Sect. 3.3.2 to some meaningful examples. Then, in Sect. 3.5, we show that (under minimal expressiveness requirements on $\mathbb{M}$) the results of Marcelino and Caleiro (2019); Caleiro and Marcelino (2019) can be used to provide analytic multiple-conclusion calculi to the strengthened logics by exploring the semantics obtained by our method and provide illustrative examples. We close the paper in Sect. 3.6, with some concluding remarks and topics for future work.

3.2 Preliminaries

For the sake of self-containment, and in order to fix notation and terminology, we start by recalling (or suitably adapting, or generalizing) a number of useful notions and results. Instead of going through this material sequentially, the reader could as well jump this section for the moment and refer back here whenever necessary.

A propositional *signature* Σ is a family $\{\Sigma^{(k)}\}_{k\in\mathbb{N}}$ of sets, where each $\Sigma^{(k)}$ contains the *k-place connectives* of Σ. To simplify notation, we express the fact that $\copyright \in \Sigma^{(k)}$ for some $k \in \mathbb{N}$ by simply writing $\copyright \in \Sigma$, and we write $\Sigma' \cup \Sigma$ or $\Sigma' \subseteq \Sigma$ to denote the union or the inclusion, respectively, if Σ' is also a signature. Given a signature Σ, the language $L_\Sigma(P)$ is the carrier of the absolutely free Σ-algebra generated over a given denumerable set of sentential variables P. Elements of $L_\Sigma(P)$ are called *formulas*. Given a formula $A \in L_\Sigma(P)$, we denote by $\mathsf{var}(A)$ (resp. $\mathsf{sub}(A)$) the set of variables (resp. subformulas) of A, defined as usual; the extension of var and sub, and other similar functions, from formulas to sets thereof is defined as expected. A *substitution* is a member $\sigma \in L_\Sigma(P)^P$, that is, a function $\sigma : P \to L_\Sigma(P)$, uniquely extendable into an endomorphism $\cdot^\sigma : L_\Sigma(P) \to L_\Sigma(P)$. Given $\Gamma \subseteq L_\Sigma(P)$, we denote by Γ^σ the set $\{A^\sigma : A \in \Gamma\}$. For $A \in L_\Sigma(P)$, define $A^{\mathsf{inst}} = \{A^\sigma : \sigma \in L_\Sigma(P)^P\}$ and $\Gamma^{\mathsf{inst}} = \bigcup_{A \in \Gamma} A^{\mathsf{inst}}$.

Given formulas $A, A_1, \dots, A_n \in L_\Sigma(P)$ with $\mathsf{var}(A) \subseteq \{p_1, \dots, p_n\}$, we write $A(A_1, \dots, A_n)$ to denote the formula A^σ where $\sigma(p_i) = A_i$ for $1 \le i \le n$.

Given a signature Σ, a Σ-*PNmatrix (partial non-deterministic matrix)* is a structure $\mathbb{M} = \langle V, D, \cdot_\mathbb{M} \rangle$ such that V is a set (of *truth-values*), $D \subseteq V$ is the set of *designated* values, and $©_\mathbb{M} : V^k \to \wp(V)$ is a function (*truth-table*) for each $k \in \mathbb{N}$ and each k-place connective $© \in \Sigma$. When $©_\mathbb{M}(x_1, \dots, x_k) \neq \emptyset$ for all $x_1, \dots, x_k \in V$ we say that the truth-table of © in $\mathbb{M}$ is *total*. When $©_\mathbb{M}(x_1, \dots, x_k)$ has at most one element for all $x_1, \dots, x_k \in V$ we say that the truth-table of © in $\mathbb{M}$ is *deterministic*. Of course, deterministic does not imply total. Given $\Sigma' \subseteq \Sigma$, we say that $\mathbb{M}$ is Σ'-*total* if the truth-tables in $\mathbb{M}$ of the connectives $© \in \Sigma'$ are all total. Analogously, we say that $\mathbb{M}$ is Σ'-*deterministic* if the truth-tables in $\mathbb{M}$ of the connectives $© \in \Sigma'$ are all deterministic. When the Σ-PNmatrix $\mathbb{M}$ is Σ-total, or just total, it is simply called a Σ-*Nmatrix, or Nmatrix (non-deterministic matrix).* When a Σ-Nmatrix $\mathbb{M}$ is Σ-deterministic, or just deterministic, it is simply called a Σ-*matrix*, or a logical matrix. For the sake of completing the picture, when a Σ-PNmatrix $\mathbb{M}$ is deterministic we call it a Σ-*Pmatrix*, or Pmatrix.

Granted a Σ-PNmatrix $\mathbb{M} = \langle V, D, \cdot_\mathbb{M} \rangle$, a $\mathbb{M}$-*valuation* is a function $v : L_\Sigma(P) \to V$ such that $v(©(A_1, \dots, A_k)) \in ©_\mathbb{M}(v(A_1), \dots, v(A_k))$ for every $k \in \mathbb{N}$, every k-place connective $© \in \Sigma$, and every $A_1, \dots, A_k \in L_\Sigma(P)$. We denote the set of all $\mathbb{M}$-valuations by $\mathrm{Val}_\mathbb{M}$. Given a formula $A \in L_\Sigma(\{p_1, \dots, p_n\})$, we extend the usual notation for connectives and use $A_\mathbb{M} : V^n \to \wp(V)$ to denote the function defined by $A_\mathbb{M}(x_1, \dots, x_n) = \{v(A) : v \in \mathrm{Val}_\mathbb{M} \text{ with } v(p_i) = x_i \text{ for } 1 \le i \le n\}$ for every $x_1, \dots, x_n \in V$.

As is well known, if $\mathbb{M} = \langle V, D, \cdot_\mathbb{M} \rangle$ is a matrix then every function $f : Q \to V$ with $Q \subseteq P$ can be extended to a $\mathbb{M}$-valuation (in an essentially unique way for all formulas A with $\mathsf{var}(A) \subseteq Q$). As a consequence, $A_\mathbb{M}(x_1, \dots, x_n)$ is a singleton when $\mathbb{M}$ is a matrix, or more generally when there is $\Sigma' \subseteq \Sigma$ such that $A \in L_{\Sigma'}(P)$ and $\mathbb{M}$ is Σ'-deterministic and Σ'-total. If $\mathbb{M}$ is only known to be Σ'-deterministic, we can at least guarantee that $A_\mathbb{M}(x_1, \dots, x_n)$ has at most one element. When $\mathbb{M}$ is a Nmatrix, however, $A_\mathbb{M}(x_1, \dots, x_n)$ can be a large (non-empty) set. Still, we know from Avron and Zamansky (2011) that a function $f : \Gamma \to V$ with $\Gamma \subseteq L_\Sigma(P)$ can be extended to a $\mathbb{M}$-valuation provided that $\mathsf{sub}(\Gamma) \subseteq \Gamma$ and that $f(©(A_1, \dots, A_k)) \in ©_\mathbb{M}(f(A_1), \dots, f(A_n))$ whenever $©(A_1, \dots, A_k) \in \Gamma$. In case $\mathbb{M}$ is a PNmatrix, in general, one does not even have such a guarantee (Baaz et al. 2013), unless $f(\Gamma) \in$

$\mathcal{T}_{\mathbb{M}} = \bigcup_{v \in \mathrm{Val}_{\mathbb{M}}} \wp(v(L_\Sigma(P)))$. In other words, given $X \subseteq V$, we have $X \in \mathcal{T}_{\mathbb{M}}$ if the values in X are all together compatible in some valuation of $\mathbb{M}$. Of course, $A_{\mathbb{M}}(x_1, \dots, x_n) \neq \emptyset$ if $\{x_1, \dots, x_n\} \in \mathcal{T}_{\mathbb{M}}$.

A set of valuations $\mathcal{V} \subseteq \mathrm{Val}_{\mathbb{M}}$ characterizes a generalized (multiple-conclusion) consequence relation $\rhd_{\mathcal{V}} \subseteq \wp(L_\Sigma(P)) \times \wp(L_\Sigma(P))$ defined by $\Gamma \rhd_{\mathcal{V}} \Delta$ when for every $v \in \mathcal{V}$ if $v(\Gamma) \subseteq D$ then $v(\Delta) \cap D \neq \emptyset$. Of course, it also defines the more usual (single conclusion) consequence relation $\vdash_{\mathcal{V}} \subseteq \wp(L_\Sigma(P)) \times L_\Sigma(P)$ such that $\Gamma \vdash_{\mathcal{V}} A$ when $\Gamma \rhd_{\mathcal{V}} \{A\}$. In both cases, $\rhd_{\mathcal{V}}$ and $\vdash_{\mathcal{V}}$ are *substitution invariant*, and respectively a Scott (Scott 1974) and Shoesmith and Smiley (Shoesmith and Smiley 1978) consequence relation, or else a Tarskian consequence relation, when $\mathcal{V}$ is closed for substitutions, that is, if $v \in \mathcal{V}$ and $\sigma \in L_\Sigma(P)^P$ then $v \circ (\cdot^\sigma) \in \mathcal{V}$.

We simply write $\rhd_{\mathbb{M}}$ or $\vdash_{\mathbb{M}}$, instead of $\rhd_{\mathrm{Val}_{\mathbb{M}}}$ or $\vdash_{\mathrm{Val}_{\mathbb{M}}}$, respectively, and say that the consequences are characterized by $\mathbb{M}$. With respect to given consequence relations $\rhd$ or $\vdash$, we say that $\mathbb{M}$ is *sound* if $\rhd \subseteq \rhd_{\mathbb{M}}$ or $\vdash \subseteq \vdash_{\mathbb{M}}$, and we say that $\mathbb{M}$ is *complete* if $\rhd_{\mathbb{M}} \subseteq \rhd$ or $\vdash_{\mathbb{M}} \subseteq \vdash$.

A *refinement* of a Σ-PNmatrix $\mathbb{M} = \langle V, D, \cdot_{\mathbb{M}} \rangle$ is any Σ-PNmatrix $\mathbb{M}' = \langle V', D', \cdot_{\mathbb{M}'} \rangle$ with $V' \subseteq V$, $D' = D \cap V'$, and $©_{\mathbb{M}'}(x_1, \dots, x_k) \subseteq ©_{\mathbb{M}}(x_1, \dots, x_k)$ for every $k \in \mathbb{N}$, every k-place connective $© \in \Sigma$, and every $x_1, \dots, x_k \in V'$. It is clear, almost by definition, that $\mathrm{Val}_{\mathbb{M}'} \subseteq \mathrm{Val}_{\mathbb{M}}$. When it is always the case that $©_{\mathbb{M}'}(x_1, \dots, x_k) = ©_{\mathbb{M}}(x_1, \dots, x_k) \cap V'$ then the refinement is called *simple* and $\mathbb{M}'$ is denoted by $\mathbb{M}_{V'}$. Clearly, $v \in \mathrm{Val}_{\mathbb{M}}$ implies that $v \in \mathrm{Val}_{\mathbb{M}_{V'}}$ with $V' = v(L_\Sigma(P))$, and also that $\mathbb{M}_{V'}$ is a non-empty total refinement of $\mathbb{M}$. This observation justifies the equivalent definition of $\mathcal{T}_{\mathbb{M}}$ put forth in Caleiro and Marcelino (2019).

$\mathcal{E} : V \to \wp(U)$ is an *expansion function* if $\mathcal{E}(x) \neq \emptyset$ for every $x \in V$, and $\mathcal{E}(x) \cap \mathcal{E}(x') = \emptyset$ if $x' \in V$ is distinct from x. Given $X \subseteq V$, we abuse notation and use $\mathcal{E}(X)$ to denote $\bigcup_{x \in X} \mathcal{E}(x)$. One associates to $\mathcal{E}$ its *contraction* $\widetilde{\mathcal{E}} : \mathcal{E}(V) \to V$ such that, for each $y \in \mathcal{E}(V)$, $\widetilde{\mathcal{E}}(y) \in V$ is the unique such that $y \in \mathcal{E}(\widetilde{\mathcal{E}}(y))$. The *$\mathcal{E}$-expansion* of a Σ-PNmatrix $\mathbb{M} = \langle V, D, \cdot_{\mathbb{M}} \rangle$ is the Σ-PNmatrix $\mathcal{E}(\mathbb{M}) = \langle \mathcal{E}(V), \mathcal{E}(D), \cdot_{\mathcal{E}(\mathbb{M})} \rangle$ such that $©_{\mathcal{E}(\mathbb{M})}(y_1, \dots, y_k) = \mathcal{E}(©_{\mathbb{M}}(\widetilde{\mathcal{E}}(y_1), \dots, \widetilde{\mathcal{E}}(y_k)))$ for every $k \in \mathbb{N}$, every k-place connective $© \in \Sigma$, and every $y_1, \dots, y_k \in \mathcal{E}(V)$. By construction, it is clear that $\widetilde{\mathcal{E}}$ preserves and reflects designated values, i.e., $\widetilde{\mathcal{E}}(y) \in D$ if and only if $y \in \mathcal{E}(D)$. Further, given a function $f : L_\Sigma(P) \to \mathcal{E}(V)$, $f \in \mathrm{Val}_{\mathcal{E}(\mathbb{M})}$ if and only if $\widetilde{\mathcal{E}} \circ f \in \mathrm{Val}_{\mathbb{M}}$.

A *rexpansion* of a Σ-PNmatrix $\mathbb{M} = \langle V, D, \cdot_{\mathbb{M}} \rangle$ is a refinement of some $\mathcal{E}$-expansion of $\mathbb{M}$. When $\mathbb{M}^\dagger = \langle V^\dagger, D^\dagger, \cdot_{\mathbb{M}^\dagger} \rangle$ is a rexpansion of $\mathbb{M}$, we still have that if $v^\dagger \in \mathrm{Val}_{\mathbb{M}^\dagger}$ then $\widetilde{\mathcal{E}} \circ v^\dagger \in \mathrm{Val}_{\mathbb{M}}$. Consequently, we have that $\widetilde{\mathcal{E}}(A_{\mathbb{M}^\dagger}(x_1, \dots, x_n)) \subseteq A_{\mathbb{M}}(\widetilde{\mathcal{E}}(x_1), \dots, \widetilde{\mathcal{E}}(x_n))$, for every $A \in L_\Sigma(\{p_1, \dots, p_n\})$ and $x_1, \dots, x_n \in V^\dagger$.

It is easy to see that the refinement relation, the expansion relation, and thus also the rexpansion relation, are all transitive.

We end this section with a very simple but useful lemma.

Lemma 3.2.1 *Let $\Sigma' \subseteq \Sigma$ and $\mathbb{M} = \langle V, D, \cdot_{\mathbb{M}} \rangle$ be a Σ'-deterministic Σ-PNmatrix.*
If $\mathbb{M}^\dagger = \langle V^\dagger, D^\dagger, \cdot_{\mathbb{M}^\dagger} \rangle$ is a rexpansion of $\mathbb{M}$, $A \in L_{\Sigma'}(\{p_1, \dots, p_n\})$, and $y, z \in A_{\mathbb{M}^\dagger}(x_1, \dots, x_n)$ then $y \in D^\dagger$ if and only if $z \in D^\dagger$.

Proof Assume that $\mathbb{M}^\dagger$ is a refinement of the expansion of $\mathbb{M}$ with $\mathcal{E}$. If $y, z \in A_{\mathbb{M}^\dagger}(x_1, \ldots, x_n)$ then $\widetilde{\mathcal{E}}(y), \widetilde{\mathcal{E}}(z) \in \widetilde{\mathcal{E}}(A_{\mathbb{M}^\dagger}(x_1, \ldots, x_n)) \subseteq A_{\mathbb{M}}(\widetilde{\mathcal{E}}(x_1), \ldots, \widetilde{\mathcal{E}}(x_n))$. Since $\mathbb{M}$ is Σ'-deterministic and $A \in L_{\Sigma'}(P)$ it follows that $A_{\mathbb{M}}(\widetilde{\mathcal{E}}(x_1), \ldots, \widetilde{\mathcal{E}}(x_n))$ has at most one element, and thus $\widetilde{\mathcal{E}}(y) = \widetilde{\mathcal{E}}(z)$. Therefore, $y \in D^\dagger$ iff $\widetilde{\mathcal{E}}(y) \in D$ iff $\widetilde{\mathcal{E}}(z) \in D$ iff $z \in D^\dagger$. □

3.3 Adding Axioms

Given a signature Σ, a Tarskian consequence relation $\vdash$ over Σ, and $\mathsf{Ax} \subseteq L_\Sigma(P)$, the *strengthening of* $\vdash$ *with (schema) axioms* Ax is the consequence relation $\vdash^{\mathsf{Ax}}$ defined by $\Gamma \vdash^{\mathsf{Ax}} A$ if and only if $\Gamma \cup \mathsf{Ax}^{\mathsf{inst}} \vdash A$.

Our aim is to provide an adequate (and usable) semantics for $\vdash^{\mathsf{Ax}}$, given a semantic characterization of $\vdash$, a task that is well within the general effort of characterizing combined logics (Caleiro et al. 2005; Marcelino and Caleiro 2016, 2017). The following simple result, whose (simple) proof we omit, is a corollary of Lemma 2.7 of Caleiro et al. (2019).

Proposition 3.3.1 *Let* $\mathbb{M} = \langle V, D, \cdot_{\mathbb{M}} \rangle$ *be a* Σ*-PNmatrix and* $\mathsf{Ax} \subseteq L_\Sigma(P)$. *The consequence relation* $\vdash^{\mathsf{Ax}}_{\mathbb{M}}$ *is characterized by* $Val^{\mathsf{Ax}}_{\mathbb{M}} = \{v \in Val_{\mathbb{M}} : v(\mathsf{Ax}^{\mathsf{inst}}) \subseteq D\}$.

Our aim in the forthcoming subsections is to design some systematic way of using the ideas behind rexpansions for transforming $\mathbb{M}$ into a PNmatrix whose valuations somehow coincide with $\mathrm{Val}^{\mathsf{Ax}}_{\mathbb{M}}$.

3.3.1 A General Construction

As a first attempt, we employ a general technique from the theory of combining logics (Caleiro et al. 2005; Marcelino and Caleiro 2016, 2017). The overall idea, when starting from a given PNmatrix and a set of strengthening axioms, is to pair each formula of the logic with its possible values but guaranteeing that instances of axioms can only be paired with designated values.

Theorem 3.3.2 *Let* $\mathbb{M} = \langle V, D, \cdot_{\mathbb{M}} \rangle$ *be a* Σ*-PNmatrix and* $\mathsf{Ax} \subseteq L_\Sigma(P)$.
The consequence $\vdash^{\mathsf{Ax}}_{\mathbb{M}}$ *is characterized by the rexpansion* $\mathbb{M}^\flat_{\mathsf{Ax}} = \langle V^\flat_{\mathsf{Ax}}, D^\flat_{\mathsf{Ax}}, \cdot_{\mathbb{M}^\flat_{\mathsf{Ax}}} \rangle$ *of* $\mathbb{M}$ *defined by*

- $V^\flat_{\mathsf{Ax}} = \{(x, A) \in V \times L_\Sigma(P) : \text{ if } A \in \mathsf{Ax}^{\mathsf{inst}} \text{ then } x \in D\}$,
- $D^\flat_{\mathsf{Ax}} = D \times L_\Sigma(P)$,
- *for each* $k \in \mathbb{N}$ *and* $© \in \Sigma^{(k)}$,

$$©_{\mathbb{M}^\flat_{\mathsf{Ax}}}((x_1, A_1), \ldots, (x_k, A_k)) =$$

$$\{(x, \copyright(A_1, \ldots, A_k)) \in V^{\flat}_{\mathsf{Ax}} : x \in \copyright_{\mathbb{M}}(x_1, \ldots, x_n)\}.$$

Proof We prove, in turn, that $\mathbb{M}^{\flat}_{\mathsf{Ax}}$ is a rexpansion of $\mathbb{M}$, and then the soundness and completeness of $\mathbb{M}^{\flat}_{\mathsf{Ax}}$ with respect to $\vdash^{\mathsf{Ax}}_{\mathbb{M}}$.

Rexpansion. It is easy to see that the PNmatrix $\mathbb{M}^{\flat}_{\mathsf{Ax}}$ is a refinement of the expansion of $\mathbb{M}$ with $\mathcal{E}(x) = \{x\} \times L_{\Sigma}(P)$. $\widetilde{\mathcal{E}} : V^{\flat}_{\mathsf{Ax}} \to V$ is such that $\widetilde{\mathcal{E}}(x, A) = x$, and clearly preserves and reflects designated values. Using Proposition 3.3.1, it suffices to show that $\{\widetilde{\mathcal{E}} \circ v^{\flat} : v^{\flat} \in \mathrm{Val}_{\mathbb{M}^{\flat}_{\mathsf{Ax}}}\} = \mathrm{Val}^{\mathsf{Ax}}_{\mathbb{M}}$.

Note that if $v^{\flat} \in \mathrm{Val}_{\mathbb{M}^{\flat}_{\mathsf{Ax}}}$ and $v^{\flat}(A) = (x, B)$ then $B \in A^{\mathsf{inst}}$. Namely, we have $B = A^{\sigma}$ where $\sigma \in L_{\Sigma}(P)^P$ is such that $\sigma(p) = C$ if $v^{\flat}(p) = (y, C)$.

Soundness. Since $\mathbb{M}^{\flat}_{\mathsf{Ax}}$ is a rexpansion of $\mathbb{M}$ with $\mathcal{E}$, we know that if $v^{\flat} \in \mathrm{Val}_{\mathbb{M}^{\flat}_{\mathsf{Ax}}}$ then $\widetilde{\mathcal{E}} \circ v^{\flat} \in \mathrm{Val}_{\mathbb{M}}$. Further, if $A \in \mathsf{Ax}^{\mathsf{inst}}$ and $v^{\flat}(A) = (x, B)$ then $B \in (\mathsf{Ax}^{\mathsf{inst}})^{\mathsf{inst}} = \mathsf{Ax}^{\mathsf{inst}}$ and $\widetilde{\mathcal{E}}(v^{\flat}(A)) = x \subset D$. We conclude that $\{\widetilde{\mathcal{E}} \circ v^{\flat} : v^{\flat} \in \mathrm{Val}_{\mathbb{M}^{\flat}_{\mathsf{Ax}}}\} \subseteq \mathrm{Val}^{\mathsf{Ax}}_{\mathbb{M}}$ and thus that $\vdash^{\mathsf{Ax}}_{\mathbb{M}} \subseteq \vdash_{\mathbb{M}^{\flat}_{\mathsf{Ax}}}$.

Completeness. Reciprocally, if $v \in \mathrm{Val}_{\mathbb{M}}$ and $v(\mathsf{Ax}^{\mathsf{inst}}) \subseteq D$ then $v = \widetilde{\mathcal{E}} \circ v^{\flat}$ with $v^{\flat}(A) = (v(A), A)$ for each $A \in L_{\Sigma}(P)$. Since $v \in \mathrm{Val}_{\mathbb{M}}$, the fact that $v(\mathsf{Ax}^{\mathsf{inst}}) \subseteq D$ guarantees that $v^{\flat} \in \mathrm{Val}_{\mathbb{M}^{\flat}_{\mathsf{Ax}}}$. We conclude that $\mathrm{Val}^{\mathsf{Ax}}_{\mathbb{M}} \subseteq \{\widetilde{\mathcal{E}} \circ v^{\flat} : v^{\flat} \in \mathrm{Val}_{\mathbb{M}^{\flat}_{\mathsf{Ax}}}\}$ and thus that $\vdash_{\mathbb{M}^{\flat}_{\mathsf{Ax}}} \subseteq \vdash^{\mathsf{Ax}}_{\mathbb{M}}$. □

In the definition of $\mathbb{M}^{\flat}_{\mathsf{Ax}}$, if $\copyright_{\mathbb{M}}(x_1, \ldots, x_k) \cap D = \emptyset$ and moreover one has $\copyright(A_1, \ldots, A_k) \in \mathsf{Ax}^{\mathsf{inst}}$ then $\copyright_{\mathbb{M}^{\flat}_{\mathsf{Ax}}}((x_1, A_1), \ldots, (x_k, A_k)) = \emptyset$, which in general explains why the resulting PNmatrix may fail to be total. Still, $\mathbb{M}^{\flat}_{\mathsf{Ax}}$ is deterministic (actually a Pmatrix) when $\mathbb{M}$ is a (P)matrix. These two observations mean that the construction actually uses partiality in a most relevant way, but not non-determinism, which is simply imported from the starting PNmatrix. Note also that the construction, though fully illustrative of the power of rexpansions (generalized to PNmatrices) to accommodate new axioms, has other drawbacks. In fact, $\mathbb{M}^{\flat}_{\mathsf{Ax}}$ is always infinite, even if starting from a finite $\mathbb{M}$. Further, the structure of $\mathbb{M}^{\flat}_{\mathsf{Ax}}$ is quite syntactic, as it incorporates an obvious pattern-matching mechanism for recognizing instances of axioms into the received structure of $\mathbb{M}$.

In general, it is not possible to do much better, as it may happen that $\vdash^{\mathsf{Ax}}_{\mathbb{M}}$ cannot be characterized by a finite PNmatrix. For instance, as noted in Avron and Zohar (2019), Avron and coauthors show in Avron (2007) that the logic resulting from strengthening the Nmatrix characterizing the basic paraconsistent logic $\mathcal{BK}$ of Avron et al. (2012) with the axiom $\neg(p_1 \wedge \neg p_1) \to \circ p_1$ yields a logic that cannot be characterized by a finite Nmatrix. Thus, in order to improve on our result, it can be useful to look for suitable ways of controlling the shape of the axioms considered, as many other examples are known to have finite characterizations (Avron 2005a, 2007; Avron et al. 2012; Ciabattoni et al. 2014; Avron and Zohar 2019; Carnielli and Coniglio 2016).

On the other hand, the construction of Theorem 3.3.2 unveils a very interesting property of PNmatrices: every axiomatic extension of the logic of a finite (or denumerable) PNmatrix can be characterized by a denumerable PNmatrix. Just by itself, the result entails that *intuitionistic propositional logic* ($\mathcal{IPL}$) can be given by a single denumerable PNmatrix, sharply contrasting with the known fact that a characteristic matrix for $\mathcal{IPL}$ needs to be non-denumerable (see Gödel (1932), Wroński (1974), Wójcicki (1998)).

Example 3.3.3 Fix a suitable signature containing the two-place connective $\to$, and use the method above for strengthening with the usual axioms Int of intuitionistic logic the consequence relation characterized by the Nmatrix $\mathbb{MP} = \langle\{0, 1\}, \{1\}, \cdot_{\mathbb{MP}}\rangle$ where $©_{\mathbb{MP}}(x_1, \ldots, x_k) = \{0, 1\}$ for every k-place $© \in \Sigma$ such that $© \neq \to$, and $\to_{\mathbb{MP}}$ has the truth-table below.[1]

$\to_{\mathbb{MP}}$	0	1
0	0, 1	0, 1
1	0	0, 1

It is easy to see that $\vdash_{\mathbb{MP}}$ is precisely the consequence determined by the single rule $\frac{p \quad p\to q}{q}$ of *modus ponens*, and so $\vdash_{\mathbb{MP}^\flat_{\mathsf{Int}}}$ is precisely $\mathcal{IPL}$. △

This idea applies also to *propositional normal (global) modal logic* $\mathcal{K}$.

Example 3.3.4 For simplicity, take a signature containing only the 1-place modality $\Box$, and the 2-place connective $\to$. The logic determined by the rules of *modus ponens* and *necessitation*, i.e., $\frac{p}{\Box p}$, is easily seen to be characterized by the Nmatrix $\mathbb{MP}_\Box = \langle\{0, 1\}, \{1\}, \cdot_{\mathbb{MP}_\Box}\rangle$ given by the truth-tables below.

$\to_{\mathbb{MP}_\Box}$	0	1
0	0, 1	0, 1
1	0	0, 1

	$\Box_{\mathbb{MP}_\Box}$
0	0, 1
1	1

Collecting in Norm the usual axioms of classical implication plus the *normalization axiom* $\Box(p \to q) \to (\Box p \to \Box q)$ and applying Theorem 3.3.2, we get a denumerable PNmatrix $(\mathbb{MP}_\Box)^\flat_{\mathsf{Norm}}$ characterizing $\mathcal{K}$. △

These cases suggest another possible obstacle to improving our result, namely when the received PNmatrix is not deterministic and actually mixes designated with undesignated values in some entry of its truth-tables. When the basis is deterministic (enough) many examples are known to be finitely characterizable.

[1] For simplicity, in this and other examples, we omit the usual brackets of set notation when describing the truth-tables.

3.3.2 A Better (Less General) Construction

In order to improve on the construction presented in the previous subsection, we will borrow full inspiration from the construction in Ciabattoni et al. (2014), and try to push the boundaries of the scope of application of the underlying ideas.

Let Σ be a signature, fix $\Sigma^d \subseteq \Sigma$ and set $\mathcal{U} \subseteq (\Sigma \setminus \Sigma^d)^{(1)}$ to be the set of all 1-place connectives not in Σ^d. We shall consider the set $\mathcal{U}^*$ of all finite strings of elements of $\mathcal{U}$ (the Kleene closure of $\mathcal{U}$). We shall use ε to denote the *empty string*, and $uw \in \mathcal{U}^*$ to denote the *concatenation* of strings $u, w \in \mathcal{U}^*$. We use $\mathsf{prfx}(w)$ to denote the set of all *prefixes* of string w, including ε. Given $w \in \mathcal{U}^*$ and $A \in L_\Sigma(P)$ we will use wA to denote the formula defined inductively by $\varepsilon A = A$, and $\bullet wA = \bullet(wA)$ if $\bullet \in \mathcal{U}$.

Definition 3.3.5 Let $© \in \Sigma$ be a k-place connective. Σ^d*-simple formulas based on* © are formulas $B \in L_\Sigma(\{p_1, \dots, p_k\})$ such that $B = A^\sigma$ for some *structure formula* $A \in L_{\Sigma^d}(\{q_1, \dots, q_n, r_1, \dots, r_m\})$ and some substitution σ for which

- $\sigma(q_i) = w_i p_j$ with $w_i \in \mathcal{U}^*$ and $1 \leq j \leq k$, for each $1 \leq i \leq n$, and
- $\sigma(r_l) = u_l©(p_1, \dots, p_k)$ with $u_l \in \mathcal{U}^*$, for each $1 \leq l \leq m$.

For ease of notation, we will simply write

$$A(\dots w_i p_j \dots u_l©(p_1, \dots, p_k) \dots)$$

for a generic Σ^d-simple formula based on ©.

The *look-ahead* set induced by B is $\Theta_B = (\cup_{i=1}^n \mathsf{prfx}(w_i)) \cup (\cup_{l=1}^m \mathsf{prfx}(u_l))$.

We call Σ^d-simple formula to any formula which is Σ^d-simple based on some[2] connective of Σ. The *look-ahead* set induced by a set Γ of Σ^d-simple formulas is[3] $\Theta_\Gamma = \{\varepsilon\} \cup (\cup_{B\in\Gamma}\Theta_B)$. △

Σ^d-simple formulas will be the allowed shapes of our (schema) axioms. Comparing with Ciabattoni et al. (2014), our setup is strictly more general in that it allows for an arbitrary base signature Σ. If we set Σ^d to consist of the usual 2-place connectives of positive logic $\wedge, \vee, \to$, and let $\Sigma = \Sigma^d \cup \mathcal{U}$ where $\mathcal{U}$ collects a number of additional 1-place connectives (e.g., $\neg, \circ$), we recover the setup of Ciabattoni et al. (2014).

---

[2] Since not all the variables $q_1, \dots, q_n, r_1, \dots, r_m$ need to occur in A, it may well happen that the subformula $©(p_1, \dots, p_k)$ ends up not appearing in the Σ^d-simple formula B based on ©. For this reason, such a Σ^d-simple formula can also be based on any available k'-place connective distinct from ©, as long as $k' \geq k$ (more precisely, k' needs to be at least as big as the number of distinct variables p_j occurring in B).

[3] Note that, in our definition, Θ_Γ is not simply the union of the look-ahead sets of each formula in Γ. We not only want Θ_Γ to be closed for taking prefixes, but we want $\varepsilon \in \Theta_\Gamma$ even if $\Gamma = \emptyset$ (a rather pathological case).

For instance, axiom $B = \circ\neg(p_1 \wedge p_2) \to (\neg\circ p_1 \vee \neg\circ p_2)$ is Σ^d-simple in this setting, as can be seen by taking $A = r_1 \to (q_1 \vee q_2)$, $\copyright = \wedge$, and $\sigma(q_1) = w_1 p_1 = \neg\circ p_1$, $\sigma(q_2) = w_2 p_2 = \neg\circ p_2$, thus with $w_1 = w_2 = \neg\circ$, and $\sigma(r_1) = u_1(p_1 \wedge p_2) = \circ\neg(p_1 \wedge p_2)$, thus with $u_1 = \circ\neg$.

Easily, all axioms covered in Ciabattoni et al. (2014) are Σ^d-simple. However, $p_1 \wedge \neg p_1$ or $p_1 \to (\neg p_1 \to \neg p_2)$ fall outside the scope of Ciabattoni et al. (2014) but are still Σ^d-simple (based on any of the 2-place connectives, as the r_l variables are not necessary). Axioms like $\neg(p_1 \wedge \neg p_1) \to \circ p_1$ are not Σ^d-simple, due to the interleaved nesting of $\neg$ and $\wedge$, and fall outside the scope of both methods.

Having set up our syntactic restriction on the set of allowed axioms, we will still need to match them with appropriate semantic restrictions. Before we do it, we need to shape up another crucial idea from Ciabattoni et al. (2014): when strengthening with a set of axioms Ax, the truth-values of the intended PNmatrix will correspond to suitable functions $f : \Theta_{\mathsf{Ax}} \to V$ where V is the set of truth-values of the given PNmatrix; when the value of a formula A is f this does not only settle its face value to $f(\varepsilon)$ but also gives as look-ahead information the value $f(w)$ for the value of formulas wA with $w \in \Theta_{\mathsf{Ax}}$.

Definition 3.3.6 Let $\mathbb{M} = \langle V, D, \cdot_{\mathbb{M}}\rangle$ be a Σ-PNmatrix and Ax a set of Σ^d-simple formulas. For each $v \in \mathrm{Val}_{\mathbb{M}}$ and $A \in L_\Sigma(P)$, we define $f_v^A \in V^{\Theta_{\mathsf{Ax}}}$ by letting $f_v^A(w) = v(wA)$ for each $w \in \Theta_{\mathsf{Ax}}$. △.

It is worth noting that, by definition, $f_v^A(uw) = f_v^{wA}(u)$ whenever $uw \in \Theta_{\mathsf{Ax}}$.

We can finally put forth our improved construction, taking Σ^d-simple axioms. In order to make it work it will suffice to require that the given PNmatrix is Σ^d-deterministic (not necessarily Σ^d-total). The more general condition, though, will be to require that the PNmatrix is a rexpansion of a Σ^d-deterministic PNmatrix, as the crucial necessary property is granted by Lemma 3.2.1.

Theorem 3.3.7 *Let $\mathbb{M} = \langle V, D, \cdot_{\mathbb{M}}\rangle$ be a Σ-PNmatrix and $\mathsf{Ax} \subseteq L_\Sigma(P)$.*

If there exists $\Sigma^d \subseteq \Sigma$ such that $\mathbb{M}$ is a rexpansion of some Σ^d-deterministic PNmatrix, and the formulas in Ax are all Σ^d-simple, then the consequence $\vdash_{\mathbb{M}}^{\mathsf{Ax}}$ is characterized by the rexpansion $\mathbb{M}_{\mathsf{Ax}}^\sharp = \langle V_{\mathsf{Ax}}^\sharp, D_{\mathsf{Ax}}^\sharp, \cdot_{\mathbb{M}_{\mathsf{Ax}}^\sharp}\rangle$ of $\mathbb{M}$ defined by

- $V_{\mathsf{Ax}}^\sharp = \bigcup_{v \in Val_{\mathbb{M}}^{\mathsf{Ax}}} \{f_v^A : A \in L_\Sigma(P)\}$,
- $D_{\mathsf{Ax}}^\sharp = \{f \in V_{\mathsf{Ax}}^\sharp : f(\varepsilon) \in D\}$,
- *for each $k \in \mathbb{N}$ and $\copyright \in \Sigma^{(k)}$,*

$$\copyright_{\mathbb{M}_{\mathsf{Ax}}^\sharp}(f_1, \dots, f_k) =$$

$$\bigcup_{v \in Val_{\mathbb{M}}^{\mathsf{Ax}}} \{f_v^{\copyright(A_1,\dots,A_k)} : A_i \in L_\Sigma(P) \text{ with } f_v^{A_i} = f_i \text{ for } 1 \le i \le k\}.$$

Proof We prove that $\mathbb{M}^{\sharp}_{\mathsf{Ax}}$ is a rexpansion of $\mathbb{M}$, and then its soundness and completeness with respect to $\vdash^{\mathsf{Ax}}_{\mathbb{M}}$.

Rexpansion. It is simple to check that the PNmatrix $\mathbb{M}^{\sharp}_{\mathsf{Ax}}$ is a refinement of the expansion of $\mathbb{M}$ with $\mathcal{E}(x) = \{f \in V^{\Theta_{\mathsf{Ax}}} : f(\varepsilon) = x\}$. Just note that one has $f_v^{©(A_1,\ldots,A_k)}(\varepsilon) = v(©(A_1, \ldots, A_k)) \in ©_{\mathbb{M}}(v(A_1), \ldots, v(A_k)) = ©_{\mathbb{M}}(f_v^{A_1}(\varepsilon), \ldots, f_v^{A_k}(\varepsilon))$ whenever it is the case that $v \in \mathrm{Val}_{\mathbb{M}}$, $k \in \mathbb{N}$, $© \in \Sigma^{(k)}$ and $A_1, \ldots, A_k \in L_\Sigma(P)$. $\widetilde{\mathcal{E}} : V^{\sharp}_{\mathsf{Ax}} \to V$ is such that $\widetilde{\mathcal{E}}(f) = f(\varepsilon)$, and clearly preserves and reflects designated values. As before, using Proposition 3.3.1, it suffices to show that $\{\widetilde{\mathcal{E}} \circ v^{\sharp} : v^{\sharp} \in \mathrm{Val}_{\mathbb{M}^{\sharp}_{\mathsf{Ax}}}\} = \mathrm{Val}^{\mathsf{Ax}}_{\mathbb{M}}$.

For a 1-place connective $\bullet \in \mathcal{U}$ and $u \in \mathcal{U}^*$ such that $u\bullet \in \Theta_{\mathsf{Ax}}$, given a valuation $v^{\sharp} \in \mathrm{Val}_{\mathbb{M}^{\sharp}_{\mathsf{Ax}}}$, we have that $v^{\sharp}(\bullet A)(u) = v^{\sharp}(A)(u\bullet)$, simply because $v^{\sharp}(\bullet A) \in \bullet_{\mathbb{M}^{\sharp}_{\mathsf{Ax}}}(v^{\sharp}(A))$ and by definition of $\bullet_{\mathbb{M}^{\sharp}_{\mathsf{Ax}}}$ there must exist $v \in \mathrm{Val}^{\mathsf{Ax}}_{\mathbb{M}}$ such that $v^{\sharp}(\bullet A) = f_v^{\bullet B}$ and $v^{\sharp}(A) = f_v^{B}$. It easily follows, by induction, that if $w \in \mathcal{U}^*$ is such that $uw \in \Theta_{\mathsf{Ax}}$ then also $v^{\sharp}(wA)(u) = v^{\sharp}(A)(uw)$.

Soundness. Since $\mathbb{M}^{\sharp}_{\mathsf{Ax}}$ is a rexpansion of $\mathbb{M}$ with $\mathcal{E}$, if $v^{\sharp} \in \mathrm{Val}_{\mathbb{M}^{\sharp}_{\mathsf{Ax}}}$ then $\widetilde{\mathcal{E}} \circ v^{\sharp} \in \mathrm{Val}_{\mathbb{M}}$. Hence, when $B = A(\ldots w_i A_j \ldots u_n ©(A_1, \ldots, A_k) \ldots) \in \mathsf{Ax}^{\mathsf{inst}}$ then setting $y = v^{\sharp}(B)(\varepsilon)$ we have

$$y \in A_{\mathbb{M}}(\ldots v^{\sharp}(w_i A_j)(\varepsilon) \ldots v^{\sharp}(u_n ©(A_1, \ldots, A_k))(\varepsilon) \ldots) =$$

$$A_{\mathbb{M}}(\ldots v^{\sharp}(A_j)(w_i) \ldots v^{\sharp}(©(A_1, \ldots, A_k))(u_n) \ldots).$$

By definition of $©_{\mathbb{M}^{\sharp}_{\mathsf{Ax}}}$, we know there exist $v \in \mathrm{Val}^{\mathsf{Ax}}_{\mathbb{M}}$ and $B_1, \ldots, B_k \in L_\Sigma(P)$ such that $v^{\sharp}(©(A_1, \ldots, A_k)) = f_v^{©(B_1,\ldots,B_k)}$ and $v^{\sharp}(A_j) = f_v^{B_j}$ for $1 \leq j \leq k$. Thus, we have

$$y \in A_{\mathbb{M}}(\ldots f_v^{B_j}(w_i) \ldots f_v^{©(B_1,\ldots,B_k)}(u_n) \ldots) =$$

$$A_{\mathbb{M}}(\ldots v(w_i B_j) \ldots v(u_n ©(B_1, \ldots, B_k)) \ldots).$$

Clearly, setting $z = v(A(\ldots w_i B_j \ldots u_n ©(B_1, \ldots, B_k) \ldots))$ we also have

$$z \in A_{\mathbb{M}}(\ldots v(w_i B_j) \ldots v(u_n ©(B_1, \ldots, B_k)) \ldots).$$

Using Lemma 3.2.1, since $A \in L_{\Sigma^d}(P)$ and $\mathbb{M}$ is a rexpansion of a Σ^d-deterministic PNmatrix, we conclude that $y \in D$ iff $z \in D$. Now, it is also the case that $A(\ldots w_i B_j \ldots u_n ©(B_1, \ldots, B_k) \ldots) \in \mathsf{Ax}^{\mathsf{inst}}$ and we know that $v \in \mathrm{Val}^{\mathsf{Ax}}_{\mathbb{M}}$, so we conclude that $z \in D$. Therefore, $y \in D$ and $v^{\sharp}(B) \in D^{\sharp}_{\mathsf{Ax}}$. We conclude $\{\widetilde{\mathcal{E}} \circ v^{\sharp} : v^{\sharp} \in \mathrm{Val}_{\mathbb{M}^{\sharp}_{\mathsf{Ax}}}\} \subseteq \mathrm{Val}^{\mathsf{Ax}}_{\mathbb{M}}$ and $\vdash^{\mathsf{Ax}}_{\mathbb{M}} \subseteq \vdash_{\mathbb{M}^{\sharp}_{\mathsf{Ax}}}$.

Completeness. Reciprocally, if $v \in \mathrm{Val}^{\mathsf{Ax}}_{\mathbb{M}}$ then $v = \widetilde{\mathcal{E}} \circ v^{\sharp}$ with $v^{\sharp}(A) = f_v^{A}$ for each $A \in L_\Sigma(P)$. It is immediate, by definition of $©_{\mathbb{M}^{\sharp}_{\mathsf{Ax}}}$, that $f_v^{©(A_1,\ldots,A_k)} \in$

$\copyright_{\mathbb{M}^{\sharp}_{\mathsf{Ax}}}(f_v^{A_1}, \ldots, f_v^{A_k})$ for every k-place connective $\copyright \in \Sigma$ and formulas $A_1, \ldots, A_k \in L_\Sigma(P)$. We conclude that $v^\sharp \in \mathrm{Val}_{\mathbb{M}^{\sharp}_{\mathsf{Ax}}}$. Therefore, we have $\mathrm{Val}^{\mathsf{Ax}}_{\mathbb{M}} \subseteq \{\widetilde{\mathcal{E}} \circ v^\flat : v^\flat \in \mathrm{Val}_{\mathbb{M}^{\flat}_{\mathsf{Ax}}}\}$ and $\vdash_{\mathbb{M}^{\flat}_{\mathsf{Ax}}} \subseteq \vdash^{\mathsf{Ax}}_{\mathbb{M}}$. □.

As intended, we have pushed the boundaries of the method in Ciabattoni et al. (2014) as much as we could. Beyond the arbitrariness of the signature, and the more permissive syntactic restrictions on the axioms, we also allow a more general PNmatrix to start with. Instead of demanding it to be the two-valued Boolean matrix on the Σ^d-connectives, we simply require that it be a rexpansion of any Pmatrix. This has the advantage of applying to a large range of non-classical base logics, but also of making the method incremental, allowing us to add axioms one by one and not necessarily all at once. Further, in our method, the interpretation of the connectives not in Σ^d is completely unrestricted, which contrasts with Ciabattoni et al. (2014), where the remaining (1-place) connectives are implicitly forced to be fully non-deterministic. This additional degree of freedom allowed by our method applies not only to the connectives in $\mathcal{U}$, but also to any other connectives not appearing in the structural formulas of the axioms.

3.4 Worked Examples

In order to show the workings and scope of the method we have put forth in Sect. 3.3.2, we shall now consider a few meaningful illustrative examples.

Example 3.4.1 Suppose that we want to add to the logic of classical implication a negation connective satisfying the *explosion* axiom $p_1 \to (\neg p_1 \to p_2)$.

We consider the signature Σ with a single 2-place connective $\to$, and a single 1-place connective $\neg$, and we start from the two-valued (P)Nmatrix $\mathbb{B} = \langle \{0, 1\}, \{1\}, \cdot_{\mathbb{B}} \rangle$ given by the truth-tables below.

$\to_{\mathbb{B}}$	0	1
0	1	1
1	0	1

	$\neg_{\mathbb{B}}$
0	0, 1
1	0, 1

Clearly, $\to_{\mathbb{B}}$ corresponds to the usual matrix truth-table of classical implication. The truth-table of $\neg_{\mathbb{B}}$ is fully non-deterministic.

Setting Σ^d to contain only $\to$, and $\mathcal{U} = \{\neg\}$ it is clear that $\mathbb{B}$ is Σ^d-deterministic and that the axiom is Σ^d-simple. With $\mathsf{Exp} = \{p_1 \to (\neg p_1 \to p_2)\}$, we have that $\Theta_{\mathsf{Exp}} = \{\varepsilon, \neg\}$. From Theorem 3.3.7, the strengthening of $\vdash_{\mathbb{B}}$ with Exp is characterized by the PNmatrix $\mathbb{B}^{\sharp}_{\mathsf{Exp}} = \langle \{00, 01, 10, 11\}, \{10, 11\}, \cdot_{\mathbb{B}^{\sharp}_{\mathsf{Exp}}} \rangle$, where

$\to_{\mathbb{B}^\sharp_{\mathsf{Exp}}}$	00	01	10	11
00	10	10	10	$\emptyset$
01	10	10	10	$\emptyset$
10	00, 01	00, 01	10	$\emptyset$
11	$\emptyset$	$\emptyset$	$\emptyset$	11

	$\neg_{\mathbb{B}^\sharp_{\mathsf{Exp}}}$
00	00, 01
01	10
10	00, 01
11	11

Note that, for ease of notation, we are denoting a function $f \in V^\sharp_{\mathsf{Exp}}$ simply by the string $f(\varepsilon)f(\neg)$. For instance, the value 01 corresponds to the function such that $f(\varepsilon) = 0$ and $f(\neg) = 1$. In this example, all four possibilities correspond to truth-values of the resulting PNmatrix. The reader may refer to Example 3.4.3 below, for a situation where this does not happen.

For illustration purposes, let us clarify why $\neg_{\mathbb{B}^\sharp_{\mathsf{Exp}}}(10) = \{00, 01\}$. Easily, if $xy \in \neg_{\mathbb{B}^\sharp_{\mathsf{Exp}}}(10)$ it is clear that $x = 0$ as this is the value of the $\neg$ look-ahead provided by the value 10. The fact that y can be either 0 or 1 boils down to noting that $\neg_{\mathbb{B}}(0) = \{0, 1\}$, none of these choices being incompatible with satisfying the axiom. Namely, $10 = f_v^{p_1}$ and $00 = f_v^{\neg p_1}$ for any $\mathbb{B}$-valuation v with $v(p_1) = 1$ and $v(\neg p_1) = v(\neg\neg p_1) = 0$ and classical for other formulas, whereas $10 = f_v^{p_1}$ and $01 = f_v^{\neg p_1}$ would result from any fully classical $\mathbb{B}$-valuation with $v(p_1) = 1$, both valuations clearly in $\mathrm{Val}_{\mathbb{B}}^{\mathsf{Exp}}$. Another interesting case is $\neg_{\mathbb{B}^\sharp_{\mathsf{Exp}}}(01) = \{10\}$. Easily, the 1 on the left of 10 is explained by the 1 on the right of 01. Once again, $\neg_{\mathbb{B}}(1) = \{0, 1\}$. However, we must exclude 11 because $01 = f_v^A$ and $11 = f_v^{\neg A}$ would jointly imply that $v(\neg A \to (\neg\neg A \to A)) = 0$ and therefore $v \notin \mathrm{Val}_{\mathbb{B}}^{\mathsf{Exp}}$. Similar justifications can be given, for instance, to explain why $11 \to_{\mathbb{B}^\sharp_{\mathsf{Exp}}} 00 = \emptyset$.

The PNmatrix $\mathbb{B}^\sharp_{\mathsf{Exp}}$ obtained is slightly more complex than one could expect. Note, however, that the value 11 is isolated from the others in the sense that a valuation that assigns 11 to some formula must assign 11 to all formulas. Concretely, $\mathbb{B}^\sharp_{\mathsf{Exp}}$ has two maximal total refinements: the three-valued Nmatrix $(\mathbb{B}^\sharp_{\mathsf{Exp}})_{\{00,01,10\}}$ one would expect, plus the trivial one-valued matrix $(\mathbb{B}^\sharp_{\mathsf{Exp}})_{\{11\}}$ (whose only trivial valuation is irrelevant for the definition of $\vdash_{\mathbb{B}}^{\mathsf{Exp}}$). △

Let us now consider a slight variation on this theme.

Example 3.4.2 To see the contrast with the previous example, suppose now that we want to add to the logic of classical implication a negation connective satisfying the weaker *partial explosion* axiom $p_1 \to (\neg p_1 \to \neg p_2)$. This is a case that is out of the scope of the method in Ciabattoni et al. (2014).

The setting up we need to consider is the same used in Example 3.4.1: the same Σ, Σ^d and $\mathcal{U}$, and the same starting PNmatrix $\mathbb{B}$. Setting now $\mathsf{Exp}_\neg = \{p_1 \to (\neg p_1 \to \neg p_2)\}$, we still have that $\Theta_{\mathsf{Exp}_\neg} = \{\varepsilon, \neg\}$. From Theorem 3.3.7, the strengthening of $\vdash_{\mathbb{B}}$ with $\mathsf{Exp}_\neg$ is now characterized by the PNmatrix $\mathbb{B}^\sharp_{\mathsf{Exp}_\neg} = \langle\{00, 01, 10, 11\}, \{10, 11\}, \cdot_{\mathbb{B}^\sharp_{\mathsf{Exp}_\neg}}\rangle$, where using the same notation convention used in Example 3.4.1, we have

$\to_{\mathbb{B}^\sharp_{\mathsf{Exp}_\neg}}$	00	01	10	11
00	10	10	10	$\emptyset$
01	10	10, 11	10	11
10	00, 01	00, 01	10	$\emptyset$
11	$\emptyset$	01	$\emptyset$	11

	$\neg_{\mathbb{B}^\sharp_{\mathsf{Exp}_\neg}}$
00	00, 01
01	10, 11
10	00, 01
11	11

The PNmatrix $\mathbb{B}^\sharp_{\mathsf{Exp}_\neg}$ is more interesting than before. Note that it also has two maximal total refinements: the three-valued Nmatrix $(\mathbb{B}^\sharp_{\mathsf{Exp}_\neg})_{\{00,01,10\}}$ (which is precisely the same as the one obtained in Example 3.4.1), plus the two-valued matrix $(\mathbb{B}^\sharp_{\mathsf{Exp}_\neg})_{\{01,11\}}$ (whose implication is classical but whose negation is always designated). △

Next, we will analyze a number of examples that appear scattered in the literature and show how our method can be systematically used in all of them. We start by revisiting an example from Avron (2005b), paradigmatic of many similar examples considered by Avron and coauthors.

Example 3.4.3 Let us consider strengthening the logic $\mathcal{CLuN}$ from Batens (2000, 1980) with the *double negation elimination* axiom $\neg\neg p_1 \to p_1$. Actually, for the sake of simplicity, we shall consider only the $\{\neg, \to\}$-fragment of the logic.

Let Σ_d contain a single 2-place connective $\to$, $\mathcal{U}$ contain a 1-place connective $\neg$. The (fragment of the) logic $\mathcal{CLuN}$ is characterized by the Nmatrix $\mathbb{M} = \langle\{0,1\},\{1\},\cdot_{\mathbb{M}}\rangle$ with:

$\to_{\mathbb{M}}$	0	1
0	1	1
1	0	1

	$\neg_{\mathbb{M}}$
0	1
1	0, 1

It is clear that $\mathbb{M}$ is Σ_d-deterministic and that the axiom is Σ^d-simple. If we let $\mathsf{DNe} = \{\neg\neg p_1 \to p_1\}$, we have that $\Theta_{\mathsf{DNe}} = \{\varepsilon, \neg, \neg\neg\}$. From Theorem 3.3.7, the strengthening of $\vdash_{\mathbb{M}}$ with DNe, which is well known to coincide with the logic $\mathcal{C}_{\min}$ of Carnielli and Marcos (1999, 2002), is characterized by the four-valued Nmatrix $\mathbb{M}^\sharp_{\mathsf{DNe}} = \langle\{010, 101, 110, 111\}, \{101, 110, 111\}, \cdot_{\mathbb{M}^\sharp_{\mathsf{DNe}}}\rangle$, where

$\to_{\mathbb{M}^\sharp_{\mathsf{DNe}}}$	010	101	110	111
010	$D^\sharp_{\mathsf{DNe}}$	$D^\sharp_{\mathsf{DNe}}$	$D^\sharp_{\mathsf{DNe}}$	$D^\sharp_{\mathsf{DNe}}$
101	010	$D^\sharp_{\mathsf{DNe}}$	$D^\sharp_{\mathsf{DNe}}$	$D^\sharp_{\mathsf{DNe}}$
110	010	$D^\sharp_{\mathsf{DNe}}$	$D^\sharp_{\mathsf{DNe}}$	$D^\sharp_{\mathsf{DNe}}$
111	010	$D^\sharp_{\mathsf{DNe}}$	$D^\sharp_{\mathsf{DNe}}$	$D^\sharp_{\mathsf{DNe}}$

	$\neg_{\mathbb{M}^\sharp_{\mathsf{DNe}}}$
010	101
101	010
110	101
111	110, 111

Above, for ease of notation, we are denoting a function $f \in V^\sharp_{\mathsf{DNe}}$ simply by the string $f(\varepsilon)f(\neg)f(\neg\neg)$. As this is a new feature in our row of examples, it is worth explaining why only four of the eight possible such functions appear as truth-values of the resulting Nmatrix. Namely, 000, 001, 100 are all unattainable as f_v^A in the Nmatrix $\mathbb{M}$ since $\neg_{\mathbb{M}^\sharp_{\mathsf{DNe}}}(0) = 1$. The remaining string 011 is excluded for more interesting reasons, as any $\mathbb{M}$-valuation v with $001 = f_v^A$ makes $v(\neg\neg A \to A) = 0$ and thus $v \notin \mathrm{Val}_{\mathbb{M}}^{\mathsf{DNe}}$.

This example shows that our method, though very general, may not be as tight as possible. It is a mandatory topic for further research to best understand how to equate the equivalence between this Nmatrix and the three-valued Nmatrix from Avron (2005b).

If we want to strengthen the resulting logic, $\mathcal{C}_{\min}$, with the *double negation introduction* axiom $p_1 \to \neg\neg p_1$, we can readily apply Theorem 3.3.7 to $\mathbb{M}^\sharp_{\mathsf{DNe}}$ and $\mathsf{DNi} = \{p_1 \to \neg\neg p_1\}$, obtaining (up to renaming of the truth-values) the three-valued Nmatrix $\mathbb{N} = (\mathbb{M}^\sharp_{\mathsf{DNe}})^\sharp_{\mathsf{DNi}} = \langle\{01, 10, 11\}, \{10, 11\}, \cdot_{\mathbb{N}}\rangle$, where

$\to_{\mathbb{N}}$	01	10	11
01	10, 11	10, 11	10, 11
10	01	10, 11	10, 11
11	01	10, 11	10, 11

	$\neg_{\mathbb{N}}$
01	10
10	01
11	11

Note that, by construction, the Nmatrix $\mathbb{N}$ has three values $g : \Theta_{\mathsf{DNi}} \to V^\sharp_{\mathsf{DNe}}$ which, given that $\Theta_{\mathsf{DNi}} = \{\varepsilon, \neg, \neg\neg\}$, can be written in string notation as $g(\varepsilon)g(\neg)g(\neg\neg)$, corresponding to the strings 010101010, 101010101, 111111111. Clearly, each of them can be named simply by their first two symbols.

It is interesting to further note that this Nmatrix is isomorphic to $\mathbb{M}^\sharp_{\mathsf{DNe}\cup\mathsf{DNi}}$. This is a particularly happy case as, in general, adding axioms incrementally, instead of all at once (as in Ciabattoni et al. (2014)), will yield an equivalent PNmatrix but not necessarily the same, often with more truth-values. △

We now consider a more elaborate example in the family of paraconsistent logics, as also tackled by Avron and coauthors, which is developed in detail in Ciabattoni et al. (2014).

Example 3.4.4 As in Example 5.1 of Ciabattoni et al. (2014), we want to characterize the logic obtained by adding two additional 1-place connectives $\neg$, $\circ$ to positive classical logic, subject to the set of axioms Ax containing:

$$p_1 \vee \neg p_1$$

$$p_1 \to (\neg p_1 \to (\circ p_1 \to p_2))$$

$$\circ p_1 \vee (p_1 \wedge \neg p_1)$$

$$\circ p_1 \to \circ(p_1 \wedge p_2)$$

$$(\neg p_1 \vee \neg p_2) \to \neg(p_1 \wedge p_2)$$

Let Σ_d contain the three 2-place connectives $\wedge, \vee, \to$, and $\mathcal{U}$ contain the two 1-place connectives $\neg, \circ$ and consider the Nmatrix $\mathbb{C} = \langle \{0, 1\}, \{1\}, \cdot_{\mathbb{C}} \rangle$ with

$\wedge_{\mathbb{C}}$	0	1
0	0	0
1	0	1

$\vee_{\mathbb{C}}$	0	1
0	0	1
1	1	1

$\to_{\mathbb{C}}$	0	1
0	1	1
1	0	1

	$\neg_{\mathbb{C}}$	$\circ_{\mathbb{C}}$
0	0, 1	0, 1
1	0, 1	0, 1

It is clear that $\mathbb{C}$ is Σ_d-deterministic and that the axioms are all Σ^d-simple. Further, we get $\Theta_{\mathsf{Ax}} = \{\varepsilon, \neg, \circ\}$. From Theorem 3.3.7, the strengthening $\vdash_{\mathbb{C}}^{\mathsf{Ax}}$ is characterized by the PNmatrix $\mathbb{C}_{\mathsf{Ax}}^{\sharp} = \langle \{011, 101, 110, 111\}, \{101, 110, 111\}, \cdot_{\mathbb{C}_{\mathsf{Ax}}^{\sharp}} \rangle$, where

$\wedge_{\mathbb{C}_{\mathsf{Ax}}^{\sharp}}$	011	101	110	111
011	011	011	011	$\emptyset$
101	011	101	$\emptyset$	$\emptyset$
110	011	$\emptyset$	110	$\emptyset$
111	$\emptyset$	$\emptyset$	$\emptyset$	111

$\vee_{\mathbb{C}_{\mathsf{Ax}}^{\sharp}}$	011	101	110	111
011	011	101	110	$\emptyset$
101	101	101	$\emptyset$	$\emptyset$
110	110	$\emptyset$	110	$\emptyset$
111	$\emptyset$	$\emptyset$	$\emptyset$	111

$\to_{\mathbb{C}_{\mathsf{Ax}}^{\sharp}}$	011	101	110	111
011	101, 110	101	110	$\emptyset$
101	011	101	$\emptyset$	$\emptyset$
110	011	$\emptyset$	110	$\emptyset$
111	$\emptyset$	$\emptyset$	$\emptyset$	111

	$\neg_{\mathbb{C}_{\mathsf{Ax}}^{\sharp}}$	$\circ_{\mathbb{C}_{\mathsf{Ax}}^{\sharp}}$
011	101, 110	101, 110
101	011	101
110	110	011
111	111	111

For ease of notation, once again, we are denoting a function $f \in V_{\mathsf{Ax}}^{\sharp}$ simply by the string $f(\varepsilon)f(\neg)f(\circ)$.

Notably, the PNmatrix $\mathbb{C}_{\mathsf{Ax}}^{\sharp}$ is slightly different from the PNmatrix obtained using the method in Ciabattoni et al. (2014). Still, it is easy to see that $\mathbb{C}_{\mathsf{Ax}}^{\sharp}$ has two maximal total refinements: the three-valued PNmatrix $(\mathbb{C}_{\mathsf{Ax}}^{\sharp})_{\{011,101,110\}}$ (which is an equivalent refinement of the PNmatrix in Ciabattoni et al. (2014) maximizing the partiality), plus the trivial one-valued matrix $(\mathbb{C}_{\mathsf{Ax}}^{\sharp})_{\{111\}}$. △

Our next example deals with Nelson-like logics and twist-structures.

Example 3.4.5 The addition of a paraconsistent Nelson-like (Nelson 1948; Vakarelov 1977; Odintsov 2008) *strong negation* $\sim$ to a given intermediate logic (as in Kracht (1998)) can be easily captured by our construction.

Let Σ_d be a signature containing binary connectives $\wedge, \vee, \to$, and $\mathcal{U}$ contain the 1-place connective $\sim$, and consider an Nmatrix $\mathbb{M} = \langle V, \{1\}, \cdot_{\mathbb{M}} \rangle$ whose $\{\wedge, \vee, \to\}$-reduct of $\mathbb{M}$, dubbed $\mathbb{N}$, is an *implicative lattice* (Odintsov 2008), and such that $\sim_{\mathbb{M}}(x) = V$ for every $x \in V$, and let Ax contain:

$$\sim\sim p_1 \to p_1 \qquad\qquad p_1 \to \sim\sim p_1$$

$$\sim(p_1 \vee p_2) \to (\sim p_1 \wedge \sim p_2) \qquad\qquad (\sim p_1 \wedge \sim p_2) \to \sim(p_1 \vee p_2)$$

$$\sim (p_1 \wedge p_2) \to (\sim p_1 \vee \sim p_2) \qquad (\sim p_1 \vee \sim p_2) \to \sim (p_1 \wedge p_2)$$

$$\sim (p_1 \to p_2) \to (p_1 \wedge \sim p_2) \qquad (p_1 \wedge \sim p_2) \to \sim (p_1 \to p_2)$$

Clearly, the axioms in Ax are Σ_d-simple and $\Theta_{\mathsf{Ax}} = \{\varepsilon, \sim, \sim\sim\}$. From Theorem 3.3.7, $\vdash_{\mathbb{M}}^{\mathsf{Ax}}$ is characterized by the matrix $\mathbb{M}^\sharp_{\mathsf{Ax}} = \langle V^\sharp_{\mathsf{Ax}}, D^\sharp_{\mathsf{Ax}}, \cdot_{\mathbb{M}^\sharp_{\mathsf{Ax}}}\rangle$ isomorphic to the well-known full twist-structure $\mathbb{N}^{\bowtie}$ over $\mathbb{N}$ (see Odintsov (2008)). Namely, we have $V^\sharp_{\mathsf{Ax}} = \{f \in V^{\{\varepsilon,\sim,\sim\sim\}} : f(\varepsilon) = f(\sim\sim)\}$. For simplicity, we can represent each such function $f \in V^\sharp_{\mathsf{Ax}}$ simply by the pair $(f(\varepsilon), f(\sim))$. Hence, we have

- $V^\sharp_{\mathsf{Ax}} = V \times V$ and $D^\sharp_{\mathsf{Ax}} = \{1\} \times V$,
- $(x_1, y_1) \wedge_{\mathbb{M}^\sharp_{\mathsf{Ax}}} (x_1, y_1) = (x_1 \wedge_{\mathbb{M}} x_2, y_1 \vee_{\mathbb{M}} y_2)$,
- $(x_1, y_1) \vee_{\mathbb{M}^\sharp_{\mathsf{Ax}}} (x_1, y_1) = (x_1 \vee_{\mathbb{M}} x_2, y_1 \wedge_{\mathbb{M}} y_2)$,
- $(x_1, y_1) \to_{\mathbb{M}^\sharp_{\mathsf{Ax}}} (x_2, y_2) = (x_1 \to_{\mathbb{M}} x_2, x_1 \wedge_{\mathbb{M}} y_2)$, and
- $\sim_{\mathbb{M}^\sharp_{\mathsf{Ax}}} (x, y) = (y, x)$.

When we take $\mathbb{N}$ to be the two-valued Boolean matrix, and using now xy instead of (x, y), we obtain, $\mathbb{M}^\sharp_{\mathsf{Ax}} = \langle\{00, 01, 10, 11\}, \{10, 11\}, \cdot_{\mathbb{M}^\sharp_{\mathsf{Ax}}}\rangle$, where

$\wedge_{\mathbb{M}^\sharp_{\mathsf{Ax}}}$	00	01	10	11
00	00	01	00	01
01	01	01	01	01
10	00	01	10	11
11	01	01	11	11

$\vee_{\mathbb{M}^\sharp_{\mathsf{Ax}}}$	00	01	10	11
00	00	00	10	10
01	00	01	10	11
10	10	10	10	10
11	10	11	10	11

$\to_{\mathbb{M}^\sharp_{\mathsf{Ax}}}$	00	01	10	11
00	10	10	10	10
01	10	10	10	10
10	00	01	10	11
11	00	01	10	11

	$\sim_{\mathbb{M}^\sharp_{\mathsf{Ax}}}$
00	00
01	10
10	01
11	11

Note this semantics coincides precisely with the semantic extension of Belnap's four-valued logic (Belnap 1977b, a) with *true implication* of Avron (Arieli and Avron 1998).

If we further impose the axiom

$$\sim p_1 \to (p_1 \to p_2)$$

we obtain corresponding explosive versions of Nelson's construction. Making $\mathsf{Ax}' = \mathsf{Ax} \cup \{\sim p_1 \to (p_1 \to p_2)\}$, the resulting twist-structure is now a refinement resulting from isolating the truth-value $(1, 1)$, i.e., such that for $* \in \{\wedge, \vee, \to\}$ we have $(1, 1) *_{\mathbb{M}^\sharp_{\mathsf{Ax}}} (x, y) = (x, y) *_{\mathbb{M}^\sharp_{\mathsf{Ax}}} (1, 1) = \emptyset$ if $(x, y) \neq (1, 1)$. Concretely, if we take $\mathbb{N}$ to be the two-valued Boolean matrix, again, we obtain the Pmatrix $\mathbb{M}^{\sharp}_{\mathsf{Ax}'} = \langle \{00, 01, 10, 11\}, \{10, 11\}, \cdot_{\mathbb{M}^{\sharp}_{\mathsf{Ax}'}} \rangle$, where

$\wedge_{\mathbb{M}^{\sharp}_{\mathsf{Ax}'}}$	00	01	10	11
00	00	01	00	∅
01	01	01	01	∅
10	00	01	10	∅
11	∅	∅	∅	11

$\vee_{\mathbb{M}^{\sharp}_{\mathsf{Ax}'}}$	00	01	10	11
00	00	00	10	∅
01	00	01	10	∅
10	10	10	10	∅
11	∅	∅	∅	11

$\to_{\mathbb{M}^{\sharp}_{\mathsf{Ax}'}}$	00	01	10	11
00	10	10	10	∅
01	10	10	10	∅
10	00	01	10	∅
11	∅	∅	∅	11

	$\neg_{\mathbb{M}^{\sharp}_{\mathsf{Ax}'}}$
00	00
01	10
10	01
11	11

Easily, $\mathbb{M}^{\sharp}_{\mathsf{Ax}'}$ has two maximal total refinements: the three-valued matrix $(\mathbb{M}^{\sharp}_{\mathsf{Ax}'})_{\{00,01,10\}}$, plus the trivial one-valued matrix $(\mathbb{M}^{\sharp}_{\mathsf{Ax}'})_{\{11\}}$. Expectedly, we have that $(\mathbb{M}^{\sharp}_{\mathsf{Ax}'})_{\{00,01,10\}}$ is precisely the matrix characterizing the three-valued logic of Vakarelov (Vakarelov 1977; Kracht 1998) (which coincides with $\vdash^{\mathsf{Ax}'}_{\mathbb{M}}$, and is known to be translationally equivalent to Łukasiewicz's three-valued logic). △

Next, we will show, by means of an example, that our method subsumes the idea of *swap-structure semantics* put forth in Carnielli and Coniglio (2016), Coniglio and Golzio (2019).

Example 3.4.6 As in Coniglio and Golzio (2019), we consider obtaining a semantic characterization of the non-normal modal logic $\mathcal{T}$ of Kearns (Kearns 1981), which coincides with the logic $\mathcal{S}_a+$ of Ivlev (Ivlev 1988). This can be done by using our method to characterize the logic obtained by a 1-place connective $\Box$ to the $\{\neg, \to\}$-fragment of classical logic, further demanding the Tm axioms of Coniglio and Golzio (2019), namely,

$$\Box(p_1 \to p_2) \to (\Box p_1 \to \Box p_2)$$

$$\Box(p_1 \to p_2) \to (\Box\neg p_2 \to \Box\neg p_1)$$

$$\neg\Box\neg(p_1 \to p_2) \to (\Box p_1 \to \neg\Box\neg p_2)$$

$$\Box\neg p_1 \to \Box(p_1 \to p_2)$$

$$\Box p_2 \to \Box(p_1 \to p_2)$$

$$\Box\neg(p_1 \to p_2) \to \Box\neg p_2$$

$$\Box\neg(p_1 \to p_2) \to \Box p_1$$

$$\Box p_1 \to p_1$$

$$\Box p_1 \to \Box\neg\neg p_1$$

$$\Box\neg\neg p_1 \to \Box p_1$$

Let Σ_d contain $\to$, and $\mathcal{U} = \{\neg, \Box\}$. Take the Nmatrix $\mathbb{D} = \langle\{0, 1\}, \{1\}, \cdot_{\mathbb{D}}\rangle$ with

$\to_{\mathbb{D}}$	0	1
0	1	1
1	0	1

	$\neg_{\mathbb{D}}$	$\Box_{\mathbb{D}}$
0	1	0, 1
1	0	0, 1

Clearly the axioms in Tm are Σ_d-simple. Furthermore, now, we have that $\Theta_{\mathsf{Tm}} = \{\varepsilon\} \cup \mathsf{prfx}(\{\Box, \Box\neg, \neg\Box\neg, \Box\neg\neg\}) = \{\varepsilon, \neg, \neg\Box, \neg\Box\neg, \Box, \Box\neg, \Box\neg\neg\}$. Note that for any $f \in V^{\sharp}_{\mathsf{Tm}}$ and $\neg w \in \Theta_{\mathsf{Tm}}$ we have $f(\neg w) = 1 - f(w)$. Note also that due to the last two axioms of Tm, it follows that $f(\Box\neg\neg) = f(\Box)$ for any $f \in V^{\sharp}_{\mathsf{Tm}}$. Hence, we can represent each f simply by the string $f(\varepsilon)f(\Box)f(\Box\neg)$. Further, note that the antepenultimate axiom $\Box p_1 \to p_1$ guarantees both that $f(\Box) \leq f(\varepsilon)$ and $f(\Box\neg) \leq f(\neg) = 1 - f(\varepsilon)$. Now, applying Theorem 3.3.7, we conclude that the strengthening $\vdash^{\mathsf{Tm}}_{\mathbb{D}}$ is characterized by the four-valued Nmatrix given by $\mathbb{D}^{\sharp}_{\mathsf{Tm}} = \langle\{000, 001, 100, 110\}, \{100, 110\}, \cdot_{\mathbb{D}^{\sharp}_{\mathsf{Tm}}}\rangle$, where

$\to_{\mathbb{D}^{\sharp}_{\mathsf{Tm}}}$	000	001	100	110
000	100, 110	100	100, 110	110
001	110	110	110	110
100	000	000	100, 110	110
110	000	001	100	110

	$\neg_{\mathbb{D}^{\sharp}_{\mathsf{Tm}}}$	$\Box_{\mathbb{D}^{\sharp}_{\mathsf{Tm}}}$
000	100	000, 001
001	110	000, 001
100	000	000, 001
110	001	100, 110

It is straightforward to check that this Nmatrix is isomorphic to the Kearns and Ivlev semantics (Kearns 1981; Ivlev 1988), also recovered in Coniglio and Golzio (2019), by renaming the truth-values 000, 001, 100, 110 by f, F, t, T, respectively. △

We finish this section with another example, starting from a non-classical base, namely, Łukasiewicz's five-valued logic.

Example 3.4.7 We start from Łukasiewicz's logic $\mathcal{L}_5$ and strengthen it by axiom $((p_1 \to \neg p_1) \to p_1) \to p_1$ in order to obtain Łukasiewicz's three-valued logic $\mathcal{L}_3$ (see, for instance, (Wójcicki 1998; Gottwald 2001)). In this case, no new connectives are added.

Let Σ_d contain the 2-place connective $\to$, and also the 1-place connective $\neg$, and let $\mathcal{U} = \emptyset$. Let also $\mathsf{Ax} = \{((p_1 \to \neg p_1) \to p_1) \to p_1\}$. Consider the five-valued matrix $\mathbb{L}_5 = \langle\{0, \frac{1}{4}, \frac{1}{2}, \frac{3}{4}, 1\}, \{1\}, \cdot_{\mathbb{L}_5}\rangle$ with:

$\to_{\mathbb{L}_5}$	0	$\frac{1}{4}$	$\frac{1}{2}$	$\frac{3}{4}$	1
0	1	1	1	1	1
$\frac{1}{4}$	$\frac{3}{4}$	1	1	1	1
$\frac{1}{2}$	$\frac{1}{2}$	$\frac{3}{4}$	1	1	1
$\frac{3}{4}$	$\frac{1}{4}$	$\frac{1}{2}$	$\frac{3}{4}$	1	1
1	0	$\frac{1}{4}$	$\frac{1}{2}$	$\frac{3}{4}$	1

	$\neg_{\mathbb{L}_5}$
0	1
$\frac{1}{4}$	$\frac{3}{4}$
$\frac{1}{2}$	$\frac{1}{2}$
$\frac{3}{4}$	$\frac{1}{4}$
1	0

Clearly the axiom is Σ_d-simple and $\Theta_{\mathsf{Ax}} = \{\varepsilon\}$. Hence we represent any $f \in V_{\mathsf{Ax}}^{\natural}$ simply by $f(\varepsilon)$. From Theorem 3.3.7, the strengthening $\vdash_{\mathbb{L}_5}^{\mathsf{Ax}}$ is characterized by the well-known three-valued matrix $(\mathbb{L}_5)_{\mathsf{Ax}}^{\natural} = \mathbb{L}_3 = \langle\{0, \frac{1}{2}, 1\}, \{1\}, \cdot_{\mathbb{L}_3}\rangle$, where

$\to_{\mathbb{L}_3}$	0	$\frac{1}{2}$	1
0	1	1	1
$\frac{1}{2}$	$\frac{1}{2}$	1	1
1	0	$\frac{1}{2}$	1

	$\neg_{\mathbb{L}_3}$
0	1
$\frac{1}{2}$	$\frac{1}{2}$
1	0

△

Examples 3.4.1, 3.4.4, 3.4.6 are also covered by the method in Ciabattoni et al. (2014). The two-valued based case of Example 3.4.5 could also be obtained using Ciabattoni et al. (2014), but not the general case we deal with, over an arbitrary implicative lattice. Example 3.4.3, the way it is formulated, is outside the scope of Ciabattoni et al. (2014), not only because it starts from a Nmatrix where negation is not fully non-deterministic, but also because we are adding one axiom and then another. Examples 3.4.2, 3.4.7 are also not covered by Ciabattoni et al. (2014). Namely, Example 3.4.2 uses an axiom which does not respect their syntactic criteria, and Example 3.4.7 uses a five-valued non-classical matrix.

3.5 Analytic Multiple-Conclusion Calculi

In the work of Arnon Avron on Nmatrices and rexpansions, obtaining a concise semantics for a logic (typically in the form of a Nmatrix) is not an end in itself but a means for obtaining (sequent-like) analytic calculi for that logic (Avron et al. 2007, 2012, 2013). In other works (e.g., Ciabattoni et al. (2014), Baaz et al. (2013)), the semantics (typically in the form of a PNmatrix) is not a basis for obtaining a calculus but it is still instrumental in proving its analyticity (when the PNmatrix is total). In this paper, so far, we have not worried about proof-theoretic aspects. Therefore, this is a good point for applying to our previous construction the techniques developed in Marcelino and Caleiro (2019); Caleiro and Marcelino (2019) for obtaining analytic multiple-conclusion calculi for logics defined by finite PNmatrices, under a reasonable expressiveness proviso. This contrasts with the above-mentioned results for sequent-like calculi (Avron et al. 2013; Baaz et al. 2013; Ciabattoni et al. 2014), for which partiality seems to devoid them of a usable (even if generalized) subformula property capable of guaranteeing analyticity (and elimination of non-analytic cuts).

In what follows, we will consider so-called *multiple-conclusion calculi*, a simple generalization of Hilbert-style calculi with (schematic) inference rules of the form $\frac{\Gamma}{\Delta}$, where Γ (*premises* read conjunctively, as usual) and Δ (*conclusions* read disjunctively) are sets of formulas. Such calculi were studied by Shoesmith and Smiley in Shoesmith and Smiley (1978), and have very interesting properties. A set R of such multiple-conclusion rules induces a consequence relation $\rhd_R$ by means of an adequate notion of proof, simply defined as a tree-like version of Hilbert-style proofs. We shall show some illustrative examples later, but refer the reader to Shoesmith and Smiley (1978), Marcelino and Caleiro (2019, 2017) for details. As usual, we say that R constitutes a calculus for a consequence relation $\rhd$ if $\rhd_R = \rhd$.

A set $\mathcal{S} \subseteq L_\Sigma(\{p\})$ induces a simple notion of a generalized subformula: A is a $\mathcal{S}$-*subformula of* B if $A \in \mathsf{sub}_\mathcal{S}(B) = \mathsf{sub}(B) \cup \{S(B') : S \in \mathcal{S}, B' \in \mathsf{sub}(B)\}$. We say that R *is an* $\mathcal{S}$*-analytic calculus* if whenever $\Gamma \rhd_R \Delta$ then there exists a proof of Δ from Γ using only formulas in $\mathsf{sub}_\mathcal{S}(\Gamma \cup \Delta)$. For finite $\mathcal{S}$, we have shown in Marcelino and Caleiro (2019, 2017) that $\mathcal{S}$-analyticity implies that deciding $\rhd_R$ is in coNP, and that proof-search can be implemented in $\mathsf{EXPTIME}$.

Producing analytic calculi for logics characterized by finite PNmatrices is possible, as long as the syntax of the logic is sufficiently expressive (a notion intimately connected with the methods in Shoesmith and Smiley (1978), Avron et al. (2007), Avron et al. (2013), Caleiro et al. (2015), Ciabattoni et al. (2014)). Fix a Σ-PNmatrix $\mathbb{M} = \langle V, D, \cdot_\mathbb{M}\rangle$. A pair of non-empty sets of elements $\emptyset \neq X, Y \subseteq V$ are *separated*, $X\#Y$, if $X \subseteq D$ and $Y \subseteq V \setminus D$, or vice versa. A formula S with $\mathsf{var}(S) \subseteq \{p\}$ with $S_\mathbb{M}(z) \neq \emptyset$ for every $z \in V$, and such that $S_\mathbb{M}(x)\#S_\mathbb{M}(y)$ is said to *separate* x and y, and called a *(monadic) separator*. The PNmatrix $\mathbb{M}$ is said to be *monadic* if there is a separator for every pair of distinct truth-values.

Granted a monadic PNmatrix $\mathbb{M} = \langle V, D, \cdot_\mathbb{M}\rangle$ and some set $\mathcal{S} = \{S^{xy} : x, y \in V, x \neq y\}$ of monadic separators for $\mathbb{M}$ such that each S^{xy} separates x and y, a *discriminator* for $\mathbb{M}$ is the V-indexed family $\widetilde{\mathcal{S}} = \{\widetilde{\mathcal{S}}_x\}_{x\in V}$, with each $\widetilde{\mathcal{S}}_x = \{S^{xy} : y \in V \setminus \{x\}\}$. Each $\widetilde{\mathcal{S}}_x$ is naturally partitioned into $\Omega_x = \{S \in \widetilde{\mathcal{S}}_x : S_\mathbb{M}(x) \subseteq D\}$ and $\mho_x = \{S \in \widetilde{\mathcal{S}}_x : S_\mathbb{M}(x) \subseteq V \setminus D\}$. This partition is easily seen to characterize precisely each of the truth-values of $\mathbb{M}$.

Given $X \subseteq V$, we denote by Ω^*_X any of the possible sets built by choosing one element from each Ω_x for $x \in X$, that is, $\Omega^*_X \subseteq \bigcup_{x\in X}\Omega_x$ is such that $\Omega^*_X \cap \Omega_x \neq \emptyset$ for each $x \in X$. Analogously, we let $\mho^*_X$ denote any of the possible sets built by choosing one element from each $\mho_x$ for $x \in X$, that is, $\mho^*_X \subseteq \bigcup_{x\in X}\mho_x$ is such that $\mho^*_X \cap \mho_x \neq \emptyset$ for each $x \in X$. The following result is taken from Caleiro and Marcelino (2019).

Theorem 3.5.1 *Let* $\mathbb{M} = \langle V, D, \cdot_\mathbb{M}\rangle$ *be a monadic PNmatrix with discriminator* $\widetilde{\mathcal{S}}$. *Then,* $R^{\widetilde{\mathcal{S}}}_\mathbb{M} = R_\exists \cup R_\mathsf{D} \cup R_\Sigma \cup R_\mathcal{T}$ *is an* $\mathcal{S}$*-analytic calculus for* $\rhd_\mathbb{M}$, *where*

- $R_\exists$ *contains, for each* $X \subseteq V$ *and each possible* $\mho^*_X$ *and* $\Omega^*_{V\setminus X}$, *the rule*

$$\frac{\mho^*_X(p)}{\Omega^*_{V\setminus X}(p)}.$$

- R_D *contains, for each* $x \in V$, *the rule*

$$\frac{\Omega_x(p)}{p, \mho_x(p)} \text{ if } x \in D \quad or \quad \frac{\Omega_x(p), p}{\mho_x(p)} \text{ if } x \notin D$$

- $R_\Sigma = \bigcup_{© \in \Sigma} R_©$ *where, for* $© \in \Sigma^{(k)}$, $R_©$ *contains, for each* $x_1, \ldots, x_k \in V$ *and* $y \notin ©_\mathbb{M}(x_1, \ldots, x_k)$, *the rule*

$$\frac{\bigcup\limits_{1\leq i\leq k} \Omega_{x_i}(p_i)\,,\; \Omega_y(©(p_1 \ldots, p_k))}{\bigcup\limits_{1\leq i\leq k} \mho_{x_i}(p_i)\,,\; \mho_y(©(p_1 \ldots, p_k))}.$$

- $R_\mathcal{T}$ *contains, for each* $X \subseteq V$ *with* $X \notin \mathcal{T}_\mathbb{M}$, *the rule*

$$\frac{\bigcup\limits_{x_i\in X} \Omega_{x_i}(p_i)}{\bigcup\limits_{x_i\in X} \mho_{x_i}(p_i)}.$$

It is worth understanding the role of each of the rules proposed, as they fully capture the behavior of $\mathbb{M}$. Namely, $R_\exists$ allows one to exclude combinations of separators that do not correspond to truth-values. Actually, in examples where the separators S are such that, in all cases, $S_\mathbb{M}(z) \subseteq D$ or $S_\mathbb{M}(z) \subseteq V \setminus D$, one can always in practice set up the discriminator in a way that makes all $R_\exists$ rules trivial, in the sense that they will necessarily have a formula that appears both as a premise and as a conclusion. Rules in R_D distinguish those combinations of separators that characterize designated values from those that characterize undesignated values. Again, in practice, whenever $\mathbb{M}$ has both designated and undesignated values and $S(p) = p$ is used to separate them, all R_D rules are also trivial. The most operational rules are perhaps R_Σ, as they completely determine the interpretation of connectives in $\mathbb{M}$. The rules in $R = R_\exists \cup R_\mathsf{D} \cup R_\Sigma$ already guarantee that $\rhd_R = \rhd_\mathbb{M}$, but not necessarily analyticity. The rules in $R_\mathcal{T}$ are crucial in proving analyticity (they are already derivable from the previous rules, but with seemingly non-analytic proofs). Indeed, rules in $R_\mathcal{T}$ guarantee that one deals with combinations of separators that correspond to values taken within a total refinement of $\mathbb{M}$.

In order to be able to apply this general result to obtain analytic calculi for the logics characterized by the PNmatrices produced by the method we have devised in Sect. 3.3.2, we need to make sure that the PNmatrices are monadic. Of course, not every PNmatrix is monadic, but we can easily show that our construction preserves monadicity.

Proposition 3.5.2 *Let* $\mathbb{M} = \langle V, D, \cdot_{\mathbb{M}}\rangle$ *be a* Σ*-PNmatrix and* $\mathsf{Ax} \subseteq L_\Sigma(P)$ *that fulfill the conditions of Theorem 3.3.7. If* $\mathbb{M}$ *is monadic then* $\mathbb{M}^\sharp_{\mathsf{Ax}}$ *is also monadic.*

Proof Let $f^{A_1}_{v_1}, f^{A_2}_{v_2} \in V^\sharp_{\mathsf{Ax}}$ with $f^{A_1}_{v_1} \neq f^{A_2}_{v_2}$. This means that there exists $w \in \Theta_{\mathsf{Ax}}$ such that $x_1 = f^{A_1}_{v_1}(w) \neq f^{A_2}_{v_2}(w) = x_2$. Given that $\mathbb{M}$ is monadic, we know that there exists $S \in L_\Sigma(\{p\})$ which separates x_1 from x_2 in $\mathbb{M}$, that is, $S_{\mathbb{M}}(x_1)\# S_{\mathbb{M}}(x_2)$. We show that $R(p) = S(w\, p)$ separates $f^{A_1}_{v_1}$ from $f^{A_2}_{v_2}$ in $\mathbb{M}^\sharp_{\mathsf{Ax}}$.

Given $f^B_v \in V^\sharp_{\mathsf{Ax}}$ we know (from the completeness part of the proof of Theorem 3.3.7) that $v^\sharp(C) = f^C_v$ for each $C \in L_\Sigma(P)$ defines a valuation $v^\sharp \in \mathrm{Val}_{V^\sharp_{\mathsf{Ax}}}$. Easily, then, $v^\sharp(R(B)) \in R_{\mathbb{M}^\sharp_{\mathsf{Ax}}}(v^\sharp(B)) = R_{\mathbb{M}^\sharp_{\mathsf{Ax}}}(f^B_v)$, and therefore $R_{\mathbb{M}^\sharp_{\mathsf{Ax}}}(f^B_v) \neq \emptyset$.

In order to show that $R_{\mathbb{M}^\sharp_{\mathsf{Ax}}}(f^{A_1}_{v_1})\# R_{\mathbb{M}^\sharp_{\mathsf{Ax}}}(f^{A_2}_{v_2})$ we just need to show that $R_{\mathbb{M}^\sharp_{\mathsf{Ax}}}(f^{A_1}_{v_1})(\varepsilon) \subseteq S_{\mathbb{M}}(x_1)\# S_{\mathbb{M}}(x_2) \supseteq R_{\mathbb{M}^\sharp_{\mathsf{Ax}}}(f^{A_2}_{v_2})(\varepsilon)$, and use the fact that in a rexpansion designated values are preserved and reflected.

Take $i \in \{1, 2\}$ and any valuation $v^\sharp \in \mathrm{Val}_{\mathbb{M}^\sharp_{\mathsf{Ax}}}$ with $v^\sharp(p) = f^{A_i}_{v_i}$. We have that $v^\sharp(R(p)) = v^\sharp(S(w\, p)) \in S_{\mathbb{M}^\sharp_{\mathsf{Ax}}}(v^\sharp(w\, p))$. Thus, it follows that $v^\sharp(R(p))(\varepsilon) \in S_{\mathbb{M}^\sharp_{\mathsf{Ax}}}(v^\sharp(w\, p))(\varepsilon) \subseteq S_{\mathbb{M}}(v^\sharp(w\, p)(\varepsilon)) = S_{\mathbb{M}}(v^\sharp(p)(w)) = S_{\mathbb{M}}(f^{A_i}_{v_i}(w)) = S_{\mathbb{M}}(x_i)$. □

Note that this result encompasses the *sufficient expressiveness* preservation result of Ciabattoni et al. (2014), as the two-valued Boolean matrix is trivially separable using just $S(p) = p$.

We now illustrate the powerful result of Theorem 3.5.1 by producing suitably analytic calculi for the resulting logics in each of the examples of Sect. 3.4. In some cases, we also take the opportunity to illustrate the (obvious) notion of proof in multiple-conclusion calculi. In each of the examples, rules $R_\exists$ and R_{D} are omitted, as they are all trivial, as discussed before. We refer the reader to Marcelino and Caleiro (2019); Caleiro and Marcelino (2019) for further details.

Example 3.4.1**, revisited.** In Example 3.4.1 we have obtained a four-valued PNmatrix characterizing the strengthening of the logic of classical implication with the additional axiom $p_1 \to (\neg p_1 \to p_2)$. Easily, $\mathcal{S} = \{p, \neg p\}$ is a corresponding set of monadic separators, which yields the discriminator $\widetilde{\mathcal{S}}$ with $\widetilde{\mathcal{S}}_x = \mathcal{S}$ for each truth-value x. This gives rise to the following partitions.

x	Ω_x	$\mho_x$
00	$\emptyset$	$\{p, \neg p\}$
01	$\{\neg p\}$	$\{p\}$
10	$\{p\}$	$\{\neg p\}$
11	$\{p, \neg p\}$	$\emptyset$

Using Theorem 3.5.1, the following rules constitute an $\mathcal{S}$-analytic calculus R for the logic.

$$\frac{}{p\,,\;p \to q}\;r_1 \qquad \frac{p\,,\;p \to q}{q}\;r_2 \qquad \frac{q}{p \to q}\;r_3 \qquad \frac{p\,,\;\neg p}{q}\;r_{\mathsf{Exp}}$$

After simplifications, the rules r_1–r_3 correspond to $R_{\to}$, and r_{Exp} to $R_{\mathcal{T}}$ with $X = \{00, 11\}$, $X = \{01, 11\}$, and $X = \{10, 11\}$.

For illustration, we next depict an analytic proof of $\rhd_R\; p_1 \to (\neg p_1 \to p_2)$. Note that rules with multiple conclusions give rise to branching in the proof-tree, which makes it necessary for the target formula $p_1 \to (\neg p_1 \to p_2)$ to appear in all the branches.

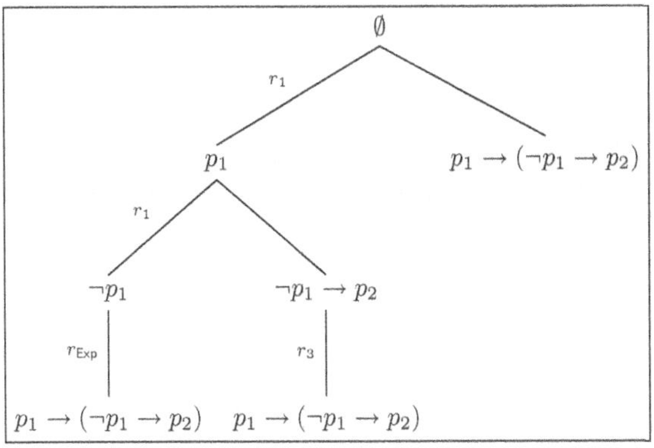

△

Example 3.4.2**, revisited.** In Example 3.4.2 we have obtained a four-valued PNmatrix characterizing the strengthening of the logic of classical implication with the additional axiom $p_1 \to (\neg p_1 \to \neg p_2)$. Easily, one can reuse the set of monadic separators, and the discriminator, from the previous example.

Using Theorem 3.5.1, an $\mathcal{S}$-analytic calculus R for the logic can be obtained by replacing the rule r_{Exp} of Example 3.4.1 with the rule below.

$$\frac{p\,,\;\neg p}{\neg q}\;r_{\mathsf{Exp}_\neg}$$

Expectedly, rule $r_{\mathsf{Exp}_\neg}$ corresponds to $R_{\mathcal{T}}$ with $X = \{00, 11\}$, and $X = \{10, 11\}$. △

Example 3.4.3**, revisited.** In Example 3.4.3 we have obtained a four-valued Nmatrix characterizing $\mathcal{C}_{\min}$, the strengthening of the logic $\mathcal{CLuN}$ with the additional axiom $\neg\neg p_1 \to p_1$. Easily, $\mathcal{S} = \{p, \neg p, \neg\neg p\}$ is a corresponding set of monadic separators, which allows for the discriminator $\widetilde{\mathcal{S}}$ with $\widetilde{\mathcal{S}}_{010} = \{p\}$, $\widetilde{\mathcal{S}}_{101} = \{p, \neg p\}$, and $\widetilde{\mathcal{S}}_{110} = \widetilde{\mathcal{S}}_{111} = \{p, \neg p, \neg\neg p\}$, giving rise to the following partitions.

x	Ω_x	$\mho_x$
010	$\emptyset$	$\{p\}$
101	$\{p\}$	$\{\neg p\}$
110	$\{p, \neg p\}$	$\{\neg\neg p\}$
111	$\{p, \neg p, \neg\neg p\}$	$\emptyset$

Using Theorem 3.5.1, the following rules constitute an $\mathcal{S}$-analytic calculus R for $\mathcal{C}_{\min}$.

$$\frac{}{p\,,\;p \to q}\,r_1 \qquad \frac{p\,,\;p \to q}{q}\,r_2 \qquad \frac{q}{p \to q}\,r_3 \qquad \frac{}{p\,,\;\neg p}\,r_4 \qquad \frac{\neg\neg p}{p}\,r_5$$

After simplifications, the rules r_1–r_3 correspond to $R_{\to}$, and r_4, r_5 to $R_{\neg}$.

We then obtained a three-valued Nmatrix characterizing the strengthening of $\mathcal{C}_{\min}$ with the axiom $p_1 \to \neg\neg p_1$. Easily, $\mathcal{S}' = \{p, \neg p\}$ is a corresponding set of monadic separators, which allows for the discriminator $\widetilde{\mathcal{S}}'$ with $\widetilde{\mathcal{S}}'_{01} = \{p\}$, and $\widetilde{\mathcal{S}}_{10} = \widetilde{\mathcal{S}}_{11} = \{p, \neg p\}$, giving rise to the following partitions.

x	Ω_x	$\mho_x$
01	$\emptyset$	$\{p\}$
10	$\{p\}$	$\{\neg p\}$
11	$\{p, \neg p\}$	$\emptyset$

Using Theorem 3.5.1, an $\mathcal{S}'$-analytic calculus R' for the logic can be obtained by joining to the calculus R obtained above the new $R_{\neg}$ rule:

$$\frac{p}{\neg\neg p}$$

△

Example 3.4.4**, revisited.** In Example 3.4.4 we have obtained a four-valued PNmatrix characterizing the strengthening of positive classical logic with axioms

$$p_1 \vee \neg p_1$$

$$p_1 \to (\neg p_1 \to (\circ p_1 \to p_2))$$

$$\circ p_1 \vee (p_1 \wedge \neg p_1)$$

$$\circ p_1 \to \circ(p_1 \wedge p_2)$$

$$(\neg p_1 \vee \neg p_2) \to \neg(p_1 \wedge p_2)$$

It is easy to see that $\mathcal{S} = \{p, \neg p, \circ p\}$ is a corresponding set of monadic separators, which allows for the discriminator $\widetilde{\mathcal{S}}$ with $\widetilde{\mathcal{S}}_{011} = \{p\}$, $\widetilde{\mathcal{S}}_{101} = \{p, \neg p\}$, and $\widetilde{\mathcal{S}}_{110} = \widetilde{\mathcal{S}}_{111} = \{p, \neg p, \circ p\}$. This gives rise to the following partitions.

x	Ω_x	$\mho_x$
011	$\emptyset$	$\{p\}$
101	$\{p\}$	$\{\neg p\}$
110	$\{p, \neg p\}$	$\{\circ p\}$
111	$\{p, \neg p, \circ p\}$	$\emptyset$

Using Theorem 3.5.1, the following rules constitute an $\mathcal{S}$-analytic calculus R for the logic.

$$\frac{p\,,\,q}{p \wedge q}\,r_1 \qquad \frac{p \wedge q}{p}\,r_2 \qquad \frac{p \wedge q}{q}\,r_3 \qquad \frac{\neg p}{\neg(p \wedge q)}\,r_4$$

$$\frac{p}{p \vee q}\,r_5 \qquad \frac{q}{p \vee q}\,r_6 \qquad \frac{p \vee q}{p\,,\,q}\,r_7 \qquad \frac{p\,,\,p \to q}{q}\,r_8 \qquad \frac{q}{p \to q}\,r_9 \qquad \frac{}{p\,,\,p \to q}\,r_{10}$$

$$\frac{}{p\,,\,\neg p}\,r_{11} \qquad \frac{}{p\,,\,\circ p}\,r_{12} \qquad \frac{p}{\neg p\,,\,\circ p}\,r_{13} \qquad \frac{p\,,\,q\,,\,\neg q}{\neg p}\,r_{14} \qquad \frac{p\,,\,\neg p\,,\,\circ p}{q}\,r_{15}$$

After simplifications, the rules r_1–r_4 correspond to $R_\wedge$, r_5–r_7 to $R_\vee$, r_8–r_{10} to $R_\to$, r_{11} to $R_\neg$, r_{12} and r_{13} to $R_\circ$. Finally, r_{14} and r_{15} result from $R_{\mathcal{T}}$, with $X = \{101, 110\}$ and $X = \{111, 011\}$, respectively.

Sample proofs, namely, for some of the axioms, with a very similar calculus can be found in Caleiro and Marcelino (2019). △

Example 3.4.5, **revisited.** In Example 3.4.5 we have obtained a four-valued twist-structure characterizing the addition of a paraconsistent Nelson-like strong negation to positive classical logic. Easily, $\mathcal{S} = \{p, \sim p\}$ is a corresponding set of monadic separators, yielding the discriminator $\widetilde{\mathcal{S}}$ with $\widetilde{\mathcal{S}}_x = \mathcal{S}$ for each truth-value x. This gives rise to the following partitions.

x	Ω_x	$\mho_x$
00	$\emptyset$	$\{p, \sim p\}$
01	$\{\sim p\}$	$\{p\}$
10	$\{p\}$	$\{\sim p\}$
11	$\{p, \sim p\}$	$\emptyset$

Using Theorem 3.5.1, the following rules constitute an $\mathcal{S}$-analytic calculus R for the logic.

$$\frac{p \wedge q}{p}\,r_1 \qquad \frac{p \wedge q}{q}\,r_2 \qquad \frac{p\,,\,q}{p \wedge q}\,r_3 \qquad \frac{\sim p}{\sim(p \wedge q)}\,r_4 \qquad \frac{\sim q}{\sim(p \wedge q)}\,r_5 \qquad \frac{\sim(p \wedge q)}{\sim p\,,\,\sim q}\,r_6$$

$$\frac{p}{p \vee q}\,r_7 \quad \frac{q}{p \vee q}\,r_8 \quad \frac{p \vee q}{p\,,\,q}\,r_9 \quad \frac{\sim(p \vee q)}{\sim q}\,r_{10} \quad \frac{\sim(p \vee q)}{\sim q}\,r_{11} \quad \frac{\sim p\,,\sim q}{\sim(p \vee q)}\,r_{12}$$

$$\frac{p\,,\,p \to q}{q}\,r_{13} \quad \frac{q}{p \to q}\,r_{14} \quad \frac{}{p\,,\,p \to q}\,r_{15}$$

$$\frac{\sim(p \to q)}{p}\,r_{16} \quad \frac{\sim(p \to q)}{\sim q}\,r_{17} \quad \frac{p\,,\sim q}{\sim(p \to q)}\,r_{18}$$

$$\frac{p}{\sim\sim p}\,r_{19} \quad \frac{\sim\sim p}{p}\,r_{20}$$

After simplifications, the rules r_1–r_6 correspond to $R_\wedge$, r_7–r_{12} to $R_\vee$, r_{13}–r_{18} to $R_\to$, r_{19} and r_{18} to $R_\sim$.

A strengthening with an additional (explosion) axiom $\sim p_1 \to (p_1 \to p_2)$ was then shown to be characterized by a four-valued Pmatrix. It is straightforward to see that one can reuse the set of monadic separators, and the discriminator, from above. Using Theorem 3.5.1, an $\mathcal{S}$-analytic calculus R' for the logic can be obtained by simply adding to R the new $R_\mathcal{T}$ rule

$$\frac{p\,,\sim p}{q}$$

obtained by considering $X = \{11, 00\}$, $X = \{11, 01\}$, and $X = \{11, 10\}$. △

Example 3.4.6, **revisited.** In Example 3.4.6 we obtained a four-valued Nmatrix characterizing the non-normal modal logic of Kearns and Ivlev (Kearns 1981; Ivlev 1988). It is not difficult to check (namely, using Proposition 3.5.2) that $\mathcal{S} = \{p, \Box p, \Box\neg p\}$ is a set of monadic separators for the Nmatrix. This allows for the discriminator $\widetilde{\mathcal{S}}$ with $\widetilde{\mathcal{S}}_{000} = \widetilde{\mathcal{S}}_{001} = \{p, \Box\neg p\}$, and $\widetilde{\mathcal{S}}_{100} = \widetilde{\mathcal{S}}_{110} = \{p, \Box p\}$, which gives rise to the following partitions.

x	Ω_x	$\mho_x$
000	$\emptyset$	$\{p, \Box\neg p\}$
001	$\{\Box\neg p\}$	$\{p\}$
100	$\{p\}$	$\{\Box p\}$
110	$\{p, \Box p\}$	$\emptyset$

Using Theorem 3.5.1, we get an $\mathcal{S}$-analytic calculus R for the logic.

$$\frac{}{p\,,\,p \to q}\,r_1 \quad \frac{p\,,\,p \to q}{q}\,r_2 \quad \frac{q}{p \to q}\,r_3 \quad \frac{p\,,\,\neg p}{}\,r_4 \quad \frac{}{p\,,\,\neg p}\,r_5$$

$$\frac{\Box(p \to q)\,,\,\Box p}{\Box q}\,k \quad \frac{\Box(p \to q)\,,\,\Box\neg q}{\Box\neg p}\,k_1 \quad \frac{\Box p\,,\,\Box\neg q}{\Box\neg(p \to q)}\,k_2$$

$$\frac{\Box\neg p}{\Box(p \to q)}\, m_1 \qquad \frac{\Box q}{\Box(p \to q)}\, m_2 \qquad \frac{\Box\neg(p \to q)}{\Box\neg q}\, m_3 \qquad \frac{\Box\neg(p \to q)}{\Box p}\, m_4$$

$$\frac{\Box p}{p}\, T \qquad \frac{\Box p}{\Box\neg\neg p}\, dn_1 \qquad \frac{\Box\neg\neg p}{\Box p}\, dn_2$$

After simplifications, the rules r_1–r_3, k, k_1–k_2, m_1–m_4 correspond to $R_{\to}$, r_4–r_5 and dn_1–dn_2 to $R_{\neg}$, and T to $R_{\Box}$. It is interesting to note that rules r_1–r_5 characterize classical logic, and the remaining rules are in a one-to-one correspondence with the axioms considered (see Coniglio and Golzio (2019)). The only less obvious case is the rule k_2. For this reason we present below an analytic proof of the corresponding axiom $K_2 = \neg\Box\neg(p \to q) \to (\Box p \to \neg\Box\neg q)$, i.e., $\rhd_R K_2$. Note that K_2 is obtained in all the branches of the proof-tree, except for the leftmost one, which is discontinued due to rule r_4 (as signaled by the use of $*$).

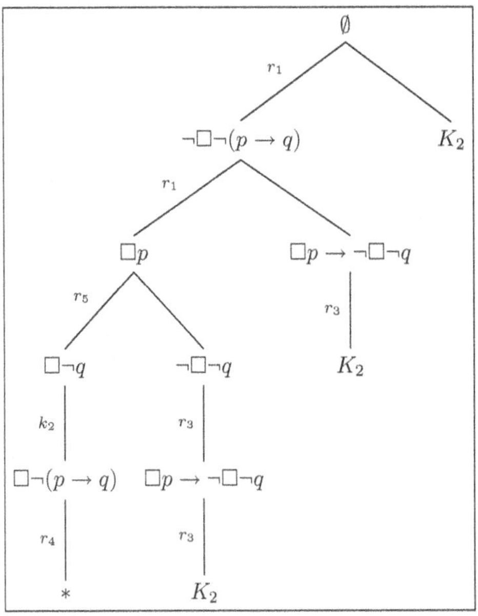

△

Example 3.4.7, **revisited.** In Example 3.4.7 we have obtained the usual three-valued Łukasiewicz's matrix (by strengthening the five-valued Łukasiewicz logic with an additional axiom). Easily, $S = \{p, \neg p\}$ is a set of monadic separators, yielding the discriminator $\widetilde{S}$ with $\widetilde{S}_0 = \widetilde{S}_{\frac{1}{2}} = \{p, \neg p\}$, and $\widetilde{S}_1 = \{p\}$, which gives rise to the following partitions.

x	Ω_x	$\mho_x$
0	$\{\neg p\}$	$\{p\}$
$\frac{1}{2}$	$\emptyset$	$\{p, \neg p\}$
1	$\{p\}$	$\emptyset$

Using Theorem 3.5.1, the following rules constitute an $\mathcal{S}$-analytic calculus R for $\mathcal{L}_3$.

$$\frac{p\,,\,\neg p}{\quad}\,r_1 \qquad \frac{p}{\neg\neg p}\,r_2 \qquad \frac{\neg\neg p}{p}\,r_3$$

$$\frac{\quad}{p\,,\,p \to q\,,\,\neg q}\,r_4 \qquad \frac{p\,,\,p \to q}{q}\,r_5 \qquad \frac{q}{p \to q}\,r_6$$

$$\frac{\neg p}{p \to q}\,r_7 \qquad \frac{\neg q\,,\,p \to q}{\neg p}\,r_8$$

$$\frac{\neg(p \to q)}{p}\,r_9 \qquad \frac{\neg(p \to q)}{\neg q}\,r_{10} \qquad \frac{p\,,\,\neg q}{\neg(p \to q)}\,r_{11}$$

After simplifications, the rules r_1–r_3 correspond to $R_\neg$, and r_4–r_{11} to $R_\to$. For illustration, we depict an analytic proof of the added axiom $A = ((p \to \neg p) \to p) \to p)$, i.e., $\rhd_R A$.

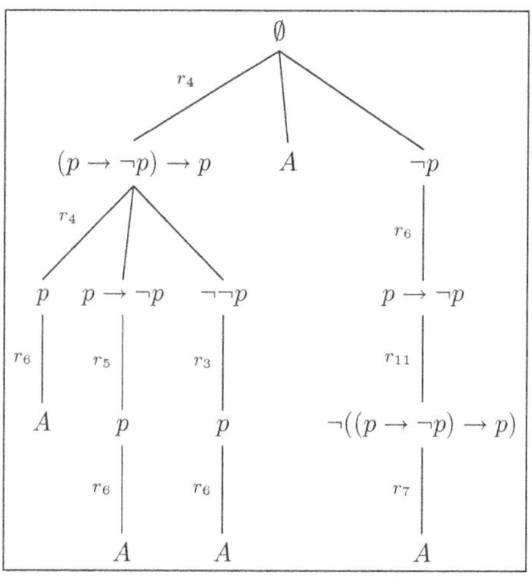

△

3.6 Concluding Remarks

In this paper we have shown that rexpansions of (P)(N)matrices are a universal tool for explaining the strengthening of logics with additional axioms. This does not come as a surprise, as non-determinism and partiality are well known for enabling a plethora of compositionality results in logic. Our general method in Theorem 3.3.2 is not effective, but it still brings about some interesting phenomena, such as the possibility of building a denumerable semantics for intuitionistic propositional logic (where the precise roles of non-determinism and partiality need further clarification). More practical, though, is our less general method in Theorem 3.3.7 as, despite the necessary restrictions on its scope, it brings about an effective method for producing finite semantic characterizations whenever starting from a finite basis. Our results cover a myriad of examples in the literature, namely, those motivated by the study of logics of formal inconsistency, which played an important role in the work of Arnon Avron. Besides, our effective method, while more general and incremental, is fully inspired by the fundamental ideas in Ciabattoni et al. (2014). It is also worth noting that our results apply not just to the Tarskian notion of consequence relation, but also to the multiple-conclusion case. An obvious topic for further work is to provide a usable tool implementing these methods.

Other opportunities for further research, aimed at generalizing the results presented, would be to find more general syntactic conditions on the set of allowed axioms. For instance, the number of sentential variables occurring in an axiom seems to be easy to flexibilize by artificially extending the logic with big-arity connectives. Beyond axioms, one could think even further away, and consider strengthening logics with fully fledged inference rules. In any case, such extensions will expectedly need more sophisticated techniques than the simple idea behind *look-aheads*.

These results reinforce the need to better understand the conditions under which two (P)(N)matrices characterize the same logic. This is by no means a trivial question, but we believe that the notion of rexpansion can be a useful tool in that direction.

If not for its own sake, this line of research aimed at providing effective semantic characterizations for combined logics is quite well justified by another recurring goal of many of the papers that inspired us: ultimately obtaining suitably analytic calculi for the resulting logics.

References

Arieli, O., & Avron, A. (1998). The value of the four values. *Artificial Intelligence*, *102*(1), 97–141.

Avron, A. (2005a). Non-deterministic matrices and modular semantics of rules. In J.-Y. Béziau (Ed.), *Logica Universalis* (pp. 149–167). Birkhäuser.

Avron, A. (2005b). A non-deterministic view on non-classical negations. *Studia Logica*, *80*(2–3), 159–194.

Avron, A. (2007). Non-deterministic semantics for logics with a consistency operator. *International Journal of Approximate Reasoning*, *45*(2), 271–287.

Avron, A., Ben-Naim, J., & Konikowska, B. (2007). Cut-free ordinary sequent calculi for logics having generalized finite-valued semantics. *Logica Universalis*, *1*(1), 41–70.
Avron, A., Konikowska, B., & Zamansky, A. (2012). Modular construction of cut-free sequent calculi for paraconsistent logics. In *Proceedings of the 27th Annual IEEE Symposium on Logic in Computer Science (LICS 2012)* (pp. 85–94).
Avron, A., Konikowska, B., & Zamansky, A. (2013). Cut-free sequent calculi for C-systems with generalized finite-valued semantics. *Journal of Logic and Computation*, *23*(3), 517–540.
Avron, A., & Lev, I. (2005). Non-deterministic multiple-valued structures. *Journal of Logic and Computation*, *15*(3), 241–261.
Avron, A., & Zamansky, A. (2011). Non-deterministic semantics for logical systems: A survey. In D. Gabbay & F. Guenthner (Eds.), *Handbook of Philosophical Logic* (Vol. 16, pp. 227–304). Springer.
Avron, A., & Zohar, Y. (2019). Rexpansions of non-deterministic matrices and their applications in non-classical logics. *Review of Symbolic Logic*, *12*(1), 173–200.
Baaz, M., Lahav, O., & Zamansky, A. (2013). Finite-valued semantics for canonical labelled calculi. *Journal of Automated Reasoning*, *51*(4), 401–430.
Batens, D. (1980). Paraconsistent extensional propositional logics. *Logique et Analyse*, *23*(90–91), 195–234.
Batens, D. (2000). A survey of inconsistency-adaptive logics. In *Frontiers of Paraconsistent Logic* (pp. 49–73). Research Studies Press.
Belnap, N. (1977a). How a computer should think. In G. Ryle (Ed.), *Contemporary Aspects of Philosophy* (Vol. 2, pp. 30–55), Episteme. Oriel Press.
Belnap, N. (1977b). A useful four-valued logic. In G. Epstein & J. M. Dunn (Eds.), *Modern Uses of Multiple-Valued Logic* (Vol. 2, pp. 5–37), Episteme. Oriel Press.
Caleiro, C., Carnielli, W., Rasga, J., & Sernadas, C. (2005). Fibring of logics as a universal construction. In D. Gabbay & F. Guenthner (Eds.), *Handbook of Philosophical Logic* (2nd ed., Vol. 13, pp. 123–187). Springer.
Caleiro, C., & Marcelino, S. (2019). Analytic calculi for monadic PNmatrices. *Logic, Language, Information and Computation (WoLLIC 2019)*, LNCS. Springer. http://sqig.math.ist.utl.pt/pub/CaleiroC/19-CM-axiomPNmatrices.pdf.
Caleiro, C., Marcelino, S., & Marcos, J. (2019). Combining fragments of classical logic: When are interaction principles needed? *Soft Computing*, *23*(7), 2213–2231.
Caleiro, C., Marcos, J., & Volpe, M. (2015). Bivalent semantics, generalized compositionality and analytic classic-like tableaux for finite-valued logics. *Theoretical Computer Science*, *603*, 84–110.
Carnielli, W., & Coniglio, M. (2016). Paraconsistent logic: consistency, contradiction and negation. In *Logic, Epistemology, and the Unity of Science* (Vol. 40). Springer.
Carnielli, W., & Marcos, J. (1999). Limits for paraconsistent calculi. *Notre Dame Journal of Formal Logic*, *40*, 375–390.
Carnielli, W., & Marcos, J. (2002). A taxonomy of C-systems. In W. Carnielli, M. Coniglio, & I. D'Ottaviano (Eds.), *Paraconsistency: The Logical Way to the Inconsistent* (Vol. 228, pp. 1–94), Lecture Notes in Pure and Applied Mathematics. Marcel Dekker.
Ciabattoni, A., Lahav, O., Spendier, L., & Zamansky, A. (2014). Taming paraconsistent (and other) logics: An algorithmic approach. *ACM Transactions on Computational Logic*, *16*(1), 5:1–5:23.
Coniglio, M., & Golzio, A. (2019). Swap structures semantics for Ivlev-like modal logics. *Soft Computing*, *23*(7), 2243–2254.
Gödel, K. (1932). Zum intuitionistischen aussagenkalkül. In *Mathematisch - naturwissenschaftliche klasse* (Vol. 69, pp. 65–66). Anzeiger Wien: Akademie der Wissenschaften.
Gottwald, S. (2001). *A Treatise on Many-Valued Logics* (Vol. 9), Studies in Logic and Computation. Research Studies Press.
Ivlev, J. (1988). A semantics for modal calculi. *Bulletin of the Section of Logic*, *17*(3–4), 114–121.
Kearns, J. (1981). Modal semantics without possible worlds. *Journal of Symbolic Logic*, *46*(1), 77–86.

Kracht, M. (1998). On extensions of intermediate logics by strong negation. *Journal of Philosophical Logic*, *27*, 49–73.
Marcelino, S., & Caleiro, C. (2016). Decidability and complexity of fibred logics without shared connectives. *Logic Journal of the IGPL*, *24*(5), 673–707.
Marcelino, S., & Caleiro, C. (2017). Disjoint fibring of non-deterministic matrices. In R. de Queiroz & J. Kennedy (Eds.), *Logic, Language, Information and Computation (WoLLIC 2017)* (Vol. 10388, pp. 242–255), LNCS. Springer.
Marcelino, S., & Caleiro, C. (2019). Axiomatizing non-deterministic many-valued generalized consequence relations. *Synthese*,. https://doi.org/10.1007/s11229-019-02142-8.
Nelson, D. (1948). Constructible falsity. *Journal of Symbolic Logic*, *14*, 247–257.
Odintsov, S. (2008). *Constructive Negations and Paraconsistency* (Vol. 26), Trends in Logic. Netherlands: Springer.
Scott, D. (1974). Completeness and axiomatizability in many-valued logic. In L. Henkin, J. Addison, C. Chang, W. Craig, D. Scott, & R. Vaught (Eds.), *Proceedings of the Tarski Symposium* (Vol. XXV, pp. 411–435), Proceedings of Symposia in Pure Mathematics. American Mathematical Society.
Shoesmith, D., & Smiley, T. (1978). *Multiple-Conclusion Logic*. Cambridge: Cambridge University Press.
Vakarelov, D. (1977). Notes on N-lattices and constructive logic with strong negation. *Studia Logica: An International Journal for Symbolic Logic*, *36*(1/2), 109–125.
Wójcicki, R. (1998). *Theory of Logical Calculi* (Vol. 199), Synthese Library. Kluwer.
Wroński, A. (1974). On the cardinality of matrices strongly adequate for the intuitionistic propositional logic. *Reports on Mathematical Logic*, *3*, 67–72.

Carlos Caleiro is an Associate Professor of Logic and Computation in the Department of Mathematics at Instituto Superior Técnico of Universidade de Lisboa, and a Senior Researcher at Instituto de Telecomunicações, in Portugal. He received his Ph.D. in Mathematics in 2000 from Universidade Técnica de Lisboa. His research focuses on compositionality in logic, namely, regarding algebraic and computational aspects of combined logics and applications, including information security. He published or edited close to 100 works in refereed journals, serials, book chapters, and conferences, often with international collaboration. He has supervised postdocs, Ph.D., and M.Sc. students, and participated in national and European research projects, namely, having coordinated the FP7 project GeTFun between 2013 and 2016. He serves on the editorial board of the journal Logica Universalis, and on several program committees, including the World Congress and School on Universal Logic and the International Joint Conference on Artificial Intelligence (IJCAI).

Sérgio Marcelino graduated in Mathematics from the University of Lisbon (Instituto Superior Técnico - IST) in 2006. In 2011, he defended his Ph.D. thesis Modal Logics for Changing Systems from King's College London, UK, under the supervision of Dov Gabbay and Agi Kurucz. Currently, he is a Researcher at Instituto de Telecomunicações and member of the Security and Quantum Information Group, a research unit of the Mathematics Department of IST, at Universidade de Lisboa, in Portugal. His research focuses primarily on compositionally issues in logic, namely, semantics, axiomatizability, computability, and complexity for combined logics. He (co-)authored around 20 publications, including in top-tier journals such as the Journal of Logic and Computation, Fuzzy Sets and Systems, Soft Computing, Synthese, Journal of Symbolic Logic, and Annals of Pure and Applied Logic.

Chapter 4
Credal Calculi, Evidence, and Consistency

Walter Carnielli and Juliana Bueno-Soler

Abstract This paper defends the use of possibility and necessity models based on the Logics of Formal Inconsistency, taking advantage of their expressivity in terms of the notions of *consistency* ($\circ$) and *inconsistency* ($\bullet$). The present proposal directly generalizes the approach of Besnard and Lang (Proceedings of 10th Conference on Uncertainty in Artificial Intelligence. Morgan Kaufmann, San Francisco, pp. 69–76 1994), whose main guidelines we borrow here. Some basic properties of possibility and necessity functions over the Logics of Formal Inconsistency are obtained and it is shown, by revisiting a paradigmatic example, how paraconsistent possibility and necessity reasoning can, in general, attain realistic models for artificial judgement. We will call such models *credal calculi*, emphasizing some of their appealing consequences.

Keywords Credal calculi · Paraconsistency · Contradiction · Consistency · Logics of Formal Inconsistency

4.1 How Should Logic, Probability, and Their Generalizations Be Related?

The term *probabilistic logic* (also referred to as probability logic) first appeared in print in Nilsson (1986), where Nilsson specifies that the (generalized) truth value of a sentence, defined in the interval [0, 1] is taken to be the probability of that sentence in ordinary first-order logic.

W. Carnielli (✉)
Centre for Logic, Epistemology and the History of Science and Department of Philosophy, University of Campinas - Unicamp, Campinas, SP, Brazil
e-mail: walter.carnielli@cle.unicamp.br

J. Bueno-Soler
School of Technology, University of Campinas - Unicamp, Limeira, SP, Brazil
e-mail: juliana@ft.unicamp.br

O. Arieli and A. Zamansky (eds.), *Arnon Avron on Semantics and Proof Theory of Non-Classical Logics*, Outstanding Contributions to Logic 21,
https://doi.org/10.1007/978-3-030-71258-7_4

The aim of a probabilistic logic is to combine the capacity of probability theory to handle uncertainty with the capacity of deductive logic to exploit the structure of deductions. As van Benthem wittily remarks in van Benthem (2017), there is an inescapable trade off between these two modes of reasoning: while quantitative probability produces less certain conclusions, but is applicable to all of life around us, deductive logic (*qua* qualitative) produces absolute certainty, but in a limited range—its 'greatest triumphs' to be found in mathematics and automated deduction. It is agreed, as confirmed by Leitgeb (2016), that probability measures in probabilistic logic are typically defined on formulas rather than on sets (events), as is usually done in standard probability theory.

The agreement between the trends of thought in probabilistic logic ends there, however. For Leitgeb (2016), specifying a probabilistic logic for a logical consequence relation is semantically determined by quantifying over probability measures (declaring, for instance, 'for all probability measures P'). Leitgeb classifies probabilistic logics along two dimensions: those which do not involve reference to (or quantification over) probability measures on the object level, and those which do. The underlying base logic for probability measures in Leitgeb (2016) is assumed to be classical, although he admits that there are also probability measures for which classical logic is not presupposed.

This already divides the field of probabilistic logic into two halves, but there is more: the notion of probability measure can itself be generalized in several ways. More than that, specifying a probabilistic logic by quantifying over probability measures is not the only way to build probabilistic logics: instead of quantifying over probability measures, the probabilistic measure itself can be intrinsically connected with (and dependent to) the deduction of a certain logic. Thus, for instance, if the logic lacks the Law of Excluded Middle (as in intuitionistic logic) or lacks the Law of Explosion (as in paraconsistent logics) the probability measure P can be such that $P(\alpha) + P(\neg\alpha)$ does not add to 1, or $P(\alpha) + P(\neg\alpha)$ can exceed 1.

We can say, in a simplified way, that probability is logic generalized, or that probability is a special case of generalized logic, or even that logic and probability are two extreme ways of reasoning governed by certain metamathematical (or metaphysical) laws.

The view that probability is a logic generalized (or the logical interpretation of probability), regarding probability as an epistemic notion concerned with degrees of belief, can be represented by Leibniz, but also Boole and later Carnap align with the same tradition which can be viewed as a logicist program in probability.

Ramsey plays an important role in this alignment, representing a mixture of the first and third views. His paper Ramsey (1990) conceived a theory that regarded probability and degrees of belief governed by consistency (or coherence) under certain laws—those laws being the 'laws of probability'. This idea founded the modern theory of subjective probability by proposing a way to measure people's beliefs through a betting method. In this way, someone's degree of belief will satisfy the laws of probability as much as this person behaves rationally. From this perspective, Ramsey can be considered the first one to conceive the well-known Dutch book theorem.

But what is this ‘consistency’ advocated by Ramsey? It’s certainly not just a reduction to the current idea of ‘lack of contradiction’. He believed that probability is the object of logic, but not merely of classical logic. His idea of consistency can be understood as including persistence in obeying certain rational principles; as he puts it in Ramsey (1990), p. 78:

> We find, therefore, that a precise account of the nature of partial belief reveals that the laws of probability are laws of consistency. [...] Having any definite degree of belief implies a certain measure of consistency, namely willingness to bet on a given proposition at the same odds for any stake, the stakes being measured in terms of ultimate values.

By relying on some intuitive rules of rational behavior, Ramsey could give a joint axiomatization of probability and utility showing that the measure of our ‘degrees of belief’ satisfies such laws of probability. We can summarize Ramsey’s position by saying that the logic of consistency for probability does not coincide with a logic of truth.

In this sense, probability expands and generalizes logic, but not necessarily only classical logic, as Adams (1998) implies. Unless one completely rules out logical pluralism (the view that there is more than one correct logic), Adam’s opinion is clearly a philosophical mistake, as there are several cases of probability calculi based on non-classical logics.

From this point of view, probability is a special case of generalized logic, in the sense that probability laws are dependent on the logic that we are assuming. This is the approach we take in this paper, investigating the interest of measures generalizing probabilities, where such measures are subject to non-standard logics.

A logicist view relating logic and probability was defended by Jan Łukasiewicz in 1913. In a publication in German (cf. Łukasiewicz 1970) Łukasiewicz maintained that probability theory would gain from being dependent on purely logical concepts, and claimed that this could save probability from its ‘obscure philosophical connotation’. Łukasiewicz’s idea was to replace the concept of probability measures by the concept of truth value (in turn regarded as degrees of truth). He was able to show that all laws of probability could be obtained from an (infinite) many-valued logical calculus, thus boldly linking probability and non-classical logic. Although (apparently) Łukasiewicz never endorsed the idea that different logics would give rise to different notions of probability (as he was more concerned with the foundations of the ‘right’ probability theory), he was nonetheless just one step away from that.

Going a bit further, as we do here and have argued elsewhere, one can regard probability as directly logic-dependent (see Bueno-Soler and Carnielli 2016), and this is the same for possibility and necessity measures. Possibility theory is an approach to uncertainty that can handle incomplete information. It is comparable to (and can be thought as generalizing) probability theory, albeit differing from probability theory as it uses a pair of dual mappings (possibility and necessity measures) instead of only one. Similar to probability measures, necessity and possibility measures can profitably be generalized to non-classical logics.

The aim of this paper is to show how necessity and possibility measures can be defined on the basis of the Logics of Formal Inconsistency (LFIs), extending previous work by Besnard and Lang (1994) in a rather natural way.

In previous papers (Bueno-Soler and Carnielli 2016 and Carnielli and Bueno-Soler 2017), we have investigated probability theories based on LFIs, emphasizing their interest and applicability. This paper goes a step further, generalizing our previous work towards possibility and necessity functions, and at the same time generalizing the work in Besnard and Lang (1994). That paper opened an area of study by proposing possibility and necessity functions over non-classical logics, concentrating on the da Costa calculus C_1. Although this was a natural choice at that time, and the paper offers valuable insights on the use of paraconsistent calculus with regard to possibility and necessity functions, there are some improvements that can be implemented in the light of new theories that extend the hierarchy C_n.

The present approach is based on the Logics of Formal Inconsistency, a generalization of da Costa's original hierarchy that takes into account operators for consistency ($\circ$) and inconsistency ($\bullet$), in a certain sense materializing the intuition of F. Ramsey. LFIs turn out to be highly flexible logic systems, as well as fixing some definitional problems of the original logics C_n (see, e.g., Carnielli et al. 2019 for references and discussion). This point will be clarified in the next section.

Since necessity measures and possibility measures can, respectively, be regarded as belief measures and plausibility measures, we refer (in an informal and generalist way) to the logic systems employing them as *credal calculi*, considering that belief can be regarded as generalized probability.

4.2 Paraconsistency, in the Guise of LFIs

Paraconsistency is the investigation of logic systems endowed with a negation $\neg$, such that not every contradiction of the form α and $\neg\alpha$ entails everything; in other words, a paraconsistent logic does not suffer from *deductive trivialism*, in the sense that a contradiction does not necessarily trivialize the deductive machinery of the system by proving everything.[1]

Deductive trivialism stems from the fact that classical logic abhors contradictions, since it endorses the inference rule of *Ex Contradictione Sequitur Quodlibet*, or *Principle of Explosion*:

(PEx) $\alpha, \neg\alpha \vdash \beta$

which authorizes one to derive anything from a pair of contradictory propositions $\alpha, \neg\alpha$.[2] The big challenge for paraconsistent logics is to avoid entertaining such

[1] There is another sense of trivialism, according to which everything is true. This should not be confused with deductive trivialism.

[2] This is independent from the fact that classical logic endorses the validity of the *Principle of Non-Contradiction*: (PNC) $\vdash \neg(\alpha \wedge \neg\alpha)$, see Carnielli et al. (2018).

an explosive negation, while still preserving resources for designing an expressive logic.

The Logics of Formal Inconsistency are a broad family of paraconsistent logics whose language, as mentioned, internalizes a notion of *consistency* independent of (but related to) negation, usually formalized by a new connective $\circ$. In this setting, the notion of inconsistency ($\bullet$) is not necessarily the negation of consistency ($\neg\circ$).

Consistent statements are those too rigid to admit contradictions, errors or vagueness, as well as exemplified by passwords. While a password does not permit any single mistake, an Internet search with typos usually returns acceptable results. Passwords are rigid, or 'consistent', in the sense that their effect is destroyed by a contradiction, while an Internet search can cope with the contradictions imposed by a typo. Starting from these basic intuitions that contradictions do not affect everything in the same way, LFIs do not validate the Principle of Explosion in its draconian form, that is, it is not the case that from any pair of contradictory sentences everything follows. The Principle of Explosion is, instead, taken into a tractable form, restricted to consistent sentences. Therefore, a contradictory theory is not necessarily trivial, provided the contradiction does not refer to something consistent.

This flexibility of the LFIs is expressed in the following law, termed the *Principle of Gentle Explosion*:

(PGE) $\circ\alpha, \alpha, \neg\alpha \vdash \beta$, for every β, although $\alpha, \neg\alpha \nvdash \beta$, for some β.

This feature of the LFIs leads to the notion that not all contradictions are equivalent, and that not all contradictions may cause deductive triviality. However, some special contradictions, the 'consistent contradictions', may indeed cause deductive explosion. LFIs expand the realm of classical logic, which may be recovered (if one so decides) by simply supposing that all kinds of propositions are consistent.

So, differently from standard logic, consistency is not synonymous with freedom from contradiction. The meaning of consistency in the LFIs is dictated by its axioms, as occurs with negation (and all other connectives). Some linguistic approximations of the idea of 'consistent' are 'coherent', 'orderly', 'tidy', etc., as in Williams (1978), where it is argued that inconsistency and contradiction should not be confused, and that one may be justified in believing inconsistent propositions.

It should be clear that the notions of consistency and non-contradiction are not coincident in the LFIs, and that the same holds for the notions of inconsistency and contradiction. There is, however, a fully fledged hierarchy of LFIs where consistency is gradually connected to non-contradiction; the reader is referred to Carnielli et al. (2007) and to Carnielli and Coniglio (2016) for a detailed treatment of the LFIs, along with conceptual motivations.

This paper, as mentioned, is devoted to redesigning the proposal of Besnard and Lang (1994) by carefully substituting the base logic C_1 by **Cie**, a logic much weaker than C_1 but one that boasts the same relevant properties with regard to the logical treatment of possibility and necessity functions, with some significant advantages.

First, as it is well known, the logics in the hierarchy C_n (C_1 included, obviously) are hardly algebraizable: there is little, if any, hope of achieving a Kolmogorovian-like approach to probabilistic theories (and their generalizations), based on C_n systems.

This is due to the non-validity of a replacement theorem (which would establish the validity of intersubstitutivity of provable equivalents, (IpE)), for such logics. Indeed, Theorem 3.51 in Carnielli and Marcos (2002) shows that (IpE) cannot hold in any paraconsistent extension of **Ci** (or, for that matter, in any LFI) in which either $(\neg\alpha \vee \neg\beta) \vdash \neg(\alpha \wedge \beta)$ or $\neg(\alpha \wedge \beta) \vdash (\neg\alpha \vee \neg\beta)$ hold.

Nevertheless, there is still a chance of obtaining (IpE) in extensions of **Ci** by the addition of weaker forms of contraposition deduction rules, as discussed in Sect. 3.7 of Carnielli and Marcos (2002). Recent (unpublished) results obtained with the help of interactive theorem provers encourage us to conjecture that this is the case (see Carnielli et al. 2020). This is one of the strongest reasons to prefer **Cie** (a slight extension of **Ci**) instead of C_1. On the other hand, a logic system weaker than **Cie** would be further away from classical logic, limiting real applications.

A second reason to prefer an LFI is the unnaturalness of the well-behavedness operator α^o of da Costa's C_1, and analogous definitions for the whole hierarchy C_n. The fact that α^o is defined by $\neg(\alpha \wedge \neg\alpha)$, together with the axiomatic propagation of this operator through other connectives ($(\alpha^o \wedge \beta^o)$ entailing $(\alpha\#\beta)^o$ for all binary connectives #) makes the semantics for C_1 (and even worse for general C_n) quite clumsy. Quoting from Carnielli et al. (2019), p. 4:

> At first glance, it may seem that the consistency operator of LFIs and the well-behavedness operator of da Costa's C_n hierarchy ... are the same thing when applied to a proposition α. This view, however, is mistaken. LFIs are a generalization of da Costa's idea of expressing the metalogical notion of consistency inside the object language. Even though the logics of C_n hierarchy (for $1 \leq n < \omega$) end up being a special case of LFIs, an important point distinguishes LFIs from da Costa's C_n. In the latter, as we have just seen, α^o is an abbreviation of $\neg(\alpha \wedge \neg\alpha)$, while in LFIs the unary connective $\circ$ may be primitive and logically independent from non-contradiction. So, in some LFIs, the equivalence between $\circ\alpha$ and $\neg(\alpha \wedge \neg\alpha)$ does not hold.

A third reason to prefer **Cie** over C_1 is that it has a nice quasi-finite valued semantics. Indeed, Avron (2007) proved that **Cie** (which he calls **Bcie**) is sound and complete with respect to a non-deterministic three-valued matrix, in this way confirming the decidability of **Cie** by a simple algorithm. Although a possible-translations semantics based on 3-valued tables and the corresponding decidability had already been obtained for **Cie** (see Marcos 2008), the non-deterministic semantics of Avron (2007) is the simplest possible and makes **Cie** semantically as close as possible to an elementary three-valued logic.

To sum up, the logic C_1 is unnecessarily heavy and bulky for the task of supporting a good and useful possibilistic theory, and it is our purpose to contend that this can be better achieved by starting from an LFI.

4.3 Formal Consistency, Possibilistic Measures, and Knowledge Representation

As widely recognized (see, e.g., Besnard and Laenens 1994 and Dubois and Prade 2015) reasoning from contradictory premises is a critical issue, since large knowledge bases are inexorably prone to incorporate contradictions. Contradictory information comes from the fact that data is provided by different sources, or by a single source that delivers contradictory data as certain.

The connections between the possibilistic and the paraconsistent paradigms are deep and complex; Dubois and Prade (2015), for instance, gives an overview of the various forms of contradiction that can be accommodated into possibilistic logic, defining concepts such as 'paraconsistency degree' and 'paraconsistent completion'. We believe, however, that defining possibility and necessity measures directly over the LFIs helps this connection, for the reasons given above. Paraconsistent logics offer simple and effective models for reasoning in the presence of contradictions, as they avoid collapsing into deductive trivialism by an uncomplicated logic machinery. Taking into consideration that it is more natural and efficient to reason from a contradictory information scenario than trying to remove the contradictions involved, the investigation of credal calculi concerned with necessity and possibility measures is well justified.

Although Besnard and Laenens (1994) refers to the 'inferential weakness' of paraconsistent logics, that paper is concerned with the calculus C_ω, a lower deductive bound of the hierarchy C_n which is considerably distinct from the elements of the hierarchy, and is not the deductive limit to this hierarchy (see Carnielli and Marcos 1999 in this respect). We intend to show that the LFIs are better candidates, and in particular that **Cie**, a Logic of Formal Inconsistency which, as said, differs very little from standard logic, can be used as a basis for paraconsistent necessity and possibility measures with an appealing potential for applications. Employing **Cie** is suggestive (as it is a sub-classical logic which is close in spirit to classical logic) but other choices can be thought of.

4.3.1 Possibility and Necessity Measures over Cie

Definition 4.3.1 The system **Cie** is composed by

1. Axioms

(PC$^+$)	all positive axioms of **PC**
(PI)	$p \vee \neg p$
(bC1)	$\circ p \supset [p \supset (\neg p \supset q)]$
(cf)	$\neg\neg p \supset p$
(ce)	$p \supset \neg\neg p$
(Ci)	$\neg \circ p \supset (p \wedge \neg p)$

2. Rule

(MP) From α and $\alpha \supset \beta$ it follows β.

Axiom **bC1** embodies the Principle of Gentle Explosion. Although the negation $\neg$ of **Cie** is more subtle than classical negation, a 'strong' or classical negation $\sim$ can be recovered within **Cie**, thus making **Cie** able to regain any classical reasoning (since all other connectives will act classically in the presence of a classical negation):

$$\sim\alpha \stackrel{\text{Def}}{=} \alpha \supset [p \wedge (\neg p \wedge \circ p)]$$

From the semantic viewpoint, **Cie** can be semantically characterized by a two-valued non-truth functional valuation semantics. As most logics of the family of LFIs, **Cie** cannot be characterizable by finite matrices, see, e.g., Theorem 4.29 of Carnielli et al. (2007). Alternatively, it can also be proved to be sound and complete by possible-translations semantics based on three-valued ingredients (Marcos 2008; Carnielli and Marcos 2002). As mentioned above, A. Avron provided in Avron (2007) an elegant sound and complete non-deterministic three-valued matrix semantics for **Cie**. All this makes **Cie** close to a many-valued logic, as it can be regarded as a special combination of three-valued logics, and at the same time minimally (but essentially) deviating from classical logic.

The idea of a graded notion of possibility in the form of a relation between possible worlds was introduced by David Lewis in 1963, in what he called 'comparative possibility'. The idea evolved progressively to possibility theory and possibilistic measures in the hands of Lotfi Zadeh, Didier Dubois and Henri Prade (cf. Dubois and Prade 2006). Possibility measures and their dual counterpart, necessity measures, can be understood respectively as plausibility and belief functions, or even as imprecise (approximate) probabilities.

The main point of this paper, as was pointed out, is to show why is it interesting to combine possibility theory with paraconsistency in the manner of the LFIs. On one hand, possibility theory based on classical logic is able to handle contradictions, but at the cost of complicated maneuvers (Dubois and Prade 2015). On the other hand, paraconsistent logics cannot easily express uncertainty in a gradual way. The blend of both via the LFIs, in view of the operators of consistency and inconsistency, offers a simple and natural qualitative and quantitative tool to reason with uncertainty. In view of the above rationale, the particular LFI we have chosen to carry out this combination, **Cie**, has some significant advantages,

Contrary to what happens in many LFIs, the logic **Cie** does not distinguish between inconsistency and contradiction as a consequence of its axiomatic presentation, and also allows for reduction of double negations. It is convenient to emphasize such properties, notably the ones concerned with consistency and inconsistency, as a theorem:

Theorem 4.3.2 *The following hold in*

1. $\vdash_{\mathbf{Cie}} \bullet\neg\alpha \equiv \bullet\alpha \equiv \neg\circ\alpha \equiv \alpha \wedge \neg\alpha$
2. $\vdash_{\mathbf{Cie}} \alpha \equiv \neg\neg\alpha$
3. $\vdash_{\mathbf{Cie}} \circ\circ\alpha$

4. $\vdash_{\mathbf{Cie}} \circ \bullet \alpha$
5. $(\alpha \supset \circ\beta) \vdash_{\mathbf{Cie}} (\neg \circ \beta \supset \neg\alpha)$
6. $(\alpha \supset \neg \circ \beta) \vdash_{\mathbf{Cie}} (\circ\beta \supset \neg\alpha)$
7. $(\neg\alpha \supset \circ\beta) \vdash_{\mathbf{Cie}} (\neg \circ \beta \supset \alpha)$
8. $(\neg\alpha \supset \neg \circ \beta) \vdash_{\mathbf{Cie}} (\circ\beta \supset \alpha)$
9. $\vdash_{\mathbf{Cie}} \alpha \vee \circ\alpha, \quad \vdash_{\mathbf{Cie}} \neg\alpha \vee \circ\alpha$

Proof 1. Item 1: Theorems 3.31, 4.2.7, and 4.28 of Carnielli et al. (2007),
2. Item 2: from axioms (ce) and (cf)) (Definition 4.3.1).
3. Items 3 to 9: from Fact 3.33, Fact 3.34, and Lemma 3.43 of Carnielli and Marcos (2002).

□

Items (3) and (4) above show a simple yet relevant attribute of the axiomatized notions of consistency and inconsistency: they are themselves consistent in their own sense. Items (5) to (9) show how the presence of consistent parts help to recover bits of standard reasoning. This makes **Cie** minimally deviating from classical logic, and easier to understand the role of the paraconsistent possibility and necessity measures. Moreover, as mentioned, classical logic can be encoded within **Cie**, as well as in most LFIs,[3] so making the credal calculi based on LFIs a legitimate expansion of standard reasoning.

A generic notion of logic-dependent necessity measures is given by the conditions below, inspired by Lewis Lewis (1976) (to the best of our knowledge, the first to consider, after Carnap, probability measures for logical sentences).

Definition 4.3.3 A necessity function (or measure) for the language $\mathcal{L}$ of **Cie**, or a **Cie**-necessity function, is a function $N : \mathcal{L} \mapsto \mathbb{R}$ satisfying the following conditions, where $\vdash_{Cie}$ stands for the syntactic derivability relation of **Cie**:

1. Non-negativity: $0 \leq N(\varphi) \leq 1$ for all $\varphi \in \mathcal{L}$
2. Tautologicity: If $\vdash_{\mathbf{Cie}} \varphi$, then $N(\varphi) = 1$
3. Anti-Tautologicity: If $\varphi \vdash_{\mathbf{Cie}}$, then $N(\varphi) = 0$
4. Comparison: If $\psi \vdash_{\mathbf{Cie}} \varphi$, then $N(\psi) \leq N(\varphi)$
5. Conjunction: $N(\varphi \wedge \psi) = min\{N(\varphi), N(\psi)\}$

A condition $N(\alpha) = \lambda$ can be understood as expressing that 'α is certain to degree λ' (in all normal states of affairs).

Possibilistic measures are also useful when representing preferences expressed as sets of prioritized goals, as, e.g., some lattice-valued possibility measures studied in the literature instead of real-valued possibility measures.

We do not want to imply that **Cie** is the only basis for defining useful necessity or possibility functions; other LFIs sharing the same signature can be good candidates. A particularly apt candidate is the three-valued logic **LFI1**. This logic has several properties that justify its role as one of the most natural three-valued paraconsistent logics, as argued in Avron (1991). In addition to having been specially designed to

[3] See discussions on the Derivability Adjustment Theorems in Carnielli et al. (2007).

be used in databases (cf. Carnielli et al. 2000), all their binary connectives behave classically, and **LFI1** is maximal with respect to classical sentential logic.

The properties of **Cie**-necessity functions permit us to derive some immediate, but useful consequences, that attest that we do not need the full force of C_1:

Theorem 4.3.4 *1. If $\phi_1, \ldots, \phi_n \vdash_{\mathbf{Cie}} \psi$ then $min_{i=1,n} N(\phi_i) \leq N(\psi)$.*
2. $N(\bullet\neg\alpha) = N(\bullet\alpha) = N(\neg\circ\alpha) = min\{N(\alpha), N(\neg\alpha)\}$
3. $N(\bullet\alpha) = N(\alpha)$ *or* $N(\bullet\alpha) = N(\neg\alpha)$
4. $N(\circ\varphi \wedge \varphi \wedge \neg\varphi) = 0$
5. $min\{N(\bullet\alpha), N(\circ\alpha)\} = 0$ *(Metaconsistency)*

Proof 1. An iterated application of Comparison.
2. A consequence of Theorem 4.3.2 (1) and the Conjunctive property of necessity functions.
3. Immediate after (2).
4. A consequence of Anti-Tautologicity and Conjunction.
5. Again, a consequence of Anti-Tautologicity and Conjunction since $\circ\varphi \wedge \bullet\varphi$ is a bottom particle.

□

Analogously, a generic notion of logic-dependent possibility measures (dual to a necessity function) is defined as follows:

Definition 4.3.5 A possibility function (or measure) for the language $\mathcal{L}$ of **Cie**, or a **Cie**- possibility function, is a function $\Pi : \mathcal{L} \mapsto \mathbb{R}$ satisfying the following conditions:

1. Non-negativity: $0 \leq \Pi(\varphi) \leq 1$ for all $\varphi \in \mathcal{L}$
2. Tautologicity: If $\vdash_{\mathbf{Cie}} \varphi$, then $\Pi(\varphi) = 1$
3. Anti-Tautologicity: If $\varphi \vdash_{\mathbf{Cie}}$, then $\Pi(\varphi) = 0$
4. Comparison: If $\psi \vdash_{\mathbf{Cie}} \varphi$, then $\Pi(\psi) \leq \Pi(\varphi)$
5. Disjunction: $\Pi(\varphi \vee \psi) = max\{\Pi(\varphi), \Pi(\psi)\}$

A condition $\Pi(\alpha) = \lambda$ can be understood as expressing that 'α is possible to degree λ'. If **L** is the standard classical logic, an **L**-possibility or -necessity function is a classical possibility or -necessity function (with the understanding that $\circ\alpha$ holds for any classical sentence α).

It is also instructive to note that analogous properties of necessity measures as in Theorem 4.3.4 hold for possibilistic measures, and that in both cases the values for the negation of a sentence are independent of the values of that sentence, their relationship being expressed by the operators of consistency (or equivalently inconsistency); in other words, $N(\neg\alpha)$ and $N(\alpha)$ are in principle independent, and somehow $N(\bullet\alpha) = N(\neg\circ\alpha)$ expresses their relationship (similarly for Π).

The meaning of such axioms is clear in showing how possibility and necessity functions can be perceived as logic-dependent: Non-Negativity, Disjunction and Conjunction are independent of any underlying logic, whereas Tautologicity, Anti-Tautologicity, and Comparison are clearly logic-dependent axioms.

By reviewing more carefully the intuition behind such measures, we can say that a value for a necessity function $N(\alpha) > 0$ measures how much α holds in the most normal situations, or how much α is a 'commonly accepted' belief. In other words, one may act as if α were 'almost' true. A value for a possibility function $\Pi(\alpha) > 0$ measures how much α is plausible, though perhaps unexpected.

The main difference between probability and possibility is that (standard) probability is self-dual, in the sense that 'it is not probable that not-α' means that 'it is probable that α', while possibility is not: 'it is not possible that not-α' does not mean 'it is possible that α', but 'it is necessary that α'.

An assertion like 'it is not possible that α' does not entail anything about the possibility, nor about the impossibility of not-α. Also, It may happen that $\Pi(\alpha) = \Pi(\neg\alpha)$, as well as $N(\alpha) = N(\neg\alpha)$.

There is a significant distinction at this point between standard and paraconsistent possibility and necessity measures. In the standard case one cannot be doubtful, that is, one cannot be both 'somewhat certain' about a proposition and about its negation, in the sense of maintaining $N(\varphi) > 0$ and $N(\neg\varphi) > 0$ since $N(\varphi \wedge \neg\varphi) = 0 = min\{N(\varphi), N(\neg\varphi)\}$. This kind of reasoning, however, can be expressed in the present theory: $N(\varphi)$ and $N(\neg\varphi)$ can be both positive. Notice, notwithstanding, that one cannot have $N(\varphi) > 0$, $N(\neg\varphi) > 0$ and $N(\circ\varphi) > 0$. That is, one cannot be both 'somewhat certain' about a consistent proposition and its negation in view of Theorem 4.3.4, item (4).

As much as in the standard case, one may have both $\Pi(\varphi) = 1$ and $\Pi(\neg\varphi) = 1$ unproblematically, as this merely acknowledges a state of total ignorance about the truth value of φ. What one cannot have, either in the standard theory or in the present theory, is $\Pi(\varphi) = 0$ and $\Pi(\neg\varphi) = 0$ (in view of consequences of Tautologicity and Disjunction for Π-measures; just recall that $\Pi(\varphi \vee \neg\varphi) = 1$ because $\vdash_{\mathbf{Cie}} \varphi \vee \neg\varphi$). This justifies investigating possibility and necessity theories based on Logics of Formal Inconsistency and Undeterminateness (LFIUs), where this would be possible (see Sect. 4.5) in view of the non-universal validity of the Law of Excluded Middle.

The relationship between possibility and necessity measures depends on the strength of the logical connectives involved. Under certain conditions (e.g., for classical logic, or under sufficiently strong properties for a negation $\backsim$, as in Proposition 4 of Besnard and Lang (1994)) the following dualities hold:

NecPos N is a necessity function iff $\Pi_N(\varphi) = 1 - N(\backsim \varphi)$ is a possibility function
PosNec Π is a possibility function iff $N_\Pi(\varphi) = 1 - \Pi(\backsim \varphi)$ is a necessity function

It should be clear that the strong negation $\sim$ definable in the logic **Cie** fulfills the requirements of Proposition 4 in Besnard and Lang (1994), which means that in principle we can restrict ourselves to elect one of the measures as preferential in **Cie**, taking the other as defined.

4.3.2 The Principle of Minimum Specificity (PMS) and Paraconsistency

Although quite flexible, standard necessity and possibility measures do not cope well with contradictions, since they treat contradictions in a global form (see a detailed explanation in Besnard and Lang (1994)), even if in a gradual way. This is the main reason to define them based upon paraconsistent logics; although they lack graduality, LFIs offer a tool for handling contradictions in knowledge bases in a local form, by locating the contradictions on critical sentences. Yet, the combination of them reaches a good balance: the paraconsistent paradigm by itself does not allow for any fine-grained graduality in the treatment of contradictions, which may lead to some loss of information when contradictions appear in a knowledge base. But when enriched with possibility and necessity functions, a new reasoning tool emerges.

At this point, it may be convenient to address a minimal information-theoretical treatment, just enough to discuss an illustrative example of the potentialities of the credal calculi when endowed with the consistency operators granted by the LFIs. We will be adapting some definitions from Besnard and Lang (1994), also revisiting the same example in that paper (itself borrowed from [Cho 94]) in order to highlight the potentialities and naturalness of the necessity measures built over the LFIs.

Consider a finite collection of items of information about a certain event X represented as pairs $\langle \phi_i, \alpha_i \rangle$, $1 \leq i \leq n$, for $\phi_i \in \mathcal{L}$, $\alpha_i \in [0, 1]$, interpreted as uncertain pieces of information about X whose meaning is 'My belief that X is ϕ_i is α_i'. $\langle \phi_i, 1 \rangle$ means 'My belief that X is ϕ_i is certain', while $\langle \phi_i, 0 \rangle$ means 'total ignorance about X'.

In general, $\langle \phi_i, \alpha_i \rangle$ is defined by the grade of credibility deriving from some body of evidence as defined by the mathematical concept of belief function in the way of Shafer [12] (but other measures are also accepted[4]).

The items of information about a certain event can be collected into a finite **Cie**-*knowledge base* $KB_0 = \{\langle \phi_i, \rho_i \rangle : 1 \leq i \leq n\}$ where $\phi_i \in \mathcal{L}$, $\rho_i \in [0, 1]$ and ρ_i respect the conditions of a necessity measure (i.e., there is some necessity function N such that $N(\phi_i) = \rho_i$ when defined), under the following provisos:

1. if φ is a tautology and $\varphi \in KB_0$ then $\langle \varphi, 1 \rangle \in KB_0$;
2. KB_0 is non-trivial, that is, KB_0 does not deduce (by means of the logic **Cie**) any bottom particle $\bot$.

The second proviso reflects an important presumption of the LFIs (and of paraconsistent logics in general): the distinction between contradictoriness and triviality. A **Cie**-knowledge base can be contradictory, but not deductively trivial. Consistency is the concept that makes the bridge between them: only a consistency contradiction

[4] Shafer's proposal as a theory of probable reasoning (and the Dempster-Shafer theory for that matter) is open to a number of objections, as discussed in Williams (1978). This aspect is not particularly relevant to our approach, since it is robust enough to correct distortions caused by a less precise belief function.

leads to deductive triviality. This is granted by Principle of Gentle Explosion (see axiom **bC1** in Sect. 4.3.1).

We consider two information-theoretic principles, one of them to complete a **Cie**-knowledge base: the *Principle of Evidence*, and another one the *Closed-World Assumption*, to be specified below.

Since sentences in **Cie**-knowledge bases belong to the language $\mathcal{L}$ of **Cie** without $\bullet$ and $\circ$ (as they are sentences about facts, and not about consistency or inconsistency of facts) the *Principle of Evidence* requires that we add to KB_0 a sentence $\bullet\alpha$ any time α and $\neg\alpha$ belong to KB_0, as well as all tautologies in **Cie**, thus stipulating an extended knowledge base:

$$KB = KB_0 \cup \{\langle \bullet\alpha, \rho\rangle : \langle \alpha, \rho_1\rangle \in KB_0 \text{ and } \langle \neg\alpha, \rho_2\rangle \in KB_0\}$$

where $\rho = min\{\rho_1, \rho_2\}$.

Definition 4.3.6 A necessity function N *satisfies* KB iff for every $\langle \phi_i, \rho_i\rangle \in KB$, $N(\phi_i) \geq \rho_i$.

The *Principle of Minimum Specificity*, **PMS** (Dubois and Prade 1987; Kraus et al. 1990) is a guiding provision of possibility theory. Also referred to as the minimum compact ranking and rational closure, it states that no hypothesis not known to be impossible should be left out. In this way, it induces a canonical necessity function which can be shown to be the smallest among all necessity functions satisfying the knowledge base.

Definition 4.3.7 Given a **Cie**-possibilistic knowledge base KB, the λ-*cut* KB_λ of KB is defined as $KB_\lambda = \{\phi : \langle \phi, \rho\rangle \in KB$ and æ $\geq$ ˘}.

A λ-cut singles out the items of information about a certain event X such that 'My belief that X is ϕ' is at least λ.

Definition 4.3.8 The minimum specificity closure of a **Cie**-possibilistic knowledge base KB is the function N^*_{KB} (called a pre-necessity measure) defined by
$N^*_{KB}(\phi) = sup\{\lambda : KB_\lambda \vdash_{\mathbf{Cie}} \phi\}$.

We now state our second information-theoretic principle, the *Closed-World Assumption* applicable to extend the pre-necessity measure N^*_{KB} in the following way: when N^*_{KB} is not defined (the piece of information ϕ does not appear in the database KB) then $N^*_{KB}(\phi) = 0$ and $N^*_{KB}(\neg\phi) = 0$. That is, when there is no evidence, either favorable or contrary, the pieces of information are considered to be consistent, since in this case $N^*_{KB}(\bullet\phi) = 0$, we can take $N^*_{KB}(\circ\phi) = 1$ (what means, in other words, assuming that consistency is maximal by lack of evidence to the contrary).

The next result sketches a proof showing how a pre-necessity measure N^*_{KB} can be extended to a necessity measure.

Theorem 4.3.9 *N^*_{KB} can be extended to a* **Cie**-*necessity function N_{KB}.*

Proof We first show some fundamental properties of N^*_{KB}:

1. Non-negativity: obvious by the definition of KB_0,
2. Conditional Tautologicity: If $\vdash_{\mathbf{Cie}} \varphi$ and $N^*_{KB}(\varphi)$ is defined, i.e., $\varphi \in KB$, then $N^*_{KB}(\varphi) = 1$ by construction of KB.
3. Anti-Tautologicity: If $\varphi \vdash_{\mathbf{Cie}}$, then $KB_\lambda \nvdash_{\mathbf{Cie}} \varphi$, for every λ, as KB does not contain any bottom particle. Then $sup\{\lambda : KB_\lambda \vdash_{\mathbf{Cie}} \varphi\} = sup\ \emptyset = 0$.[5] Thus $N^*_{KB}(\varphi) = 0$ if φ is a bottom particle.
4. Conditional Comparison: If $\psi \vdash_{\mathbf{Cie}} \varphi$ and $N^*_{KB}(\psi)$ and $N^*_{KB}(\varphi)$ are defined (that is, $\varphi, \psi \in KB$) then $N^*_{KB}(\psi) \leq N^*_{KB}(\varphi)$, since $sup\{\lambda : KB_\lambda \vdash_{\mathbf{Cie}} \psi\} \leq sup\{\lambda : KB_\lambda \vdash_{\mathbf{Cie}} \varphi\}$.
5. Quasi-Conditional Conjunction: If $N^*_{KB}(\varphi \wedge \psi)$, $N^*_{KB}(\psi)$, and $N^*_{KB}(\varphi)$ are defined (that is, all sentences belong to KB) then $N^*_{KB}(\varphi \wedge \psi) \leq min\{N^*_{KB}(\psi), N^*_{KB}(\varphi)\}$, since $sup\{\lambda : KB_\lambda \vdash_{\mathbf{Cie}} \varphi \wedge \psi\} \leq sup\{\lambda : KB_\lambda \vdash_{\mathbf{Cie}} \varphi\}$ and $sup\{\lambda : KB_\lambda \vdash_{\mathbf{Cie}} \varphi \wedge \psi\} \leq sup\{\lambda : KB_\lambda \vdash_{\mathbf{Cie}} \psi\}$.
   Now, if $N^*_{KB}(\varphi \wedge \psi) < min\{N^*_{KB}(\psi), N^*_{KB}(\varphi)\}$, change the value of $N^*_{KB}(\varphi \wedge \psi)$ in KB to $\langle \varphi \wedge \psi, \rho \rangle$ where $\rho = min\{N^*_{KB}(\psi), N^*_{KB}(\varphi)\}$. This establishes Conjunction.
   The above items show that N^*_{KB} conditionally satisfies the clauses of Definition 4.3.3. N^*_{KB} can be now extended to a **Cie**-necessity function N_{KB} in the following way, for the sentences α where $N^*_{KB}(\alpha)$ is not defined (that is, $\varphi \notin KB$) under the proviso that $\alpha \neq \bot$:

   a. If $\vdash_{\mathbf{Cie}} \varphi$ and $N^*_{KB}(\varphi)$ is not defined set $N^*_{KB}(\varphi) = 1$. This establishes Tautologicity.
   b. If $\psi \vdash_{\mathbf{Cie}} \varphi$, and $N^*_{KB}(\psi)$ or $N^*_{KB}(\varphi)$ is undefined, set the value of $N^*_{KB}(\psi)$ to be the same as $N^*_{KB}(\varphi)$. Vice versa for the other case. If neither are defined, set the value of both as 1. This establishes Comparison.
   c. If $N^*_{KB}(\varphi \wedge \psi)$, $N^*_{KB}(\psi)$, and $N^*_{KB}(\varphi)$ are all undefined, set $N^*_{KB}(\varphi \wedge \psi) = N^*_{KB}(\varphi) = N^*_{KB}(\psi) = 1$
      If two of them are undefined, set them as the defined one.
      If just one of them is undefined, there are two subcases:
      – If $N^*_{KB}(\varphi \wedge \psi)$ is undefined, set $N^*_{KB}(\varphi \wedge \psi) = min\{N^*_{KB}(\psi), N^*_{KB}(\varphi)\}$.
      – If either $N^*_{KB}(\psi)$ or $N^*_{KB}(\varphi)$ is undefined, set its value as the same as $N^*_{KB}(\varphi \wedge \psi)$.
   d. No sentences of the form $\bullet\varphi$ and $\circ\varphi$ belong simultaneously to KB, since KB is non trivial by hypothesis. If $N^*_{KB}(\bullet\alpha)$ or $N^*_{KB}(\circ\alpha)$ is defined, set the other one such that $N^*_{KB}(\bullet\alpha) + N^*_{KB}(\circ\alpha) = 1$. If neither are defined, set $N^*_{KB}(\bullet\alpha) = 0$ and $N^*_{KB}(\circ\alpha) = 1$.
      The pre-necessity measure N^*_{KB} is therefore extended to N_{KB} which enjoys the properties of Definition 4.3.3 for all sentences in the language.

□

[5] We are supported by the following properties: (1) every real number is both an upper and a lower bound of an empty set, and (2) If $A \subset B$ are sets of real numbers and sup exists, then $supA \leq supB$, see, e.g., Harzheim (2005).

The intuition behind N_{KB} is to provide a necessity function with a 'higher degree of generality', and therefore with a 'lower degree of specificity'. We can now prove that N_{KB} is indeed minimal:

Theorem 4.3.10 *For any* **Cie**-*necessity function* N, N *satisfies* KB *iff* $N(\varphi) \geq N_{KB}(\varphi)$ *for every* φ *such that* $\langle \varphi, \lambda \rangle \in KB$ *for some* λ.

Proof N *satisfies* KB by Definition 4.3.6 iff for every $\langle \phi, \rho \rangle \in KB$, $N(\phi) \geq \rho$. As $N_{KB} = N^*_{KB}$ for sentences in KB and $N^*_{KB} \leq N$ by Definition 4.3.8, the result follows. □

Non-monotonic and paraconsistent logics are no strangers to each other. In certain cases, we may not need non-monotonic reasoning if we are within a paraconsistent domain. Non-monotonic logic is structurally closed to the internal reasoning of belief revision, as argued in Gärdenfors (1991), where it is shown that the formal structures of the two theories are similar. Following a suggestion in Besnard and Lang (1994), in light of Kraus et al. (1990), it is possible to define a natural non-monotonic consequence relation[6] on databases acting under the logic **Cie** as follows:

Definition 4.3.11 $KB \mid\sim \phi$ iff $N_{KB}(\phi) > N_{KB}(\bullet\phi)$

The definition above can be equivalently written as the following:

Theorem 4.3.12 $KB \mid\sim \phi$ *iff* $N_{KB}(\phi) > N_{KB}(\neg\phi)$ *and* $KB \mid\sim \neg\phi$ *iff* $N_{KB}(\neg\phi) > N_{KB}(\phi)$.

Proof Consequence of Theorem 4.3.4. □

The general non-monotonic consequence relation $\phi \mid\sim_{KB} \psi$ is defined by $\phi \mid\sim_{KB} \psi$ iff $N_{KB}(\phi \supset \psi) > N_{KB}(\phi \supset \bullet\psi)$ (or equivalently $\phi \mid\sim_{KB} \psi$ iff $N_{KB}(\phi \supset \psi) > N_{KB}(\phi \supset \neg\psi)$).

Intuitively, $\phi \mid\sim_{KB} \psi$ means 'If ϕ, then normally ψ', or 'ψ is a plausible consequence of ϕ'. In the particular case of KB, ϕ is deducible from KB iff the certainty of ϕ is higher than the degree of inconsistency of ϕ, or equivalently, iff the certainty of ϕ is higher than the certainty of $\neg\phi$.

4.4 Contradictory Evidence via Multi-source Reasoning

By revisiting a paradigmatic example of Cholvy (1994) and Besnard and Lang (1994) we can appraise the suitability of general paraconsistent credal calculi, exemplified by the present case of **Cie** possibility and necessity measures in formalizing artificial reasoning with discernment, producing sensible judgements.

Two witnesses, W_1 and W_2, report a crime, under the following pieces of testimonial evidence:

---

[6] Formally, the fact that the consequence relation $KB \mid\sim \phi$ is monotonic follows from results in Dubois and Prade (1991) but it can be checked that if $KB \subset KB'$ and $KB \mid\sim \phi$ it does not necessarily follow that $KB' \mid\sim \phi$, in view of the role of the *sup* operator.

1. W_1 is certain the suspect was a woman with blond hair, and believes (with different degrees of uncertainty) that she was using glasses, wearing a hat and driving a BMW.
2. W_2 is also certain the suspect was a indeed a woman, but with brown hair and that she was not using glasses, and believes (with some degree of uncertainty) she was not driving a BMW. W_2 cannot say anything about a hat.
   Now, let us assign a necessity function N to this information provided by W_1 and W_2 (where $\lambda, \lambda', \lambda'', \lambda''' < 1$):
3. For W_1:
   $N(female) = 1; N(blond) = 1; N(glasses) = \lambda; N(hat) = \lambda'; N(BMW) = \lambda''$.
4. For W_2:
   $N(female) = 1;\ N(\neg blond) = 1;\ N(\neg glasses) = 1;\ N(\neg BMW) = \lambda'''$.

Let KB be a possibilistic knowledge base containing the credal information extracted from the witnesses, that is, KB contains the items of information $\{\langle female, 1\rangle, \langle blond, 1\rangle, \langle \neg blond, 1\rangle, \langle glasses, \lambda\rangle, \langle \neg glasses, 1\rangle, \langle hat, \lambda'\rangle, \langle BMW, \lambda''\rangle, \langle \neg BMW, \lambda'''\rangle\}$. The canonical necessity function N_{KB} now gives the following values, in line with Theorem 4.3.4 and applying the Closed-World Assumption:

- $N_{KB}(female) = 1;\ N_{KB}(\neg female) = 0;\ N_{KB}(\bullet female) = 0$
- $N_{KB}(blond) = 1;\ N_{KB}(\neg blond) = 1;\ N_{KB}(\bullet blond) = 1$
- $N_{KB}(BMW) = \lambda'';\ N_{KB}(\neg BMW) = \lambda''';\ N_{KB}(\bullet BMW) = min\{\lambda'', \lambda'''\}$
- $N_{KB}(hat) = \lambda';\ N_{KB}(\neg hat) = 0;\ N_{KB}(\bullet hat) = 0$
- $N_{KB}(glasses) = \lambda;\ N_{KB}(\neg glasses) = 1;\ N_{KB}(\bullet glasses) = \lambda$

From this setting the non-monotonic associated logic $KB \mid\!\sim$ is able to conclude that:

$KB \mid\!\sim female$, $KB \mid\!\sim hat$, $KB \mid\!\sim \neg glasses$, $KB \not\mid\!\sim blond$, $KB \not\mid\!\sim \neg blond$, $KB \mid\!\sim BMW$ if $\lambda''' < \lambda''$ (respectively, $KB \mid\!\sim \neg BMW$ if $\lambda''' > \lambda''$, or $KB \not\mid\!\sim BMW$, $KB \not\mid\!\sim \neg BMW$ if $\lambda''' = \lambda''$).

The results are not substantially different from previous examples (compare to Besnard and Lang 1994) but the prospects on being based on LFIs (and LFIUs in general) are wider, for the following reasons: first, the logic **Cie** is lighter than C_1 (and of C_n) in not having to carry the weight of the axioms for propagation of consistency, unnecessary for the context of possibility and necessity measures (see discussion in Sect. 5.2 of Carnielli and Marcos 2002). **Cie**, however, enjoys all the relevant properties for dealing with useful credal calculi, and the treatment initiated here paves the way for using other LFIs. Second, **Cie** has more appealing semantics and is a candidate for having a better algebraic counterpart (see discussion in Sect. 4.2).

The resulting logic is able to perform artificial reasoning that not only automatically practices suspension of judgement, avoiding cognitive bias, but at the same time respects the beliefs of agents, even when contradictory. At the same time, it is performing an automatic procedure of belief revision, moving from a paraconsistent to a classicized configuration. It is to be noted that the paraconsistent stage is crucial, permitting to avoid trivialism and taking rational profit of the whole situation. But more than performing judicious artificial reasoning, the logic is qualified to produce explanations by analyzing the conclusions from the credal database.

4.5 Conclusion and Further Challenges

Some naive criticisms of paraconsistency seem to imply that while it is a good theory which makes it possible to reason with contradictory opinions, beliefs and other situations without falling into trivialism, an ideal missing step would be to 'restore consistency' in the sense of solving the problems caused by contradictions. This is, of course, not the role of a paraconsistent logic (and that is the reason why such criticisms are misplaced). Nonetheless, the above paraconsistent non-monotonic logic built over LFI-based possibility and necessity measures does perform this 'second step', acting as a belief revision procedure, with clear interest for applications.

The consequence relation $\mid\!\sim$ is non-monotonic and paracomplete (it may be possible that $KB \not\mid\!\sim \alpha$ and $KB \not\mid\!\sim \neg\alpha$) and works on the basis of paraconsistent credal measures, weighing contradictory evidence, partial evidence and missing evidence. It is also related to the notions of conclusive and non-conclusive evidence, as well as of preservation of evidence.

The paper Rodrigues et al. (2020) introduces the logic of evidence and truth LET_F as an extension of the Belnap-Dunn four-valued logic *FDE*. LET_F is equipped with a classicality operator $\circ$ and its dual to non-classicality operator $\bullet$. It would be interesting to define possibility and necessity measures over LET_F, generalizing the probability measures defined over LET_F and to further investigate the connections between the formal notions of evidence and the graded notions of possibility and necessity.

Both LET_J (introduced in Carnielli and Rodrigues 2017) and LET_F are part of the family of the Logics of Formal Inconsistency and Undeterminedness or LFIUs (cf. Carnielli et al. 2019 for references and results on duality). In the LFIUs, not only the Principle of Gentle Explosion holds (principle 2) $\circ\alpha, \alpha, \neg\alpha \vdash \beta$, but also $\circ\alpha \vdash \alpha \vee \neg\alpha$. What this means is that neither $\alpha \wedge \neg\alpha$ unrestrictedly causes deductive trivialism nor that $\alpha \vee \neg\alpha$ holds unrestrictedly: they both depend on the consistency of α ($\circ\alpha$). This has a deep effect on the possibility and necessity measures, as we have suggested in previous sections. This encourages the development of possibility and necessity systems based on the LFIUs, a project that we leave for future work.

There are some other research lines to be investigated concerning LFIUs and necessity measures. For instance, Benferhat et al. (1995) study three consequence relations with the ability to infer non-trivial conclusions from contradictory knowledge bases. In each case a certain 'level of paraconsistency' is computed for each conclusion, and the consequence relations treat inconsistency in a local way, in contrast with the more frequent global approaches developed in the literature. This approach can be promptly generalized to LFIUs.

As several authors have pointed out (e.g., Pearl 1988; Nilsson 1993), and famously proved in Lewis (1976), there is a remarkable discrepancy between the probability of material condition versus conditional probability: $P(\neg\alpha \vee \beta)$ seems not to reflect the proper meaning of 'if α then β' especially when $P(\alpha)$ is very small compared to $P(\neg\alpha \vee \beta) < 1$. So the proper modeling of a rule 'if α then β' which is not certain but likely, applied to a rare event α seems to be $P(\beta/\alpha)$, and the same for possibilistic measures. This suggests that it would be interesting to investigate probability in logics with no 'naive' implication, as in Rodrigues et al. (2020).

It wouldn't be too bold to say that this kind of cautious logic machinery that balances the partial, missing or even contradictory evidence could help to mitigate some criticisms raised by the abusive use of algorithms in today's society. It is argued in O'Neil (2016), for instance, by giving several examples, that mathematical models can work as ideological tools which contribute to exacerbate oppression and inequality. We should thus worry about invisible failures by algorithms used in social media and daily life. The present proposal, by emphasizing credal calculi based on formal consistency and inconsistency that produces prudent decisions and offers explanations, could hopefully be helpful to develop less oppressive algorithms.

Acknowledgements Both authors acknowledge support from the National Council for Scientific and Technological Development (CNPq), Brazil under research grants 307376/2018-4 and 308077/2018-0. We are indebted to Jérôme Lang and Philippe Besnard for personal communications that encouraged this paper to appear, and to Lluís Godo for some sharp remarks on a previous version of this paper.

References

Adams, E. W. (1998). *A Primer of Probability Logic*. Stanford: CSLI Publications.

Avron, A. (1991). Natural 3-valued logics - characterization and proof theory. *Journal of Symbolic Logic*, *56*(1), 276–294.

Avron, A. (2007). Non-deterministic semantics for logics with a consistency operator. *International Journal of Approximate Reasoning*, *45*(2), 271–287. Eighth European Conference on Symbolic and Quantitative Approaches to Reasoning with Uncertainty (ECSQARU 2005).

Benferhat, S., Dubois, D., & Prade, H. (1995). A local approach to reasoning under inconsistency in stratified knowledge bases. In C. Froidevaux & J. Kohlas (Eds.), *Symbolic and Quantitative Approaches to Reasoning and Uncertainty. ECSQARU 1995* (Vol. 946, pp. 36–43), Lecture Notes in Computer Science (Lecture Notes in Artificial Intelligence). Springer.

Besnard, Ph., & Laenens, E. (1994). A knowledge representation perspective: logics for paraconsistent reasoning. *International Journal of Intelligent Systems*, *9*, 153–168.

Besnard, Ph., & Lang, J. (1994). Possibility and necessity functions over non-classical logics. In R. Lopez de Mantaras & P. Poole (Eds.), *Proceedings of 10th Conference on Uncertainty in Artificial Intelligence* (Vol. 229, pp. 69–76). San Francisco: Morgan Kaufmann.

Bueno-Soler, J., & Carnielli, W. A. (2016). Paraconsistent probabilities: consistency, contradictions and Bayes' theorem. In J. Stern (Ed.), *Statistical Significance and the Logic of Hypothesis Testing* (Vol. Entropy 18(9)). MDPI Publications. Open acess at http://www.mdpi.com/1099-4300/18/9/325/htm.

Carnielli, W. A., & Bueno-Soler, J. (2017). Paraconsistent probabilities, their significance and their uses. In C. Caleiro, F. Dionisio, P. Gouveia, P. Mateus, & J. Rasga (Eds.), *Essays in Honour of Amilcar Sernadas* (Vol. 10500, pp. 197–230). London: College Publications.
Carnielli, W. A., & Coniglio, M. E. (2016). *Paraconsistent Logic: Consistency, Contradiction and Negation*. Springer.
Carnielli, W. A., Coniglio, M. E., & Fuenmayor, D. (2020). Formal inconsistency enriched with replacement: an algebraic and modal account. Submitted for publication, preprint available from https://www.cle.unicamp.br/eprints/index.php/CLE_e-Prints.
Carnielli, W. A., Coniglio, M. E., & Marcos, J. (2007). Logics of formal inconsistency. In D. Gabbay & F. Guenthner (Eds.), *Handbook of Philosophical Logic* (Vol. 14, pp. 1–93). Amsterdam: Springer.
Carnielli, W. A., Coniglio, M. E., & Rodrigues, A. (2018). On formal aspects of the epistemic approach to paraconsistency. In M. Freund, M. Fernandez de Castro, & M. Ruffino (Eds.), *Logic and Philosophy of Logic: Recent Trends in Latin America and Spain* (pp. 48–74). College Publications.
Carnielli, W. A., Coniglio, M. E., & Rodrigues, A. (2019). Recovery operators, paraconsistency and duality. *Logic Journal of the IGPL*. First published online at https://doi.org/10.1093/jigpal/jzy054.
Carnielli, W. A., de Amo, S., & Marcos, J. (2000). Formal inconsistency and evolutionary databases. *Logic and Logical Philosophy*, *8*, 115–152.
Carnielli, W. A., & Marcos, J. (1999). Limits for paraconsistent calculi. *Notre Dame Journal of Formal Logic*, *40*(3), 375–390.
Carnielli, W. A., & Marcos, J. (2002). A taxonomy of C-systems. In W. A. Carnielli, M. E. Coniglio, & I. M. L. D'Ottaviano (Eds.), *Paraconsistency - The Logical Way to the Inconsistent* (Vol. 228, pp. 1–94). Lecture Notes in Pure and Applied Mathematics. New York: Marcel Dekker.
Carnielli, W. A., & Rodrigues, A. (2017). An epistemic approach to paraconsistency: a logic of evidence and truth. In *Synthese*. https://link.springer.com/article/10.1007/s11229-017-1621-7.
Cholvy, L. (1994). Fusion de sources d'information ordonnees en fonction des themes. In *Proceedings of RFIA'94* (pp. 487–494). Paris.
Dubois, D., & Prade, H. (1987). The principle of minimum specificity as a basis for evidential reasoning. In B. Bouchon & R. R. Yager (Eds.), *Uncertainty in Knowledge-Based Systems. IPMU 1986* (Vol. 286, pp. 75–84). Springer Lecture Notes in Computer Science.
Dubois, D., & Prade, H. (1991). Possibilistic logic, preferential models, non-monotonicity and related issues. In *Proceedings of the 12th International Joint Conference on Artificial Intelligence (IJCAI-91) Sydney, Australia, Aug. 24–30 1991* (pp. 419–424). Paris.
Dubois, D., & Prade, H. (2006). Possibility theory and its applications: a retrospective and prospective view. In G. Della Riccia, D. Dubois, R. Kruse, & H. J. Lenz (Eds.), *Decision Theory and Multi-Agent Planning* (Vol. 482, pp. 89–109). CISM International Centre for Mechanical Sciences (Courses and Lectures).
Dubois, D., & Prade, H. (2015). Inconsistency management from the standpoint of possibilistic logic. *International Journal of Uncertainty, Fuzziness and Knowledge-Based Systems*, *23*, 15–30.
Gärdenfors, P. (1991). Belief revision and nonmonotonic logic: Two sides of the same coin? In J. van Eijck (Ed.), *Logics in AI. JELIA 1990. Lecture Notes* (Vol. 478, pp. 52–54). Springer Lecture Notes in Computer Science (Lecture Notes in Artificial Intelligence).
Harzheim, E. (2005). *Ordered Sets (Advances in Mathematics)*. Springer.
Kraus, S., Lehmann, D., & Magidor, M. (1990). Nonmonotonic reasoning, preferential models and cumulative logics. *Artificial Intelligence*, *44*, 167–207.
Leitgeb, H. (2016). Probability in logic. In A. Hájek & C. Hitchcock (Eds.), *The Oxford Handbook of Probability and Philosophy* (pp. 681–704). Oxford- Oxford Handbooks in Philosophy.
Lewis, D. (1976). Probabilities of conditionals and conditional probabilities. *The Philosophical Review*, *85*(3), 297–315.

Łukasiewicz, J. (1970). Die logischen Grundlagen der Wahrscheinlichkeitsrechnung. Krakow: Spółka Wydawnicza Polska, 1913. Translated as Logical Foundations of Probability Theory. In L. Borkowski (Ed.), *Jan Łukasiewicz-Selected Works* (pp. 16–63). Amsterdam, London: North Holland Publishing Company.
Marcos, J. (2008). Possible-translations semantics for some weak classically-based paraconsistent logics. *Journal of Applied Non-Classical Logics*, *18*(1), 7–28.
Nilsson, N. J. (1986). Probabilistic logic. *Artificial Intelligence*, *28*, 71–87.
Nilsson, N. J. (1993). Probabilistic logic revisited. *Artificial Intelligence*, *39*, 39–42.
O'Neil, C. (2016). *Weapons of Math Destruction: How Big Data Increases Inequality and Threatens Democracy*. New York, NY: Crown Publishing Group.
Pearl, J. (1988). *Probabilistic Reasoning in Intelligent Systems: Networks of Plausible Inference* (2nd ed.). Morgan Kaufmann.
Ramsey, F. (1990). Truth and probability. In D. H. Mellor (Ed.), *Philosophical Papers* (pp. 52–109). Cambridge: Cambridge University Press.
Rodrigues, A., Bueno-Soler, J., & Carnielli, W. A. (2020). Measuring evidence: a probabilistic approach to an extension of Belnap-Dunn logic. In *Synthese*.
van Benthem, J. (2017). Against all odds: when logic meets probability. In J. P. Katoen, R. Langerak, & A. Rensink (Eds.), *ModelEd, TestEd, TrustEd* (Vol. 10500, pp. 239–253). Springer Lecture Notes in Computer Science.
Williams, P. M. (1978). On a new theory of epistemic probability. *The British Journal for the Philosophy of Science*, *29*(4), 375–387.

Walter Carnielli is a Full Professor of Logic at the University of Campinas (Unicamp), where he served as Director of the Centre for Logic for several years. With relevant contributions to both technical results and philosophical aspects of logic, he is one of the pioneers of the Brazilian School of Paraconsistent Logic, and was also President of the Brazilian Society of Logic. He published several books in the fields of Computability, Logic, and Philosophy of Logic as well as books for the general public. He is Grantee of the Alexander von Humboldt Foundation, and he received national and international awards, such as the 2013 Telesio Galilei Academy of Science Gold Medal, the 2007 Jabuti Prize, and the 2019 CAPES Prize as the supervisor for the best Ph.D. thesis in philosophy. He has interests in the area of Artificial Intelligence and Rationality, and is currently a Vice-President of the Advanced Institute for Artificial Intelligence (AI2).

Juliana Bueno-Soler is a Bachelor and Licentiate in Mathematics from the São Paulo State University "Julio de Mesquita Filho" (Unesp) and earned her Ph.D. in Philosophy and Logic (2009) from the University of Campinas. She has contributed to Logic and Philosophy, mainly in the subjects of algebraization of logics, paraconsistent logics, foundations of probability, multimodal logics, and semantics for non-classical logics. She is a member of the Centre for Logic (Unicamp), and she is currently working as an Associate Professor in the area of Statistics and Differential and Integral Calculus at the School of Technology of the University of Campinas (Unicamp), Campus Limeira. She has been currently actively researching the foundations of probability theory based on paraconsistent logics.

Chapter 5
Theorems of Alternatives for Substructural Logics

Almudena Colacito, Nikolaos Galatos, and George Metcalfe

Abstract A theorem of alternatives provides a reduction of validity in a substructural logic to validity in its multiplicative fragment. Notable examples include a theorem of Arnon Avron that reduces the validity of a disjunction of multiplicative formulas in the "R-mingle" logic RM to the validity of a linear combination of these formulas, and Gordan's theorem for solutions of linear systems over the real numbers that yields an analogous reduction for validity in Abelian logic A. In this paper, general conditions are provided for axiomatic extensions of involutive uninorm logic without additive constants to admit a theorem of alternatives. It is also shown that a theorem of alternatives for a logic can be used to establish (uniform) deductive interpolation and completeness with respect to a class of dense totally ordered residuated lattices.

5.1 Introduction

In Avron (1987), Arnon Avron proved a remarkable theorem relating derivability in the "R-mingle" logic RM (see, e.g., Dunn 1970; Anderson and Belnap 1975; Avron 1986) formulated with connectives $+$, $\neg$, $\wedge$, and $\vee$, to derivability in its multiplicative fragment with connectives $+$ and $\neg$. More precisely, Avron proved that a disjunction of multiplicative formulas $\varphi_1 \vee \ldots \vee \varphi_n$ is derivable in RM if and only if $\varphi_{j_1} + \ldots + \varphi_{j_k}$ is derivable in RM for some $1 \leq j_1 < \ldots < j_k \leq n$. Indeed,

The research reported here was funded by the Swiss National Science Foundation (SNF) grant 200021_184693 and the EU Horizon 2020 research and innovation programme under the Marie Skłodowska-Curie grant agreement no. 689176.

A. Colacito · G. Metcalfe (✉)
Mathematical Institute, University of Bern, Sidlerstrasse 5, 3012 Bern, Switzerland
e-mail: george.metcalfe@math.unibe.ch

A. Colacito
e-mail: almudena.colacito@math.unibe.ch

N. Galatos
Department of Mathematics, University of Denver, 2390 S. York St., Denver, CO 80208, USA
e-mail: ngalatos@du.edu

O. Arieli and A. Zamansky (eds.), *Arnon Avron on Semantics and Proof Theory of Non-Classical Logics*, Outstanding Contributions to Logic 21,
https://doi.org/10.1007/978-3-030-71258-7_5

two proofs are given of this result. The first is an easy consequence of a (quite hard) cut elimination proof for a proof system for RM defined in the framework of hypersequents, introduced in the paper as sequences of sequents. The second proof is semantic and makes use of the completeness of RM and its multiplicative fragment with respect to an infinite-valued and a three-valued matrix, respectively.

The central aim of the first part of this paper is to show that Avron's theorem belongs to a family of results that may be understood as "theorems of alternatives" for substructural logics. Such theorems in the field of linear programming are duality principles stating that either one or another linear system has a solution over the real numbers, but not both (see, e.g., Dantzig 1963). In particular, Gordan's theorem (replacing real numbers with integers) asserts that for any $M \in \mathbb{Z}^{m \times n}$,

$$\textit{either } y^T M > \mathbf{0} \text{ for some } y \in \mathbb{Z}^m \textit{ or } Mx = \mathbf{0} \text{ for some } x \in \mathbb{N}^n \backslash \{\mathbf{0}\}.$$

This version of Gordan's theorem is proved in Colacito and Metcalfe (2017) by extending partial orders on free abelian groups to total orders and formulated as a correspondence between derivability in Abelian logic A (see, e.g., Meyer and Slaney 1989; Casari 1989; Metcalfe et al. 2005) and derivability in its multiplicative fragment. That is, a disjunction of multiplicative formulas $\varphi_1 \vee \ldots \vee \varphi_n$ is derivable in A if and only if $\lambda_1\varphi_1 + \cdots + \lambda_n\varphi_n$ is derivable in A for some $\lambda_1, \ldots, \lambda_n \in \mathbb{N}$ not all 0. In Sect. 5.3, we provide a sufficient condition for an axiomatic extension of involutive uninorm logic without additive constants IUL^- (see Metcalfe and Montagna 2007) to satisfy such an equivalence. The condition is based on derivability and determines a family of substructural logics admitting a theorem of alternatives that includes RM^{t} (RM with an additional truth constant), involutive uninorm mingle logic without additive constants IUML^- (axiomatized relative to RM^{t} by $1 \to 0$, see Metcalfe and Montagna (2007)), Abelian logic A, and the "balanced" extension BIUL^- of IUL^- with additional axioms $n\varphi \to \varphi^n$ and $\varphi^n \to n\varphi$ for each $n \in \mathbb{N}$.

The second part of the paper focuses on applications of theorems of alternatives. In Sect. 5.4, we show that if an extension of IUL^- with a theorem of alternatives admits deductive interpolation or right uniform deductive interpolation (see, e.g., Metcalfe et al. 2014; van Gool et al. 2017; Kowalski and Metcalfe 2019) for its multiplicative fragment, then the full logic admits the property. For example, this provides an alternative proof that RM^{t} admits deductive interpolation (and hence right uniform deductive interpolation), first proved in Meyer (1980) (see also Avron 1986; Marchioni and Metcalfe 2012). In Sect. 5.5, we show that any extension of IUL^- that derives $1 \to 0$ and has a theorem of alternatives is complete with respect to a class of dense totally ordered residuated lattices. Obtaining such a "dense chain completeness" result is important in the field of mathematical fuzzy logic as a key intermediate step toward proving that an axiom system is "standard complete," that is, complete with respect to a class of algebras with lattice reduct [0, 1] (see, e.g., Jenei and Montagna 2002; Metcalfe and Montagna 2007; Ciabattoni and Metcalfe 2008; Baldi and Terui 2016; Metcalfe and Tsinakis 2017; Galatos and Horčik 2021). Although theorems of alternatives hold only for a fairly narrow class of substructural logics, we obtain here new dense chain completeness proofs for IUML^- and A, and a dense chain completeness result for BIUL^- that does not seem to be easily proved using other methods developed in the literature.

5.2 Preliminaries

Let $\mathcal{L}$ be any propositional language and let $\mathrm{Fm}_{\mathcal{L}}$ denote the set of formulas of this language over a fixed countably infinite set of variables, denoting arbitrary variables and formulas by $p, q, r, \ldots$ and $\varphi, \psi, \chi, \ldots$, respectively. Given $\Sigma \subseteq \mathrm{Fm}_{\mathcal{L}}$, we let $\mathrm{Var}(\Sigma)$ denote the set of variables occurring in Σ, shortening $\mathrm{Var}(\{\varphi\})$ to $\mathrm{Var}(\varphi)$. We also denote the formula algebra of $\mathcal{L}$ by $\mathbf{Fm}_{\mathcal{L}}$ and recall that a substitution for $\mathcal{L}$ is a homomorphism $\sigma : \mathbf{Fm}_{\mathcal{L}} \to \mathbf{Fm}_{\mathcal{L}}$.

A *substitution-invariant consequence relation* over $\mathcal{L}$ is a set $\vdash_{\mathrm{L}} \subseteq \mathcal{P}(\mathrm{Fm}_{\mathcal{L}}) \times \mathrm{Fm}_{\mathcal{L}}$ that satisfies the following conditions for all $\Sigma \cup \Sigma' \cup \{\varphi\} \subseteq \mathrm{Fm}_{\mathcal{L}}$ (writing $\Sigma \vdash_{\mathrm{L}} \varphi$ to denote $\langle \Sigma, \varphi \rangle \in \vdash_{\mathrm{L}}$):

(i) if $\varphi \in \Sigma$, then $\Sigma \vdash_{\mathrm{L}} \varphi$ *(reflexivity)*;
(ii) if $\Sigma \vdash_{\mathrm{L}} \varphi$ and $\Sigma \subseteq \Sigma'$, then $\Sigma' \vdash_{\mathrm{L}} \varphi$ *(monotonicity)*;
(iii) if $\Sigma \vdash_{\mathrm{L}} \varphi$ and $\Sigma' \vdash_{\mathrm{L}} \psi$ for every $\psi \in \Sigma$, then $\Sigma' \vdash_{\mathrm{L}} \varphi$ *(cut)*;
(iv) if $\Sigma \vdash_{\mathrm{L}} \varphi$, then $\sigma[\Sigma] \vdash_{\mathrm{L}} \sigma(\varphi)$ for any substitution σ for $\mathcal{L}$ *(substitution-invariance)*.

If also $\Sigma \vdash_{\mathrm{L}} \varphi$ implies $\Sigma' \vdash_{\mathrm{L}} \varphi$ for some finite $\Sigma' \subseteq \Sigma$, then $\vdash_{\mathrm{L}}$ is called *finitary*.

An ordered pair $\mathrm{L} = \langle \mathcal{L}, \vdash_{\mathrm{L}} \rangle$, where $\vdash_{\mathrm{L}}$ is a substitution-invariant consequence relation over a propositional language $\mathcal{L}$, is called a *logic* over $\mathcal{L}$. We call another logic L' an *extension* of L if $\vdash_{\mathrm{L}} \subseteq \vdash_{\mathrm{L}'}$. Given $X \subseteq \mathrm{Fm}_{\mathcal{L}}$, we also call the smallest extension of L that includes X an *axiomatic extension* of L and denote it by $\mathrm{L} \oplus X$.

Let us consider a propositional language $\mathcal{L}_m$ with binary connectives $\to, \cdot$ and constants $1, 0$, defining $\neg\varphi := \varphi \to 0$, $\varphi + \psi := \neg\varphi \to \psi$, and, inductively, $0\varphi := 0$, $\varphi^0 := 1$, $(n+1)\varphi = n\varphi + \varphi$, and $\varphi^{n+1} = \varphi^n \cdot \varphi$ for $n \in \mathbb{N}$. We shorten $\mathrm{Fm}_{\mathcal{L}_m}$ to Fm_m and call its members *multiplicative* formulas. *Multiplicative linear logic* MLL over $\mathcal{L}_m$ can be defined via derivability in the axiom system:

$(\varphi \to \psi) \to ((\psi \to \chi) \to (\varphi \to \chi))$	$(\varphi \to (\psi \to \chi)) \to ((\varphi \cdot \psi) \to \chi)$
$(\varphi \to (\psi \to \chi)) \to (\psi \to (\varphi \to \chi))$	$\varphi \to (\psi \to (\varphi \cdot \psi))$
$\varphi \to \varphi$	$\varphi \to (1 \to \varphi)$
$\neg\neg\varphi \to \varphi$	1

$$\frac{\varphi \quad \varphi \to \psi}{\psi}\ (\mathrm{mp})$$

The following useful deduction theorem is proved by an easy induction on the height of a derivation in an axiomatic extension of MLL.

Lemma 5.2.1 (cf. Avron 1988) *Let* L *be an axiomatic extension of* MLL. *Then for any* $\Sigma \cup \{\varphi, \psi\} \subseteq \mathrm{Fm}_m$,

$$\Sigma \cup \{\varphi\} \vdash_{\mathrm{L}} \psi \iff \Sigma \vdash_{\mathrm{L}} \varphi^n \to \psi \text{ for some } n \in \mathbb{N}.$$

It will also be useful later to consider the logic $\mathrm{MLL}^{\mathrm{u}}$ defined over $\mathcal{L}_m$ by the axiom system for MLL extended with the "unperforated" rule schema

$$\frac{n\varphi}{\varphi}(\mathrm{u}_n) \quad (n \in \mathbb{N}^+).$$

Let $\mathcal{L}_\ell$ be the propositional language with connectives $\wedge, \vee, \cdot, \to, 1$, and 0, shortening $\mathrm{Fm}_{\mathcal{L}_\ell}$ to Fm_ℓ. *Multiplicative additive linear logic without additive constants* MALL^- over $\mathcal{L}_\ell$ can be defined via the axiom system for MLL extended with the axiom and rule schema

$$\begin{array}{ll} (\varphi \wedge \psi) \to \varphi & \varphi \to (\varphi \vee \psi) \\ (\varphi \wedge \psi) \to \psi & \psi \to (\varphi \vee \psi) \\ ((\varphi \to \psi) \wedge (\varphi \to \chi)) \to (\varphi \to (\psi \wedge \chi)) & ((\varphi \to \chi) \wedge (\psi \to \chi)) \to ((\varphi \vee \psi) \to \chi) \end{array}$$

$$\frac{\varphi \quad \psi}{\varphi \wedge \psi}\ (\mathrm{adj})$$

Appropriate algebraic semantics for MALL^- and other substructural logics are provided by classes of residuated lattices (Blount and Tsinakis 2003; Jipsen and Tsinakis 2002; Galatos et al. 2007). An *involutive commutative residuated lattice* is an algebraic structure $\mathbf{A} = \langle A, \wedge, \vee, \cdot, \to, 1, 0\rangle$ such that $\langle A, \wedge, \vee\rangle$ is a lattice (where $a \leq b :\iff a \wedge b = a$), $\langle A, \cdot, 1\rangle$ is a monoid, $\neg\neg a = a$ for all $a \in A$, and $\to$ is the residual of $\cdot$, i.e., $b \leq a \to c \iff a \cdot b \leq c$ for all $a, b, c \in A$. It is easily shown (see, e.g., Galatos et al. 2007) that the class of all involutive commutative residuated lattices can be defined by equations and hence forms a variety that we denote by $\mathcal{I}n\mathcal{CRL}$.

Let $\mathcal{K}$ be any class of involutive commutative residuated lattices. We define for $\Sigma \cup \{\varphi\} \subseteq \mathrm{Fm}_\ell$,

$$\Sigma \models_{\mathcal{K}} \varphi \quad :\iff \quad \begin{array}{c} \text{for any } \mathbf{A} \in \mathcal{K} \text{ and homomorphism } e\colon \mathbf{Fm}_\ell \to \mathbf{A}, \\ 1 \leq e(\psi) \text{ for all } \psi \in \Sigma \implies 1 \leq e(\varphi). \end{array}$$

It is easily checked that $\models_{\mathcal{K}}$ is a substitution-invariant consequence relation over $\mathcal{L}_\ell$; moreover, if $\mathcal{K}$ is a variety (equational class), this consequence relation will be finitary (see, e.g., Metcalfe et al. 2014).

For any logic $\mathrm{L} = \mathrm{MALL}^- \oplus \mathcal{A}$ for some $\mathcal{A} \subseteq \mathrm{Fm}_\ell$, we obtain a variety

$$\mathcal{V}_{\mathrm{L}} := \{\mathbf{A} \in \mathcal{I}n\mathcal{CRL} \mid \models_{\mathbf{A}} \psi \text{ for all } \psi \in \mathcal{A}\}.$$

The following algebraic completeness theorem is then standard.

Proposition 5.2.2 (cf. Galatos et al. 2007) *If* $\mathrm{L} = \mathrm{MALL}^- \oplus \mathcal{A}$ *for some* $\mathcal{A} \subseteq \mathrm{Fm}_\ell$, *then for all* $\Sigma \cup \{\varphi\} \subseteq \mathrm{Fm}_\ell$,

$$\Sigma \vdash_{\mathrm{L}} \varphi \iff \Sigma \models_{\mathcal{V}_{\mathrm{L}}} \varphi.$$

The *multiplicative fragment* of an extension L of MALL^- is the logic $\mathrm{L_m}$ defined over $\mathcal{L}_m$ with $\vdash_{\mathrm{L_m}} := \vdash_{\mathrm{L}} \cap (\mathcal{P}(\mathrm{Fm}_m) \times \mathrm{Fm}_m)$. In order to reduce consequence in L to consequence in $\mathrm{L_m}$, we require distributivity properties that are satisfied, in particular, when L is complete with respect to a class of totally ordered algebras. We therefore consider *involutive uninorm logic without additive constants* (see Metcalfe and Montagna 2007), which may be defined as

$$\mathrm{IUL}^- := \mathrm{MALL}^- \oplus \{((p \to q) \wedge 1) \vee ((q \to p) \wedge 1)\}.$$

For any variety $\mathcal{V}$ of involutive commutative residuated lattices, let us denote the class of totally ordered members of $\mathcal{V}$ by $\mathcal{V}^c$. For axiomatic extensions of IUL^-, we obtain the following more specialized completeness result.

Proposition 5.2.3 (cf. Metcalfe and Montagna 2007; Galatos et al. 2007) *If* $\mathrm{L} = \mathrm{IUL}^- \oplus \mathcal{A}$ *for some* $\mathcal{A} \subseteq \mathrm{Fm}_\ell$, *then for all* $\Sigma \cup \{\varphi\} \subseteq \mathrm{Fm}_\ell$,

$$\Sigma \vdash_{\mathrm{L}} \varphi \iff \Sigma \models_{\mathcal{V}^c_{\mathrm{L}}} \varphi.$$

Using the previous proposition, it is straightforward to prove that in any axiomatic extension L of IUL^-, each formula $\varphi \in \mathrm{Fm}_\ell$ is equivalent both to a conjunction of disjunctions of multiplicative formulas and a disjunction of conjunctions of multiplicative formulas. It is also straightforward to establish the following equivalences:

$$\begin{aligned} \Sigma \vdash_{\mathrm{L}} \varphi_1 \wedge \varphi_2 &\iff \Sigma \vdash_{\mathrm{L}} \varphi_1 \text{ and } \Sigma \vdash_{\mathrm{L}} \varphi_2 \\ \Sigma \cup \{\psi_1 \wedge \psi_2\} \vdash_{\mathrm{L}} \varphi &\iff \Sigma \cup \{\psi_1, \psi_2\} \vdash_{\mathrm{L}} \varphi \\ \Sigma \cup \{\psi_1 \vee \psi_2\} \vdash_{\mathrm{L}} \varphi &\iff \Sigma \cup \{\psi_1\} \vdash_{\mathrm{L}} \varphi \text{ and } \Sigma \cup \{\psi_2\} \vdash_{\mathrm{L}} \varphi. \end{aligned}$$

Consequence in L can therefore be reduced to consequences of the form $\Sigma \vdash_{\mathrm{L}} \varphi_1 \vee \ldots \vee \varphi_n$ where $\Sigma \cup \{\varphi_1, \ldots, \varphi_n\} \subseteq \mathrm{Fm}_m$.

The following axiomatic extensions of IUL^- will be of particular interest in this paper:

$$\begin{aligned} \mathrm{A} &:= \mathrm{IUL}^- \oplus \{(p \to p) \to 0, 0 \to 1\} \\ \mathrm{RM}^{\mathrm{t}} &:= \mathrm{IUL}^- \oplus \{p \to (p + p), (p + p) \to p\} \\ \mathrm{IUML}^- &:= \mathrm{RM}^{\mathrm{t}} \oplus \{1 \to 0\} \\ \mathrm{BIUL}^- &:= \mathrm{IUL}^- \oplus \{np \to p^n, p^n \to np \mid n \in \mathbb{N}\}. \end{aligned}$$

The varieties $\mathcal{V}_{\mathrm{A}}$, $\mathcal{V}_{\mathrm{RM^t}}$, and $\mathcal{V}_{\mathrm{IUML^-}}$ are term-equivalent to lattice-ordered abelian groups, Sugihara monoids, and odd Sugihara monoids, respectively, while $\mathcal{V}_{\mathrm{BIUL^-}}$ (where "B" stands for "balanced") consists of involutive commutative residuated lattices satisfying $x^n \approx nx$ for all $n \in \mathbb{N}$.

5.3 Theorems of Alternatives

We will say that an extension L of IUL^- admits a *theorem of alternatives* if for any multiplicative formulas $\Sigma \cup \{\varphi_1, \dots, \varphi_n\} \subseteq \mathrm{Fm}_m$,

$$\Sigma \vdash_{\mathrm{L}} \varphi_1 \vee \dots \vee \varphi_n \iff \Sigma \vdash_{\mathrm{L}} \lambda_1\varphi_1 + \dots + \lambda_n\varphi_n \text{ for some } \lambda_1, \dots, \lambda_n \in \mathbb{N} \text{ not all } 0.$$

Such logics must satisfy the law of excluded middle and a "mingle" axiom.

Lemma 5.3.1 *Let* L *be an extension of* IUL^- *that admits a theorem of alternatives. Then*

(i) $\vdash_{\mathrm{L}} p \vee \neg p$.
(ii) $\vdash_{\mathrm{L}} 0 \to 1$.

Proof For (i), it suffices to observe that $\vdash_{\mathrm{L}} \neg p + p$ and hence, by the theorem of alternatives for L, also $\vdash_{\mathrm{L}} p \vee \neg p$. For (ii), we note first that $\vdash_{\mathrm{IUL^-}} (\neg p + \neg p) \vee (p + p)$. Hence, by the theorem of alternatives for L, we obtain $\vdash_{\mathrm{L}} \lambda(\neg p + \neg p) + \mu(p + p)$ for some $\lambda, \mu \in \mathbb{N}$ not both 0. If $\mu \neq 0$, then substituting 1 for p yields $\vdash_{\mathrm{L}} \mu(1 + 1)$ and, by a further application of the theorem of alternatives, $\vdash_{\mathrm{L}} 1 + 1$. Similarly, if $\lambda \neq 0$, then substituting 0 for p yields $\vdash_{\mathrm{L}} \lambda(1 + 1)$ and, again by the theorem of alternatives, $\vdash_{\mathrm{L}} 1 + 1$. In both cases, it follows that $\vdash_{\mathrm{L}} 0 \to 1$. □

We therefore define $\mathrm{IUL}^\star = \mathrm{IUL}^- \oplus \{p \vee \neg p,\ 0 \to 1\}$ and note that one direction of the theorem of alternatives holds for any extension of this logic.

Lemma 5.3.2 *For any extension* L *of* $\mathrm{IUL}^\star$ *and* $\Sigma \cup \{\varphi_1, \dots, \varphi_n\} \subseteq \mathrm{Fm}_m$,

$$\Sigma \vdash_{\mathrm{L}} \lambda_1\varphi_1 + \dots + \lambda_n\varphi_n \text{ for some } \lambda_1, \dots, \lambda_n \in \mathbb{N} \text{ not all } 0 \implies \Sigma \vdash_{\mathrm{L}} \varphi_1 \vee \dots \vee \varphi_n.$$

Proof Observe that for any $\Sigma \cup \{\varphi, \psi, \chi\} \subseteq \mathrm{Fm}_\ell$, if $\Sigma \vdash_{\mathrm{L}} (\varphi + \psi) \vee \chi$, then $\Sigma \vdash_{\mathrm{L}} (\neg\varphi \to \psi) \vee \chi$ and, since $\vdash_{\mathrm{L}} \varphi \vee \neg\varphi$, also $\Sigma \vdash_{\mathrm{L}} \varphi \vee \psi \vee \chi$. The claim follows by a simple inductive argument. □

We now establish a sufficient condition for extensions of $\mathrm{IUL}^\star$ that are axiomatized via additional multiplicative formulas by considering corresponding axiomatic extensions of MLL. Since $0 \to 1$ is derivable in the multiplicative fragment of any extension of $\mathrm{IUL}^\star$, we let $\mathrm{MLL}_0 := \mathrm{MLL} \oplus \{0 \to 1\}$ and $\mathrm{MLL}_0^{\mathrm{u}} := \mathrm{MLL}^{\mathrm{u}} \oplus \{0 \to 1\}$.

Lemma 5.3.3 *Let* $\mathcal{A} \subseteq \mathrm{Fm}_m$ *and suppose that for any* $\Sigma \cup \{\varphi\} \subseteq \mathrm{Fm}_m$,

$$\Sigma \vdash_{\mathrm{IUL}^\star \oplus \mathcal{A}} \varphi \iff \Sigma \vdash_{\mathrm{MLL}_0 \oplus \mathcal{A}} \lambda\varphi \textit{ for some } \lambda \in \mathbb{N}^+.$$

Then IUL$^\star \oplus \mathcal{A}$ *admits a theorem of alternatives and its multiplicative fragment is* $\mathrm{MLL}_0^{\mathrm{u}} \oplus \mathcal{A}$.

Proof By Lemma 5.3.2, it suffices to prove the left-to-right direction of the theorem of alternatives for IUL$^\star \oplus \mathcal{A}$. Suppose that $\Sigma \vdash_{\mathrm{IUL}^\star \oplus \mathcal{A}} \varphi_1 \vee \ldots \vee \varphi_n$ for some $\Sigma \cup \{\varphi_1, \ldots, \varphi_n\} \subseteq \mathrm{Fm}_m$ and let p be a variable such that $p \notin \mathrm{Var}(\Sigma \cup \{\varphi_1, \ldots, \varphi_n\})$. Observe that

$$\{\varphi_1 \to p, \ldots, \varphi_n \to p, \varphi_1 \vee \ldots \vee \varphi_n\} \vdash_{\mathrm{IUL}^\star \oplus \mathcal{A}} p$$

and hence

$$\Sigma \cup \{\varphi_1 \to p, \ldots, \varphi_n \to p\} \vdash_{\mathrm{IUL}^\star \oplus \mathcal{A}} p.$$

By assumption, there exists $\lambda \in \mathbb{N}^+$ such that

$$\Sigma \cup \{\varphi_1 \to p, \ldots, \varphi_n \to p\} \vdash_{\mathrm{MLL}_0 \oplus \mathcal{A}} \lambda p.$$

But then, using Lemma 5.2.1, there exist $\lambda_1, \ldots, \lambda_n \in \mathbb{N}$ such that

$$\Sigma \vdash_{\mathrm{MLL}_0 \oplus \mathcal{A}} (\varphi_1 \to p)^{\lambda_1} \to \ldots \to (\varphi_n \to p)^{\lambda_n} \to \lambda p.$$

If all the $\lambda_1, \ldots, \lambda_n$ are 0, then we can substitute p with φ_1 and obtain $\Sigma \vdash_{\mathrm{MLL}_0 \oplus \mathcal{A}} \lambda\varphi_1$. Otherwise, we substitute p with 0 and obtain

$$\Sigma \vdash_{\mathrm{MLL}_0 \oplus \mathcal{A}} \lambda_1\varphi_1 + \cdots + \lambda_n\varphi_n.$$

So clearly also $\Sigma \vdash_{\mathrm{IUL}^\star \oplus \mathcal{A}} \lambda_1\varphi_1 + \cdots + \lambda_n\varphi_n$.

Finally, it follows directly from the assumption and the fact that $\{\lambda\varphi\} \vdash_{\mathrm{IUL}^\star} \varphi$ that $\mathrm{MLL}_0^{\mathrm{u}} \oplus \mathcal{A}$ is the multiplicative fragment of IUL$^\star \oplus \mathcal{A}$. □

We are now able to formulate a sufficient condition for admitting a theorem of alternatives for logics axiomatized relative to IUL* by multiplicative formulas.

Theorem 5.3.4 *Let* $\mathcal{A} \subseteq \mathrm{Fm}_m$ *and suppose that for some* $n_0 \in \mathbb{N}$, *whenever* $n \geq n_0$, *there exist* $m \in \mathbb{N}^+, k \in \mathbb{N}$ *such that* $\vdash_{\mathrm{MLL}_0 \oplus \mathcal{A}} (np)^k \to m(p^n)$. *Then* IUL$^\star \oplus \mathcal{A}$ *admits a theorem of alternatives and its multiplicative fragment is* $\mathrm{MLL}_0^{\mathrm{u}} \oplus \mathcal{A}$.

Proof Assume that for some $n_0 \in \mathbb{N}$, whenever $n \geq n_0$, there exist $m \in \mathbb{N}^+, k \in \mathbb{N}$ such that $\vdash_{\mathrm{IUL}^\star \oplus \mathcal{A}} (np)^k \to m(p^n)$. By Lemma 5.3.3, to show that IUL$^\star \oplus \mathcal{A}$ admits a theorem of alternatives and its multiplicative fragment is $\mathrm{MLL}_0^{\mathrm{u}} \oplus \mathcal{A}$, it suffices to prove that for any $\Sigma \cup \{\varphi\} \subseteq \mathrm{Fm}_m$,

$$\Sigma \vdash_{\mathrm{IUL}^\star \oplus \mathcal{A}} \varphi \iff \Sigma \vdash_{\mathrm{MLL}_0 \oplus \mathcal{A}} \lambda\varphi \text{ for some } \lambda \in \mathbb{N}^+.$$

Suppose first that $\Sigma \vdash_{\mathrm{MLL}_0 \oplus \mathcal{A}} \lambda\varphi$ for some $\lambda \in \mathbb{N}^+$. Then also $\Sigma \vdash_{\mathrm{IUL}^\star \oplus \mathcal{A}} \lambda\varphi$ and, since $\{\lambda\varphi\} \vdash_{\mathrm{IUL}^\star \oplus \mathcal{A}} \varphi$, it follows that $\Sigma \vdash_{\mathrm{IUL}^\star \oplus \mathcal{A}} \varphi$. To prove the converse, we assume contrapositively that $\Sigma_0 := \Sigma$ satisfies

$$(\star)\ \ \Sigma_0 \nvdash_{\mathrm{MLL}_0 \oplus \mathcal{A}} \lambda\varphi \ \text{ for all } \lambda \in \mathbb{N}^+.$$

We enumerate Fm_m as $(\psi_i)_{i \in \mathbb{N}}$. Suppose now that Σ_N for some $N \in \mathbb{N}$ contains ψ_i or $\neg\psi_i$ for all $i < N$ and satisfies $(\star)$. Consider ψ_N and suppose for a contradiction that for some $\lambda, \mu \in \mathbb{N}^+$,

$$\Sigma_N \cup \{\psi_N\} \vdash_{\mathrm{MLL}_0 \oplus \mathcal{A}} \lambda\varphi \ \text{ and } \ \Sigma_N \cup \{\neg\psi_N\} \vdash_{\mathrm{MLL}_0 \oplus \mathcal{A}} \mu\varphi.$$

By Lemma 5.2.1, there exist $r, s \in \mathbb{N}$ such that

$$\Sigma_N \vdash_{\mathrm{MLL}_0 \oplus \mathcal{A}} (\psi_N)^r \to \lambda\varphi \ \text{ and } \ \Sigma_N \vdash_{\mathrm{MLL}_0 \oplus \mathcal{A}} (\neg\psi_N)^s \to \mu\varphi.$$

Using the fact that $\vdash_{\mathrm{MLL}_0} 0 \to 1$, it follows easily that also

$$\text{(i) } \Sigma_N \vdash_{\mathrm{MLL}_0 \oplus \mathcal{A}} (\psi_N)^{rs} \to s\lambda\varphi \ \text{ and (ii) } \ \Sigma_N \vdash_{\mathrm{MLL}_0 \oplus \mathcal{A}} (\neg\psi_N)^{rs} \to r\mu\varphi,$$

where (ii) can be rewritten, more conveniently, as

$$\text{(ii')}\ \ \Sigma_N \vdash_{\mathrm{MLL}_0 \oplus \mathcal{A}} rs\psi_N + r\mu\varphi.$$

By assumption, with $n = rst$ for some large $t \in \mathbb{N}^+$, there exist $m \in \mathbb{N}^+$, $k \in \mathbb{N}$ such that

$$\vdash_{\mathrm{MLL}_0 \oplus \mathcal{A}} (rst(\psi_N))^k \to m(\psi_N)^{rst}.$$

Observe also that, using (i) and (ii'),

$$\Sigma_N \vdash_{\mathrm{MLL}_0 \oplus \mathcal{A}} m(\psi_N)^{rst} \to mst\lambda\varphi \ \text{ and } \ \Sigma_N \vdash_{\mathrm{MLL}_0 \oplus \mathcal{A}} (rst)\psi_N + (rt\mu)\varphi.$$

Hence $\Sigma_N \vdash_{\mathrm{MLL}_0 \oplus \mathcal{A}} (rst(\psi_N))^k \to mst\lambda\varphi$, yielding

$$\Sigma_N \vdash_{\mathrm{MLL}_0 \oplus \mathcal{A}} (mst\lambda + krt\mu)\varphi,$$

which contradicts the assumption that Σ_N satisfies $(\star)$. So Σ_N can be extended with either ψ_N or $\neg\psi_N$ to obtain $\Sigma_{N+1} \subseteq \mathrm{Fm}_m$ that satisfies $(\star)$. We then let

$$\Sigma^* := \bigcup_{i \in \mathbb{N}} \Sigma_i,$$

noting that, by a simple argument using the fact that $\vdash_{\mathrm{MLL}_0 \oplus \mathcal{A}}$ is finitary, Σ^* also satisfies $(\star)$.

Next, we define a binary relation Θ on Fm_m by

$$\psi \,\Theta\, \chi \;:\Longleftrightarrow\; \Sigma^* \vdash_{\mathrm{MLL_0} \oplus \mathcal{A}} \psi \to \chi \text{ and } \Sigma^* \vdash_{\mathrm{MLL_0} \oplus \mathcal{A}} \chi \to \psi.$$

It is then straightforward to show that Θ is, in fact, a congruence on $\mathbf{Fm_m}$ and hence that the set of equivalence classes $\mathrm{Fm}_m/\Theta = \{[\psi] \mid \psi \in \mathrm{Fm}_m\}$, where $[\psi] = \{\chi \in \mathrm{Fm}_m \mid \psi \,\Theta\, \chi\}$ can be equipped with well-defined binary operations $\cdot$ and $\to$ and constants $[1]$ and $[0]$. Now define also

$$[\psi] \leq [\chi] \;:\Longleftrightarrow\; \Sigma^* \vdash_{\mathrm{MLL_0} \oplus \mathcal{A}} \psi \to \chi.$$

This is a total order by construction and hence we can equip $\langle \mathrm{Fm}_m/\Theta, \cdot, \to, [1], [0] \rangle$ also with meet and join operations $\wedge$ and $\vee$. It is then straightforward to show that the resulting algebra belongs to $\mathcal{I}n\mathcal{CRL}$ and satisfies each member of $\mathcal{A} \cup \{p \vee \neg p, 0 \to 1\}$. Finally, we consider a homomorphism e mapping each formula χ to its equivalence class $[\chi]$, obtaining $[1] \leq e(\psi)$ for all $\psi \in \Sigma^\star$ and $[1] \not\leq e(\varphi)$. Hence $\Sigma \nvdash_{\mathrm{IUL}^\star \oplus \mathcal{A}} \varphi$ as required. □

Clearly, the logics BIUL$^-$, A, RM^{t}, and IUML$^-$ defined in Sect. 5.2 satisfy the condition of the previous theorem and admit a theorem of alternatives. More generally, we obtain the following result for extensions of BIUL$^-$.

Corollary 5.3.5 *For any $\mathcal{A} \subseteq \mathrm{Fm}_m$, the logic* BIUL$^-$ $\oplus\, \mathcal{A}$ *admits a theorem of alternatives and its multiplicative fragment is* $\mathrm{MLL}_0^{\mathrm{u}} \oplus \{np \to p^n, p^n \to np \mid n \in \mathbb{N}\} \oplus \mathcal{A}$.

Note that in the statement of Theorem 5.3.4, the $m \in \mathbb{N}^+, k \in \mathbb{N}$ satisfying the condition $\vdash_{\mathrm{MLL_0} \oplus \mathcal{A}} (np)^k \to m(p^n)$ depend, in general, on the particular $n \geq n_0$. If the logic proves a knotted axiom of the form $p^t \to p^{t+1}$ $(t \in \mathbb{N}^+)$, however, these parameters can be fixed.

Corollary 5.3.6 *For any $\mathcal{A} \subseteq \mathrm{Fm}_m$ and $k, m, r, s, t \in \mathbb{N}^+$, with $r, s \geq t$, the logic* $\mathrm{IUL}^\star \oplus \mathcal{A} \cup \{p^t \to p^{t+1}, (rp)^k \to m(p^s)\}$ *admits a theorem of alternatives and its multiplicative fragment is* $\mathrm{MLL}_0^{\mathrm{u}} \oplus \mathcal{A} \cup \{p^t \to p^{t+1}, (rp)^k \to m(p^s)\}$.

For example, the logic $\mathrm{IUL}^\star \oplus \{p^2 \to p^3, (4p)^5 \to 6(p^7)\}$ has a theorem of alternatives. More generally, for any $\mathcal{A} \subseteq \mathrm{Fm}_m$ and $t, u, r_0, k_0, m_0, s_0, \ldots, r_{u-1}, k_{u-1}, m_{u-1}, s_{u-1} \in \mathbb{N}^+$, where all the r_i and s_i are congruent to i modulo u and greater or equal to t, the logic $\mathrm{IUL}^\star \oplus \mathcal{A} \cup \{p^t \to p^{t+u}\} \cup \{(r_i p)^{k_i} \to m_i(p^{s_i}) \mid 0 \leq i < u\}$ admits a theorem of alternatives. To see this, we apply Theorem 5.3.4 with $n_0 = \max(\{r_i \mid i < u\} \cup \{s_i \mid i < u\})$. For each $n \geq n_0$, we let i be the remainder of dividing n by u and choose $k = k_i, m = m_i$. That $(np)^k \to mp^n$ is derivable in this logic can be shown by reasoning that in the corresponding variety of residuated lattices, $(np)^{k_i} \leq (r_i p)^{k_i} \leq m_i(p^{s_i}) \leq m_i(p^n)$. (The last inequality follows from repeated applications of $p^t \leq p^{t+u}$ and the first by repeated applications of $(t+u)p \leq tp$, which follows from $p^t \leq p^{t+u}$ and involutivity.)

Let us note that in the special cases of A, RM^{t}, and IUML$^-$, the theorem of alternatives can be established à la Avron (1987) using either the completeness of the logic and its multiplicative fragment with respect to certain algebras or a hypersequent calculus that admits cut elimination. However, in the case of BIUL^- and other logics covered by the above results, suitable algebras and hypersequent calculi are not available, so these methods cannot be followed. What can be said is that if an extension of $\mathrm{IUL}^\star$ admits a theorem of alternatives, then any analytic calculus for its multiplicative fragment can be extended to an analytic calculus for the full logic using versions of the mix and split rules.

5.4 Interpolation

A logic L over a propositional language $\mathcal{L}$ is said to have the *deductive interpolation property* if for any finite $\Sigma \cup \{\varphi\} \subseteq \mathrm{Fm}_\mathcal{L}$ satisfying $\Sigma \vdash_\mathrm{L} \varphi$, there exists $\Pi \subseteq \mathrm{Fm}_\mathcal{L}$ with $\mathrm{Var}(\Pi) \subseteq \mathrm{Var}(\Sigma) \cap \mathrm{Var}(\varphi)$ such that $\Pi \vdash_\mathrm{L} \varphi$ and $\Sigma \vdash_\mathrm{L} \psi$ for all $\psi \in \Pi$. It is easily shown (see, e.g., van Gool et al. 2017) that this is equivalent to the following condition:

(†) For any finite $\Sigma \subseteq \mathrm{Fm}_\mathcal{L}$ and $X \subseteq \mathrm{Var}(\Sigma)$, there exists $\Pi \subseteq \mathrm{Fm}_\mathcal{L}$ with $\mathrm{Var}(\Pi) \subseteq X$ such that for any $\varphi \in \mathrm{Fm}_\mathcal{L}$ satisfying $\mathrm{Var}(\Sigma) \cap \mathrm{Var}(\varphi) \subseteq X$,

$$\Sigma \vdash_\mathrm{L} \varphi \iff \Pi \vdash_\mathrm{L} \varphi.$$

If Π in (†) can always be finite, then L is said to have the *right uniform deductive interpolation property*. If Π can always be finite, but (†) is restricted to formulas $\varphi \in \mathrm{Fm}_\mathcal{L}$ with $\mathrm{Var}(\varphi) \subseteq X$, then L is said to be *coherent*. It is proved in Kowalski and Metcalfe (2019) that L has the right uniform deductive interpolation property if and only if it has the deductive interpolation property and is coherent.

Recall that the *multiplicative fragment* of an extension L of MALL^- is the logic $\mathrm{L_m}$ defined over $\mathcal{L}_m$ with consequence relation $\vdash_{\mathrm{L_m}} := \vdash_\mathrm{L} \cap (\mathcal{P}(\mathrm{Fm}_m) \times \mathrm{Fm}_m)$. We show now that an extension of IUL^- that admits a theorem of alternatives inherits deductive interpolation and coherence properties from its multiplicative fragment.

Theorem 5.4.1 *Let* L *be an extension of* IUL^- *that admits a theorem of alternatives.*

(a) *If* $\mathrm{L_m}$ *has the deductive interpolation property, then so does* L.
(b) *If* $\mathrm{L_m}$ *is coherent, then so is* L.
(c) *If* $\mathrm{L_m}$ *has the right uniform deductive interpolation property, then so does* L.

Proof Suppose for (a) that $\mathrm{L_m}$ has the deductive interpolation property. We consider first any finite $\Sigma \subseteq \mathrm{Fm}_m$ and $X \subseteq \mathrm{Var}(\Sigma)$. By assumption, there exists $\Pi \subseteq \mathrm{Fm}_m$ such that $\mathrm{Var}(\Pi) \subseteq X$ and for any $\varphi \in \mathrm{Fm}_m$ satisfying $\mathrm{Var}(\Sigma) \cap \mathrm{Var}(\varphi) \subseteq X$,

$$\Sigma \vdash_\mathrm{L} \varphi \iff \Pi \vdash_\mathrm{L} \varphi.$$

Hence also for any $\varphi_1, \ldots, \varphi_n \in \mathrm{Fm}_m$ satisfying $\mathrm{Var}(\Sigma) \cap \mathrm{Var}(\{\varphi_1, \ldots, \varphi_n\}) \subseteq X$, by the theorem of alternatives,

$$\begin{aligned}\Sigma \vdash_{\mathrm{L}} \varphi_1 \vee \ldots \vee \varphi_n &\iff \Sigma \vdash_{\mathrm{L}} \lambda_1\varphi_1 + \cdots + \lambda_n\varphi_n \text{ for some } \lambda_1, \ldots, \lambda_n \in \mathbb{N} \text{ not all } 0\\ &\iff \Pi \vdash_{\mathrm{L}} \lambda_1\varphi_1 + \cdots + \lambda_n\varphi_n \text{ for some } \lambda_1, \ldots, \lambda_n \in \mathbb{N} \text{ not all } 0\\ &\iff \Pi \vdash_{\mathrm{L}} \varphi_1 \vee \ldots \vee \varphi_n.\end{aligned}$$

Moreover, recalling that every $\varphi \in \mathrm{Fm}_\ell$ is equivalent in L to a conjunction of disjunctions of formulas in Fm_m and that for any $\Delta \cup \{\psi_1, \psi_2\} \subseteq \mathrm{Fm}_\ell$,

$$\Delta \vdash_{\mathrm{L}} \psi_1 \wedge \psi_2 \iff \Delta \vdash_{\mathrm{L}} \psi_1 \text{ and } \Delta \vdash_{\mathrm{L}} \psi_2,$$

it follows that for any $\varphi \in \mathrm{Fm}_\ell$ satisfying $\mathrm{Var}(\Sigma) \cap \mathrm{Var}(\varphi) \subseteq X$,

$$\Sigma \vdash_{\mathrm{L}} \varphi \iff \Pi \vdash_{\mathrm{L}} \varphi.$$

Now consider any finite $\Sigma \subseteq \mathrm{Fm}_\ell$ and $X \subseteq \mathrm{Var}(\Sigma)$. Since for any $\Delta \cup \{\psi_1, \psi_2\} \subseteq \mathrm{Fm}_\ell$,

$$\Delta \cup \{\psi_1 \wedge \psi_2\} \vdash_{\mathrm{L}} \varphi \iff \Delta \cup \{\psi_1, \psi_2\} \vdash_{\mathrm{L}} \varphi,$$

we may assume that Σ consists of disjunctions of formulas in Fm_m. Suppose that $\Sigma = \Sigma' \cup \{\psi_1 \vee \psi_2\}$ and there exist $\Pi_1 \cup \Pi_2 \subseteq \mathrm{Fm}_\ell$ such that $\mathrm{Var}(\Pi_1 \cup \Pi_2) \subseteq X$ and for any $\varphi \in \mathrm{Fm}_\ell$ satisfying $\mathrm{Var}(\Sigma) \cap \mathrm{Var}(\varphi) \subseteq X$,

$$\Sigma' \cup \{\psi_1\} \vdash_{\mathrm{L}} \varphi \iff \Pi_1 \vdash_{\mathrm{L}} \varphi \quad \text{and} \quad \Sigma' \cup \{\psi_2\} \vdash_{\mathrm{L}} \varphi \iff \Pi_2 \vdash_{\mathrm{L}} \varphi.$$

We define

$$\Pi := \{(\chi_1 \wedge \ldots \wedge \chi_n) \vee (\chi_1' \wedge \ldots \wedge \chi_m') \mid \chi_1, \ldots, \chi_n \in \Pi_1,\ \chi_1', \ldots, \chi_m' \in \Pi_2\}.$$

Then $\mathrm{Var}(\Pi) \subseteq X$ and for any $\varphi \in \mathrm{Fm}_\ell$ satisfying $\mathrm{Var}(\Sigma) \cap \mathrm{Var}(\varphi) \subseteq X$,

$$\begin{aligned}\Sigma \vdash_{\mathrm{L}} \varphi &\iff \Sigma' \cup \{\psi_1 \vee \psi_2\} \vdash_{\mathrm{L}} \varphi\\ &\iff \Sigma' \cup \{\psi_1\} \vdash_{\mathrm{L}} \varphi \text{ and } \Sigma' \cup \{\psi_2\} \vdash_{\mathrm{L}} \varphi\\ &\iff \Pi_1 \vdash_{\mathrm{L}} \varphi \text{ and } \Pi_2 \vdash_{\mathrm{L}} \varphi\\ &\iff \Pi \vdash_{\mathrm{L}} \varphi.\end{aligned}$$

Hence, it follows by induction on the number of occurrences of $\vee$ in Σ that there exists $\Pi \subseteq \mathrm{Fm}_{\mathcal{L}}$ such that $\mathrm{Var}(\Pi) \subseteq X$ and for any $\varphi \in \mathrm{Fm}_{\mathcal{L}}$ satisfying $\mathrm{Var}(\Sigma) \cap \mathrm{Var}(\varphi) \subseteq X$,

$$\Sigma \vdash_{\mathrm{L}} \varphi \iff \Pi \vdash_{\mathrm{L}} \varphi.$$

For (b) and (c), we just note that if $\mathrm{L_m}$ is coherent or has the right uniform deductive interpolation property, then the construction described above will also yield a set Π that is finite and hence L will be coherent or have the right uniform deductive interpolation property, respectively. □

For example, deductive interpolation for the multiplicative fragment of $\mathrm{RM^t}$ can be established proof theoretically via a Maehara lemma argument for the sequent calculus for this fragment defined in Avron (1987). This yields a further proof of the fact that $\mathrm{RM^t}$ has the deductive interpolation property, first proved in Meyer (1980) (see also Avron 1986; Marchioni and Metcalfe 2012). Indeed, since the variety of Sugihara algebras corresponding to $\mathrm{RM^t}$ is locally finite, this logic is coherent and has the right uniform deductive interpolation property. Similarly, it can be shown semantically that the multiplicative fragment of A has the right uniform deductive interpolation property and hence that the same holds for the full logic (see Metcalfe et al. 2014 for a proof that proceeds along these lines without mentioning a theorem of alternatives).

5.5 Density

Recall from Sect. 5.2 that any axiomatic extension L of IUL^- is complete both with respect to a variety $\mathcal{V}_\mathrm{L}$ of involutive commutative residuated lattices and to the class $\mathcal{V}_\mathrm{L}^c$ of totally ordered members of $\mathcal{V}_\mathrm{L}$. In this section, we show that if L has a theorem of alternatives and $\vdash_\mathrm{L} 1 \to 0$, then L is also complete with respect to the class $\mathcal{V}_\mathrm{L}^d$ of *dense* totally ordered members of $\mathcal{V}_\mathrm{L}$. From an algebraic perspective, such a completeness result corresponds to $\mathcal{V}_\mathrm{L}$ being generated as a generalized quasivariety by the class $\mathcal{V}_\mathrm{L}^d$ (i.e., $\mathcal{V}_\mathrm{L} = \mathbb{ISP}(\mathcal{V}_\mathrm{L}^d)$) or, equivalently, the property that each member of $\mathcal{V}_\mathrm{L}^c$ embeds into a member of $\mathcal{V}_\mathrm{L}^d$ (see Metcalfe and Tsinakis 2017 for details).

Let us say that an extension L of IUL^- is *dense chain complete* if for any $\Sigma \cup \{\varphi\} \subseteq \mathrm{Fm}_\ell$,

$$\Sigma \vdash_\mathrm{L} \varphi \iff \Sigma \models_{\mathcal{V}_\mathrm{L}^d} \varphi.$$

If L has a theorem of alternatives and is dense chain complete, then $\vdash_\mathrm{L} 1 \to 0$. Just consider any $\mathbf{A} \in \mathcal{V}_\mathrm{L}^d$ and observe that $\vdash_\mathrm{L} (1 \to x) \vee (x \to 0)$ by part (i) of Lemma 5.3.1, so $1 \leq x$ or $x \leq 0$ for all $x \in A$, and, since $\mathbf{A}$ is dense, $1 = 0$ and, by dense chain completeness, $\vdash_\mathrm{L} 1 \to 0$. It follows, for example, that $\mathrm{RM^t}$ is not dense chain complete, although, as shown below (or see Metcalfe and Montagna 2007), $\mathrm{IUML}^- = \mathrm{RM^t} \oplus 1 \to 0$ does have this property.

One method for establishing dense chain completeness for a logic is to establish the admissibility of a certain "density rule" (see Metcalfe and Montagna 2007; Ciabattoni and Metcalfe 2008). We say that an extension L of IUL^- has the *density property* if for any $\Sigma \cup \{\varphi, \psi, \chi\} \subseteq \mathrm{Fm}_\ell$ and $p \notin \mathrm{Var}(\Sigma \cup \{\varphi, \psi, \chi\})$,

$$\Sigma \vdash_\mathrm{L} (\varphi \to p) \vee (p \to \psi) \vee \chi \iff \Sigma \vdash_\mathrm{L} (\varphi \to \psi) \vee \chi.$$

We make use of the following result, proved in a more general setting in Metcalfe and Montagna (2007) (see also Ciabattoni and Metcalfe 2008; Baldi and Terui 2016; Metcalfe and Tsinakis 2017; Galatos and Horčík 2021).

Theorem 5.5.1 (Metcalfe and Montagna 2007) *Any extension of* IUL$^-$ *that has the density property is dense chain complete.*

Theorem 5.5.2 *Any extension of* IUL$^- \oplus \{1 \to 0\}$ *that admits a theorem of alternatives is dense chain complete.*

Proof Let L be an extension of IUL$^- \oplus \{1 \to 0\}$ that admits a theorem of alternatives. By Theorem 5.5.1, it suffices to prove that L has the density property. Suppose first that for some $\Sigma \cup \{\varphi, \psi, \chi\} \subseteq \mathrm{Fm}_m$ and $p \notin \mathrm{Var}(\Sigma \cup \{\varphi, \psi, \chi\})$,

$$\Sigma \vdash_{\mathrm{L}} (\varphi \to p) \vee (p \to \psi) \vee \chi.$$

Since L admits a theorem of alternatives, there exist $\lambda, \mu, \gamma \in \mathbb{N}$ not all 0 such that

$$\Sigma \vdash_{\mathrm{L}} \lambda(\varphi \to p) + \mu(p \to \psi) + \gamma\chi.$$

Substituting p with 0, and separately all other variables with 0, yields the consequences

$$\Sigma \vdash_{\mathrm{L}} \lambda(\neg\varphi) + \mu\psi + \gamma\chi \text{ and } \vdash_{\mathrm{L}} \lambda p + \mu(\neg p).$$

Multiplying the conclusion in the first consequence by λ and substituting p with φ^λ in the second consequence produce the consequences

$$\Sigma \vdash_{\mathrm{L}} \lambda(\varphi^\lambda) \to (\lambda\mu\psi + \lambda\gamma\chi) \text{ and } \vdash_{\mathrm{L}} \varphi^{\lambda\mu} \to \lambda(\varphi^\lambda).$$

By transitivity of implication, we obtain

$$\Sigma \vdash_{\mathrm{L}} \varphi^{\lambda\mu} \to (\lambda\mu\psi + \lambda\gamma\chi),$$

which can be rewritten as

$$\Sigma \vdash_{\mathrm{L}} \lambda\mu(\varphi \to \psi) + \lambda\gamma\chi.$$

By the theorem of alternatives again, $\Sigma \vdash_{\mathrm{L}} (\varphi \to \psi) \vee \chi$.

Now consider the more general case where $\Sigma \vdash_{\mathrm{L}} (\varphi \to p) \vee (p \to \psi) \vee \chi$ for some $\Sigma \cup \{\varphi, \psi, \chi\} \subseteq \mathrm{Fm}_\ell$ and $p \notin \mathrm{Var}(\Sigma \cup \{\varphi, \psi, \chi\})$. If $\varphi, \psi, \chi \in \mathrm{Fm}_m$, then using the equivalences presented in Sect. 5.2 and the multiplicative case just established, we obtain again $\Sigma \vdash_{\mathrm{L}} (\varphi \to \psi) \vee \chi$ as required. Otherwise, for $q, r, s \notin \{p\} \cup \mathrm{Var}(\Sigma \cup \{\varphi, \psi, \chi\})$, we obtain

$$\Sigma \cup \{q \to \varphi, \psi \to r, \chi \to s\} \vdash_{\mathrm{L}} (q \to p) \vee (p \to r) \vee s.$$

But then also using the equivalences presented in Sect. 5.2 and the multiplicative case just established,

$$\Sigma \cup \{q \to \varphi, \psi \to r, \chi \to s\} \vdash_{\mathrm{L}} (q \to r) \vee s,$$

and finally, substituting φ for q, ψ for r, and χ for s yields $\Sigma \vdash_{\mathrm{L}} (\varphi \to \psi) \vee \chi$. □

Corollary 5.5.3 *Let* $\mathcal{A} \subseteq \mathrm{Fm}_m$ *and* $\mathrm{L} = \mathrm{BIUL}^- \oplus \mathcal{A}$. *Then* L *is dense chain complete.*

Let us remark finally that dense chain completeness can be established for A and IUML^- via a direct semantic argument or proof theoretically using an analytic hypersequent calculus as described in Metcalfe and Montagna (2007), but these methods do not seem to be available for BIUL^- or other logics admitting a theorem of alternatives. It may be hoped also that this new approach provides a first step toward addressing the open standard completeness problem for involutive uninorm logic posed in Metcalfe and Montagna (2007), possibly by introducing a weaker theorem of alternatives property.[1]

References

Anderson, A. R., & Belnap, N. D. (1975). *Entailment* (Vol. 1). Princeton University Press.

Avron, A. (1986). On an implication connective of RM. *Notre Dame Journal of Formal Logic*, *27*, 201–209.

Avron, A. (1987). A constructive analysis of RM. *Journal of Symbolic Logic*, *52*(4), 939–951.

Avron, A. (1988). The semantics and proof theory of linear logic. *Theoretical Computer Science*, *57*(2–3), 161–184.

Baldi, P., & Terui, K. (2016). Densification of FL-chains via residuated frames. *Algebra Universalis*, *75*, 169–195.

Blount, K., & Tsinakis, C. (2003). The structure of residuated lattices. *International Journal of Algebra and Computation*, *13*(4), 437–461.

Casari, E. (1989). Comparative logics and abelian ℓ-groups. In C. Bonotto, R. Ferro, S. Valentini, & A. Zanardo (Eds.), *Logic Colloquium '88* (pp. 161–190). Elsevier.

Ciabattoni, A., & Metcalfe, G. (2008). Density elimination. *Theoretical Computer Science*, *403*, 328–346.

Colacito, A., & Metcalfe, G. (2017). Proof theory and ordered groups. In *Proceedings of WoLLIC 2017* (Vol. 10388, pp. 80–91). LNCS. Springer.

Dantzig, G. B. (1963). *Linear Programming and Extensions*. Princeton: Princeton University Press.

Dunn, J. M. (1970). Algebraic completeness for R-mingle and its extensions. *Journal of Symbolic Logic*, *35*, 1–13.

Galatos, N., & Horčík, R. (2021). Densification via polynomials, languages, and frames. Journal of Pure and Applied Algebra.

Galatos, N., Jipsen, P., Kowalski, T., & Ono, H. (2007). *Residuated Lattices: An Algebraic Glimpse at Substructural Logics*. Elsevier.

Jenei, S. (2018). Group representation, Hahn-type embedding, and the FSSC for IUL^{fp}. *Proceedings of LATD*, *2018*, 63–66.

Jenei, S., & Montagna, F. (2002). A proof of standard completeness for Esteva and Godo's MTL logic. *Studia Logica*, *70*(2), 183–192.

Jipsen, P., & Tsinakis, C. (2002). A survey of residuated lattices. In J. Martinez (Ed.), *Ordered Algebraic Structures* (pp. 19–56). xx: Kluwer.

[1] Note that a very different approach to tackling this problem, via representations of totally ordered involutive commutative residuated lattices using ordered groups, has been described recently in Jenei (2018).

Kowalski, T., & Metcalfe, G. (2019). Uniform interpolation and coherence. *Annals of Pure and Applied Logic*, *170*(7), 825–841.
Marchioni, E., & Metcalfe, G. (2012). Craig interpolation for semilinear substructural logics. *Mathematical Logic Quarterly*, *58*(6), 468–481.
Metcalfe, G., & Montagna, F. (2007). Substructural fuzzy logics. *Journal of Symbolic Logic*, *72*(3), 834–864.
Metcalfe, G., Montagna, F., & Tsinakis, C. (2014). Amalgamation and interpolation in ordered algebras. *Journal of Algebra*, *402*, 21–82.
Metcalfe, G., Olivetti, N., & Gabbay, D. (2005). Sequent and hypersequent calculi for abelian and Łukasiewicz logics. *ACM Transactions on Computational Logic*, *6*(3), 578–613.
Metcalfe, G., & Tsinakis, C. (2017). Density revisited. *Soft Computing*, *21*, 175–189.
Meyer, R. K. (1980). Relevantly interpolating in RM. Technical Report 9, Logic Group, RSSS, Australian National University.
Meyer, R. K., & Slaney, J. K. (1989). Abelian logic from A to Z. In *Paraconsistent Logic: Essays on the Inconsistent* (pp. 245–288). Philosophia Verlag.
van Gool, S., Metcalfe, G., & Tsinakis, C. (2017). Uniform interpolation and compact congruences. *Annals of Pure and Applied Logic*, *168*, 1827–1948.

Almudena Colacito is a Ph.D. student at the University of Bern, supervised by George Metcalfe. She graduated with a Master of Logic degree from the University of Amsterdam, writing a thesis on subminimal logics with weak negations supervised by Dick de Jongh and Marta Bílková. She has worked mostly on non-classical and substructural logics, focusing on their algebraic semantics.

Nikolaos Galatos is a Professor of Mathematics at the University of Denver. After receiving his Ph.D. from Vanderbilt University in 2003, he held a Postdoctoral appointment and then a Visiting Assistant Professor position at the Japan Advanced Institute of Science and Technology, as well as a Visiting Assistant Professor position at Vanderbilt University. He has been at the University of Denver since 2007. His research interests include universal algebra, ordered algebraic structures and their applications to logic. He has co-authored a book on these topics, titled "Residuated lattices: an algebraic glimpse at substructutral logics" published by Elsevier.

George Metcalfe is a Professor of Mathematics at the University of Bern. He obtained his Ph.D. from King's College London in 2004 and subsequently held a Marie Curie Fellowship at the Vienna University of Technology and a Postdoctoral Position at Vanderbilt University. His research focuses on interactions between universal algebra and proof theory, and he is an author of the book "Proof Theory for Fuzzy Logics" published in the Springer Applied Logic series.

Chapter 6
Degree-Preserving Gödel Logics with an Involution: Intermediate Logics and (Ideal) Paraconsistency

Marcelo E. Coniglio, Francesc Esteva, Joan Gispert, and Lluis Godo

Abstract In this paper, we study intermediate logics between the logic $G^{\leq}_{\sim}$, the degree-preserving companion of Gödel fuzzy logic with involution $G_{\sim}$, and classical propositional logic CPL, as well as the intermediate logics of their finite-valued counterparts $G^{\leq}_{n\sim}$. Although $G^{\leq}_{\sim}$ and $G^{\leq}_{n\sim}$ are explosive w.r.t. Gödel negation $\neg$, they are paraconsistent w.r.t. the involutive negation $\sim$. We introduce the notion of saturated paraconsistency, a weaker notion than ideal paraconsistency, and we fully characterize the ideal and the saturated paraconsistent logics between $G^{\leq}_{n\sim}$ and CPL. We also identify a large family of saturated paraconsistent logics in the family of intermediate logics for degree-preserving finite-valued Łukasiewicz logics.

6.1 Introduction

Contradictions frequently arise in scientific theories, as well as in philosophical argumentation. In computer science, techniques for dealing with contradictory information need to be developed in areas such as logic programming, belief revision, the

---

This paper is a humble tribute to our friend and colleague Arnon Avron and his outstanding contributions on nonclassical logics, proof theory, and foundations of mathematics.

M. E. Coniglio
Dept. of Philosophy - IFCH and Centre for Logic, Epistemology and the History of Science, University of Campinas, Campinas, Brazil
e-mail: coniglio@unicamp.br

F. Esteva · L. Godo (✉)
Artificial Intelligence Research Institute (IIIA) - CSIC, Barcelona, Spain
e-mail: godo@iiia.csic.es

F. Esteva
e-mail: esteva@iiia.csic.es

J. Gispert
Departament de Matemàtiques i Informàtica, Universitat de Barcelona, Barcelona, Spain
e-mail: jgispertb@ub.edu

O. Arieli and A. Zamansky (eds.), *Arnon Avron on Semantics and Proof Theory of Non-Classical Logics*, Outstanding Contributions to Logic 21,
https://doi.org/10.1007/978-3-030-71258-7_6

semantic web, and artificial intelligence in general. Since classical logic—as well as many other nonclassical logics– trivialize in the presence of inconsistencies, it can be useful to consider logical systems tolerant to contradictions in order to formalize such situations.

A logic L is said to be *paraconsistent* with respect to a negation connective $\neg$ when it contains a $\neg$-contradictory but not trivial theory. Assuming that L is (at least) Tarskian, this is equivalent to say that the $\neg$-explosion rule

$$\frac{\varphi \quad \neg\varphi}{\psi}$$

is not valid in L. The main challenge for paraconsistent logicians is defining logic systems in which not only a contradiction does not necessarily trivialize, but also allowing that useful conclusions can be derived from such inconsistent information.

The first systematic study of paraconsistency from the point of view of formal logic is due to da Costa, which introduces in 1963 a hierarchy of paraconsistent systems called C_n. This is why da Costa is considered one of the founders of paraconsistency. Under his perspective, propositions in a paraconsistent setting are "dubious" in the sense that, in general, a sentence and its negation can be hold simultaneously without trivialization. That is, it is possible to consider contradictory but non-trivial theories. Moreover, it is possible to express (in every system C_n) the fact that a given sentence φ has a classical behavior w.r.t. the explosion law. This approach to paraconsistency, in which the explosion law is recovered in a controlled way, was generalized by Carnielli and Marcos (2000) by means of the notion of *Logics of Formal Inconsistency* (**LFI**s, in short). An **LFI** is a paraconsistent logic (w.r.t. a negation $\neg$) having, in addition, an unary connective $\circ$ (a *consistency operator*), primitive or defined, such that any theory of the form $\{\varphi, \neg\varphi, \circ\varphi\}$ is trivial, despite $\{\varphi, \neg\varphi\}$ not being necessarily so. Of course, the main novelty with respect to da Costa's systems C_n is that the consistency operator (which corresponds to the well-behavior operator) can now be a primitive connective, which allows to consider a more general and expressive theory of paraconsistency. **LFI**s have been extensively studied since then (for general references, consult (Carnielli and Marcos 2000; Carnielli et al. 2007; Carnielli and Coniglio 2016)). Avron, together with his collaborators, has significantly contributed to the development of **LFI**s by introducing several new systems, besides the ones proposed in Carnielli and Marcos (2000), Carnielli et al. (2007), Carnielli and Coniglio (2016), and by providing simple, effective, and modular semantics based on non-deterministic matrices (N-matrices) as well as elegant Gentzen-style proof methods for **LFI**s, see, for instance, (Arieli and Avron 2017; Arieli et al. 2010, 2011a, b; Avron et al. 2018; Avron 2005, 2007, 2009, 2017, 2019; Avron et al. 2010; Avron and Lev 2001; Avron and Zamansky 2011).

According to da Costa, one of the main properties that a paraconsistent logic should have is being as close as possible to classical logic. That is, a paraconsistent logic should retain as much as possible the classical inferences, and still allowing to have non-trivial, contradictory theories. A natural way to formalize this *desideratum* is by means of the notion of maximality of a logic w.r.t. another one. A (Tarskian and structural) logic L_1 is said to be *maximal* w.r.t. another logic L_2 if both are defined over the same signature, the consequence relation of L_1 is contained in that of L_2 (i.e., L_2 is an extension of L_1) and, if φ is a theorem of L_2 which is not derivable in L_1, then the extension of L_1 obtained by adding φ (and all of its instances under uniform substitutions) as a theorem coincides with L_2. Hence, a "good" paraconsistent logic L should be maximal w.r.t. classical logic CPL (presented over the same signature as L). As observed in Coniglio et al. (2019), the notion of maximality can be vacuously satisfied when both logics (L_1 and L_2) have the same theorems.

In Arieli et al. (2010), Arieli, Avron, and Zamansky propose an interesting notion of maximality w.r.t. paraconsistency: a paraconsistent logic is *maximally paraconsistent* if no proper extension of it is paraconsistent. Thus, they prove that several well-known three-valued logics such as Sette's P1 and da Costa and D'Ottaviano's J_3 are maximally paraconsistent. Note that both P1 and J_3 are also maximal w.r.t. CPL.

These strong features satisfied by logics such as P1 and J_3 led Arieli, Avron, and Zamansky to introduce in Arieli et al. (2011b) the notion of ideal paraconsistent logics. Briefly, a logic L is called *ideal paraconsistent* when it is maximally paraconsistent and maximal w.r.t. to classical logic CPL (the formal definition of ideal paraconsistency will we recalled in Sect. 6.5). One interesting problem is to find ideal paraconsistent logics, and in this sense (Arieli et al. 2011b) provides a vast variety of examples of ideal paraconsistent finite-valued logics, aside from P1 and J_3.

As mentioned above, one of da Costa's requirements for defining reasonable paraconsistent logics is maximality w.r.t. CPL. Many paraconsistent logicians (probably including Avron and his collaborators) would agree with the relevance of this feature. However, this position is by no means uncontroversial. In Wansing and Odintsov (2016), Wansing and Odintsov extensively criticized that requirement. According to these authors, maximality w.r.t. classical logic is not a good choice. On the one hand, the phenomenon of paraconsistency should be interpreted from an informational perspective instead of considering epistemological or ontological terms. Indeed, the authors claim that "logic should avoid as many ontological commitments as possible."[1] To this end, they argue that, by definition, logic "is committed to the existence of languages but not necessarily to the existence of language users."[2] This means that, despite the models for logics cannot avoid linguistic entities, valid inferences should not refer to notions such as "knowledge," "belief states" of any other epistemic or doxastic subjects. Thus, it would be preferable to motivate a system of paraconsistent logic in terms of *information*, without appealing to epistemological or ontological commitments such as language users, epistemic subjects possessing mental states, etc. For instance, by considering that formulas in an inference process are pieces of information, the fact that in a paraconsistent logic $\{A, \neg A\}$ does not entail B can be read as "it is just not the case that $\{A, \neg A\}$ provides the information that B." The following are some excerpts from Wansing and Odintsov (2016):

---

[1] Wansing and Odintsov (2016), p. 179.

[2] Ibid., p. 180.

> "classical logic is not at all a natural reference logic for reasoning about information and information structures. On the other hand, it is reasoning about information that suggests paraconsistent reasoning."[3]

> "one may wonder why exactly a *nonclassical* paraconsistent logic, if correct, should have a distinguished status in virtue of being faithful to classical logic "as much as possible"."[4]

> "Paraconsistency does deviate from logical orthodoxy, but it is not at all clear that classical logic indeed is the logical orthodoxy from which paraconsistent logics ought to deviate only minimally."[5]

Although it could be argued against this emphatic perspective, it also seems that being maximal w.r.t. CPL should not be a necessary requirement for being an "ideal" (meaning "optimal") paraconsistent logic.[6] This is why we propose in this paper the notion of *saturated paraconsistent* logic, which is just a weakening of the concept of ideal paraconsistent logic, by dropping the requirement of maximality w.r.t. CPL. As we shall see along this paper, there are several interesting examples of saturated paraconsistent logics.

While paraconsistency deals with excessive or dubious information, fuzzy logics were designed for reasoning with imprecise information, in particular, for reasoning with propositions containing vague predicates. Given that both paradigms are able to deal with information—unreliable, in the case of paraconsistent logics, and imprecise, in the case of fuzzy logics—it seems reasonable to consider logics which combine both features, namely, paraconsistent fuzzy logic. The first steps along this way were taken in Ertola et al. (2015), where a consistency operator $\circ$ was defined in terms of the other connectives (for instance, by using the Monteiro-Baaz Δ-operator) in several fuzzy logics. In Coniglio et al. (2014), this approach was generalized to fuzzy **LFI**s extending the logic MTL of pre-linear (integral, commutative, bounded) residuated lattices, in which the consistency operator is primitive.

We have studied in different papers the paraconsistent logics arising from the family of mathematical fuzzy logics, see, e.g., Ertola et al. (2015); Coniglio et al. (2014, 2016, 2019). In particular, in Ertola et al. (2015), the authors observe that even though all truth-preserving fuzzy logics L are explosive, their degree-preserving companions $L^{\leq}$ (as introduced in Bou et al. (2009)) are paraconsistent in many cases. This provides a large family of paraconsistent fuzzy logics. In Coniglio et al. (2016), the authors studied the lattice of logics between the n-valued Łukasiewicz logics $Ł_n$ and their degree-preserving companions $Ł_n^{\leq}$. Although there are many paraconsistent logics for each n, no one of them is ideal. However, in Coniglio et al. (2019), the authors of this paper consider a wide class of logics between $Ł_n^{\leq}$ and CPL, and in that

[3] Ibid., p. 181.

[4] Ibid., p. 181.

[5] Ibid., p. 183.

[6] It is worth noting that, more recently, the authors have changed the terminology "ideal paraconsistent logic" in Arieli et al. (2011b) to "fully maximal and normal paraconsistent logic," e.g., in Avron et al. (2018). According to them, they choose the latter "to use a more neutral terminology" (see Avron et al. (2018, Footnote 9, p. 57)).

case they indeed find and axiomatically characterize a family of ideal paraconsistent logics.

In this paper, we study paraconsistent logics arising from Gödel fuzzy logic expanded with an involutive negation $G_{\sim}$, introduced in Esteva et al. (2000), as well as from its finite-valued extensions $G_{n\sim}$. It is well known that Gödel logic G coincides with its degree-preserving companion (since G has the deduction-detachment theorem), but this is not the case for $G_{\sim}$. In fact, $G_{\sim}$ and $G_{\sim}^{\leq}$ are different logics, and moreover, while $G_{\sim}^{\leq}$ is explosive w.r.t. Gödel negation $\neg$, it is paraconsistent w.r.t. the involutive negation $\sim$.[7] We also study the logics between $G_{n\sim}^{\leq}$ (the finite-valued Gödel logic with an involutive negation) and CPL, and we find that the ideal paraconsistent logics of this family are only the abovementioned three-valued logic J_3 and its four-valued version J_4, introduced in Coniglio et al. (2019). Moreover, we fully characterize the ideal and the saturated paraconsistent logics between $G_{n\sim}^{\leq}$ and CPL.

The paper is structured as follows. After this introduction, some basic definitions and known results to be used along the paper will be presented. In Sect. 6.3, we show that the logics between $G_{\sim}^{\leq}$ and CPL are defined by matrices over a $G_{\sim}$-algebra with lattice filters, and, in particular, we study the logics defined by matrices over $[0, 1]_{\sim}$ with order filters. In Sect. 6.4, we study the case of finite-valued Gödel logics with involution $G_{n\sim}$, and we observe that $G_{3\sim}$ and $G_{4\sim}$ coincide with $Ł_3$ and $Ł_4$ (the three- and four-valued Łukasiewicz logics) already studied in Coniglio et al. (2019). We prove that, in the general case, each finite $G_{n\sim}$-algebra is a direct product of subalgebras of $\mathbf{GV_{n\sim}}$, the Gödel chain of n elements with the unique involution $\sim$ one can define on it. This result allows us to characterize the logics between $G_{n\sim}^{\leq}$ and CPL. In Sect. 6.5, the definition of saturated paraconsistent logic is formally introduced, and it is proved that between $G_{n\sim}^{\leq}$ and CPL there are only three saturated paraconsistent logics: two of them (J_3 and J_4) are already known and are in fact ideal paraconsistent, and there is only one that is saturated but not ideal paraconsistent, which we call $J_3 \times J_4$. Finally, in Sect. 6.6, we return to the study of finite-valued Łukasiewicz logic and prove that in this framework there is a large family of saturated paraconsistent logics that are not ideal paraconsistent. Some concluding remarks are discussed in the final section.

[7] In fact, $G_{\sim}^{\leq}$ is then a *paradefinite* logic (w.r.t. $\sim$) in the sense of Arieli and Avron (2017), as it is both paraconsistent and paracomplete, since the law of excluded middle $\varphi \vee \sim\varphi$ fails, as in all fuzzy logics. Logics with a negation which is both paraconsistent and paracomplete were already considered in the literature under different names: *non-alethic* logics (da Costa) and *paranormal* logics (Beziau).

6.2 Preliminaries

6.2.1 Truth-Preserving Gödel Logics

This section is devoted to needed preliminaries on the Gödel fuzzy logic G, its axiomatic extensions, as well as their expansions with an involutive negation. We present their syntax and semantics, their main logical properties, and the notation we use throughout the article.

The language of Gödel propositional logic is built as usual from a countable set of propositional variables V, the constant $\bot$, and the binary connectives $\wedge$ and $\rightarrow$. Disjunction and negation are, respectively, defined as $\varphi \vee \psi := ((\varphi \rightarrow \psi) \rightarrow \psi) \wedge ((\psi \rightarrow \varphi) \rightarrow \varphi)$ and $\neg\varphi := \varphi \rightarrow \bot$, equivalence is defined as $\varphi \leftrightarrow \psi := (\varphi \rightarrow \psi) \wedge (\psi \rightarrow \varphi)$, and the constant $\top$ is taken as $\bot \rightarrow \bot$.

The following are the *axioms* of G[8]:

(A1) $(\varphi \rightarrow \psi) \rightarrow ((\psi \rightarrow \chi) \rightarrow (\varphi \rightarrow \chi))$
(A2) $(\varphi \wedge \psi) \rightarrow \varphi$
(A3) $(\varphi \wedge \psi) \rightarrow (\psi \wedge \varphi)$
(A4a) $(\varphi \rightarrow (\psi \rightarrow \chi)) \rightarrow ((\varphi \wedge \psi) \rightarrow \chi)$
(A4b) $((\varphi \wedge \psi) \rightarrow \chi) \rightarrow (\varphi \rightarrow (\psi \rightarrow \chi))$
(A5) $((\varphi \rightarrow \psi) \rightarrow \chi) \rightarrow (((\psi \rightarrow \varphi) \rightarrow \chi) \rightarrow \chi)$
(A6) $\bot \rightarrow \varphi$
(A7) $\varphi \rightarrow (\varphi \wedge \varphi)$.

The *deduction rule* of G is modus ponens.

As a many-valued logic, Gödel logic is the axiomatic extension of Hájek's basic fuzzy logic BL (Hájek 1998) (which is the logic of continuous t-norms and their residua) by means of the contraction axiom (A7).

Since the unique idempotent continuous t-norm is the minimum, this yields that Gödel logic is strongly complete with respect to its standard fuzzy semantics that interprets formulas over the structure $[0, 1]_G = ([0, 1], \min, \Rightarrow_G, 0, 1)$,[9] i.e., semantics defined by truth evaluations of formulas e on $[0, 1]$, where 1 is the only designated truth value, such that $e(\varphi \wedge \psi) = \min(e(\varphi), e(\psi))$, $e(\varphi \rightarrow \psi) = e(\varphi) \Rightarrow_G e(\psi)$, and $e(\bot) = 0$, where $\Rightarrow_G$ is the binary operation on $[0, 1]$ defined as

$$x \Rightarrow_G y = \begin{cases} 1, & \text{if } x \leq y \\ y, & \text{otherwise.} \end{cases}$$

[8] This axiomatization comes from adding axiom (A7) to the axioms of Hájek's BL logic (Hájek 1998). Later it was shown that axioms (A2) and (A3) were, in fact, redundant, see Běhounek (2011) for a detailed exposition and the references therein.

[9] Called *standard* Gödel algebra.

As a consequence, $e(\varphi \vee \psi) = \max(e(\varphi), e(\psi))$ and $e(\neg\varphi) = \neg_G e(\varphi) = e(\varphi) \Rightarrow_G 0$. By definition, $\Gamma \models_G \varphi$ iff, for every evaluation e over $[0, 1]_G$, if $e(\gamma) = 1$ for every $\gamma \in \Gamma$ then $e(\varphi) = 1$.

Gödel logic can also be seen as the axiomatic extension of intuitionistic propositional logic by the pre-linearity axiom

$$(\varphi \to \psi) \vee (\psi \to \varphi).$$

Its algebraic semantics is, therefore, given by the variety of pre-linear Heyting algebras, also known as Gödel algebras. A Gödel algebra is thus a (bounded, integral, commutative) residuated lattice $\mathbf{A} = (A, \wedge, \vee, *, \Rightarrow, 0, 1)$ such that the monoidal operation $*$ coincides with the lattice meet $\wedge$, and such that the pre-linearity equation

$$(x \Rightarrow y) \vee (y \Rightarrow x) = 1$$

is satisfied, where $x \vee y = ((x \Rightarrow y) \Rightarrow y) * ((y \Rightarrow x) \Rightarrow x))$. Gödel algebras are locally finite, i.e., given a Gödel algebra $\mathbf{A}$ and a finite set $F \subseteq A$, the Gödel subalgebra generated by F is finite as well.

It is also well known that the axiomatic extensions of Gödel logic correspond to its finite-valued counterparts. If we replace the unit interval [0, 1] by the truth-value set $GV_n = \{0, 1/(n-1), \ldots, (n-2)/(n-1), 1\}$ in the standard Gödel algebra $[0, 1]_G$ then the structure $\mathbf{GV_n} = (GV_n, \min, \Rightarrow_G, 0, 1)$ becomes the "standard" algebra for the n-valued Gödel logic G_n, that is, the axiomatic extension of G with the axiom

$$(A_{G_n})\ (\varphi_1 \to \varphi_2) \vee \ldots \vee (\varphi_n \to \varphi_{n+1}).$$

By definition, $\Gamma \models_{G_n} \varphi$ iff, for every evaluation e over GV_n, if $e(\gamma) = 1$ for every $\gamma \in \Gamma$ then $e(\varphi) = 1$. In fact, the logics G_n are all the axiomatic extensions of G, and for each n, G_n is an axiomatic extension of G_{n+1}, where G_2 coincides with CPL. Thus, the set of axiomatic extensions of G form a chain of logics (and of the corresponding varieties of algebras) of strictly increasing strength:

$$G < \ldots < G_{n+1} < G_n < \ldots < G_3 < G_2 = \text{CPL},$$

where $L < L'$ denotes that L' is an axiomatic extension of L.

Since the negation in Gödel logics is a pseudo-complementation and not an involution, in Esteva et al. (2000), the authors investigate the residuated fuzzy logics arising from continuous t-norms without non-trivial zero divisors and extended with an involutive negation. In particular, they consider the extension of Gödel logic G with an involutive negation $\sim$, denoted as $G_\sim$, and axiomatize it.

The intended semantics of the $\sim$ connective on the real unit interval [0, 1] is an arbitrary order-reversing involution $n : [0, 1] \to [0, 1]$, i.e., satisfying $n(n(x)) = x$ and $n(x) \leq n(y)$ whenever $x \geq y$.

It turns out that in $G_\sim$, with both negations, $\neg$ and $\sim$, the projection Monteiro-Baaz connective Δ is definable as

$$\Delta\varphi := \neg \sim \varphi,$$

and whose semantics on [0, 1] is given by the mapping $\delta : [0, 1] \to [0, 1]$ defined as $\delta(1) = 1$ and $\delta(x) = 0$ for $x < 1$.

Axioms of $G_{\sim}$ are those of G plus[10]:

(~ 1) $\sim\sim\varphi \leftrightarrow \varphi$ *(Involution)*
(~ 2) $\neg\varphi \to \sim \varphi$
(~ 3) $\Delta(\varphi \to \psi) \to \Delta(\sim \psi \to \sim \varphi)$ *(Order Reversing)*
$(\Delta 1)$ $\Delta\varphi \vee \neg\Delta\varphi$
$(\Delta 2)$ $\Delta(\varphi \vee \psi) \to (\Delta\varphi \vee \Delta\psi)$
$(\Delta 5)$ $\Delta(\varphi \to \psi) \to (\Delta\varphi \to \Delta\psi)$,

and inference rules of $G_{\sim}$ are *modus ponens* and *necessitation* for Δ:

$$\frac{\varphi, \varphi \to \psi}{\psi} \qquad \frac{\varphi}{\Delta\varphi}.$$

$G_{\sim}$ is an algebraizable logic, whose equivalent algebraic semantics is the quasi-variety of $G_{\sim}$-algebras, defined in the natural way, and generated by the class of its linearly ordered members. Among them, the so-called *standard* $G_{\sim}$-algebra, denoted $[0, 1]_{G_{\sim}}$, is the algebra on the real interval [0, 1] with Gödel truth functions extended by the involutive negation $\sim x = 1 - x$. This $G_{\sim}$-chain generates the whole quasi-variety of $G_{\sim}$-algebras. In fact, we have a strong *standard* completeness result for $G_{\sim}$, see Esteva et al. (2000, 2011): for any set $\Gamma \cup \{\varphi\}$ of $G_{\sim}$-formulas, $\Gamma \vdash_{G_{\sim}} \varphi$ iff $\Gamma \models_{G_{\sim}} \varphi$, where the latter means: for every evaluation e over $[0, 1]_{G_{\sim}}$, if $e(\gamma) = 1$ for every $\gamma \in \Gamma$ then $e(\varphi) = 1$.

Finally, let us mention that, while G enjoys the usual deduction-detachment theorem (i.e., $\Gamma \cup \{\varphi\} \vdash_G \psi$ iff $\Gamma \vdash_G \varphi \to \psi$), this is not the case for $G_{\sim}$, which has only the following form of Δ-deduction theorem: $\Gamma \cup \{\varphi\} \vdash_{G_{\sim}} \psi$ iff $\Gamma \vdash_{G_{\sim}} \Delta\varphi \to \psi$. See also the handbook chapter (Esteva et al. 2011) for further properties of $G_{\sim}$.

On the other hand, as in the case of Gödel logic, one can also consider the logics $G_{n\sim}$ for each $n \geq 2$, the finite-valued counterparts of $G_{\sim}$. Namely, $G_{n\sim}$ can be obtained as the axiomatic extension of $G_{\sim}$ with the axiom (A_{G_n}),[11] and can be shown to be complete with respect to its intended algebraic semantics, the variety of algebras generated by the linearly ordered algebra $\mathbf{GV_{n\sim}}$ obtained in turn by expanding $\mathbf{GV_n}$ with the involutive negation $\sim x = 1 - x$, the only involutive order-reversing mapping that can be defined on GV_n. Thus, for any set $\Gamma \cup \{\varphi\}$ of $G_{n\sim}$-formulas, $\Gamma \vdash_{G_{n\sim}} \varphi$ iff $\Gamma \models_{G_{n\sim}} \varphi$, where the latter means: for every evaluation e over the expansion of $\mathbf{GV_n}$ by $\sim$, if $e(\gamma) = 1$ for every $\gamma \in \Gamma$ then $e(\varphi) = 1$. Clearly, $G_{2\sim} = \text{CPL}$. The graph of axiomatic extensions of $G_{n\sim}$ is depicted in Fig. 6.1, where edges denote

---

[10] These are the original axioms from Esteva et al. (2000), see again Běhounek (2011) and the references therein for a shorter axiomatization.

[11] Equivalently, as the expansion of G_n with $\sim$ along with the axioms (~ 1)-(~ 3), $(\Delta 1)$-$(\Delta 3)$, and the necessitation rule for Δ.

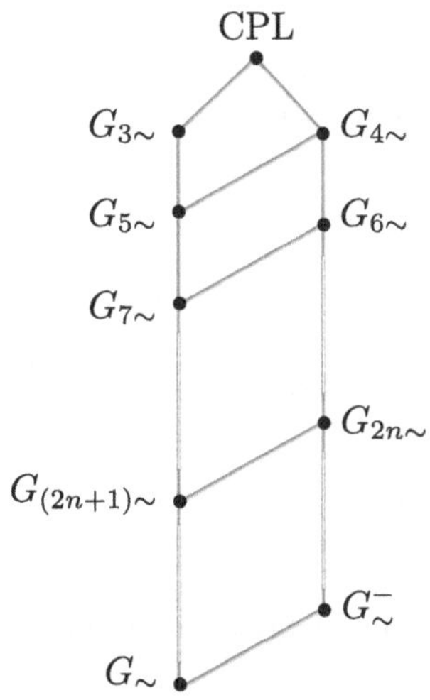

Fig. 6.1 Graph of axiomatic extensions of $G_{\sim}$

extensions. It can be observed that, if n is even then $G_{n\sim}$ is an extension of $G_{m\sim}$ for any $m > n$, while if n is odd, $G_{n\sim}$ is an extension of $G_{m\sim}$ only for those $m > n$ being odd as well. Also, note that, in the figure, $G^{-}_{\sim}$ denotes the extension of $G_{\sim}$ with the axiom
(NFP) $\sim\Delta(\varphi \leftrightarrow \sim \varphi)$

that captures the condition that the involutive negation does not have a fixed point, a condition satisfied by all the logics $G_{n\sim}$ with n even.

6.2.2 *Degree-Preserving Gödel Logics with Involution*

Main logics studied in Mathematical Fuzzy Logic are (full) truth-preserving fuzzy logics, like the Gödel logics introduced in the previous section. But we can also find in the literature companion logics that preserve degrees of truth, see, e.g., Font et al. (2006), Bou et al. (2009). It has been argued in Font (2009) that this approach is more coherent with the commitment of many-valued logics to truth-degree semantics because all values play an equally important role in the corresponding notion of consequence. Namely, given a fuzzy logic L,[12] one can introduce a variant of L that is usually denoted $L^{\leq}$, whose associated deducibility relation has the following semantics: for every set of formulas $\Gamma \cup \{\varphi\}$,

[12] For practical purposes, we can assume in this paper that L is an axiomatic extension of Hájek's BL logic.

$\Gamma \vdash_{L^{\leq}} \varphi$ iff for every L-chain $\boldsymbol{A}$, every $a \in A$, and every $\boldsymbol{A}$-evaluation e,
if $a \leq e(\psi)$ for every $\psi \in \Gamma$, then $a \leq e(\varphi)$.

For this reason, $L^{\leq}$ is known as a fuzzy logic *preserving degrees of truth* or the *degree-preserving companion* of L. It is not difficult to show that L and $L^{\leq}$ have the same theorems and also that for every finite set of formulas $\Gamma \cup \{\varphi\}$:

$$\Gamma \vdash_{L^{\leq}} \varphi \text{ iff } \vdash_L \Gamma^{\wedge} \to \varphi,$$

where $\Gamma^{\wedge}$ means $\gamma_1 \wedge \ldots \wedge \gamma_k$ for $\Gamma = \{\gamma_1, \ldots, \gamma_k\}$ (when Γ is empty then $\Gamma^{\wedge}$ is $\top$).

Remark 6.1 It is worth noting that the idea of degree-preserving consequence relations is already present in the context of (classical) modal logic. As it is well known, under the usual Kripke relational semantics one can consider in modal logic two notions of consequence relation: a *local* and a *global* one.[13] But modal logics have also been given algebraic semantics by means of the so-called modal algebras. Given such a modal algebra, one can define associated truth-preserving and degree-preserving consequence relations, in an analogous way as done above for a given linearly ordered L-algebra. It is immediate to see that, for modal logics, local Kripke semantics corresponds to degree-preserving algebraic semantics, while global semantics corresponds to truth-preserving semantics, see, e.g., Blackburn et al. (2002, Defs. 1.35 and 1.37).

As regard to axiomatization, the logic $L^{\leq}$ admits a Hilbert-style axiomatization having the same axioms as L and the following deduction rules (Bou et al. 2009):

(**Adj**-$\wedge$) from φ and ψ derive $\varphi \wedge \psi$;
(**MP**-r) if $\vdash_L \varphi \to \psi$, then from φ and $\varphi \to \psi$, derive ψ.

Note that (MP-r) is a restricted form of the Modus Ponens rule, it is only applicable when $\varphi \to \psi$ is a theorem of L.

Since Gödel logic G enjoys the deduction-detachment theorem, a key observation is that $G^{\leq} = G$. However, the case is different for the expansion of G with an involutive negation, since $G_{\sim}$ does not satisfy the usual deduction-detachment theorem, and hence $G_{\sim}$ and $G_{\sim}^{\leq}$ are different logics. Moreover, while $G_{\sim}^{\leq}$ keeps being $\neg$-explosive, it is $\sim$-paraconsistent. Indeed, there are φ, ψ such that $\varphi \wedge {\sim}\varphi \nvdash_{G_{\sim}^{\leq}} \psi$. Take, for instance, φ and ψ as being two different propositional variables, and e a truth evaluation over $[0, 1]_{G_{\sim}}$ such that $e(\varphi) = \frac{1}{2}$ and $e(\psi) < \frac{1}{2}$.

As for the axiomatization of $G_{\sim}^{\leq}$, we need to consider an extra rule regarding Δ. As shown in Ertola et al. (2015), a complete Hilbert-style axiomatization for $G_{\sim}^{\leq}$ can be obtained by the axioms of $G_{\sim}$, the previous rules (**Adj**-$\wedge$) and (**MP**-r),[14] together with the following restricted form of the usual necessitation rule for Δ:

[13] Given a class of Kripke models, a formula φ follows *locally* from a set Γ of formulas if, for any Kripke model M in the class and every world w in M, φ is true in $\langle M, w\rangle$ whenever every formula in Γ is true in $\langle M, w\rangle$ as well. On the other hand, φ follows *globally* from Γ in the class if, for every Kripke model M, φ is true in $\langle M, w\rangle$ for every w whenever every formula in Γ is true in $\langle M, w\rangle$ for every w.

[14] For $L = G_{\sim}$.

(ΔNec-r) if $\vdash_{G_\sim} \varphi$, then from φ derive $\Delta\varphi$.

Finally, let us consider the logics $G^{\leq}_{n\sim}$, the degree-preserving companions of the finite-valued logics $G^{\leq}_{\sim}$, defined in the obvious way as above for $L = G_{n\sim}$. Similar to $G^{\leq}_{\sim}$, $G^{\leq}_{n\sim}$ also admits the following Hilbert-style axiomatization: $G^{\leq}_{n\sim}$ has as axioms those of $G_{n\sim}$, and as rules, the rule **(Adj-$\wedge$)** and the following restricted rules:

(**MP-**r) if $\vdash_{G_{n\sim}} \varphi \to \psi$, then from φ and $\varphi \to \psi$, derive ψ;
(Δ**Nec-**r) if $\vdash_{G_{n\sim}} \varphi$, then from φ derive $\Delta\varphi$.

6.3 Logics Defined by Matrices Over $[0, 1]_{G\sim}$ by Means of Order Filters

By a *logical matrix* we understand a pair $\langle \mathbf{A}, F\rangle$ where $\mathbf{A}$ is an algebra and F is a subset of the domain A of $\mathbf{A}$. The logic $L(M)$ defined by the matrix $M = \langle \mathbf{A}, F\rangle$ is obtained by stipulating, for any set of formulas $\Gamma \cup \{\varphi\}$,

$$\Gamma \vdash_{L(M)} \varphi \text{ if for every evaluation } e \text{ on } \mathbf{A}, \\ \text{if } e(\gamma) \in F \text{ for every } \gamma \in \Gamma, \text{ then } e(\varphi) \in F.$$

On the other hand, the logic $L(\mathcal{M})$ determined by a class of matrices $\mathcal{M}$ is defined as the intersection of the logics defined by all the matrices in the family. A logic is said to be a *matrix logic* if it is of the form $L(\mathcal{M})$ for some class of matrices $\mathcal{M}$.

Notation: *In the rest of the paper, without danger of confusion and for the sake of a lighter notation, we will often identify a matrix M or a set of matrices $\mathcal{M}$ with their corresponding logics $L(M)$ and $L(\mathcal{M})$.*

As proved in Bou et al. (2009) for logics of residuated lattices, one can show that $G^{\leq}_{\sim}$, the degree-preserving companion of $G_\sim$, is not algebraizable in the sense of Block and Pigozzi and thus it has no algebraic semantics. But it has a semantics via matrices. Indeed, $G^{\leq}_{\sim}$ is the logic defined by the set of matrices

$$\mathcal{M}_{G_\sim} = \{\langle \mathbf{A}, F\rangle : \mathbf{A} \text{ is a } G_\sim\text{-algebra and } F \text{ is a lattice filter of } \mathbf{A}\}.$$

Using similar arguments as in the proof of Bou et al. (2009, Theorem 2.12), in fact, we can also prove that $G^{\leq}_{\sim}$ is complete with respect to a subset of $\mathcal{M}_{G_\sim}$, namely, the set of matrices over the standard $G_\sim$-algebra

$$\mathcal{M}_{[0,1]} = \{\langle [0, 1]_{G_\sim}, F\rangle : F \text{ is an order filter of } [0, 1]\}.$$

Next, we study the relationships among all the logics defined by matrices from $\mathcal{M}_{[0,1]}$, i.e., matrices over the algebra $[0, 1]_{G_\sim}$ by order filters, identifying which ones are paraconsistent. Actually, the order filters on $[0, 1]_{G_\sim}$ are the following sets:

$F_{[a} = \{x \in [0,1] : x \geq a\}$ for all $a \in (0,1]$ and $F_{(a} = \{x \in [0,1] : x > a\}$ for all $a \in [0,1)$. Abusing the notation, we will denote the corresponding logics as

$$\mathrm{G}_{\sim}^{[a} = \langle [0,1]_{\mathrm{G}_{\sim}}, F_{[a} \rangle \text{ and } \mathrm{G}_{\sim}^{(a} = \langle [0,1]_{\mathrm{G}_{\sim}}, F_{(a} \rangle.$$

The consequence relations corresponding to these logics will be, respectively, denoted by $\vdash_{[a}$ and $\vdash_{(a}$, while $\vdash_{[a}^{f}$ and $\vdash_{(a}^{f}$ will denote the finitary companions of $\vdash_{[a}$ and $\vdash_{(a}$, respectively.[15] We will write $\vdash_1$ and $\vdash_1^f$ instead of $\vdash_{[1}$ and $\vdash_{[1}^f$, respectively.

Some of the logics $\vdash_{[a}$ and $\vdash_{(a}$ are, in fact, finitary. This is shown in the next lemma.

Lemma 6.1 *The logic $\vdash_1$ is finitary. Moreover, the logics $\vdash_{(1/2}, \vdash_{[1/2}$ and $\vdash_{(0}$ are equivalent, as deductive systems, to $\vdash_1$ and hence they are finitary as well. Therefore, all these logics coincide with their finitary companions $\vdash_1^f, \vdash_{(1/2}^f, \vdash_{[1/2}^f$, and $\vdash_{(0}^f$, respectively.*

Proof • Since $\mathrm{G}_{\sim}$ has a finitary axiomatization (see previous Sect. 6.2.1) and it is strongly standard complete w.r.t. to $\vdash_1$, then $\vdash_1$ is finitary and coincides with $\vdash_1^f$.

• We prove that, in fact, $\vdash_{(1/2}, \vdash_{[1/2}$, and $\vdash_{(0}$ are all of them equivalent to $\vdash_1$, in the sense of Blok and Pigozzi (Blok and Pigozzi 2001). Indeed, for each formula φ, define the following transformations:

- $\varphi^{*_1} := (\sim\varphi \to \varphi) \wedge \neg\Delta(\varphi \leftrightarrow \sim\varphi)$, $\varphi^{*_2} := \sim\varphi \to \varphi$, $\varphi^{*_3} := \neg\neg\varphi$.

Further, if Γ is a set of formulas, define $\Gamma^* := \{\psi^* \mid \psi \in \Gamma\}$ for $* \in \{*_1, *_2, *_3\}$. It is easy to check that for any $\mathrm{G}_{\sim}$-evaluation e, we have

- $e(\varphi) > 1/2$ iff $e(\varphi^{*_1}) = 1$,
- $e(\varphi) \geq 1/2$ iff $e(\varphi^{*_2}) = 1$,
- $e(\varphi) > 0$ iff $e(\varphi^{*_3}) = 1$.

Then one can check that the following three conditions are satisfied:

(i) The logics $\vdash_{(1/2}, \vdash_{[1/2}$, and $\vdash_{(0}$ can be faithfully interpreted in $\vdash_1$ as the following equivalences hold:
$\Gamma \vdash_{(1/2} \varphi$ iff $\Gamma^{*_1} \vdash_1 \varphi^{*_1}$, $\Gamma \vdash_{[1/2} \varphi$ iff $\Gamma^{*_2} \vdash_1 \varphi^{*_2}$, and
$\Gamma \vdash_{(0} \varphi$ iff $\Gamma^{*_3} \vdash_1 \varphi^{*_3}$.

(ii) We can also interpret $\vdash_1$ in any of the other consequence relations by using the Δ operator, indeed, we have
$\Gamma \vdash_1 \varphi$ iff $\Gamma^{\Delta} \vdash_{(1/2} \Delta(\varphi)$ iff $\Gamma^{\Delta} \vdash_{[1/2} \Delta(\varphi)$ iff $\Gamma^{\Delta} \vdash_{>0} \Delta(\varphi)$,
where $\Gamma^{\Delta} = \{\Delta(\psi) : \psi \in \Gamma\}$.

(iii) Finally, the following inter-derivabilities show that the Δ acts as a proper converse transformation of each $*_i$ in the corresponding logic:
$\psi \dashv\vdash_{(1/2} \Delta(\psi^{*_1})$ and $\varphi \dashv\vdash_1 (\Delta\varphi)^{*_1}$,
$\psi \dashv\vdash_{[1/2} \Delta(\psi^{*_2})$ and $\varphi \dashv\vdash_1 (\Delta\varphi)^{*_2}$,
$\psi \dashv\vdash_{(0} \Delta(\psi^{*_3})$ and $\varphi \dashv\vdash_1 (\Delta\varphi)^{*_3}$.

[15] Recall that the finitary companion of a logic $(L, \vdash)$ is given by $(L, \vdash^f)$ where, for every $\Gamma \cup \{\varphi\} \subseteq L$, $\Gamma \vdash^f \varphi$ iff there exists a finite $\Gamma_0 \subseteq \Gamma$ such that $\Gamma_0 \vdash \varphi$.

As a consequence, the logics $\vdash_1$, $\vdash_{(1/2}$, $\vdash_{[1/2}$, and $\vdash_{(0}$ are equivalent, and since $\vdash_1$ is finitary, so are the other logics as well.

□

Therefore, as a consequence of previous lemma, the only cases left open are whether the logics $\vdash_{[a}$ and $\vdash_{(a}$ are finitary for $a \in (0, 1/2) \cup (1/2, 1)$.

Next proposition shows the relationships among the remaining logics defined by matrices over the algebra $[0, 1]_{G_\sim}$ by order filters.

Proposition 6.1 *The logics* $G^{[a}_{\sim} = \langle [0, 1]_{G_\sim}, F_{[a} \rangle$ *for* $a \in (0, 1]$, $G^{(a}_{\sim} = \langle [0, 1]_{G_\sim}, F_{(a} \rangle$ *for* $a \in [0, 1)$, *and their finitary companions, satisfy the following properties*[16]*:*

P1. $\vdash_{[p} = \vdash_{[p'}$ *and* $\vdash_{(p} = \vdash_{(p'}$, *for all* $p, p' \in (1/2, 1)$.
Moreover, $\vdash_{(p} \subseteq \vdash_{[p}$ *and* $\vdash^f_{[p} = \vdash^f_{(p}$ *for all* $p \in (1/2, 1)$.

P2. $\vdash_{[n} = \vdash_{[n'}$ *and* $\vdash_{(n} = \vdash_{(n'}$, *for all* $n, n' \in (0, 1/2)$.
Moreover, $\vdash_{(n} \subseteq \vdash_{[n}$ *and* $\vdash^f_{[n} = \vdash^f_{(n}$ *for all* $n \in (0, 1/2)$.

P3. $\vdash_{[p} \subsetneq \vdash_1$, *for any* $p \in (1/2, 1)$.

P4. $\vdash_1$ *and* $\vdash_{[1/2}$ *are not comparable.*

P5. $\vdash_{[p}$ *and* $\vdash_{[1/2}$, *as well as* $\vdash^f_{[p}$ *and* $\vdash_{[1/2}$, *are not comparable, for any* $p \in (1/2, 1)$.

P6. $\vdash^f_{[p} \subsetneq \vdash_{(1/2}$, *for any* $p \in (1/2, 1)$.

P7. $\vdash_{[p}$ *and* $\vdash_{[n}$ *are not comparable, for any* $p \in (1/2, 1)$ *and any* $n \in (0, 1/2)$.
The same holds for $\vdash^f_{[p}$ *and* $\vdash^f_{[n}$.

P8. $\vdash_{[n} \subsetneq \vdash_{[1/2}$, *for any* $n \in (0, 1/2)$.

P9. $\vdash_{(0}$, $\vdash_{[1/2}$, *and* $\vdash_{(1/2}$ *are not pairwise comparable.*

P10. $\vdash^f_{[n} \subsetneq \vdash_{(0}$, *for any* $n \in (0, 1/2)$.

Proof P1. We divide the proof in four steps:

(i) That $\vdash_{[p} = \vdash_{[p'}$ and $\vdash_{(p} = \vdash_{(p'}$ is an easy consequence of the fact that for every $p, p' \in (1/2, 1)$ it is possible to define an automorphism f of $[0, 1]_{G\sim}$ such that $f(p) = p'$. Let us then show that $\vdash^f_{[p} = \vdash^f_{(p}$ for every $p \in (1/2, 1)$.

(ii) Assume $\{\varphi_i : i \in I\} \vdash^f_{[p} \psi$, with I finite, for some $p \in (1/2, 1)$. Let q such that $1/2 < q < p$, and let e be an evaluation such that $e(\varphi_i) > q$ for all $i \in I$. Let $p' = \min_{i \in I} e(\varphi_i)$. Obviously $p' > q$. Then, by (i), we also have $\{\varphi_i : i \in I\} \vdash^f_{[p'} \psi$, and therefore we have $e(\psi) \geq p' > q$, and hence $\{\varphi_i : i \in I\} \vdash^f_{(q} \psi$. Therefore, we have $\vdash^f_{[p} \subseteq \vdash^f_{(q}$ for all $1/2 < q < p$.

(iii) Recall from (i) that $\Gamma \vdash_{(p} \varphi$ iff $\Gamma \vdash_{(p'} \varphi$ for all $1/2 < p' < 1$. Let $p_1, p_2, \ldots, p_n, \ldots$ be an increasing sequence of values $p_i \in (1/2, p)$ such that $\lim_n p_n = p$. Suppose $\Gamma \vdash_{(p} \varphi$, and further assume $e(\psi) \geq p$ for all $\psi \in \Gamma$. Clearly, for each p_i, $e(\psi) > p_i$ for all $\psi \in \Gamma$. Since $\Gamma \vdash_{(p_i} \varphi$ for each p_i, we have that $e(\varphi) > p_i$ for each p_i. Hence $e(\varphi) \geq p$.

(iv) Assume $\{\varphi_i : i \in I\} \vdash^f_{(q} \psi$, with I finite, for some $q \in (1/2, 1)$. Let p be such that $q < p < 1$, and let e be an evaluation such that $e(\varphi_i) \geq p$ for

[16] In the following we use p and n to denote positive and negative values in $[0, 1]$ with respect to the negation $\sim x = 1 - x$; in other words, $p > 1/2$ and $n < 1/2$.

all $i \in I$. Let $q' = \min_{i \in I} e(\varphi_i)$. Obviously $q' \geq p$. Then, by (i), we also have $\{\varphi_i : i \in I\} \vdash^f_{(q'} \psi$, and therefore we have $e(\psi) \geq q' \geq p$, and hence $\{\varphi_i : i \in I\} \vdash^f_p \psi$. Therefore, we have $\vdash^f_{(q} \subseteq \vdash^f_{[p}$ for all $1/2 < q < p$.

P2. The proofs are analogous to those of P1.

P3. Assume $\{\varphi_i : i \in I\} \vdash_{[p} \psi$ for a given $p \in (1/2, 1)$, and let e be an evaluation such that $e(\varphi_i) = 1$ for all $i \in I$. Since it is also true that $e(\varphi_i) \geq p'$ for all $p' \in (1/2, 1)$, by P1 it follows that $\{\varphi_i : i \in I\} \vdash_{[p'} \psi$ for all $p' \in (1/2, 1)$, and hence $e(\psi) \geq p'$ for all $p' \in (1/2, 1)$, and thus $e(\psi) = 1$. Therefore $\{\varphi_i : i \in I\} \vdash_1 \psi$.
The strict inclusion can be easily noticed since it holds that $\varphi \vdash_1 \Delta\varphi$ but $\varphi \nvdash_{[p} \Delta\varphi$ for any $p < 1$.

P4. It clearly holds that, on the one hand, $\Delta(\varphi \leftrightarrow {\sim}\varphi) \vdash_{[1/2} \varphi$ but $\Delta(\varphi \leftrightarrow {\sim}\varphi) \nvdash_1 \varphi$, while, on the other hand, $\varphi \vdash_1 \Delta\varphi$ but $\varphi \nvdash_{[1/2} \Delta\varphi$.

P5. It follows from noticing that $\Delta(\varphi \leftrightarrow {\sim}\varphi) \wedge \varphi \vdash_{[p} \bot$ and $\Delta(\varphi \leftrightarrow {\sim}\varphi) \wedge \varphi \nvdash_{[1/2} \bot$, while $\Delta(\varphi \leftrightarrow {\sim}\varphi) \vdash_{[1/2} \varphi$ and $\Delta(\varphi \leftrightarrow {\sim}\varphi) \nvdash_{[p} \varphi$.

P6. Assume that, for a given $p \in (1/2, 1)$, $\{\varphi_i : i \in I\} \vdash_{[p} \psi$, with I finite, and let e be an evaluation such that $e(\varphi_i) > 1/2$ for all $i \in I$. Let $p' = \min_{i \in I} e(\varphi_i)$. Obviously $p' > 1/2$. Then, from P1 we also have $\{\varphi_i : i \in I\} \vdash_{[p'} \psi$, and therefore we have $e(\psi) \geq p' > 1/2$, and hence $\{\varphi_i : i \in I\} \vdash_{(1/2} \psi$. Therefore, we have $\vdash_{[p} \subseteq \vdash_{(1/2}$.
That the inclusion is strict follows from observing that ${\sim}\Delta(\varphi \to {\sim}\varphi) \vdash_{(1/2} \varphi$, but ${\sim}\Delta(\varphi \to {\sim}\varphi) \nvdash_{[p} \varphi$.

P7. It follows from observing (i) $\Delta(\varphi \leftrightarrow {\sim}\varphi) \vdash_{[n} \varphi$ and $\Delta(\varphi \leftrightarrow {\sim}\varphi) \nvdash_{[p} \varphi$, and (ii) $\varphi \vdash_{[p} {\sim}\Delta(\varphi \to {\sim}\varphi)$ and $\varphi \nvdash_{[n} {\sim}\Delta(\varphi \to {\sim}\varphi)$.

P8. That $\vdash_{[n} \subsetneq \vdash_{[1/2}$ is proved in a similar way to P3. The strict inclusion is a consequence of the following facts:
(i) $\varphi \wedge {\sim}\varphi \vdash_{[1/2} \Delta(\varphi \leftrightarrow {\sim}\varphi)$;
(ii) $\varphi \wedge {\sim}\varphi \nvdash_{[n} \Delta(\varphi \leftrightarrow {\sim}\varphi)$.
Notice that $e(\varphi \wedge {\sim}\varphi) \geq 1/2$ iff $e(\varphi) = 1/2$ iff $e(\varphi \leftrightarrow {\sim}\varphi) = 1$, while if $e(\varphi) = n$, then $e(\varphi \wedge {\sim}\varphi) = e(\varphi \leftrightarrow {\sim}\varphi) = n$, but $e(\Delta(\varphi \leftrightarrow {\sim}\varphi)) = 0$.

P9. That $\vdash_{[1/2}$ and $\vdash_{(1/2}$ are not comparable results from noticing, e.g., (i) $\Delta(\varphi \leftrightarrow {\sim}\varphi) \vdash_{[1/2} \varphi$ but $\Delta(\varphi \leftrightarrow {\sim}\varphi) \nvdash_{(1/2} \varphi$, and (ii) $\varphi \vdash_{(1/2} {\sim}\Delta(\varphi \to {\sim}\varphi)$ but $\varphi \nvdash_{[1/2} {\sim}\Delta(\varphi \to {\sim}\varphi)$.
On the other hand, it is easy to check that $\bot$ follows from $\varphi \wedge \Delta(\varphi \to {\sim}\varphi) \wedge \neg\Delta({\sim}\varphi \to \varphi)$ in $\vdash_{(1/2}$ and $\vdash_{[1/2}$, but not in $\vdash_{(0}$. Conversely, $\neg\neg\varphi \wedge \neg\Delta\varphi \vdash_{(0} \varphi \wedge {\sim}\varphi$, but this is neither the case for $\vdash_{(1/2}$ nor for $\vdash_{[1/2}$.

P10. Assume $\{\varphi_i : i \in I\} \vdash_{[n} \psi$ for a given $n \in (0, 1/2)$ and a finite set I, and let e be an evaluation such that $e(\varphi_i) > 0$ for all $i \in I$. Let $n' = \min_{i \in I} e(\varphi_i)$. Obviously $n' > 0$ and $e(\varphi_i) \geq n'$, for all $i \in I$. Then, from P1, we also have that $\{\varphi_i : i \in I\} \vdash_{[n'} \psi$, and hence we have $e(\psi) \geq n' > 0$. This means $\{\varphi_i : i \in I\} \vdash_{(0} \psi$. Therefore, we have $\vdash^f_{[n} \subseteq \vdash_{(0}$.
On the other hand, $\neg\neg\varphi \vdash_{(0} \varphi$ but $\neg\neg\varphi \nvdash_{[n} \varphi$, hence we have proved that $\vdash^f_{[n} \subsetneq \vdash_{(0}$.

□

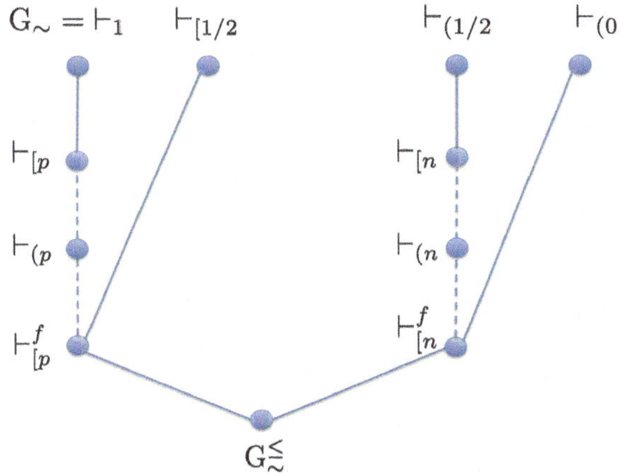

Fig. 6.2 Graph of logics over $[0, 1]_{G_\sim}$ defined by order filters, where $1/2 < p < 1$ and $0 < n < 1/2$, where edges stand for inclusions (upward sense). The dashed edges denote that it is an open problem whether the connected logics are different

A graphical representation of the different logics (consequence relations) involved in the above proposition can be seen in Fig. 6.2.

It is clear that a matrix logic $G^{[a}_{\sim} = \langle [0, 1]_{G_\sim}, F_{[a} \rangle$ (resp. $G^{(a}_{\sim} = \langle [0, 1]_{G_\sim}, F_{(a} \rangle$) is paraconsistent only in the case that $a \leq 1/2$ (resp. $a < 1/2$). As a consequence of the above classification, it turns out that there are only three different paraconsistent logics among them.

Corollary 6.1 *Among the families of logics* $\{G^{[a}_{\sim}\}_{a\in(0,1]}$ *and* $\{G^{(a}_{\sim}\}_{a\in[0,1)}$*:*

- *there are only three different paraconsistent logics:* $G^{[a}_{\sim}$ *for any* $a \in (0, 1/2)$, $G^{[1/2}_{\sim}$, *and* $G^{(0}_{\sim}$.
- *there are only three different explosive logics:* $G^{[a}_{\sim}$ *for any* $a \in (1/2, 1)$, $G^{(1/2}_{\sim}$, *and* $G^{[1}_{\sim}$.

In analogy to Coniglio et al. (2016, Theorem 2), it is easy to show that every intermediate logic L between $G^{\leq}_{\sim}$ and CPL is, in fact, the logic $L(\mathcal{M}')$ defined by a subfamily of matrices $\mathcal{M}' \subseteq \mathcal{M}_{G_\sim}$. However, note that the set of $G_\sim$-algebras and their lattice filters is very large. Then, an exhaustive analysis of the set of intermediate logics between $G^{\leq}_{\sim}$ and CPL actually seems to be a difficult task. Because of this, in the next section, we will restrict ourselves to the case of finite-valued Gödel logics with an involutive negation $G_{n\sim}$.

Remark 6.2 In the last corollary, we have shown that $G^{(0}_{\sim}$, the matrix logic defined by the standard $G_\sim$-algebra $[0, 1]_{G_\sim}$ and the filter $(0, 1]$ of designated values, is a paraconsistent logic. In Avron (2016), Avron introduces a paraconsistent extension of the logic T of Anderson and Belnap called FT. This logic, which intends to be "a paraconsistent counterpart of Łukasiewicz Logic $Ł_\infty$" (Avron (2016, pp. 75)), is firstly defined axiomatically over a propositional language with connectives $\wedge$, $\vee$, $\sim$, $\rightarrow_{FT}$,[17] and then it is proved that FT is semantically characterized by the logic

[17] In Avron (2016), the symbols $\neg$ and $\rightarrow$ were used instead of $\sim$ and $\rightarrow_{FT}$. We adopt this notation in order to keep the notation of the present paper uniform.

matrix defined by the ordered algebra $\mathbf{M}_{[0,1]} = ([0,1], \wedge, \vee, \sim, \to_{FT}, 0, 1)$ and the filter $(0, 1]$ of designated values (this is why Avron considers FT as a logic that preserves *non-falsity*). Here $\wedge$, $\vee$, and $\sim$ are defined as in $[0,1]_{G_\sim}$, while $\to_{FT}$ is defined as follows:

$$x \to_{FT} y = \begin{cases} \max(1-x, y), & \text{if } x \leq y \\ 0, & \text{otherwise.} \end{cases}$$

Now, observe that the implication $\to_{FT}$ of $\mathbf{M}_{[0,1]}$ is definable in $[0,1]_{G_\sim}$ as $x \to_{FT} y = \Delta(x \to_G y) \wedge (\sim x \vee y)$. As a consequence of this, the logic FT is interpretable in $G_\sim^{(0}$ by means of a mapping $* : Fm_{FT} \to Fm_{G_\sim}$ defined recursively as follows: $p^* = p$ if p is a propositional variable; $(\sim\varphi)^* = \sim\varphi^*$; $(\varphi \wedge \psi)^* = \varphi^* \wedge \psi^*$; $(\varphi \vee \psi)^* = \varphi^* \vee \psi^*$; and $(\varphi \to_{FT} \psi)^* = \Delta(\varphi^* \to_G \psi^*) \wedge (\sim\varphi^* \vee \psi^*)$. Then, for every $\Gamma \cup \{\varphi\} \subseteq Fm_{FT}$, we have

$$\Gamma \vdash_{FT} \varphi \text{ iff } \Gamma^* \vdash_{(0} \varphi^*,$$

where Γ^* denotes the set $\{\psi^* : \psi \in \Gamma\}$ and $\vdash_{(0}$ denotes the consequence relation of $G_\sim^{(0}$. Moreover, in Avron (2016, Example 3.4), Avron considers, for every $n > 1$, the finite subalgebra $\mathbf{M}^{\mathbf{n}}_{[0,1]}$ of $\mathbf{M}_{[0,1]}$ with domain $GV_n = \{0, 1/(n-1), \ldots, (n-2)/(n-1), 1\}$. Let FT^n be the logic characterized by the logic matrix defined by the algebra $\mathbf{M}^{\mathbf{n}}_{[0,1]}$ and the filter $F_{\frac{1}{n-1}} = \{a \in GV_n : a > 0\} = GV_n \cap (0, 1]$ of designated values. Then, the interpretation $*$ above also shows that FT^n is interpretable in $\langle \mathbf{GV_{n\sim}}, F_{\frac{1}{n-1}} \rangle$, since we also have

$$\Gamma \vdash_{FT^n} \varphi \text{ iff } \Gamma^* \vdash_{\{\frac{1}{n-1}\}} \varphi^*,$$

where $\vdash_{\{\frac{1}{n-1}\}}$ is the consequence relation of the matrix logic $\langle \mathbf{GV_{n\sim}}, F_{\frac{1}{n-1}} \rangle$. This notation will be also used in Sect. 6.4.1.

As a matter of fact, it can be observed that the Δ operator of $[\mathbf{0}, \mathbf{1}]_{G_\sim}$ is definable in $\mathbf{M}_{[0,1]}$ as $\Delta x = 1 \to_{FT} x$ and so the Gödel implication $\to_G$ of $[\mathbf{0}, \mathbf{1}]_{G_\sim}$ is also definable in $\mathbf{M}_{[0,1]}$ as $x \to_G y = \sim\Delta\sim(x \to_{FT} y) \vee y$. Observe, however, that the logic FT has neither bottom nor top,[18] hence there is no formula in FT which can express the Δ operator. Let FT_0 be the logic defined by the same matrix $\langle \mathbf{M}_{[0,1]}, (0, 1] \rangle$ of FT, but now over an expanded language Fm_{FT_0} containing a constant $\bot$ and adding the requirement that $e(\bot) = 0$ for every evaluation e. Let us denote by $\vdash_{FT_0}$ its corresponding consequence relation. Consider the mapping $\# : Fm_{G_\sim} \to Fm_{FT_0}$ defined recursively as follows: $p^\# = p$ if p is a propositional variable; $\bot^\# = \bot$; $(\sim\varphi)^\# = \sim\varphi^\#$; $(\varphi \wedge \psi)^\# = \varphi^\# \wedge \psi^\#$; and $(\varphi \to_G \psi)^\# = \sim\Delta\sim(\varphi^\# \to_{FT} \psi^\#) \vee \psi^\#$ (where $\Delta\alpha = \sim\bot \to_{FT} \alpha$ for every α). Then, $\Gamma \vdash_{(0} \varphi$ iff $\Gamma^\# \vdash_{FT_0} \varphi^\#$, showing that the logic $G_\sim^{(0}$ is interpretable in FT_0. Since φ is equivalent to $\varphi^{*\#}$ in FT_0 for every $\varphi \in Fm_{FT_0}$ and ψ is equivalent to $\psi^{\#*}$ in $G_\sim^{(0}$ for every $\psi \in Fm_{G_\sim}$ (here, $*$ is extended

[18] Indeed, every formula $\varphi \in Fm_{FT}$ gets the value $1/2$ in any evaluation e over $\mathbf{M}_{[0,1]}$ such that e assigns the value $1/2$ to any propositional variable occurring in φ.

to Fm_{FT_0} by putting $\bot^* = \bot$), the logics $\mathrm{G}^{(0}_{\sim}$ and FT_0 are the same up to language. The same relationship holds between $\langle \mathbf{GV_{n\sim}}, F_{\frac{1}{n-1}} \rangle$ and the logic $\mathrm{FT}^{\mathrm{n}}_0$ obtained from FT^{n} by adding $\bot$.

6.4 Logics Between $\mathrm{G}^{\leq}_{n\sim}$ and CPL

In this section, we will study the intermediate logics between $\mathrm{G}^{\leq}_{n\sim}$ and CPL, for a natural $n > 2$. The cases $n = 3$ and $n = 4$ are easy to analyze since $\mathrm{G}_{3\sim}$ and $\mathrm{G}_{4\sim}$ coincide, respectively, with the three-valued and four-valued Łukasiewicz logics $Ł_3$ and $Ł_4$.

Proposition 6.2 *$\mathrm{G}_{3\sim}$ and $\mathrm{G}_{4\sim}$ are logically equivalent to $Ł_3$ and $Ł_4$, respectively.*

Proof The proof is algebraic, we prove that the standard algebras $\mathbf{GV_{3\sim}}$ and $\mathbf{GV_{4\sim}}$ are termwise equivalent to the standard Łukasiewicz algebras of $Ł_3$ and $Ł_4$, respectively. First, in the algebra $\mathbf{GV_{3\sim}}$, it is possible to define the binary connective $x \rightarrow_{3Ł} y = (x \rightarrow y) \vee ({\sim}x \vee y)$, that coincides with the three-valued Łukasiewicz implication, i.e., we have $x \rightarrow_{3Ł} y = \min(1, 1 - x + y)$ for every $x, y \in GV_3$. Thus, in $\mathbf{GV_{3\sim}}$, we can define all the Łukasiewicz connectives, in other words, $(GV_{3\sim}, \rightarrow_{3Ł}, {\sim}, 0, 1)$ is, in fact, the

Second, also in the algebra $\mathbf{GV_{4\sim}}$, we can define the binary connective

$$x \rightarrow_{4Ł} y = {\sim}x \vee [\Delta({\sim}x \rightarrow x) \wedge ({\sim}\, \Delta x) \wedge (\neg\neg y) \wedge x)] \vee (x \rightarrow y)$$

which coincides again with the four-valued Łukasiewicz implication, i.e., $x \rightarrow_{4Ł} y = \min(1, 1 - x + y)$ for every $x, y \in GV_4$.

On the other hand, in any finite MV-algebra $\mathbf{ŁV_n}$ we can always define Gödel implication as $x \rightarrow_G y = \Delta(x \rightarrow_Ł y) \vee y$ and Gödel negation as $\neg_G x = \Delta({\sim}x)$.[19] □

Remark 6.3 From the last result, it follows that the logics between $\mathrm{G}^{\leq}_{3\sim}$ (resp. $\mathrm{G}^{\leq}_{4\sim}$) and CPL coincide with the logics between $Ł^{\leq}_3$ (resp. $Ł^{\leq}_4$) and CPL studied in Coniglio et al. (2016, 2019). Among them, the well-known da Costa and D'Ottaviano's three-valued logic J_3, that is, equivalent (up to language) to the matrix logic $\langle \mathbf{ŁV_3}, \{1/2, 1\} \rangle$, is a logic between $Ł^{\leq}_3$ and CPL. Analogously, its four-valued generalization J_4, defined as the matrix logic $\langle \mathbf{ŁV_4}, \{1/3, 2/3, 1\} \rangle$ in Coniglio et al. (2019), is a logic between $Ł^{\leq}_4$ and CPL. Therefore, we can consider the logics J_3 and J_4 to be equivalent as well to the intermediate logics $\langle \mathbf{GV_{3\sim}}, \{1/2, 1\} \rangle$ and $\langle \mathbf{GV_{4\sim}}, \{1/3, 2/3, 1\} \rangle$, respectively.

Observe, however, that for any $n > 4$, $\mathrm{G}_{n\sim}$ is no longer equivalent to $Ł_n$. Thus, we need to study the intermediate logics for $\mathrm{G}^{\leq}_{n\sim}$ for $n > 4$, and this is the goal of the next subsection, while in Sect. 6.4.2 we will have a closer look to the case $n = 5$.

[19] Recall that the Δ connective is definable in any algebra $ŁV_n$.

6.4.1 The Intermediate Logics of $G^{\leq}_{n\sim}$ for $n > 4$

Throughout this section n will denote a natural number such that $n > 4$.

Following the same arguments as in previous sections, it is easy to check that $G^{\leq}_{n\sim}$ is, in fact, the logic semantically defined by the class of matrices:

$$\{\langle \mathbf{A}, F\rangle : \mathbf{A} \text{ is a } G_{n\sim}\text{-algebra and } F \text{ is a lattice filter of } \mathbf{A}\}.$$

Therefore, in order to study the intermediate logics between $G^{\leq}_{n\sim}$ and CPL, we need to characterize the (finite) $G_{n\sim}$-algebras.

Proposition 6.3 *Every finite* $G_{n\sim}$*-algebra is a finite direct product of finite* $G_{n\sim}$*-chains.*

Proof Notice that for every $G_{n\sim}$-chain the term $t(x, y, z) := (\Delta(x \leftrightarrow y) \wedge z) \vee (\neg\Delta(x \leftrightarrow y) \wedge x)$ is a discriminator term,[20] hence every $G_{n\sim}$-variety is a discriminator variety. Then the result is a consequence of a result of universal algebra (see, for instance, Burris and Sankappanavar (1981, Theorem 9.4, item (d))). □

In the following, we will need to consider products of logical matrices.

Definition 6.1 Let $L_i = \langle \mathbf{A}_i, D_i\rangle$ (for $i \in I$) be a family of logical matrices, where each D_i is an order filter in $\mathbf{A_i}$. The product of these matrices is the logical matrix $L = \Pi_{i\in I} L_i = \langle \Pi_{i\in I}\mathbf{A}_i, \Pi_{i\in I} D_i\rangle$, where $\Gamma \vdash_L \varphi$ iff, for every tuple of evaluations $(e_i)_{i\in I}$, each e_i over $\mathbf{A_i}$, the following condition holds: if $e_i(\psi) \in D_i$ for every $i \in I$ and every $\psi \in \Gamma$, then $e_i(\varphi) \in D_i$ for every $i \in I$.

Remark 6.4 Obviously, a matrix logic L as above is paraconsistent iff all the components L_i are paraconsistent. For example, if one component is $\langle \mathbf{GV_{2\sim}}, F_1\rangle$, then the matrix logic is not paraconsistent.

Since every $G_{\sim}$-algebra is locally finite, every intermediate logic L between $G^{\leq}_{n\sim}$ and CPL is induced by a family of product matrices $\langle \mathbf{A}, F\rangle$ where $\mathbf{A}$ is a finite direct product of subalgebras of $\mathbf{GV_{n\sim}}$ and F is a lattice filter of A compatible[21] with L.

In Coniglio et al. (2016, 2019), products of logical matrices were considered for Łukasiewicz finite-valued logics. For instance, Coniglio et al. (2016) contains a full description of the set $Int_{\Pi}(\text{Ł}_n)$ of logics defined by (sets of) products of matrices over the standard Ł_n-algebra. Additionally, it also contains an almost full description of the set $Int(\text{Ł}_n)$ of logics defined by sets of products of matrices over subalgebras of the standard Ł_n-algebra which are sublogics of Ł_n, when $n - 1$ is a prime number.

In the rest of this section, we will consider families of intermediate logics between $G^{\leq}_{n\sim}$ and CPL of increasing generality.

---

[20] In fact, this is a discriminator term in the whole variety of $G_{\sim}$-algebras. For a definition of discriminator term and discriminator variety, see Burris and Sankappanavar (1981).

[21] A filter F of an algebra $\mathbf{A}$ is *compatible* with a logic L if, whenever $\Gamma \vdash_L \varphi$, the following holds: for every $\mathbf{A}$-evaluation e, if $e(\gamma) \in F$ for every $\gamma \in \Gamma$ then $e(\varphi) \in F$.

First of all, we study the matrix logics $\langle \mathbf{GV_n}, F\rangle$ where F is an order filter of $\mathbf{GV_n}$. In order to simplify the notation, for every nonempty subset $T \subseteq GV_n$ we denote by $L(\mathcal{M}_T)$ the logic defined by the set of matrices $\mathcal{M}_T = \{\langle \mathbf{GV_{n\sim}}, F_t\rangle : t \in T\}$, where F_t denotes the order filter in GV_n generated by $t \in GV_n$, namely, if $t = i/(n-1)$ then $F_t = \{i/(n-1), (i+1)/(n-1), \ldots, (n-2)/(n-1), 1\}$.[22] Note that $F_1 = \{1\}$. The set of all the logics $L(\mathcal{M}_T)$, for $\emptyset \neq T \subseteq GV_n \setminus \{0\}$, will be denoted by $L(\mathbf{GV_{n\sim}})$.

Proposition 6.4 *The logics $L(\mathcal{M}_{\{t\}})$, with $t \in GV_n \setminus \{0\}$, are pairwise incomparable. Moreover, $L(\mathcal{M}_T)$ is not comparable to $L(\mathcal{M}_R)$ if $\emptyset \neq T, R \subseteq GV_n \setminus \{0\}$ such that $T \neq R$ and T and R have the same cardinality. In addition, the set of logics $L(\mathbf{GV_{n\sim}})$ is a meet-semilattice where the logics $L(\mathcal{M}_{\{t\}})$, for $t \in GV_n \setminus \{0\}$, are its maximal elements.*

Proof Let $\vdash_{\{t\}}$ be the consequence relation of the logic $L(\mathcal{M}_{\{t\}})$ defined by the matrix $\langle \mathbf{GV_{n\sim}}, F_t\rangle$, with $t \in GV_n \setminus \{0\}$. Observe that in any of these logics, since we have the Δ operator, it is possible to build a propositional formula on n variables $\Phi(p_0, p_1, \ldots, p_{n-1})$ such that, for every evaluation e of formulas on $\mathbf{GV_{n\sim}}$,

$$e(\Phi(p_0, p_1, \ldots, p_n)) = \begin{cases} 1, \text{ if } e(p_i) = \frac{i}{n-1} \text{ for all } i = 0, 1, \ldots, n-1 \\ 0, \text{ otherwise.} \end{cases}$$

Let $i, j \in \{1, 2, \ldots, n-1\}$ be such that $i < j$. Then,

- $\Phi(p_0, p_1, \ldots, p_n) \wedge p_i \vdash_{\{\frac{j}{n-1}\}} \perp$ and $\Phi(p_0, p_1, \ldots, p_n) \wedge p_i \nvdash_{\{\frac{i}{n-1}\}} \perp$;
- $\Phi(p_0, p_1, \ldots, p_n) \wedge p_j \nvdash_{\{\frac{j}{n-1}\}} p_i$ and $\Phi(p_0, p_1, \ldots, p_n) \wedge p_j \vdash_{\{\frac{i}{n-1}\}} p_i$.

Therefore, $\vdash_{\{t\}}$ and $\vdash_{\{t'\}}$ are not comparable if $0 < t < t' < 1$. From this, it is easy to prove that for any subsets $\emptyset \neq T, R \subseteq GV_n \setminus \{0\}$ with the same cardinality and such that $T \neq R$, the logic $L(\mathcal{M}_T)$ is not comparable to $L(\mathcal{M}_R)$. Finally, if $\emptyset \neq T, R \subseteq GV_n \setminus \{0\}$ then $L(\mathcal{M}_T) \cap L(\mathcal{M}_R) = L(\mathcal{M}_{T \cup R})$. Hence, $L(\mathbf{GV_{n\sim}})$ is a meet-semilattice such that the maximal elements are exactly the logics $L(\mathcal{M}_{\{t\}})$, for $t \in GV_n \setminus \{0\}$. □

On the other hand, as it was done in Coniglio et al. (2016, 2019) for Łukasiewicz finite-valued logics, product matrices can be considered.

Definition 6.2 Given a nonempty set $T \subseteq GV_n \setminus \{0\}, T = \{t_1, \ldots, t_k\}$ (where $k \geq 1$ and $t_i < t_j$ if $i < j$), we will denote by $\mathbb{L}(T)$ the matrix logic $\langle (\mathbf{GV_{n\sim}})^k, \Pi_{i=1}^k F_{t_i}\rangle$ defined on a direct product of $\mathbf{GV_{n\sim}}$ by means of order filters.

[22] Strictly speaking, this notation becomes ambiguous if n is not clear from the context and we identify rational numbers such as $i/(n-1)$ and $i.k/(n-1).k$, for instance, $1/2$, $2/4$, $3/6$, and so on. In this case, the notation $F_{\frac{1}{2}}$ is problematic, since it could denote any of an infinite sequence of different filters in GV_3, GV_5, GV_7,... respectively. The right notation for order filters in GV_n should be F_t^n. However, the superscript n will be avoided when there is no risk of confusion.

Proceeding as in Coniglio et al. (2016, Prop. 11), one can show that $\mathbb{L}(T)$ can be characterized as follows:

$$\Gamma \vdash_{\mathbb{L}(T)} \varphi \text{ iff either } \Gamma \vdash_{\{t_k\}} \bot \text{ or } \Gamma \vdash_{\{t\}} \varphi \text{ for all } t \in T.$$

Note that the first condition, $\Gamma \vdash_{\{t_k\}} \bot$, amounts to a sort of graded inconsistency condition for Γ (it reads $e(\psi_i) < \max T$ for any $\mathbf{GV_n}$-evaluation e and for any $\psi_i \in \Gamma$). On the other hand, the second condition, $\Gamma \vdash_{\{t\}} \varphi$ for all $t \in T$, amounts to require that φ follows from Γ in the logic defined by the set matrices $\mathcal{M}_T = \{\langle \mathbf{GV_{n\sim}}, F_t\rangle : t \in T\}$. Hence, φ follows from Γ in $\mathbb{L}(T)$ whenever, either Γ is inconsistent or contradictory to a certain degree (the maximum of T), or φ follows from Γ in the logic of $\mathcal{M}_T$. This makes it clear that the latter is a sublogic of the product matrix logic $\mathbb{L}(T)$.

The results become different when studying the matrix logics that involve components over finite subalgebras belonging to the variety generated by $\mathbf{GV_{n\sim}}$ because even though all of them are direct products of subalgebras of $\mathbf{GV_{n\sim}}$, the number of subalgebras of $\mathbf{GV_{n\sim}}$ is significantly larger than in the Łukasiewicz case. Indeed:

- Subalgebras of $\mathbf{GV_{n\sim}}$ are those chains that can be obtained from GV_n by removing a set of pairs of elements $\{a_i, \sim a_i\}$ with $a_i \notin \{0, 1\}$. In particular, if n is odd one can remove just the fixed point.
- Therefore, the logics between $\mathrm{G}^{\leq}_{n\sim}$ and CPL are those logics defined by matrices over direct products of subalgebras of $\mathbf{GV_{n\sim}}$ and with products of order filters on the corresponding components of the product algebra. Of course, we have to avoid the repetition of components in these products.

Example 6.1 Recall the logics J_3 and J_4 from Remark 6.3. Since $\mathbf{GV}_3$ and $\mathbf{GV}_4$ are subalgebras of $\mathbf{GV}_5$, by the characterization of all extensions of $\mathrm{G}^{\leq}_{n\sim}$ we have that $\mathsf{J}_3 \times \mathsf{J}_4$ coincides (up to language) with $\langle \mathbf{GV_{3\sim}} \times \mathbf{GV_{4\sim}}, F_{\frac{1}{2}} \times F_{\frac{1}{3}}\rangle$, hence it is a paraconsistent extension of $\mathrm{G}^{\leq}_{5\sim}$ that is comparable neither to J_3 nor to J_4. Indeed, it is immediate to see that $\vdash_{\mathsf{J}_3\times\mathsf{J}_4} \varphi$ iff $\vdash_{\mathsf{J}_3} \varphi$ and $\vdash_{\mathsf{J}_4} \varphi$ for every formula φ. Thus, since the theorems of J_3 and those of J_4 are not comparable, $\mathsf{J}_3 \times \mathsf{J}_4$ is an extension neither of J_3 nor of J_4.

On the other hand, it is also easy to check that $\varphi \vdash_{\mathsf{J}_3\times\mathsf{J}_4} \bot$ iff $\varphi \vdash_{\mathsf{J}_3} \bot$ or $\varphi \vdash_{\mathsf{J}_4} \bot$. Consider now the following formulas:

- $\alpha = \Delta(p \leftrightarrow \sim p)$;
- $\beta = \neg((p_1 \to p_2) \vee (p_2 \to p_3) \vee (p_3 \to p_4))$.

It is clear that $\alpha \nvdash_{\mathsf{J}_3} \bot$ while $\alpha \vdash_{\mathsf{J}_4} \bot$, and hence $\alpha \vdash_{\mathsf{J}_3\times\mathsf{J}_4} \bot$ as well. Analogously, we also have that $\beta \nvdash_{\mathsf{J}_4} \bot$ while $\beta \vdash_{\mathsf{J}_3} \bot$, and hence $\beta \vdash_{\mathsf{J}_3\times\mathsf{J}_4} \bot$. Thus, $\mathsf{J}_3 \times \mathsf{J}_4$, J_3 and J_4 are mutually not comparable.

Finally, we can characterize the logics satisfying the explosion rule for $\sim$:

$$\frac{\varphi \quad \sim\varphi}{\bot}.$$

Indeed, we have

- Following the same reasoning as in Coniglio et al. (2016) for the n-valued Łukasiewicz case, one can show that the minimal matrix logic satisfying the explosion rule, i.e., the expansion of $G^{\leq}_{n\sim}$ with the above rule is the logic L_{exp} whose consequence relation is defined as

$$\Gamma \vdash_{L_{exp}} \varphi \text{ iff either } \Gamma \vdash_{\{\frac{i}{n-1}\}} \bot, \text{ or } \Gamma \vdash_{G^{\leq}_{n\sim}} \varphi,$$

where i is the first natural such that $\frac{i}{n-1} > 1/2$. By manipulating the right-hand side of the above condition, $\Gamma \vdash_{L_{exp}} \varphi$ turns out to be equivalent to the two further conditions:

$$\begin{aligned}\Gamma \vdash_{L_{exp}} \varphi \text{ iff either } & \Gamma \vdash_{\{\frac{i}{n-1}\}} \bot \text{ or } (\Gamma \vdash_{T_1} \varphi \text{ and } \Gamma \vdash_{T_2} \varphi) \\ \text{iff } & [\Gamma \vdash_{\{\frac{i}{n-1}\}} \bot \text{ or } \Gamma \vdash_{T_1} \varphi] \text{ and } [\Gamma \vdash_{\{\frac{i}{n-1}\}} \bot \text{ or } \Gamma \vdash_{T_2} \varphi)]\end{aligned}$$

But, according to the paragraph after Definition 6.2, the condition $[\Gamma \vdash_{\{\frac{i}{n-1}\}} \bot$ or $\Gamma \vdash_{T_1} \varphi]$ is just saying that φ follows from Γ in the logic $\mathbb{L}(T_1)$, while the condition $[\Gamma \vdash_{\{\frac{i}{n-1}\}} \bot$ or $\Gamma \vdash_{T_2} \varphi)]$ is clearly equivalent to only $\Gamma \vdash_{T_2} \varphi$. Therefore,

$$L_{exp} = \mathbb{L}(T_1) \cap L(\mathcal{M}_{T_2}),$$

or in other words, L_{exp} is the logic defined by the following set of matrices:

$$\begin{aligned}\mathcal{M}_n = \{ & \langle (\mathbf{GV_{n\sim}})^i, \Pi^i_{r=1} F_{\frac{r}{n-1}} \rangle \} \cup \\ & \{ \langle \mathbf{GV_{n\sim}}, F_1 \rangle, \langle \mathbf{GV_{n\sim}}, F_{\frac{n-2}{n-1}} \rangle, \ldots, \langle \mathbf{GV_{n\sim}}, F_{\frac{i+1}{n-1}} \rangle \}.\end{aligned}$$

- Therefore, the explosion rule is valid in all the logics extending the logic L_{exp}. Hence, all of them are explosive, while those not extending it are paraconsistent.

6.4.2 Example: the Case n = 5

As an example, we study the case of the set $Int(G^{\leq}_{5\sim})$ of matrix logics defining intermediate logics between $G^{\leq}_{5\sim}$ and CPL. Recall that GV_5 denotes the ordered set $\{0, 1/4, 1/2, 3/4, 1\}$. We start with some basic facts:

- Consider the subset $L(\mathbf{GV_{5\sim}}) \subset Int(G^{\leq}_{5\sim})$ of logics defined by the set of matrices $\mathcal{M}_T = \{\langle \mathbf{GV_{5\sim}}, F_t \rangle : t \in T\}$ for $\emptyset \neq T \subseteq GV_5 \setminus \{0\}$, as it was done in Sect. 6.4.1. According to Proposition 6.4, the logics of the matrices $\langle \mathbf{GV_{5\sim}}, F_{i/4} \rangle$ for $i \in \{1, 2, 3, 4\}$ are pairwise incomparable, and in fact they are the maximal logics in $L(\mathbf{GV_{5\sim}})$, while $\bigcap_{i \in \{1,2,3,4\}} \langle \mathbf{GV_{5\sim}}, F_{i/4} \rangle = G^{\leq}_{5\sim}$ is the minimum logic of $L(\mathbf{GV_{5\sim}})$ (and clearly of $Int(G^{\leq}_{5\sim})$ as well).
- Let $L_{\Pi}(G_{5\sim}) \subset Int(G^{\leq}_{5\sim})$ be the set of matrix logics of the form $\mathbb{L}(T)$ defined on direct products of $\mathbf{GV_{5\sim}}$ by means of products of order filters (recall Definition 6.2).

Then, these logics satisfy the following conditions (like in the case of Łukasiewicz logics):

- If $\emptyset \neq T, R \subseteq GV_5 \setminus \{0\}$ are such that $\max\ T = \max\ R$, then $\mathbb{L}(T) \cap \mathbb{L}(R) = \mathbb{L}(T \cup R)$.
- The maximal elements of $L_{\Pi}(G_{5\sim})$ are the matrix logics of the type $\langle(\mathbf{GV_{5\sim}})^2, F_{i/4} \times F_{j/4}\rangle$ with $i, j \in \{1, 2, 3, 4\}$ and $i < j$.
- The matrix logic $\langle(\mathbf{GV_{5\sim}})^2, F_i \times F_j\rangle$ for $0 < i < j$ contains $\langle\mathbf{GV_{5\sim}}, F_j\rangle$ and it is not comparable with $\langle\mathbf{GV_{5\sim}}, F_k\rangle$ for $0 < k \neq j$.

- Finally, let us consider the subset $L_{\Pi^*}(G_{5\sim}) \subset Int(G^{\leq}_{5\sim})$ of matrix logics defined on direct products of $\mathbf{GV_{5\sim}}$ and their subalgebras together with direct products of order filters. The subalgebras of $\mathbf{GV_{5\sim}}$ are (isomorphic to) $\mathbf{GV_{2\sim}}$, $\mathbf{GV_{3\sim}}$, and $\mathbf{GV_{4\sim}}$, and thus the number of matrix logics in $L_{\Pi^*}(G_{5\sim})$ proliferate in a large number. Namely, to define matrix logics we have the following components to combine: four algebras, $\mathbf{GV_{5\sim}}$, $\mathbf{GV_{4\sim}}$, $\mathbf{GV_{3\sim}}$ and $\mathbf{GV_{2\sim}}$, and ten order filters: four over $\mathbf{GV_{5\sim}}$, three over $\mathbf{GV_{4\sim}}$, two over $\mathbf{GV_{3\sim}}$, and one over $\mathbf{GV_{2\sim}}$. Therefore, we have all the possible products (without repetitions) of these ten components.

We can also characterize the minimal extension of $G^{\leq}_{5\sim}$ with the explosion rule as the logic $L(\mathcal{M}_5)$ of the set of matrices

$$\mathcal{M}_5 = \{\langle(\mathbf{GV_{5\sim}})^3, F_{3/4} \times F_{2/4} \times F_{1/4}\rangle, \langle\mathbf{GV_{5\sim}}, F_1\rangle\}.$$

Concerning axiomatization, as in case of Łukasiewicz logics, we can give an axiomatic characterization of the logics of $L\Pi(G_{5\sim})$. To see this, first of all, observe that in $G_{5\sim}$, for every value $i/4 \in GV_5 \setminus \{0\}$ there exists a formula in one variable $\varphi(p)$ characterizing the value $i/4$, i.e., such that for any evaluation e, $e(\varphi(p)) = 1$ if $e(p) = i/4$, and 0 otherwise. For example, for the value $1/2$, the formula can be $\Delta(p \leftrightarrow {\sim}p)$. It is also possible to define a formula characterizing the sets of values $\geq i/4$, $> i/4$, $\leq i/4$, and $< i/4$.

Using this observation, it is easy to see that every matrix logic of type $\langle\mathbf{GV_{5\sim}}, F_{i/4}\rangle$ or $\mathbb{L}(T) \in L\Pi(G_{5\sim})$ can be axiomatized. For instance, here we give the following example:

- The matrix logic $\langle\mathbf{GV_{5\sim}}, F_{i/4}\rangle$ is axiomatized by adding to the axioms and rules of $G^{\leq}_{5\sim}$ the following restricted inference rule:

  if $\vdash_{G_{5\sim}} (\varphi < i/4) \vee ((\varphi \geq i/4) \wedge (\psi \geq i/4)$, from φ derive ψ.

Other matrix logics of $L\Pi(G_{5\sim})$ can be axiomatized in an analogous way. Notice that these axiomatizations are possible since, in $G_{5\sim}$, for every element $a \in GV_5$ there exists a characterizing formula in one variable. This is not true in $G_{n\sim}$ for $n > 5$, and thus the previous axiomatization results are not generalizable to $G_{n\sim}$ for $n > 5$.

6.5 Ideal and Saturated Paraconsistent Extensions of $G_{n\sim}^{\leq}$

As already noticed, matrix logics over direct products of subalgebras of $\mathbf{GV_{n\sim}}$ with products of order filters are $\sim$-paraconsistent iff all the components are $\sim$-paraconsistent. In this section, using the results of the previous section, we study the status of the logics between $G_{n\sim}^{\leq}$ and CPL in relation to ideal $\sim$-paraconsistency. Namely, we show that there are only two extensions of $G_{n\sim}^{\leq}$ which are ideal $\sim$-paraconsistent. Moreover, we show that there is another $\sim$-paraconsistent extension of $G_{n\sim}^{\leq}$ which, although not being ideal $\sim$-paraconsistent, it has the remarkable property of not having any proper $\sim$-paraconsistent extension.

We have already briefly discussed in the Introduction, the concept of *ideal paraconsistent logics*, introduced by Arieli et al. (2011b).[23] We recall here this notion.

Definition 6.3 (c.f. Arieli et al. (2011b)) Let L be a propositional logic defined over a signature Θ (with consequence relation $\vdash_L$) containing at least a unary connective $\urcorner$ and a binary connective $\rightarrow$ such that

(i) L is paraconsistent w.r.t. $\urcorner$ (or simply $\urcorner$-paraconsistent), that is, there are formulas $\varphi, \psi \in \mathcal{L}(\Theta)$ such that $\varphi, \urcorner\varphi, \nvdash_L \psi$.
(ii) $\rightarrow$ is an implication for which the deduction-detachment theorem holds in L, that is, $\Gamma \cup \{\varphi\} \vdash_L \psi$ iff $\Gamma \vdash_L \varphi \rightarrow \psi$, for every set for formulas $\Gamma \cup \{\varphi, \psi\} \subseteq \mathcal{L}(\Theta)$.
(iii) There is a presentation of CPL as a matrix logic $L' = \langle \mathbf{A}, \{1\}\rangle$ over the signature Θ such that the domain of $\mathbf{A}$ is $\{0, 1\}$, and $\urcorner$ and $\rightarrow$ are interpreted as the usual two-valued negation and implication of CPL, respectively.
(iv) L is a sublogic of CPL in the sense that $\vdash_L \subseteq \vdash_{L'}$, that is, $\Gamma \vdash_L \varphi$ implies $\Gamma \vdash_{L'} \varphi$, for every set for formulas $\Gamma \cup \{\varphi\} \subseteq \mathcal{L}(\Theta)$.

Then, L is said to be an *ideal $\urcorner$-paraconsistent logic* if it is maximal w.r.t. CPL, and every proper extension of L over Θ is not $\urcorner$-paraconsistent.

An implication connective satisfying the above condition (ii) is usually called *deductive implication*.

Remark 6.5 As it has been argued in Remark 6.3, J_3 is equivalent to $\langle \mathbf{GV}_{3\sim}, F_{\frac{1}{2}}\rangle$ and therefore for every odd number $n \geq 3$, J_3 is an extension of any $G_{n\sim}^{\leq}$ (recall Fig. 2.1). Similarly, J_4 is equivalent to $\langle \mathbf{GV}_{4\sim}, F_{\frac{1}{3}}\rangle$. Thus, J_4 is an extension of $G_{n\sim}^{\leq}$ for every $n \geq 4$.

---

[23] The authors, as it was mentioned in Sect. 6.1, have changed the terminology "ideal paraconsistent logic" to "fully maximal and normal paraconsistent logic." However, it should be noticed that being normal, according to Avron et al. (2018, Definition 1.32), means that the logic L has, besides a deductive implication, a conjunction and a disjunction satisfying the usual properties in terms of consequence relations. Here we decide to keep the original definition of ideal paraconsistency. It is worth noting that all the ideal (and saturated) logics considered in this paper and in Coniglio et al. (2019) are normal in the sense of Avron et al. (2018).

In Coniglio et al. (2019, Proposition 6.3), it is shown that J_3 and J_4 are ideal $\sim$-paraconsistent logics where the deductive implication in the signature of $G_{\sim}$ is the term-defined implication $x \Rightarrow y := \neg x \vee y$.[24]

As discussed in Sect. 6.1, requiring a paraconsistent logic to be maximal w.r.t. CPL in order to be "ideal" (in the sense of being "optimal") is a debatable issue (see Wansing and Odintsov (2016)). On the other hand, the other requirements of Definition 6.3 seem interesting, and a system enjoying such features could be considered as a remarkable paraconsistent logic. This motivates the following definition.

Definition 6.4 A logic L is *saturated* $\urcorner$-paraconsistent if it satisfies all the conditions (i) to (iv) of the previous definition, and every proper extension of L over Θ is not $\urcorner$-paraconsistent.[25]

Remark 6.6 In Ribeiro and Coniglio (2012, p. 273), it was introduced the notion of maximality of a logic L w.r.t. an inference rule r. Namely, given a Tarskian and structural propositional logic L over a signature Θ, and given an inference rule r over Θ, L is *r-maximal* if r is not derivable in L, but any proper extension of L over Θ derives r.[26] Clearly, ideal and saturated paraconsistent logics are special cases of r-maximal logics, where r is the explosion rule.[27]

Proposition 6.5 $J_3 \times J_4 := \langle \mathbf{GV_{3\sim}} \times \mathbf{GV_{4\sim}}, F_{\frac{1}{2}} \times F_{\frac{1}{3}} \rangle$ *is saturated $\sim$-paraconsistent, but not ideal $\sim$-paraconsistent.*

Proof Since $\mathbf{GV_{3\sim}}$ and $VG_{4\sim}$ are subalgebras of $\mathbf{GV_{5\sim}}$, by the characterization of all extensions of $G^{\leq}_{n\sim}$ given in Sect. 6.4.1, $\langle \mathbf{GV_{3\sim}} \times \mathbf{GV_{4\sim}}, F_{\frac{1}{2}} \times F_{\frac{1}{3}} \rangle$ is an extension of $G^{\leq}_{5\sim}$ satisfying conditions (i) to (iv) because every component is $\sim$-paraconsistent and $x \Rightarrow y := \neg x \vee y$ is a term-defined deductive implication. We prove by contradiction that $J_3 \times J_4$ has no proper $\sim$-paraconsistent extensions. Assume there is a proper $\sim$-paraconsistent extension L of $J_3 \times J_4$. In this case, there is a matrix $\langle \mathbf{A_1} \times \cdots \times \mathbf{A_k}, F_{i_1} \times \cdots \times F_{i_k} \rangle$ which is an extension of L such that each $\langle \mathbf{A_j}, F_{i_j} \rangle$ is either J_3, J_4, $\langle \mathbf{GV_{5\sim}}, F_{\frac{1}{2}} \rangle$ or $\langle \mathbf{GV_{5\sim}}, F_{\frac{1}{4}} \rangle$. Since J_3 is not comparable with $J_3 \times J_4$ and J_3 is a submatrix of $\langle \mathbf{GV_5}, F_{\frac{1}{2}} \rangle$ and also a submatrix of $\langle \mathbf{GV_{5\sim}}, F_{\frac{1}{4}} \rangle$, there is a component $\langle \mathbf{A_{j0}}, F_{j0} \rangle = J_4$. Similarly, there should be a different component $\langle \mathbf{A_{j1}}, F_{j1} \rangle \neq J_4$, otherwise J_4 would be an extension of $J_3 \times J_4$. Finally, in the case $\langle \mathbf{A_1} \times \cdots \times \mathbf{A_k}, F_{i_1} \times \cdots \times F_{i_k} \rangle$ has a component equal to J_4 and another which

[24] Observe that in Coniglio et al. (2019), $\neg$ denotes the Łukasiewicz negation, while the Gödel negation for J_3 and J_4 is, respectively, denoted by $\sim_{\frac{1}{2}}$ and $\sim_{\frac{1}{3}}$.

[25] Using the terminology of Avron et al. (2018), a saturated paraconsistent logic is a logic such that: (i) it has an implication, (ii) it is **F**-contained in CPL, and (iii) it is strongly maximal.

[26] This was denoted by $L \in \mathbf{Triv}_\Theta \bot \{r\}$ in Ribeiro and Coniglio (2012), where $\mathbf{Triv}_\Theta$ denotes the trivial logic over the signature Θ.

[27] Indeed, by means of the notion of *remainder set* $L \bot R$ of a logic L w.r.t. a set of rules R introduced in Ribeiro and Coniglio (2012, Definition 7), several concepts relative to maximality proposed in the literature can be easily represented, see Ribeiro and Coniglio (2012, p. 273).

is different to J_4, then $\mathsf{J}_3 \times \mathsf{J}_4$ is a submatrix of $\langle \mathbf{A_1} \times \cdots \times \mathbf{A_k}, F_{i_1} \times \cdots \times F_{i_k} \rangle$, which contradicts the fact that L is a proper extension of $\mathsf{J}_3 \times \mathsf{J}_4$.

Let φ be a theorem of J_3 which is not a theorem of J_4. Then, the matrix logic $J_2 \times \mathsf{J}_3 := \langle \mathbf{GV_{2\sim}} \times \mathbf{GV_{3\sim}}, F_1 \times F_{\frac{1}{2}} \rangle$ is an extension of $\mathsf{J}_3 \times \mathsf{J}_4$ different from CPL such that $\vdash_{J_2 \times \mathsf{J}_3} \varphi$. Thus, $\mathsf{J}_3 \times \mathsf{J}_4$ is not maximal w.r.t. CPL. □

Theorem 6.1 *Let n be an integer number such that $n > 4$ and let L be an extension of $\mathrm{G}^{\leq}_{n\sim}$.*

1. *If n is an even number, the following are equivalent:*
   - *L is saturated $\sim$-paraconsistent;*
   - *L is ideal $\sim$-paraconsistent;*
   - *$L = \mathsf{J}_4$.*
2. *If n is an odd number, the following are equivalent:*
   - *L is saturated $\sim$-paraconsistent;*
   - *$L = \mathsf{J}_3$, $L = \mathsf{J}_4$ or $L = \mathsf{J}_3 \times \mathsf{J}_4$.*
3. *If n is an odd number, the following are equivalent:*
   - *L is ideal $\sim$-paraconsistent;*
   - *$L = \mathsf{J}_3$ or $L = \mathsf{J}_4$.*

Proof 1. Assume that n is even. After Remark 6.5 and Proposition 6.5, we only need to prove that if L is saturated $\sim$-paraconsistent then $L = \mathsf{J}_4$. Since n is even then, as observed in Sect. 6.4.1, every extension L' of $\mathrm{G}^{\leq}_{n\sim}$ is induced by a family of matrices of the form $\langle \mathbf{A}, F \rangle = \langle \mathbf{GV_{n_1\sim}} \times \cdots \times \mathbf{GV_{n_k\sim}}, F_{\frac{i_1}{n_1-1}} \times \cdots \times F_{\frac{i_k}{n_k-1}} \rangle$ where each n_j is also an even number. If L' is $\sim$-paraconsistent then there is a member in that family, say $\langle \mathbf{GV_{n_1\sim}} \times \cdots \times \mathbf{GV_{n_k\sim}}, F_{\frac{i_1}{n_1-1}} \times \cdots \times F_{\frac{i_k}{n_k-1}} \rangle$, such that $2 < n_j \leq n$ and $0 < \frac{i_j}{n_j-1} \leq \frac{1}{2}$ for every j such that $1 \leq j \leq k$. Then, J_4 is an extension of every $\sim$-paraconsistent extension of $\mathrm{G}^{\leq}_{n\sim}$. In particular, J_4 extends L. Thus, $L = \mathsf{J}_4$, since L is maximal paraconsistent.

2. The right to left implication follows from Remark 6.5 and Proposition 6.5. To prove the converse, let L be a saturated $\sim$-paraconsistent extension of $\mathrm{G}^{\leq}_{n\sim}$. Since L is $\sim$-paraconsistent and it has no proper $\sim$-paraconsistent extension, L is induced by a single $\sim$-paraconsistent matrix $\langle \mathbf{A}, F \rangle$ such that $\langle \mathbf{A}, F \rangle = \langle \mathbf{GV_{n_1\sim}} \times \cdots \times \mathbf{GV_{n_k\sim}}, F_{\frac{i_1}{n_1-1}} \times \cdots \times F_{\frac{i_k}{n_k-1}} \rangle$ where $2 < n_j \leq n$ and $0 < \frac{i_j}{n_j-1} \leq \frac{1}{2}$ for every j such that $1 \leq j \leq k$. If all n_j's are even, as in previous item $L = \mathsf{J}_4$. If all n_j's are odd, then J_3 is a $\sim$-paraconsistent extension of L, thus $L = \mathsf{J}_3$. Assume n is odd and some n_j's are even and some are odd, all of them bigger than 2. Then in that case $\mathsf{J}_3 \times \mathsf{J}_4 := \langle \mathbf{GV_{3\sim}} \times \mathbf{GV_{4\sim}}, F_{\frac{1}{2}} \times F_{\frac{1}{3}} \rangle$ is a $\sim$-paraconsistent extension of L, thus $L = \mathsf{J}_3 \times \mathsf{J}_4$.
3. Immediate after Proposition 6.5 and item 2.

□

6.6 Saturated Paraconsistency and Finite-Valued Łukasiewicz Logics

In Coniglio et al. (2019), we study maximality conditions for intermediate logics between CPL and the degree-preserving finite-valued Łukasiewicz logics. In particular, we have characterized the ideal paraconsistent logics in this family. Since in the last section we have introduced the weaker notion of saturated paraconsistency in the setting of degree-preserving Gödel logics with involution, we deem interesting to also explore this notion for the abovementioned setting of finite-valued Łukasiewicz logics. This is done in this section, after briefly recalling some basic notions about (degree-preserving) finite-valued Łukasiewicz logics.

The $(n+1)$-valued Łukasiewicz logic can be semantically defined as the matrix logic

$$Ł_{n+1} = \langle \mathbf{ŁV}_{n+1}, \{1\}\rangle,$$

where $\mathbf{ŁV}_{n+1} = (ŁV_{n+1}, \neg, \to)$ is the $n+1$-elements MV-chain with $ŁV_{n+1} = \left\{0, \frac{1}{n}, \ldots, \frac{n-1}{n}, 1\right\}$, and operations defined as follows: for every $x, y \in ŁV_{n+1}$,

$\neg x = 1 - x$ (Łukasiewicz negation);
$x \to y = \min\{1, 1 - x + y\}$ (Łukasiewicz implication).

In fact, $Ł_{n+1}$ is algebraizable and its generated quasivariety $MV_{n+1} := \mathcal{Q}(\mathbf{ŁV}_{n+1})$ is its equivalent algebraic semantics.

The (finitary) degree-preserving $(n+1)$-valued Łukasiewicz logic, denoted $Ł^{\leq}_{n+1}$, can be semantically defined the following way: For every finite set of formulas $\Gamma \cup \{\varphi\}$

$$\Gamma \models_{Ł^{\leq}_{n+1}} \varphi \text{ if for every evaluation } e \text{ over } \mathbf{ŁV}_{n+1} \text{ and every } a \in ŁV_{n+1}, \\ \text{if } e(\gamma) \geq a \text{ for every } \gamma \in \Gamma, \text{ then } e(\gamma) \geq a.$$

Following Coniglio et al. (2019), we denote by L^i_n the logic obtained by the matrix $\langle \mathbf{ŁV}_{n+1}, F_{\frac{i}{n}}\rangle$, where $F_{\frac{i}{n}}$ is the order filter $\{x \in ŁV_{n+1} : x \geq i/n\}$. Notice that with this notation the $n+1$-valued Łukasiewicz logic $Ł_{n+1}$ is also denoted by L^n_n.

As proved in Coniglio et al. (2019, Theorem 5.2), for every $1 \leq i \leq n$, L^i_n is equivalent, as a deductive system, to $Ł_{n+1}$ (see Blok and Pigozzi (2001) and also Blok and Pigozzi (1991)). Since algebraizability is preserved by equivalence, L^i_n is algebraizable and MV_{n+1} is also its equivalent algebraic semantics. Thus, the lattice of finitary extensions of L^i_n is isomorphic to the lattice of sub-quasivarieties of $MV_{n+1} = \mathcal{Q}(\mathbf{ŁV}_{n+1})$.

MV_{n+1} is a locally finite variety and, as proved in Gispert and Torrens (2014), every sub-quasivariety is also locally finite and it is generated by a finite family of critical[28] MV-algebras. Using the correspondence among sub-quasivarieties of MV_{n+1} and finitary extensions of L^i_n, in Coniglio et al. (2019) we obtain that every

[28] An algebra is said to be *critical* if it is a finite algebra not belonging to the quasivariety generated by all its proper subalgebras.

extension L of L_n^i is induced by a finite family of matrices of type $\langle \mathbf{A}, F\rangle$ where $\mathbf{A}$ is a critical MV_{n+1}-algebra and F is a lattice filter of $\mathbf{A}$ compatible with L. To be more precise, in Gispert and Torrens (2014); Coniglio et al. (2019), it is proved that $\mathbf{A}$ is isomorphic to a direct product of MV_{n+1}-chains $\mathbf{ŁV}_{n_0+1} \times \cdots \times \mathbf{ŁV}_{n_{l-1}+1}$, where

1. For every $j < l$, $n_j | n$.
2. For every $j, k < l$, $k \neq j$ implies $n_k \neq n_j$.
3. If there exists n_j, $j < l$ such that $n_k | n_j$ for some $k \neq j$, then n_j is unique.

And $F = (F_{\frac{i}{n}})^l \cap (ŁV_{n_0+1} \times \cdots \times ŁV_{n_{l-1}+1})$.

Thus, in analogy to Coniglio et al. (2016, Theorem 3), every extension of L_n^i is induced by a finite family of matrices where each matrix is a product of submatrices of $\langle \mathbf{ŁV}_{n+1}, F_{\frac{i}{n}}\rangle$.

As observed in Proposition 6.2, $\mathbf{ŁV}_3$ is termwise equivalent to $\mathbf{GV}_{3\sim}$ and $\mathbf{ŁV}_4$ is termwise equivalent to $\mathbf{GV}_{4\sim}$, where the involutive negation $\sim$ in $\mathbf{GV}_{3\sim}$ and $\mathbf{GV}_{4\sim}$ is, in fact, the MV-negation $\neg$. Then, as indicated in Remark 6.5, the matrix logics $\mathsf{J}_3 = \langle \mathbf{ŁV}_3, F_{\frac{1}{2}}\rangle$ and $\mathsf{J}_4 = \langle \mathbf{ŁV}_4, F_{\frac{1}{3}}\rangle$, expressed in the signature of Łukasiewicz logic, are ideal $\neg$-paraconsistent. We recall here that this can be generalized in the following way.

Proposition 6.6 (Coniglio et al. 2019, Proposition 6.2) *Let q be a prime number, and let $1 \leq i < q$ such that $i/q \leq 1/2$. Then, L_q^i is a $(q+1)$-valued ideal $\neg$-paraconsistent logic.*

In fact, we can also prove that the converse implication also holds under some circumstances. This result is not present in Coniglio et al. (2019).

Theorem 6.2 *Let $0 < i < n$ such that $\frac{i}{n} \leq \frac{1}{2}$. If L is an extension of L_n^i, then L is ideal $\neg$-paraconsistent iff $L = \mathsf{L}_q^j$ for some prime number q such that $q|n$ and some $1 \leq j$ such that $j/q \leq 1/2$*

Proof Let L be an ideal $\neg$-paraconsistent extension of L_n^i. Since L is maximal, it is induced by a single matrix $\langle \mathbf{A}, F\rangle$, where $\mathbf{A}$ is critical and F is compatible with L. In fact, as mentioned above, $\langle \mathbf{A}, F\rangle$ is of type $\langle \mathbf{ŁV}_{n_1+1} \times \cdots \times \mathbf{ŁV}_{n_k+1}, (F_{\frac{i}{n}})^k \cap (ŁV_{n_1+1} \times \cdots \times ŁV_{n_k+1})\rangle$ where

1. For every $1 \leq i \leq k$, $n_i | n$.
2. For every $1 \leq i, j \leq k$, $i \neq j$ implies $n_i \neq n_j$.
3. If there exists n_j, $1 \leq j \leq k$ such that $n_i | n_j$ for some $1 \leq i \neq j$, then n_j is unique.

Since L is $\neg$-paraconsistent, none of the components $\mathbf{ŁV}_{n_i+1}$ can be $\mathbf{ŁV}_2$ (otherwise L would be explosive), and hence $1 < n_i$ for all $1 \leq i \leq k$. If $k > 1$, then

- If there is n_j, with $1 \leq j \leq k$, such that $n_i | n_j$ for some $1 \leq i \neq j$, then without loss of generality assume that $j = k$. In that case, $\langle \mathbf{ŁV}_{n_1+1} \times \cdots \times \mathbf{ŁV}_{n_{k-1}+1}, (F_{\frac{i}{n}})^{k-1} \cap (ŁV_{n_1+1} \times \cdots \times ŁV_{n_{k-1}+1})\rangle$ is a $\neg$-paraconsistent extension of L which contradicts the assumption of L being ideal $\neg$-paraconsistent.

- If there is no n_j, with $1 \leq j \leq k$, such that $n_i | n_j$ for some $1 \leq i \neq j$, then $n_k \neq n$ and L is not maximal because $\langle \mathbf{ŁV}_2 \times \mathbf{ŁV}_{n_k+1}, (F_{\frac{i}{n}})^2 \cap (ŁV_2 \times ŁV_{n_k+1})\rangle$ is an extension of L different from CPL and there is a formula φ valid in $\langle \mathbf{ŁV}_2 \times \mathbf{ŁV}_{n_k+1}, (F_{\frac{i}{n}})^2 \cap (ŁV_2 \times ŁV_{n_k+1})\rangle$ and not valid in L. A contradiction again.

Thus $k = 1$. In that case, n should be prime because otherwise for any prime number p such that $p|n$, $\langle \mathbf{ŁV}_{p+1}, F_{\frac{i}{n}} \cap ŁV_{p+1}\rangle$ would be an extension of L different from CPL and there is a formula φ valid in $\langle \mathbf{ŁV}_{p+1}, F_{\frac{i}{n}} \cap ŁV_{p+1}\rangle$ and not valid in L. □

As regard to saturated paraconsistency, we have the following results:

Theorem 6.3 *Let $0 < i < n$ such that $\frac{i}{n} \leq \frac{1}{2}$. Let*

$$X = \{p \ : \ p \ prime,\ p|n,\ F_{\frac{i}{n}} \cap ŁV_{p+1} = \{\frac{m}{p} \ : \ m \geq k\} \ and \ \frac{k}{p} \leq \frac{1}{2}\}.$$

For every finite subset $\{p_1, \ldots, p_j\} \subseteq X$, the logic defined by the matrix

$$\langle \mathbf{ŁV}_{p_1+1} \times \cdots \times \mathbf{ŁV}_{p_j+1}, (F_{\frac{i}{n}})^j \cap (ŁV_{p_1+1} \times \cdots \times ŁV_{p_j+1})\rangle$$

is saturated $\neg$-paraconsistent.

Proof By the previous result, $\langle \mathbf{ŁV}_{p_1+1} \times \cdots \times \mathbf{ŁV}_{p_j+1}, (F_{\frac{i}{n}})^j \cap (ŁV_{p_1+1} \times \cdots \times ŁV_{p_j+1})\rangle$ is an extension of L^i_n. Moreover, it is $\neg$-paraconsistent, because every component is $\neg$-paraconsistent. Let $\Rightarrow^i_n$ defined as $\varphi \Rightarrow^i_n \psi := \sim^i_n \varphi \vee \psi$ where $\sim^i_n(x)$ is the single variable McNaughton term such that for every $a \in ŁV_{n+1}$,

$$\sim^i_n(a) = \begin{cases} 0, \text{ if } a \geq \frac{i}{n} \\ 1, \text{ otherwise.} \end{cases}$$

Similar to the proof of Coniglio et al. (2019, Proposition 6.2), the logic $\langle \mathbf{ŁV}_{p_1+1} \times \cdots \times \mathbf{ŁV}_{p_j+1}, (F_{\frac{i}{n}})^j \cap (ŁV_{p_1+1} \times \cdots \times ŁV_{p_j+1})\rangle$ satisfies conditions (i) to (iv) in Definition 6.3, the definition of ideal $\neg$-paraconsistency. Let L be a $\neg$-paraconsistent extension of $\langle \mathbf{ŁV}_{p_1+1} \times \cdots \times \mathbf{ŁV}_{p_j+1}, (F_{\frac{i}{n}})^j \cap (ŁV_{p_1+1} \times \cdots \times ŁV_{p_j+1})\rangle$, then L is induced by a finite family of matrices $\langle \mathbf{A}, F\rangle$, where $\mathbf{A}$ is critical and F is compatible with L. Since L is $\neg$-paraconsistent, there is at least one matrix $\langle \mathbf{ŁV}_{n_0+1} \times \cdots \times \mathbf{ŁV}_{n_{l-1}+1}, (F_{\frac{i}{n}})^l \cap (ŁV_{n_0+1} \times \cdots \times ŁV_{n_{l-1}+1})\rangle$ where

1. for every $m < l$, $n_m | n$;
2. for every $m, k < l$, $k \neq m$ implies $n_k \neq n_m$;
3. if there exists n_m with $m < l$ such that $n_m | n_k$ for some $k \neq m$, then n_k is unique;

which is $\neg$-paraconsistent. Thus, for every $m < l$, it is the case that $2 \leq n_m$. Since $\langle \mathbf{ŁV}_{n_0+1} \times \cdots \times \mathbf{ŁV}_{n_{l-1}+1}, (F_{\frac{i}{n}})^l \cap (ŁV_{n_0+1} \times \cdots \times ŁV_{n_{l-1}+1})\rangle$ is an extension of $\langle \mathbf{ŁV}_{p_1+1} \times \cdots \times \mathbf{ŁV}_{p_j+1}, (F_{\frac{i}{n}})^j \cap (ŁV_{p_1+1} \times \cdots \times ŁV_{p_j+1})\rangle$, then $\langle \mathbf{ŁV}_{n_0+1} \times \cdots \times \mathbf{ŁV}_{n_{l-1}+1}, (F_{\frac{i}{n}})^l \cap (ŁV_{n_0+1} \times \cdots \times ŁV_{n_{l-1}+1})\rangle$ is a submatrix of $\langle \mathbf{ŁV}_{p_1+1} \times \cdots \times$

$\text{Ł}\mathbf{V}_{p_j+1}, (F_{\frac{i}{n}})^j \cap (\text{Ł}V_{p_1+1} \times \cdots \times \text{Ł}V_{p_j+1})\rangle$. Therefore, by Coniglio et al. (2019, Lemma 5.6), for every $m < l$ there is a $0 < k \leq j$ such that $n_m | p_k$, since $2 \leq n_m$ and p_k is prime, then $n_m = p_k$. Moreover, for every $0 < k \leq j$, there is $m < l$ such that $n_m | p_k$. Thus, $\text{Ł}\mathbf{V}_{p_1+1} \times \cdots \times \text{Ł}\mathbf{V}_{p_j+1} \cong \text{Ł}\mathbf{V}_{n_0+1} \times \cdots \times \text{Ł}\mathbf{V}_{n_{l-1}+1}$ and $L = \langle \text{Ł}\mathbf{V}_{p_1+1} \times \cdots \times \text{Ł}\mathbf{V}_{p_j+1}, (F_{\frac{i}{n}})^j \cap (\text{Ł}V_{p_1+1} \times \cdots \times \text{Ł}V_{p_j+1})\rangle$, proving that any proper extension of $\langle \text{Ł}\mathbf{V}_{p_1+1} \times \cdots \times \text{Ł}\mathbf{V}_{p_j+1}, (F_{\frac{i}{n}})^j \cap (\text{Ł}V_{p_1+1} \times \cdots \times \text{Ł}V_{p_j+1})\rangle$ is not $\neg$-paraconsistent. □

Remark 6.7 One may wonder whether the saturated $\neg$-paraconsistent logics identified in the above theorem are, in fact, ideal paraconsistent. However, it is easy to see that this is not the case unless they are of the type described in Theorem 6.2. Indeed, this is a consequence of the fact that the logics considered in Theorem 6.3 (and in Corollary 6.2) are extensions of logics of the type L_n^i, and in Theorem 6.2 we have exactly characterized which of these extensions are ideal paraconsistent.

Corollary 6.2 *Let* $\{p_1, \ldots, p_j\}$ *be any finite set of prime numbers, then* $\langle \text{Ł}\mathbf{V}_{p_1+1} \times \cdots \times \text{Ł}\mathbf{V}_{p_j+1}, F_{\frac{1}{p_1}} \times \cdots \times F_{\frac{1}{p_j}}\rangle$ *is saturated* $\neg$*-paraconsistent.*

Contrary to the case of $\mathrm{G}_{n\sim}^{\leq}$ in Theorem 6.1, not every saturated $\neg$-paraconsistent extension of L_n^i is of the type of the above corollary. For instance, L_{15}^7 is saturated $\neg$-paraconsistent. Indeed, it is a $\neg$-paraconsistent logic where $\Rightarrow_{15}^7$ is a deductive implication and $\mathsf{L}_1^1 = \mathrm{CPL}$ is a submatrix logic of L_{15}^7. Moreover, every proper extension L of L_{15}^7 is induced by a family of proper submatrices of L_{15}^7, of type $\langle \text{Ł}\mathbf{V}_{n_0+1} \times \cdots \times \text{Ł}\mathbf{V}_{n_{l-1}+1}, (F_{\frac{7}{15}})^l \cap (\text{Ł}V_{n_0+1} \times \cdots \times \text{Ł}V_{n_{l-1}+1})\rangle$ where at least there is some $j < l$ such that $n_j | 15$ and $n_j \neq 15$. Hence, n_j is either 1, 3, or 5, in which case the j-th component $\langle \text{Ł}\mathbf{V}_{n_j+1}, F_{\frac{7}{15}} \cap \text{Ł}V_{n_j+1}\rangle$ is not $\neg$-paraconsistent. Thus, L is not $\neg$-paraconsistent and, therefore, L_{15}^7 is saturated $\neg$-paraconsistent.

6.7 A Final Remark: Relationship to Logics of Formal Inconsistency

To conclude this section, we provide an additional analysis—from the point of view of paraconsistency—of the logics discussed in this paper. Recall from Sect. 6.1 the class of paraconsistent logics known as *logics of formal inconsistency* (**LFI**s). It is easy to see that all the paraconsistent logics considered in the present paper are, in fact, **LFI**s.

Indeed, in Ertola et al. (2015), it is shown that, if $L_\sim$ is the expansion of a core fuzzy logic L with an involutive negation $\sim$ where Δ is definable in $L_\sim$,[29] then $L_\sim^{\leq}$ is an **LFI** w.r.t. $\sim$, and the consistency operator is given by $\circ\varphi = \Delta(\neg\varphi \vee \varphi)$. This shows the following.

[29] This is the case of any pseudo-complemented logic L where Δ is definable as $\Delta\varphi := \neg{\sim}\varphi$, in particular, the case of Gödel fuzzy logic G.

Proposition 6.7 *All the paraconsistent logics based on Gödel fuzzy logic with involution* $G_{\sim}$ *and its finite-valued extensions* $G_{n\sim}$ *considered in this paper are* **LFI***s w.r.t.* $\sim$.

As for the paraconsistent logics based on finite-valued Łukasiewicz logics analyzed in this section, they are also **LFI**s, as the following result states.

Proposition 6.8 *Let* L *be one of the matrix logics in Proposition 6.6, or one of the products of matrix logics in Theorem 6.3. Then,* L *is an* **LFI** *w.r.t.* $\neg$.

Proof Concerning the logics of Proposition 6.6, by Coniglio et al. (2019, Proposition 6.3) we know that each logic L^i_n for $i/n \leq 1/2$ is an **LFI** w.r.t. $\neg$, where the consistency operator is given by $\circ^i_n\alpha := \sim^i_n(\alpha \wedge \neg\alpha)$. Here, $\sim^i_n$ is the unary connective defined as in the proof of Theorem 6.3. Now, let

$$L = \langle \mathbf{ŁV}_{p_1+1} \times \cdots \times \mathbf{ŁV}_{p_j+1}, (F_{\frac{i}{n}})^j \cap (ŁV_{p_1+1} \times \cdots \times ŁV_{p_j+1})\rangle$$

be one of the logics in Theorem 6.3 given by a product of matrix logics, for some $\{p_1, \ldots, p_j\} \subseteq X$. By definition of X, for every $1 \leq s \leq j$ there exists $1 \leq k_s < p_s$ such that $k_s/p_s \leq 1/2$ and $\langle \mathbf{ŁV}_{p_s+1}, F_{\frac{i}{n}} \cap ŁV_{p_s+1}\rangle = \mathrm{L}^{k_s}_{p_s}$. This means that $L = \mathrm{L}^{k_1}_{p_1} \times \cdots \times \mathrm{L}^{k_j}_{p_j}$. Using again Coniglio et al. (2019, Proposition 6.3) it follows that each $\mathrm{L}^{k_s}_{p_s}$ is an **LFI** w.r.t. $\neg$, with consistency operator $\circ^{k_s}_{p_s}$ defined as above. It is immediate to see that $\sim^i_n$ restricted to $ŁV_{p_s+1}$ coincides with $\sim^{k_s}_{p_s}$, and so $\circ^i_n$ restricted to $ŁV_{p_s+1}$ coincides with $\circ^{k_s}_{p_s}$, for every $1 \leq s \leq j$. Therefore, L is an **LFI** w.r.t. $\neg$, with consistency operator given by $\circ\alpha := \circ^i_n\alpha$.

Indeed, it is clear that $\varphi, \neg\varphi, \circ\varphi \vdash_L \psi$ for every formulas φ, ψ. Let q and r be two different propositional variables, and let e be an evaluation over L such that $e(q) = 1$ and $e(r) = 0$. This ensures that $q, \circ q \nvdash_L r$. On the other hand, any evaluation e' over L such that $e'(q) = e'(r) = 0$ guarantees that $\neg q, \circ q \nvdash_L r$. Hence, L is an **LFI** w.r.t. $\neg$ and $\circ$ (recall the definition of **LFI**s in Carnielli and Marcos (2000); Carnielli et al. (2007); Carnielli and Coniglio (2016)). □

6.8 Conclusions

In this paper, the Gödel fuzzy logic G expanded with an involutive negation $\sim$, denoted $G_{\sim}$, together with its finite-valued extensions $G_{n\sim}$, was studied from the point of view of paraconsistency. In order to do this, the respective degree-preserving companions $G^{\leq}_{\sim}$ and $G^{\leq}_{n\sim}$ were analyzed given that, in contrast to $G_{\sim}$ and $G_{n\sim}$, these logics are $\sim$-paraconsistent. Observe that G coincides with $G^{\leq}$, since G satisfies the deduction-detachment theorem; hence, the addition of an involutive negation $\sim$ to G allows to develop such kind of study. The question of determining the lattice of intermediate logics between $G^{\leq}_{\sim}$ and CPL, as well as the logics between $G^{\leq}_{n\sim}$ and CPL, was addressed. After introducing the concept of saturated paraconsistent logic, which is weaker than the notion of ideal paraconsistency, it was shown that there are only three saturated paraconsistent logics between $G^{\leq}_{n\sim}$ and CPL, two of them (J_3 and J_4) being, in fact, ideal paraconsistent and the other (namely, $J_3 \times J_4$) being saturated but not ideal. Finally, the study of finite-valued Łukasiewicz logic we started in Coniglio et al. (2019) has been taken up again, in order to find additional interesting examples of saturated but not ideal paraconsistent logics.

As for future work, we aim at performing similar studies for other locally finite fuzzy logics, in particular, for the Nilpotent Minimum logic (NM) (Esteva et al. 2001) that combines and shares many features of both Gödel and Łukasiewicz logics. It is indeed logically equivalent to Gödel logic with involution when NM is expanded with the Baaz-Monteiro operator Δ.

Acknowledgements The authors are indebted to two anonymous reviewers for their careful and insightful suggestions and remarks that have definitively helped to improve the paper. Coniglio acknowledges support from the National Council for Scientific and Technological Development (CNPq), Brazil, under research grant 306530/2019-8. Gispert acknowledges partial support by the Spanish FEDER/MINECO projects (PID2019-110843GA-100, MTM2016-74892 and MDM-2014-044) and grant 2017-SGR-95 of Generalitat de Catalunya. Esteva and Godo acknowledge partial support by the Spanish FEDER/MINECO project TIN2015-71799-C2-1-P and the project PID2019-111544GB-C21.

References

Arieli, O., & Avron, A. (2017). Four-valued paradefinite logics. *Studia Logica, 105*(6), 1087–1122.

Arieli, O., Avron, A., & Zamansky, A. (2010). Maximally paraconsistent three-valued logics. In F. Lin, et al. (Eds.), *Principles of Knowledge Representation and Reasoning: Proc. of the Twelfth International Conference, (KR 2010)* (pp. 9–13). AAAI Press.

Arieli, O., Avron, A., & Zamansky, A. (2011a). Maximal and premaximal paraconsistency in the framework of three-valued semantics. *Studia Logica, 97*(1), 31–60.

Arieli, O., Avron, A., & Zamansky, A. (2011b). Ideal paraconsistent logics. *Studia Logica, 99*(1–3), 31–60.

Avron, A. (2005). Non-deterministic matrices and modular semantics of rules. In J.-Y. Beziau (Ed.), *Logica Universalis* (pp. 149–167). Birkhäuser Basel.

Avron, A. (2007). Non-deterministic semantics for logics with a consistency operator. *International Journal of Approximate Reasoning, 45*, 271–287.

Avron, A. (2009). Modular semantics for some basic logics of formal inconsistency. In W. A. Carnielli, M. E. Coniglio, & I. M. L. D' Ottaviano (Eds.), *The Many Sides of Logic* (Vol. 21, pp. 15–26), Studies in Logic. College Publications.

Avron, A. (2016). Paraconsistent fuzzy logic preserving non-falsity. *Fuzzy Sets and Systems, 292*, 75–84.

Avron, A. (2017). Self-extensional three-valued paraconsistent logics. *Logica Universalis, 11*(3), 297–315.

Avron, A. (2019). Paraconsistency and the need for infinite semantics. *Soft Computing, 23*, 2167–2175.

Avron, A., Arieli, O., & Zamansky, A. (2010). On strong maximality of paraconsistent finite-valued logics. In *Proceedings of the 25th Annual IEEE Symposium on Logic in Computer Science, LICS 2010, 11-14 July 2010, Edinburgh, United Kingdom* (pp. 304–313). IEEE Computer Society.

Avron, A., Arieli, O., & Zamansky, A. (2018). *Theory of Effective Propositional Paraconsistent Logics* (Vol. 75), Studies in Logic. College Publications.
Avron, A., & Lev, I. (2001). Canonical propositional Gentzen-type systems. In *Proceedings of the 1st International Joint Conference on Automated Reasoning (IJCAR 2001)* (Vol. 2083, pp. 529–544), LNAI. Springer.
Avron, A., & Zamansky, A. (2011). Non-deterministic semantics for logical systems–a survey. In D. M. Gabbay & F. Guenthner (Eds.), *Handbook of Philosophical Logic* (2nd ed., Vol. 16, pp. 227–304). Springer.
Běhounek, L., Cintula, P., & Hájek, P. (2011). Introduction to mathematical fuzzy logic. In P. Cintula, P. Hájek, & C. Noguera (Eds.), *Handbook of Mathematical Fuzzy Logic – volume 1, Studies in Logic, Mathematical Logic and Foundations* (Vol. 37, pp. 1–101). London: College Publications.
Blackburn, P., de Rijke, M., & Venema, Y. (2002). *Modal Logic*. Cambridge: Cambridge University Press.
Blok, W.J., & Pigozzi, D. (2001). Abstract algebraic logic and the deduction theorem. http://www.math.iastate.edu/dpigozzi/papers/aaldedth.pdf
Blok, W. J., & Pigozzi, D. (1991). Local deduction theorems in algebraic logic. In *Algebraic logic (Budapest, 1988)* (Vol. 54, pp. 75–109), Colloq. Math. Soc. János Bolyai. Amsterdam: North-Holland.
Bou, F., Esteva, F., Font, J. M., Gil, A., Godo, L., Torrens, A., et al. (2009). Logics preserving degrees of truth from varieties of residuated lattices. *Journal of Logic and Computation*, *19*(6), 1031–1069.
Burris, S., & Sankappanavar, H. P. (1981). *A Course in Universal Algebra*. Berlin: Springer.
Carnielli, W. A., & Coniglio, M. E. (2016). Paraconsistent logic: consistency, contradiction and negation. *Logic, Epistemology, and the Unity of Science* (Vol. 40). Springer.
Carnielli, W. A., Coniglio, M. E., & Marcos, J. (2007). Logics of formal inconsistency. In D. M. Gabbay & F. Guenthner (Eds.), *Handbook of Philosophical Logic* (2nd ed., Vol. 14, pp. 1–93). Springer.
Carnielli, W. A., & Marcos, J. (2002). A taxonomy of C-systems. In W. A. Carnielli, M. E. Coniglio, & I. M. L. D'Ottaviano (Eds.), *Paraconsistency: The Logical Way to the Inconsistent. Proceedings of the 2nd World Congress on Paraconsistency (WCP 2000)* (Vol. 228, pp. 1–94), Lecture Notes in Pure and Applied Mathematics. New York: Marcel Dekker.
Coniglio, M. E., Esteva, F., & Godo, L. (2014). Logics of formal inconsistency arising from systems of fuzzy logic. *Logic Journal of the IGPL*, *22*(6), 880–904.
Coniglio, M. E., Esteva, F., & Godo, L. (2016). On the set of intermediate logics between truth and degree preserving Łukasiewicz logic. *Logical Journal of the IGPL*, *24*(3), 288–320.
Coniglio, M. E., Esteva, F., Gispert, J., & Godo, L. (2019). Maximality in finite-valued Łukasiewicz Logics defined by order filters. *Journal of Logic and Computation*, *29*(1), 125–156.
Ertola, R. C., Esteva, F., Flaminio, T., Godo, L., & Noguera, C. (2015). Paraconsistency properties in degree-preserving fuzzy logics. *Soft Computing*, *19*(3), 531–546.
Esteva, Francesc, & Godo, Lluís. (2001). Monoidal t-norm based logic: Towards a logic for left-continuous t-norms. *Fuzzy Sets and Systems*, *124*(3), 271–288.
Esteva, F., Godo, L., Hájek, P., & Navara, M. (2000). Residuated fuzzy logics with an involutive negation. *Archive for Mathematical Logic*, *39*, 103–124.
Esteva, F., Godo, L., & Marchioni, E. (2011). Fuzzy Logics with Enriched Language. In P. Cintula, P. Hájek, & C. Noguera (Eds.), *Handbook of Mathematical Fuzzy Logic, Vol. II, Chap. VIII* (Vol. 38, pp. 627–711), Studies in Logic, Mathematical Logic and Foundations. College Publications.
Font, J. M. (2009). Taking degrees of truth seriously. *Studia Logica*, *91*(3), 383–406.
Font, J. M., Gil, A., Torrens, A., & Verdú, V. (2006). On the infinite-valued Łukasiewicz logic that preserves degrees of truth. *Archive for Mathematical Logic*, *45*, 839–868.
Gispert, J., & Torrens, A. (2014). Locally finite quasivarieties of MV-algebras (pp. 1–14). arXiv:1405.7504.
Hájek, P. (1998). *Metamathematics of Fuzzy Logic* (Vol. 4), Trends in Logic. Kluwer Academic Publishers.

Odintsov, S. P. (2008). *Constructive Negations and Paraconsistency* (Vol. 26), Trends in Logic. Springer.

Ribeiro, M. M., & Coniglio, M. E. (2012). Contracting logics. In L. Ong & R. de Queiroz (Eds.), *Logic, Language, Information and Computation. WoLLIC 2012* (Vol. 7456, pp. 268–281)., Lecture Notes in Computer Science Berlin: Springer.

Wansing, H., & Odintsov, S. P. (2016). On the Methodology of Paraconsistent Logic. In H. Andreas & P. Verdée (Eds.), *Logical Studies of Paraconsistent Reasoning in Science and Mathematics* (Vol. 45, pp. 175–204), Trends in Logic. Springer.

Marcelo E. Coniglio is a Full Professor in Logic at the Philosophy Department of the Institute of Philosophy and the Humanities (IFCH) of the University of Campinas (UNICAMP), and he is Director of the Centre for Logic, Epistemology and the History of Science (CLE) of UNICAMP. He was President of the Brazilian Logic Society (SBL) from 2014 to 2017. He obtained his Ph.D. in Mathematics from the University of S˜ao Paulo (Brazil) in 1997. He has published more than 60 refereed articles and published and edited 6 books on different subjects of Non-Classical Logic, with emphasis on Paraconsistent Logics and Combination of Logics.

Francesc Esteva obtained his Ph.D. in Mathematics in 1974 from the University of Barcelona. His research work is mainly devoted to fuzzy logic and applications. In the 90s, he participated (with the leadership of Prof. Petr H´ajek) in the creation of the field of Mathematical Fuzzy Logic. Since then, he has worked mainly in this topic but he has also worked in similarity-based reasoning and its application to case-based reasoning in modal many-valued logic, in fuzzy description logic, and in fuzzy paraconsistent logics. He has been Director of the Artificial Intelligence Research Institute (IIIA-CSIC), President of FLAT (Spanish Association for Fuzzy Logic and Technology), and First President of EUSFLAT (European Association for Fuzzy logic and Technology). He is Honorary Member of EUSFLAT and IFSA Fellow. Currently, he is Professor "at honorem" at the IIIA-CSIC.

Joan Gispert is an Associate Professor at the Mathematics and Computer Science Department of the University of Barcelona. He obtained his Ph.D. in Mathematics from the University of Barcelona in 1998 with a thesis on finitary extensions on the infinite-valued Lukasiewicz logic, under the advise of Antoni Torrens. His research publications are mainly on algebraic and non-classical logics.

Lluis Godo is a Research Professor at IIIA-CSIC, the Artificial Intelligence Research Institute (IIIA) of the Spanish National Research Council (CSIC), Barcelona, Spain. He obtained his M.Sc. degree in Mathematics from the University of Barcelona (1979) and the Ph.D. in Mathematics from the Technical University of Catalunya (1990). His main research interests include logics for Artificial Intelligence (AI), in particular, graded uncertainty reasoning formalisms, mathematical fuzzy logic, and argumentation systems. He is co-Editor-in-Chief of the Journal of Fuzzy Sets and Systems, and Associate Editor of the journals Artificial Intelligence, Soft Computing and Autonomous Agents, and Multi-Agent Systems. He is an ECCAI Fellow and an IFSA Fellow.

Chapter 7
R-Mingle is Nice, and so is Arnon Avron

J. Michael Dunn

Abstract Arnon Avron has written: "Dunn-McCall logic RM is by far the best understood and the most well-behaved in the family of logics developed by the school of Anderson and Belnap." I agree. There is the famous saying: "Do not let the perfect become the enemy of the good." I might say: "good enough." In this spirit, I will examine the logic R-Mingle, exploring how (in the terminology of Avron) it is only a "semi-relevant logic" but still a paraconsistent logic. I shall discuss the history of RM, and compare RM to Anderson and Belnap's system R of relevant implication and to classical two-valued logic. There is a "consumer's guide," evaluating these logics as "tools," in the light of my recent work on "Humans as Rational Toolmaking Animals."

Keywords Arnon Avron · R-Mingle · RM · Relevant · Semi-relevant · Paraconsistent · Logics as tools · Consumer's guide

7.1 Introduction

Arnon Avron was nice to write a nice paper "RM and its Nice Properties" (Avron 2016) as a nice chapter of the nice Springer Outstanding Contributions to Logic volume about (nice) me. (OK, so I am *moi.*) In return—there was no collusion—I have titled this chapter, "R-Mingle Has Nice Properties, and So Does Arnon Avron."

I was trying to think when I first met Arnon Avron. I can't be sure when that was, but I know that we corresponded in the mid-1980s when he was working on his Ph.D. dissertation *The Semantics and Proof Theory of Relevance Logics and Nontrivial Theories Containing Contradictions* (1985). We have continued to correspond since then, and also spent some time together on a number of different occasions

J. Michael Dunn died in April 2021. He was professor emeritus at IU.

J. M. Dunn (✉)
Luddy School of Informatics, Computing and Engineering, and Department of Philosophy, Indiana University, 901 East 10th Street, Bloomington, IN 47408–3912, USA
e-mail: dunn@indiana.edu

O. Arieli and A. Zamansky (eds.), *Arnon Avron on Semantics and Proof Theory of Non-Classical Logics*, Outstanding Contributions to Logic 21,
https://doi.org/10.1007/978-3-030-71258-7_7

(conferences, workshops, short visits to each other's universities, etc.). It is always enjoyable to talk to Arnon Avron about logic, and other things.

Arnon Avron has a long and strong connection to the logic **R-M**ingle (**RM** for short), and to relevance logics (and paraconsistent logics) more generally. Since his dissertation, I estimate that he has written over 100 articles and approximately 40 of these are directly or indirectly on relevance logic, and at least 8 are about **R-M**ingle. I do not know of a single researcher, including myself, who has done so much work on **RM**. Storrs McCall and I may be the "parents" of **RM**, but then Arnon Avron is one of its favorite uncles (Bob Meyer would certainly be another).

In the nice Abstract of his nice paper (Avron 2016), Arnon Avron says: "Dunn-McCall logic **RM** is by far the best understood and the most well-behaved in the family of logics developed by the school of Anderson and Belnap." **RM** has a simple axiomatization, simply add the "mingle" axiom $A \to (A \to A)$ to the Anderson and Belnap (1975) system **R** of relevant implication, but as we shall see there are complications in this that need to be unraveled.

I surely do not disagree with Arnon Avron's characterization of **RM**. Indeed I shall argue here that **RM** deserves more respect than it has gotten. I shall base my appraisal from the perspective of "logics as tools" as I discussed in Dunn (2018), using my list of considerations for choosing tools listed there. There does not seem to be the one perfect tool, and similarly I think that there is not one perfect logic. Some logics have a desirable feature, say meeting the strong Variable Sharing criterion of relevance of Anderson and Belnap, but they achieve this by missing another desirable feature, say decidability of their propositional fragment.

Arnon Avron and I are not the first to defend the "niceness" of**RM**. Robert K. Meyer, who first proved **RM** complete with respect to the class of Sugihara matrices, says in one of his contributions to Anderson and Belnap (1975), p. 393:

> I like Mingle, as the reader will discover, for two reasons. (There is a tension between them.) First, its theory of deduction, due to Dunn and McCall both in motivation and formulation, simplifies the Church theory, set out in sec. 3, in a reasonable way. Second, it is useful for many of the things that **R** might be good for, while being much more easily visualized. On the other hand, as noted in sec. 8, this involves the breakdown of the relevance principle, undermining the *raison d' etre* of the enterprise.
>
> As I argue, sometimes one doesn't need the whole relevance principle, and, on these occasions, **RM** is good enough, when some relevance is desirable. Indeed, it has proved very useful by illuminating in a context that is technically simple some of the relevance features shared among **RM**, **R**, and **E**. Nevertheless, one must confess that the system happened by accident; nobody thought that a mild modification of **R** would produce a system this strong.

What Meyer is referring to is that the simple addition of the seemingly relevant implication $A \to (A \to A)$ to the system **R** leads to irrelevant implications such as $(A \wedge \backsim A) \to (B \vee \backsim B)$. It turns out that these are limited, since, as Meyer showed, the following holds:

Proposition 7.1 *Weak Variable Sharing Property (WVSP): $A \to B$ is a theorem of **RM** only if either (i) A and B share a sentential variable, or (ii) $\backsim A$ and B are theorems of **RM**.*

This is to be contrasted with the stronger Variable Sharing Property of the Anderson and Belnap relevance logics **R** and **E**.

Proposition 7.2 *Variable Sharing Property (VSP): $A \to B$ is a theorem of* ***R*** *(or* ***E****) only if A and B share a sentential variable.*

Meyer goes on to say after the passages quoted above:

> Accordingly, it now seems to me that the two poles of motivation must be viewed as conflicting. Despite my affection for **RM** (I like its initials, especially), I agree in the end with the principle authors; all things considered **R** is the superior system.

I shall be arguing below that **RM** might instead be the superior system when all things are considered, and might especially be so in light of certain important considerations. I shall be doing this after first reviewing some of the "relevant" (pun intended) work on **RM,** including some of Arnon Avron's nice results. The history of **RM** is a bit scattered so I thought I would be doing a service by pulling some of it together. I refer the reader who would like to learn more to Anderson and Belnap (1975), Anderson et al. (1992), Dunn (1986) (and/or its updated version Dunn and Restall (2002). In particular Anderson and Belnap (1975) includes contributions by Robert Meyer and myself that are important to **RM**. Many technical details can be found in these publications, including proofs of theorems of **RM** and related systems that may only be sketched here.

7.2 Background. The Systems R of Relevant Implication and E of Entailment

Since **RM** is an extension of the Anderson–Belnap system **R** of Relevant Implication, we will start there. For its implicational fragment $\mathbf{R}_{\to}$, we take the rule *modus ponens* $(A, A \to B \vdash B)$ and the following axiom schemes:

$$A \to A \quad \text{Self-Implication,} \tag{7.1}$$

$$(A \to B) \to [(C \to A) \to (C \to B)] \quad \text{Prefixing,} \tag{7.2}$$

$$[A \to (A \to B)] \to (A \to B) \quad \text{Contraction,} \tag{7.3}$$

$$[A \to (B \to C)] \to [B \to (A \to C)] \quad \text{Permutation.} \tag{7.4}$$

This formulation is due to Church (1951), who called it "The weak implication calculus." Church formulated this before Anderson and Belnap even thought of relevance logic, but it is equivalent to the implicational fragment of their system **R** of Relevant Implication. He remarks that the axioms are the same as those of Hilbert's positive implicational calculus (the implicational fragment of the intuitionistic propositional calculus) except that (1) is replaced with

$$A \to (B \to A) \quad \text{Positive Paradox.}$$

There are a variety of alternative ways to axiomatize the system. Thus, (2) Prefixing may be replaced by

$$(A \to B) \to [(B \to C) \to (A \to C)] \quad \text{Suffixing.}$$

Note that from either Suffixing or Prefixing the following rule is derivable:
$A \to B, B \to C \vdash A \to C$ Transitivity.
(3) Contraction may be replaced by

$$[A \to (B \to C)] \to [(A \to B) \to (A \to C)] \quad \text{Self-Distribution,}$$

and (4) Permutation may be replaced by

$$A \to [(A \to B) \to B] \quad \text{Assertion.}$$

These choices of implicational axioms are nicely compartmentalized in that one choice does not affect another.

The implicational fragment $\mathbf{E}_{\to}$ of the Anderson–Belnap system **E** of Entailment combines necessity with relevant implication to get a strict, relevant implication. It may also be axiomatized in a number of equivalent ways. Let us start with the first axioms that were given above for $\mathbf{R}_{\to}$ and simply replace Permutation with

$$[A \to (\vec{B} \to C)] \to [\vec{B} \to (A \to C)] \quad \text{Restricted Permutation,}$$

where the arrow on top of the B indicates it must be of the form $B_1 \to B_2$. This is in analogy to a standard axiom for the Lewis system **S4** of strict implication (See Hacking (1963)). Similarly, Assertion can be replaced with

$$\vec{A} \to [(\vec{A} \to B) \to B] \quad \text{Restricted Assertion.}$$

As for negation, the implication–negation fragments of **R** and **E**, named $\mathbf{R}_{\to}^{\backsim}$ and $\mathbf{E}_{\to}^{\backsim}$, may be obtained by adding to their axioms the following:

$$(A \to \backsim A) \to \backsim A \quad \text{Reductio,} \tag{7.5}$$

$$(A \to \backsim B) \to (B \to \backsim A) \quad \text{Contraposition,} \tag{7.6}$$

$$\backsim\backsim A \to A \quad \text{Classical Double Negation.} \tag{7.7}$$

The axioms of Reductio and Contraposition are intuitionistically acceptable negation principles, but of course Classical Double Negation is not. Its converse though is intuitionistically acceptable and is proven as follows:

1. $(\backsim A \to \backsim A) \to (A \to \backsim\backsim A)$ Contraposition $(\backsim A/A, A\,/B)$
2. $\backsim A \to \backsim A$ Instance of Identity
3. $(A \to \backsim\backsim A)$ 1, 2, *modus ponens*

Using Classical Double Negation one can derive forms of reductio and contraposition that are intuitionistically unacceptable:

$$(\backsim A \to A) \to A \quad \text{Classical Reductio,}$$
$$(\backsim A \to \backsim B) \to (B \to A) \quad \text{Classical Contraposition.}$$

In the presence of both forms of Double Negation, all of the contraposition principles are equivalent.

Next, we add the positive extensional connectives $\wedge$ and $\vee$, in order to obtain **R** and **E**, with the following axioms:

$$(A \wedge B) \to A, \ (A \wedge B) \to B \quad \text{Conjunction Elimination,} \tag{7.8}$$
$$[(A \to B) \wedge (A \to C)] \to [A \to (B \wedge C)] \quad \text{Conjunction Introduction,} \tag{7.9}$$
$$A \to (A \vee B), \quad B \to (A \vee B) \quad \text{Disjunction Introduction,} \tag{7.10}$$
$$[(A \to C) \wedge (B \to C)] \to [(A \vee B) \to C] \quad \text{Disjunction Elimination,} \tag{7.11}$$
$$[A \wedge (B \vee C)] \to [(A \wedge B) \vee C] \quad \text{Distribution.} \tag{7.12}$$

plus the rule of Adjunction: $A, B \vdash A \wedge B$.

Axioms (7.8)–(7.11) can readily be seen to be encoding the usual elimination and introduction rules for conjunction and disjunction into axioms, giving $\wedge$ and $\vee$ what might be called "the lattice properties."

Distribution (Axiom 12) is continually problematic. It causes inelegancies in the natural deduction systems and is an obstacle to finding decision procedures.[1]

7.3 The Creation of R-Mingle

We must begin with a quick explanation of "multi-sets" versus sets, and their use in understanding the premises Γ in a deduction $\Gamma \vdash A$. Consider the seemingly obvious deduction $A \vdash A$. This seems to clearly state that the sentence A is a consequence of itself. But how many occurrences of itself? How does it compare to $A, A \vdash A$? Normally, this would not seem to make a difference, because Γ is often taken as a set. This is in effect what is said by the usual structural rules: Permutation, Contraction, and Expansion—a special case of Weakening where an already existing premise is duplicated:

$$\Gamma, A \vdash B \quad \text{(Expansion).}$$

[1] Dunn (1986) contains a natural amendment to the Anderson and Belnap natural deductions system allowing subproofs to have finite sequences of formulas as their lines. These are interpreted as extensional conjunctions of formulas. This is in sympathy with the amendments to Gentzen-systems made independently in Mints (1976) and Dunn (1973), where commas are replaced by two different kinds of punctuation (commas and semicolons in Dunn 1973), one interpreted as ordinary extensional conjunction and the other interpreted as "intensional conjunction" (often called "fusion").

$$\overline{\Gamma, A, A \vdash B}$$

But what if we don't have all of these? So-called "Substructural Logics" are based on the idea that one might not have all of these structural rules. Relevance logics indeed lack Thinning, but do all of them have the special case Expansion? Well, it is clear that $\mathbf{LR}_{\rightarrow}$ does not, for otherwise, we would have a proof of Mingle:

$$\dfrac{\dfrac{\dfrac{A \vdash A}{A, A \vdash A}}{A \vdash A \rightarrow A}}{\vdash A \rightarrow (A \rightarrow A)}$$

Identity
Expansion
$(\vdash\rightarrow)$
$(\vdash\rightarrow)$

I hope I have said enough that the reader can make sense of what Arnon Avron says in his nice paper (Avron 2016, p. 16):

> It somewhat looks strange to take relevant entailment as a relation between *multi-sets* of formulas and formulas, rather than between *sets* of formulas and formulas (as consequence relations are usually and most naturally taken to be). This observation motivated J. M. Dunn and S. McCall in investigating the results of adding to **R** and its fragments the mingle axiom $\varphi \rightarrow (\varphi \rightarrow \varphi)$ considered above. In the case of $\mathbf{R}_{\rightarrow}$, this addition yields $\mathbf{RM0}_{\rightarrow}$, which is the minimal system in which the above criterion for relevant entailment is met, with the latter taken as a relation between *sets* of formulas and formulas. In the case of the full system **R**, it yields a very interesting system called **RM** ("**R**-mingle").

I actually appreciate Arnon Avron's description of how **RM** came to be invented, and it resonates with me. There is just one small problem with it, namely, it is not true. The system **RM** has achieved enough fame that I think it would be useful to give a short history of how it came to be studied. The main point is that it was accidentally discovered/created like the Microwave Oven, the Post-It Note, Penicillin, etc.

It is often called, as Arnon Avron does, the "Dunn-McCall system **R**-**M**ingle." That is true enough. I took Storrs McCall's logic seminar in my first year as a graduate student at the University of Pittsburgh and I was in writing a paper for this course where I was trying to show that the implicational fragment $\mathbf{E}_{\rightarrow}$ of **E** was equivalent to the intersections of that of $\mathbf{R}_{\rightarrow}$ and that of some other logic. Saul Kripke had already communicated to Nuel Belnap that the other logic was not **S4**. This was a natural choice given that the natural deduction system for the system $\mathbf{E}_{\rightarrow}$ is the result of combining the restrictions on the rule of $\rightarrow$-introduction of $\mathbf{R}_{\rightarrow}$ with those of **S4**. This encouraged the thought that the system **E** could be regarded as "the best of both," combining in its implication both the relevant implication of **R** with the strict implication of **S4**.

The counterexample that Kripke provided was $A \rightarrow [(A \rightarrow (A \rightarrow B)) \rightarrow (A \rightarrow B)]$. Belnap went on to show that the system **S3** does not work either, giving the counterexample $[((A \rightarrow B) \rightarrow B) \rightarrow A] \rightarrow (A \rightarrow A)$.

My description above relies heavily on Anderson and Belnap (1975), p. 94, where you can find more details. There they also state the following:

> PROBLEM. There is along these lines one unsettled conjecture, raised by Storrs McCall: is $\mathbf{E}_{\rightarrow}$ the intersection of $\mathbf{R}_{\rightarrow}$ with the system obtained by adding "restricted mingle" $(A \rightarrow B) \rightarrow [(A \rightarrow B) \rightarrow (A \rightarrow B)]$ ... to $\mathbf{E}_{\rightarrow}$?

There clearly is a system such as Anderson and Belnap describe, and they (and others) often call it "**E-Mingle**," or "**EM.**" However, I recall that the system that McCall actually suggested to me was to add to **E** the unrestricted Mingle axiom $A \to (A \to A)$ (cf. McCall 1963), and out of respect for Storrs I shall call this system **E-Mingle**, or **EM**. I will call the system **E** plus Restricted Mingle **E-**$\overrightarrow{\mathbf{M}}$ingle, or $\mathbf{E}\overrightarrow{\mathbf{M}}$**.** I will not bother to make a similar distinction between **RM** and $\mathbf{R}\overrightarrow{\mathbf{M}}$, because they are equivalent.[2] Indeed, I will continue to use **RM** because of its currency in the literature. Storrs told me about the non-strict "mingle rule" of Onishi and Matsumoto and suggested that I add this to **E**. This ultimately led to my exploring the system obtained by adding this rule to the relevance logic **R** and ultimately to my paper (Dunn 1970) showing that every normal extension of the resulting logic "**R-Mingle**" has a finite characteristic matrix.[3]

So **R-Mingle** came to be, for me anyway, as a non-modal relative of **E-Mingle**, whose implicational fragment $\mathbf{EM}_{\to}$ is a possible solution to the equation $\mathbf{R}_{\to} \cap \mathbf{X} = \mathbf{E}_{\to}$. Curiously, I ended up not pursuing that question in my term paper but instead focused on proving Cut Elimination for the Gentzen system that is the result of adding the mingle rule to the Gentzen system for $\mathbf{E}_{\to}$ So 57 years or so after I wrote the paper for Storrs McCall's class in 1963, I publish now a description of that paper, and also say some more things about "mingle." I am sorry to say though that after all these years the problem about whether **E** is the intersection of **R** and **E-Mingle** still remains open.

7.4 Ohnishi and Matsumoto's System $\mathfrak{S}$ of "Strict Implication"

I begin by summarizing (Ohnishi and Matsumoto 1962). As Thomas (1970) points out in his review, "The system is designed to avoid paradoxes of irrelevance in the sense of Anderson–Belnap." I find this ironic since Ohnishi and Matsumoto title their paper "A System for Strict Implication." and yet the Lewis and Langford modal systems of strict (necessary) implication have been a nemesis of Anderson and Belnap's system **E** of entailment. Despite the name, Ohnishi and Matsumoto were seeking a system that avoided various paradoxes of relevance as well as paradoxes of necessity. This is very much in the spirit of the system **E**. They list some of these in their first sentence. Using the standard notation from Anderson and Belnap, these include:

$$A \to (B \to B), A \to (B \vee \backsim B), (A \wedge \backsim A) \to B, A \to (B \to A).$$

[2] Thus, clearly, $(A \to B) \to [(A \to B) \to (A \to B)]$ is a special case of $A \to (A \to A)$. For the other direction, note that it is straightforward to prove in $\mathbf{R}_{\to}$, and hence in $\mathbf{RM}_{\to}$, $A \leftrightarrow [(A \to A) \to A]$. Thus, if we replace B with $(A \to A) \to A$, it is easy to get the unrestricted mingle axiom from the restricted Mingle axiom.

[3] Thus, it is "pretabular" to use the term (Scroggs 1951) used when he showed the modal logic **S5** to have this property. Often it has since been known as the "Scroggs Property."

Ohnishi and Matsumoto formulate their system $\mathfrak{S}$ in a way based on Gentzen's sequent formulation of*LK*.[4] A sequent is a triple $\Gamma \vdash \Theta$, where Γ and Θ are finite sequences of formulas. The axioms are just all formulas of the form $A \vdash A$.

There are Gentzen's "structural rules": *Exchange* (we will call it *Permutation*) that allows adjacent formulas in the sequences to be permuted, *Contraction* that allows one of two adjacent duplicate formulas to be removed, and *Weakening* that allows additional formulas to be added. There are also natural "operational rules" for the connectives of negation $\backsim$, conjunction $\wedge$, and disjunction $\vee$. We also have the following rules for strict implication:

$$\frac{\Gamma \vdash \Theta, A \qquad B, \Delta \vdash \Lambda}{A \to B, \Gamma, \Delta \vdash \Theta, \Lambda} \quad (\to\vdash)$$

$$\frac{A, \overrightarrow{\Gamma} \vdash B}{\overrightarrow{\Gamma} \vdash A \to B.} \quad (\vdash\to)$$

Finally, there is of course the namesake rule:

$$\frac{\Gamma \vdash \Theta \qquad \Sigma \vdash \Pi}{\Gamma, \Sigma \vdash \Theta, \Pi} \quad \text{(Mingle)}$$

The first rule $(\to\vdash)$ is the same as the corresponding rule for *L***K**. The last of these, Mingle, might seem curious since one could obtain the conclusion from either premise alone by repeated Weakenings on both sides, together with some Permutations. However, there are two restrictions on the rule $(\vdash\to)$: (i) $\overrightarrow{\Gamma}$ is either empty or consists only of formulas of the form $A \to B$; (ii) the use of the rule Weakening (sometimes called Thinning) may not occur in the proof above the premise.[5] (i) is perfectly natural for a Gentzen system for strict (necessary) implication. But the way the system $\mathfrak{S}$ handles (ii) is what gives Mingle its power. The way that the standard Gentzen systems for substructural logics handle restrictions on the structural rules is to just drop them, or maybe uniformly modify them. Ohnishi and Matsumoto do not just drop Weakening; they rather limit its use in proving implications on the right.

---

[4] We use $\vdash$ in place of Gentzen's $\to$ because we use $\to$ as the connective that Ohnishi and Matsumoto denote using the Lewis "fishhook."

[5] Ohnishi and Matsumoto seem to state one exception to (2) in their footnote 4, which may not be immediately clear. "The rules of mingle which are of weakened form of weakening may appear." All this means is that special cases of the rule Mingle which might be taken as weakenings are still allowed since they are licensed by the rule Mingle. Consider:

$$\frac{A \vdash A \quad A \vdash A}{A, A \ \vdash A, A} \ \text{(Mingle).}$$

This allows curious proofs such as

$$\begin{array}{cl}
A \vdash A & \text{Identity} \\ \hline
\backsim A, A \vdash & (\backsim\vdash) \\ \hline
A, A \wedge \backsim A \vdash & (\wedge \vdash) \\ \hline
A \wedge \backsim A, A \wedge \backsim A \vdash & \text{Permutation, } (\wedge \vdash) \\ \hline
A \wedge \backsim A \vdash & \text{Contraction} \\ \hline
A \wedge \backsim A \vdash B & \text{Weakening}
\end{array}$$

Note that we are just one step away from writing

$$\vdash (A \wedge \backsim A) \to B),$$

but we are prevented from doing so because of the use of Weakening.

Let's mention one more rule, Gentzen's famous:

$$\frac{\Gamma \vdash A, \Theta \qquad \Sigma, A \vdash \Pi}{\Gamma, \Sigma \vdash \Theta, \Pi} \qquad \text{(Cut).}$$

This rule is extremely important because it contains a special case:

$$\frac{\vdash A \qquad A \vdash B}{\vdash B} \qquad (\text{“\textit{Modus ponens}”}).$$

The quotes are there to indicate that this is not literally *modus ponens* because it has $\vdash$ between A and B instead of $\to$, with $\vdash$ in front of A. But it is easy to show that $\vdash A \to B$ and $A \vdash B$ are interderivable. This is critical to showing that $\mathfrak{S}$ contains the classical propositional calculus **TV**.

But Cut could be an obstacle to decidability. To prove $\vdash B$, let's try to find a formula A that entails B and is such that $\vdash A$ is provable.

Hmm, but to prove $\vdash A$, let's try to find an A' such that A' entails A and such that $\vdash A'$ is provable. Etc. Gentzen's brilliant insight was to show that the Cut rule was redundant—the so-called Cut Elimination Theorem—Gentzen called it "Hauptzatz," (Chief Proposition). All of the other rules have the Subformula Property, any formula that appears in a premise also occurs in the conclusion (though perhaps embedded in another formula). All of Ohnishi and Matsumoto's rules have this property, including their rule Mingle. This contains proof searches in a good way, though there are still a few more needed tricks which we shall not get into here.

Ohnishi and Matsumoto state (correctly) that the implicational fragment $\mathbf{E}_{\to}$ of Anderson and Belnap's system **E** of entailment is contained within $\mathfrak{S}$, but they do not say anything about the whole system **E** itself. The restriction on Weakening prevents the proof of the Distribution axiom, which could otherwise go like this:

$$\begin{array}{ccl}
\dfrac{\dfrac{\dfrac{A \vdash A}{A, B \vdash A} \quad \dfrac{B \vdash B}{A, B \vdash B}}{A, B \vdash A \wedge B}}{A, B \vdash (A \wedge B) \vee C} & \dfrac{C \vdash C}{C \vdash (A \wedge B) \vee C} & \begin{array}{l}\text{Weakening (with Permutation)} \\ (\vdash \wedge) \text{ Identity} \\ (\vdash \vee) \text{ twice}\end{array} \\ \hline
\multicolumn{2}{c}{A, (B \vee C) \vdash ((A \wedge B) \vee C)} & (\vee \vdash) \\ \hline
\multicolumn{2}{c}{(A \wedge (B \vee C)), (A \wedge (B \vee C)) \vdash ((A \wedge B) \vee C)} & (\wedge \vdash) \text{ twice} \\ \hline
\multicolumn{2}{c}{(A \wedge (B \vee C)) \vdash ((A \wedge B) \vee C)} & \text{Contraction} \\ \hline
\multicolumn{2}{c}{\vdash (A \wedge (B \vee C)) \to ((A \wedge B) \vee C)} & (\vdash\to) \ ???
\end{array}$$

I have put the "???" is there to indicate that the restriction (ii) on ($\vdash\rightarrow$) has been violated by the Weakenings in the second steps, and so the rule ($\vdash\rightarrow$) cannot apply.

I wonder how to give a Hilbert-style axiomatization for $\mathfrak{S}$. The problem is the way it is formulated, which is kind of a cross-over between classical logic and relevance logic. The restrictions on the ($\vdash\rightarrow$) seem to raise problems. I leave this as an open problem.

Thomas (1970), in his review of Ohnishi and Matsumoto, says:

> It is asserted, and easily checked, that the Anderson-Belnap **E** is contained here. It seems to the reviewer that since **LE**, a Gentzen-type version of **E** used by Belnap and Wallace, contains (i) but not (ii), the latter restriction must be superfluous, and the systems are equivalent.

The (i) and (ii) refer to the two restrictions on the rule ($\vdash\rightarrow$).

Thomas is wrong about the equivalence on two counts. First, he seems to overlook the fact that $\mathfrak{S}$ has the rule Mingle, which easily leads to the proof of $\vdash (A \rightarrow B) \rightarrow [(A \rightarrow B) \rightarrow (A \rightarrow B)]$, which is not a theorem of E.

Second, he seems to overlook the subscript I on the Belnap and Wallace Gentzen system $\mathbf{LE}_I$. While there are Gentzen systems for the implication (and even the implication–negation) fragments of **E** and **R,** it has been notoriously difficult to give a Gentzen system for the whole of **E** (and **R**), and in particular to prove Distribution. And as we saw above Distribution is not provable in $\mathfrak{S}$. The best attempts to provide a Gentzen system for **R** are due independently to Mints (1976) and Dunn (1973), and involve the idea of two different ways of combining sentences, the usual "extensional conjunctive" way where the comma is interpreted on the left as extensional conjunction, and a new "intensional conjunctive" way where a semi-colon is added in addition to the comma and is interpreted as intensional conjunction (the right-hand sides are singular, i.e., consist of a single sentence). This provides a Gentzen system that is both sound and complete for the positive fragment of **R**, but no one has succeeded in adding just the usual De Morgan negation.[6] Belnap (1982) "Display Logic" might be seen as doing this, but it does so at the price of having classical (Boolean) negation as well.

Naively one might have thought that the implicational fragment of **RM** was simply the result of adding $A \rightarrow (A \rightarrow A)$ to the implicational fragment of **R**, $\mathbf{R}_{\rightarrow}$. But it turns out to be much more complicated than that.

One of my favorite sayings (attributed to Voltaire, and others) is "Do not let the perfect become the enemy of the good." If the purpose is to have an Anderson-Belnap type relevance logic, then **RM** is far from perfect. It has the following "bad" theorems:

Safety: $(A \wedge \backsim A) \rightarrow (B \vee \backsim B)$

Ex Falso Verum[7]: $\backsim A \rightarrow (B \rightarrow (A \rightarrow B))$

Weak Chain axiom: $(A \rightarrow B) \vee (B \rightarrow A)$.

---

[6] Bimbó (2009) provides a corresponding result for the positive fragment of **E** . See Bimbó (2014) for a a good exposition of this result and similar results for related systems.

[7] This name "From Falsity, Truth" is my invention and is an obvious play on words of the well-known Ex Falso Quodlibet, "From a Falsity Everything Follows."

None of these look like the sort of theorem that should result from adding the Mingle axiom $A \to (A \to A)$ to **R**, but Robert K. Meyer showed they did.

First, he showed: (*) $A \to (\backsim A \to A)$

Proof:

1. $\backsim A \to (\backsim A \to \backsim A)$ Mingle axiom
2. $\backsim A \to (A \to A)$ 1, Contraposition and Double Negation.
3. $A \to (\backsim A \to A)$ 2, Permutation

And then Meyer proved Safety:

1. $[(A \vee \backsim A) \wedge (B \vee \backsim B)] \to \{\backsim [(A \vee \backsim A) \wedge (B \vee \backsim B)] \to [(A \vee \backsim A) \wedge (B \vee \backsim B)]\}$ Substitution in (*) above
2. $(A \vee \backsim A) \wedge (B \vee \backsim B)$ Excluded Middle and Adjunction
3. $\backsim [(A \vee \backsim A) \wedge (B \vee \backsim B)] \to [(A \vee \backsim A) \wedge (B \vee \backsim B)]$ 1, 2 *modus ponens*
4. $[(A \wedge \backsim A) \vee (B \wedge \backsim B)] \to [(A \vee \backsim A) \wedge (B \vee \backsim B)]$ 3, De Morgan's Laws
5. $(A \wedge \backsim A) \to (B \vee \backsim B)$ 4, $\vee$-introduction and $\wedge$-elimination

Any other two theorems A and B can be plugged in instead of the two excluded-middles, and so we have generally

(**) If A and B are theorems then $\backsim A \to B$ is a theorem.

This is interesting because although its proof involves conjunction/disjunction, a special instance of it can involve only implication and negation. A particularly interesting case of this is:

Implication/Negation Safety : $\backsim (A \to A) \to (B \to B)$.

From this, we can prove Ex Falso Verum as follows:

1. $\backsim (B \to B) \to (A \to A)$ Implication/Negation Safety
2. $A \to (\backsim (B \to B) \to A)$ 1, Permutation
3. $A \to (\backsim A \to (B \to B))$ 2, Contraposition and Double Negation
4. $\backsim A \to (A \to (B \to B))$ 3, Permutation
5. $\backsim A \to (B \to (A \to B))$ 4, Permutation

We next prove the:

Strong Chain axiom: $\backsim (A \to B) \to (B \to A)$.

1. $\backsim A \to (B \to (A \to B)$ Ex Falso Verum.
2. $\backsim A \to (\backsim (A \to B) \to \backsim B)$
3. $\backsim (A \to B) \to (\backsim A \to \backsim B)$
4. $\backsim (A \to B) \to (B \to A)$

Given that $(\backsim C \to D) \to (C \vee D)$ is a well-know theorem of **R** (and **E**), and hence of **RM**, the Weak Chain axiom follows from the Strong Chain axiom. The question then arises as to whether theorems such as Implication/Negation Safety, Ex Falso Verum, and the Strong Chain axiom have proofs in the implication-negation fragment of **RM**, but just what is this fragment? And for that matter, what is the implicational fragment? We discuss these, and some other questions concerning fragments, in the next section.

7.5 Various Fragments of R-Mingle—What a Nightmare!

Actually, it begins as a pleasant dream and becomes a nightmare as the splendid bubbles of that dream begin to burst. But in the end, it is only a nightmare until we finally sort through the confusion of the dream and finally wake up. Then it is pleasant again. I will do my best to explain.

We begin not with a pleasant bubble but rather a firm mattress foundation for our sleep, the implicational fragment $\mathbf{R}_{\to}$ of the logic of relevant implication **R**. As we saw $\mathbf{R}_{\to}$ can be very nicely axiomatized. It is pleasant then to think that the implicational fragment $\mathbf{RM}_{\to}$ of **R**-**M**ingle could be axiomatized by just adding the Mingle axiom $A \to (A \to A)$. In fact, it is easy to show that this resulting system **RM**0 satisfies the Variable Sharing Property, as we shall show. But wait a minute! What is the "0" doing there? Let's not spoil the pleasant dream yet. Let us go on to say that we shall show that if the **R** axioms for extensional conjunction and disjunction $\wedge$ and $\vee$ are added, together with the rule of Adjunction, the VSP still holds. Still very pleasant. We get the system $\mathbf{RM}0^{+}$. But darn, there is that troubling "0" again. Let's go back and try adding negation to $\mathbf{RM}0_{\to}$, to obtain the implication–negation system $\mathbf{RM}0^{\sim}_{\to}$. Again this can straightforwardly be shown to satisfy VSP. So where is the unhappy ending? Well of course it is when you put everything together into one system, namely **RM**. This lacks the VSP (though it does have a Weak VSP). Note well that we do not say **RM**0, though I guess we could. But we are talking about the original **RM**, the system that results from adding the Mingle axiom to **R**.

The implicational system $\mathbf{RM}0_{\to}$ can be axiomatized in several different (equivalent) ways:

$$A \to (A \to A) \quad \text{Mingle} \tag{7.13}$$

$$(A \to B) \to [A \to (A \to B)] \text{Expansion} \tag{7.14}$$

$$(A \to C) \to [(B \to C) \to (A \to (B \to C))] \quad \text{O-M Mingle} \tag{7.15}$$

The first axiom is the one I played with in my graduate term paper, the second of these corresponds to the Gentzen structural rule Expansion that is the dual of Contraction (and a special case of Weakening), and the third one corresponds to Ohnishi and Matsumoto's rule "Mingle." This is why it as the label "O-M Mingle."

Meyer and Parks (1972) observes that the theorems of **RM** always take designated values in the following matrix due to Sobociński.

$\to$	0	1	2	$\sim$
0	2	2	2	2
*1	0	1	2	1
*2	0	0	2	0

$\wedge$	0	1	2
0	0	0	0
*1	0	1	1
*2	0	1	2

$\vee$	0	1	2
0	0	1	2
*1	1	1	2
*2	2	2	2

Notice that there is a natural way of seeing the elements of the matrix as being ordered. Defining $x \leq y$ iff $x \to y$ is designated," we obtain the following chain:

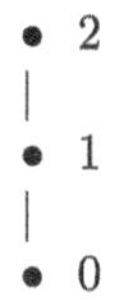

A Sugihara matrix more generally is any chain with a 1-1 order inverting mapping $\backsim$ of it onto itself. The designateds are those elements x such that $\backsim x \leq x$. The negation operation is of course just $\backsim$, and the other operations are defined as follows: $a \wedge b = \min(a, b)$; $a \vee b = \max(a, b)$; $a \to b = \backsim a \vee b$ if $a \leq b$, and $\backsim a \wedge b$ otherwise. An important example is the set of integers Z with the non-negative integers designated. Also, there is a natural sequence of finite examples: $S_1 = \{+1, -1\}$, $S_1 + 0 = \{+1, 0, -1\}$, $S_2 = \{+2, +1, -1, -2\}, \cdots$.

Sugihara matrices were introduced by Sugihara (1955) and Meyer (1968) showed that **RM** is sound and complete with respect to the class of Sugihara matrices.

Parks' matrix is isomorphic to the Sugihara matrix $S_1 + 0$:

$\to$	+1	0	−1	$\backsim$
*+1	+1	−1	−1	−1
*0	+1	0	−1	0
−1	+1	+1	+1	+1

$\wedge$	+1	0	−1
*+1	+1	0	−1
*0	0	0	−1
−1	−1	−1	−1

$\vee$	+1	0	−1
*+1	+1	+1	+1
*0	+1	0	0
−1	−1	0	−1

Anderson and Belnap (1962) use the four-valued Sugihara matrix S_2 to show the VSP for $\mathbf{E}_{\to}$. They do not say so, but it works just as well for $\mathbf{R}_{\to}$, and so does the three-valued Sugihara matrix $S_1 + 0$.

It is well known since Meyer (1968) that Sugihara matrices are sound for **RM**. It is easy to check that the axioms of **RM** are always designated (the designated values +1 and 0 are marked by *), and that *modus ponens* preserves designation. So in particular, all the theorems of **RM0** receive a designated value (either +1 or 0) for every assignment to their propositional variables. And for two sentences A and B that do not share a propositional variable, we may assign all the propositional variables in A the value +1 and all those in B the value 0. It is straightforward to see that A must take the value +1 and that B must take the value 0, and since +1→0 = 0, then $A \to B$ is undesignated and so cannot be a theorem of $\mathbf{RM}_{\to}$. So $\mathbf{RM0}_{\to}$ satisfies the VSP.

Let us next add conjunction and disjunction to $\mathbf{RM0}_{\to}$, to obtain the positive system $\mathbf{RM0}^{+}$. It is easy to see that if all the variables in A are assigned +1, then A receives the value +1, and if all the variables in B are assigned 0, then B receives the value 0, so the same argument as just above shows that $\mathbf{RM0}^{+}$ satisfies the VSP.

Now what about adding negation to $\mathbf{RM0}_{\to}$, say Reductio, Contraposition, and Classical Double Negation? Because negation "flips and flops," it is not so easy to falsify $A \to B$, even when A and B do not share a propositional variable. But as Meyer and Parks (1972) showed, we can do so if we have a negation with two fixed points (not just one as with the 0 in a Sugihara matrix). Here is Parks' matrix from Parks (1972) (note that $\backsim 1 = 1$, and $\backsim 2 = 2$).[8]

[8] The reader with some knowledge of relevance logic might think that Parks' matrix is built upon the Belnap–Dunn four-valued logic (cf. Dunn 1976d), so that if one defines $a \leq b$ iff $a \to b$ is

$\rightarrow$	0	1	2	3	$\backsim$
0	3	3	3	3	3
*1	0	1	0	3	1
*2	0	0	3	2	2
*3	0	0	0	3	0

If A and B share no propositional variables, every variable in A can be assigned 1, while every variable in B is assigned 2. It is easy to see then that A will then end up with the value 1 while B will have the value 2. And since $1 \rightarrow 2 = 0$ (undesignated), $A \rightarrow B$ will be undesignated and hence not a theorem of $\mathbf{RM0}^{\backsim}_{\rightarrow}$.

So you can add negation to $\mathbf{RM0}_{\rightarrow}$ and still retain VSP, and we already saw that you can add conjunction/disjunction to $\mathbf{RM0}_{\rightarrow}$ and still retain VSP. But what about doing both? "The third time is a charm," as the saying goes, but in this case, it is a bad charm, not a good one. Adding both gives us the axioms for **RM** and we already saw, or Meyer already saw, that this leads to Safety. We have only the weak VSP, not the full one.

There must be theorems of **RM** in just implication that are not provable in $\mathbf{RM0}_{\rightarrow}$, and again there must be theorems of **RM** in just implication and negation that are not provable in $\mathbf{RM0}^{\backsim}_{\rightarrow}$.

Meyer and Parks (1972) showed that the following axiom when added to $\mathbf{RM0}_{\rightarrow}$ gives the full implicational fragment of **RM:**

$$((((A \rightarrow B) \rightarrow B) \rightarrow A) \rightarrow C) \rightarrow (((((B \rightarrow A) \rightarrow A) \rightarrow B) \rightarrow C) \rightarrow C)$$

This is clearly related to the Chain Property.

Meyer and Parks (1972) observed that the axioms of Sobociński (1952) characterize the implication-negation fragment of **RM**. It is easy to see that these axioms can be taken to be the axioms for $\mathbf{R}^{\backsim}_{\rightarrow}$ plus Ex Falso Verum.

S1 $(A \rightarrow B) \rightarrow [(B \rightarrow C) \rightarrow (A \rightarrow C)]$ Suffixing
S2 $A \rightarrow [(A \rightarrow B) \rightarrow B]$ Assertion
S3 $[A \rightarrow (A \rightarrow B)] \rightarrow (A \rightarrow B)$ Contraction
S4 $A \rightarrow (B \rightarrow (\backsim B \rightarrow A))$ Variant of Ex Falso Verum
S5 $(\backsim A \rightarrow \backsim B) \rightarrow (B \rightarrow A)$ Classical Contraposition

S4 is a variant of Ex Falso Verum $\backsim A \rightarrow (B \rightarrow (A \rightarrow B))$—a little relettering along with with Permutation and Double Negation will get you from one to the other. It is an example of an implication-negation formula that is provable in $\mathbf{RM}^{\backsim}_{\rightarrow}$ but not provable in $\mathbf{RM0}^{\backsim}_{\rightarrow}$.

Note that $A \rightarrow A$ (Identity) is missing as an axiom. That is OK if we can somehow derive the Mingle axiom $A \rightarrow (A \rightarrow A)$, since it follows then by Contraction.

Proof of Mingle axiom $A \rightarrow (A \rightarrow A)$ from Ex Falso Verum:

1. $\backsim A \rightarrow (\backsim A \rightarrow (A \rightarrow \backsim A))$ Instance of Ex Falso Verum
2. $\backsim A \rightarrow (A \rightarrow \backsim A)$ 2, Contraction

designated one gets the familiar diamond De Morgan lattice (3 at the top, 0 on the bottom, 1 and 2 each on a different side). But the Belnap–Dunn lattice has no operation for $\rightarrow$ (just a relation) and more importantly only two of its elements are designated.

3. $A \to (\backsim A \to \backsim A)$
4. $(\backsim A \to \backsim A) \to (A \to A)$ Classical Contraposition
5. $A \to (A \to A)$ 3, 4 Transitivity

Parks' observation has an amazing consequence, namely, that the Implication–negation fragment of **RM** is characterized by the three-element Sugihara matrix $S_1 + 0$. So, all the extra work of the Sugihara matrix on the integers has to do with conjunction/disjunction. An algebraic view on the fragments of RM is expounded in Blok and Raftery (2004).

Let us now mention extensions of the system **E** with mingle-type axioms. Of course one can simply add the mingle axiom above. As we said above we shall simply call this **EM.** Adding any of the following to **E** gives the usual restricted **E**-**M**ingle, which we have chosen to name as **E**-$\overrightarrow{\mathbf{M}}$ingle, or **E**$\overrightarrow{\mathbf{M}}$.

$$\overrightarrow{A} \to (\overrightarrow{A} \to \overrightarrow{A}) \text{ Restricted Mingle,} \tag{7.16}$$

$$(\overrightarrow{A} \to B) \to [\overrightarrow{A} \to (\overrightarrow{A} \to B)] \text{ Restricted Expansion,} \tag{7.17}$$

$$(\overrightarrow{A} \to C) \to [(\overrightarrow{B} \to C) \to (\overrightarrow{A} \to (\overrightarrow{B} \to C))] \text{ Restricted O-M Mingle.} \tag{7.18}$$

7.6 Some of Arnon Avron's Contributions to the Study of R-Mingle

Arnon Avron has made a number of substantial contributions to the study of **RM** and we cannot begin to discuss all of them here. Perhaps principle among those we shall neglect are "bilattices." Dana Scott observed that the four-valued De Morgan lattice DM4 is a lattice no matter whether its diagram is seen from top to bottom (the usual) or side to side (unusual), and he communicated this to Nuel Belnap. This lattice plays a fundamental role in the semantics of First-Degree Relevant Entailments (**FDE**). Arnon Avron (with Ofer Arieli) has made fundamental contributions to extending this observation, but we shall skip this to discuss some of his contributions more relevant to our discussions in this chapter.

7.6.1 Arnon Avron and Hypersequents

As I have already stated, and as Arnon Avron clearly agrees, Distribution is a major problem for axiomatizing relevance logic. Hypersequents were his solution to this problem. Hypersequents were not his creation though. Hypersequents seem to have been first introduced by Mints (1971) and Pottinger (1983). Put quickly a hypersequent is a finite sequence of sequents. We will not go through any details here but do want to mention that Avron (1987) contains a hypersequent formulation of **RM** that allows (as it must since it really gives **RM**) a proof of the Distribution axiom. See Avron (1996) for further hypersequent systems.

7.6.2 Arnon Avron's Characterization of Semi-relevant Logics

As Arnon Avron points out, the system **RM** has been called a semi-relevant logic by many authors. This is because it satisfies the Weak Variable Sharing Property even though it fails to satisfy the Variable Sharing Property itself. Arnon Avron wants to give a deeper definition of "semi-relevance." Before he can do so with generality he has to give some abstract definitions characterizing the notion of propositional logic. These preliminary definitions are quite standard and I shall describe them somewhat informally and refer the reader to Avron (2016) if there are any questions.

Arnon Avron starts with the notion of a *propositional language* L, which is just a set of *well-formed formulas (wffs)* built up from a set of atomic formulas using finitary connectives. He calls any set of wffs a *theory*.[9] He next defines a *Tarskian consequence relation* (tcr) $T \vdash \psi$ as a binary relation between a theory T and a single formula ψ. A tcr $\vdash$ is *structural* just when it is closed under uniform substitution. A tcr is *non-trivial* (or *consistent*) just when for any two distinct atomic formulas p and q, $p \nvdash q$. A *(propositional) logic* **L** is then defined as a pair $(L\ \vdash)$ where L is a propositional language and $\vdash$ is a structural and non-trivial tcr for L.

Definition 7.3 (*Basic relevance criterion*) A logic $\mathbf{L} = (L\ \vdash)$ satisfies the *basic relevance criterion* if for every two theories T_1, T_2 and formula ψ, we have that $T_1 \vdash \psi$ whenever

$$T_1 \cup T_2 \vdash \psi \text{ and } T_2 \text{ has no atomic formulas in common with } T_1 \cup \{\psi\}.$$

Definition 7.4 (*Minimal semantic relevance criterion*) A logic **L** satisfies the minimal semantic relevance criterion if it does not have a finite weakly characteristic matrix.

Definition 7.5 A logic **L** which satisfies both the basic relevance criterion and the minimal semantic relevance criterion is called semi-relevant.

Arnon Avron shows that **RM** is a semi-relevant logic in this sense.

[9] I find this terminology non-standard since typically the word "theory" is reserved for a set of wffs closed under some given consequence relation. But this is Arnon Avron's word and so it means just what he says it means.

7.6.3 Arnon Avron's Characterization of Relevance Logics

Strictly speaking, this contribution is more about **R** and **E** than **RM**, but it helps clarify in what ways **RM** is not a "relevance logic." See also Avron (1992). Avron (2014) gives two characterizations of a "relevance logic," and also of a "strong relevance logic." The details are complicated and we shall not pursue them here. Put quickly the relevance logic **R** of Anderson and Belnap is not a *strong* relevance logic, but it is a relevance logic using Arnon Avron's weaker criterion. Arnon Avron seems to favor the stronger notion, but he realizes that the reason that **R** fails to satisfy the stronger notion is Anderson and Belnap's desire to add conjunction to the implication–negation fragment $\mathbf{R}^{\sim}_{\rightarrow}$. Arnon Avron briefly flirts with the idea that the stronger notion would be appropriate to a purely intensionalist approach to relevance logic (no extensional connectives such as $\wedge$ and $\vee$). I would suggest that in that context the intensional logic **RM0** might also be a good candidate for "the" relevance logic.

7.7 A *Consumer Report* Style Checklist

Now let us consider a *Consumer Report* style checklist for you to use to judge the desirability of various logics. We consider just three logics, the semi-relevance logic **RM**, the relevance logic **R**, and the standard two-valued classical logic **TV**. There are of course many other logics, and as a referee pointed out, just as there is no car that is the best in every respect, there is no logic that is best for every use. We shall be focusing on how these logics handle issues of relevance, as well as some other basic properties, e.g., decidability.

Relevance Consumer Checklist		**RM**	**R**	**TV**
1.	Decidable	√	*X*	√
2.	Low complexity	√	na	√
3.	Simple, easy to interpret semantics	√	*X*	√
4.	Constant domain semantics for quantifiers	√	*X*	√
5.	Variable Sharing Property	*X*	√	*X*
6.	Relevant in sense of Avron	*X*	√	*X*
7	Semi-relevant in sense of Avron	√	√	*X*
8.	Paraconsistent	√	√	*X*
9.	No finite characteristic matrix	√	√	*X*
10.	No Chain Property	*X*	√	*X*

I have to admit in honesty that this list has somewhat of a science fiction quality to it. Logics are not yet being bought by the same number of customers as say cell phones. But maybe someday they will be. The last property will be particularly important to the consumer interested in safety and who does not want everything to explode if there is the smallest conflict in the logic she is using to build her theory T (cf. Priest 1998). Of course **RM** is barely paraconsistent. Because of its theorem

Safety one small contradiction $A \wedge \backsim A$ will lead to a lot of irrelevant consequences, e.g., anything of the form $B \vee \backsim B$, and more generally any theorem C of the theory T. I have been trying to think of an analogy and the best I have been able to come up with goes something like this. Suppose I am building an electrical circuit and I want to protect against faults. Normally, a small fault will turn all the switches on. But what if I somehow insert a clever circuit that allows a switch to be turned on only if it is already on? Something like that.

Let's now go through the Checklist row by row.

1. Decidability. Avron (2016) presents a number of "nice properties" of **RM**, the first of which is decidability. Decidability was shown by Robert Meyer in an unpublished manuscript in 1968 and finally published as a contribution to Anderson and Belnap (1975). His proof was based on a completeness theorem he proved for **RM** with respect to a class of matrices created by Sugihara. He showed that a formula A containing n propositional variables is a theorem of **RM** iff A is valid in the Sugihara matrix S_n (or equivalently $S_n + 0$).

2. Complexity. This is to be contrasted strongly with Urquhart's stunning result that **RM**'s older cousin **R** is undecidable (see Urquhart 1984). So of course **RM** has lower complexity in the sense that it has a finite complexity, whereas **R** can be said to be of "infinite complexity." But in fact, its complexity is not just finite but very, very low. Recently, Urquhart, the master of complexity, has provided me with a proof that the complexity of the decidability of **RM** is co-NP-Complete, and with his permission and encouragement, I include a sketch of his proof here. We shall prove that the family of unprovable formulas of **RM** is NP-complete, from which the result showing that **RM** has essentially the same complexity as classical propositional calculus.

Formulate classical propositional logic with $\neg$, $\wedge$, and $\vee$. A *sequent* has the form $\Gamma \vdash \Delta$, where Γ and Δ are finite sets of classical propositional formulas. If the sequent contains no negation signs, then we say that it is *monotone*. A sequent $\Gamma \vdash \Delta$ is *valid* if for any Boolean assignment to the variables in it, provided all of the formulas in Γ are true, then at least one formula in Δ is true. The family of invalid sequents is NP-complete.

Lemma 7.6 *The family of invalid monotone sequents is NP-complete.*

Proof Given a classical Boolean sequent $\Gamma \vdash \Delta$, we show how to construct a monotone sequent from it that is valid if and only the classical sequent is valid.

First, by using the De Morgan laws, convert the sequent $\Gamma \vdash \Delta$ to the form where negations apply only to variables. If $\{p_1, \ldots, p_k\}$ are the variables in the sequent, let $\{q_1, \ldots, q_k\}$ be a fresh set of k variables. Second, let Γ^* and Δ^* be the result of replacing all literals $\neg p_i$ by q_i in Γ and Δ. Let $(\Gamma \vdash \Delta)^m$ be the following monotone sequent:

$$\{p_i \vee q_i : 1 \leq i \leq k\}, \Gamma^* \vdash \Delta^*, \{p_i \wedge q_i : 1 \leq i \leq k\}.$$

If $\Gamma \vdash \Delta$ is invalid under the assignment φ, then $(\Gamma \vdash \Delta)^m$ is also invalid, extending the assignment φ so that q_i has the opposite value to p_i. Conversely, if $(\Gamma \vdash \Delta)^m$ is invalid, then $\Gamma \vdash \Delta$ is also invalid, because any assignment invalidating the monotone sequent $(\Gamma \vdash \Delta)^m$ gives the variables q_i the same value as that of $\neg p_i$. □

Theorem 7.7 *The family of unprovable formulas of* ***RM*** *is NP-complete.*

Proof If a formula with n variables is unprovable in **RM**, then it can be refuted in the finite Sugihara matrix S_n. This matrix is clearly constructible in polynomial time, given n. Hence, we can define a nondeterministic procedure as follows: Given a formula with n variables, first construct the matrix S_n, then guess an assignment to the variables in the formula that invalidates it. This procedure succeeds if the formula is unprovable in **RM**, showing that the unprovable formulas of **RM** are in NP.

Conversely, the monotone sequent calculus can be embedded in **RM**, using the translation

$$\Gamma \vdash \Delta \Longrightarrow (\wedge\Gamma \to \vee\Delta).$$

The implication is provable in **RM** if and only if the monotone sequent is valid. Hence, the NP-completeness claim follows by Lemma 7.6.

3. Simple, easy-to-understand semantics. The algebraic semantics given by the Sugihara matrix is easy to grasp. But better yet, from a philosophical and intuitive point of view is the "Kripke-style semantics" in Dunn (1976a). This is a modification of Kripke (1965) semantics for intuitionistic logic (which semantics was independently discovered by Grzegorczyk 1964).

The basic idea of this semantics is to have "evidential situations" in place of the "possible worlds" in Kripke's semantics for modal logic and to understand the binary relation not as "relative possibility" but something like one evidential situation is included in another. It is required that if an atomic sentence is true in an evidential situation that is included in a second, then that sentence stays true in the second. This "hereditary condition" can be proven by induction to hold for compound sentences as well. The basic modification that I made was to allow a sentence to be both true and false (but not neither) and to require falsity preservation as well as truth preservation. Dunn (2000) presents a comprehensive overview of some variations on these ideas. This is much easier to understand that the Routley–Meyer semantics which uses a ternary accessibility relation.

One negative of the semantics for **RM**, as we will see below, is that it satisfied the Chain axiom. This reflects that its Sugihara matrices are all chains, and also that in its Kripke semantics the accessibility relation is a chain. It was shown in Dunn (1976c) that one can give a Kripke-style semantics for $E\overrightarrow{M}$ by dropping both requirement that the accessibility relation be a chain and the Hereditary Condition.

4. Constant domain semantics. The binary accessibility relation is not the only thing that makes for a simple semantics. When we add first-order quantifiers to **R** and **RM** to obtain their first-order versions **RQ** and **RMQ** the question arises as to how to treat their domains. A real deficit of **R** (shared by **E**), at least in my opinion, is the lack of a constant domain semantics for the system **RQ**. Routley and Meyer had proposed an extension of the semantics for **R** with a constant domain of individuals (the same domain for each of their "setups"—a term they use in place

of Kripke's "possible worlds," since they can be incomplete and/or inconsistent). This is not unlike the constant domain semantics in various Kripke-style semantics for modal logic (Kripke 1963). But unfortunately, Fine (1989) showed that this semantics is incomplete. Fine (1988) came up with an ingenious complete semantics that involved expanding domains of "arbitrary objects." but of course this was not a constant-domain semantics. For Fine, each setup has its own domain, which can be expanded in an arbitrary way.
This gets complicated and we refer to Fine (1988) for details.

Mares and Goldblatt (2006) have come up with a completeness result with a different semantics. I have talked to at least one person who views it as a kind of constant-domain semantics, but I do not believe that Mares and Goldblatt (2006) refers to it that way. While their semantics is certainly interesting, it is to my mind not a "constant domain" semantics. It is true that it might give that appearance since for Mares and Goldblatt a model adds just one domain I to a Routley–Meyer frame for a relevant proposition logic, say **R**. But their frames also have a set of *propositions* and *propositional functions*, where a *prop* is a set of setups satisfying certain conditions, and a propositional *function* is a function (satisfying other conditions of course) from the denumerable sequences of individuals in I that takes propositions as its values. It gets complicated. The reason I do not see their semantics as "constant domain" is that they use the propositions to in effect restrict the domain. In their own words (note they use the word "world" in place of "set-up" but they mean the same):

> A proposition is a set of worlds. As we shall see soon, not every set of worlds is a proposition. A proposition X entails a proposition Y if X is a subset of Y and a proposition X obtains at a world a if a is in X. Thus, our truth condition says that $\forall x A$ is true at a if and only if there is some proposition X that is a subset of any proposition expressed by A on any assignment of x and $a \in X$. Mares and Goldblatt (2006, p. 164).

However, Mares and Goldblatt's semantics is construed, Dunn (1976b) produced a constant domain semantics for **RM** based on his Kripke-style relational semantics, which is very analogous to constant domain semantics for quantified modal logics. $\forall x A$ is true at a iff A is true at all b in the domain. And $\forall x A$ is false at a iff A is false at some b in the domain. Remember, we allow a sentence to be both true and false (but not neither). This is directly in parallel with the treatment of conjunction, wherein a conjunction is true iff both conjuncts are true, and false if at least once conjunct is false.

The next three (5, 6, 7) all have to do with various degrees of relevance required by the implications.

5–7. The Variable Sharing Criterion is the classic Anderson and Belnap way of characterizing relevance, and their classic systems **R** and **E** satisfy it, while as we have seen, the systems **RM**, **EM**, and $\overrightarrow{\mathbf{EM}}$ do not. 6 is Arnon Avron's strengthening of 5, and while it is philosophically interesting it does not change the ranking of the systems I just mentioned. 7 gives some points to the mingle systems but does not take anything away from the relevance scores **R** and **E** achieved.

8. All of the logics **R**, **E**, **RM**, **EM**, and $\overrightarrow{\mathbf{EM}}$ are paraconsistent in the sense that contradiction does not imply every sentence whatsoever ("Explosion"). However the

mingle systems do all have the irrelevant implication "Safety," which I have already argued is safe in that, unlike Explosion leads to nothing new.

9. Is it desirable or undesirable that a logic does not have a finite characteristic matrix. Well clearly when the issues are decidability and complexity and finite characteristic matrix is good, and the smaller, the better. That is why the classical two-valued logic has a leg up here. But it, and all logics with finite characteristic matrices and sufficient expressive power, have also a serious liability. Where $\leftrightarrow$ is the classical material equivalence, the following "Dugundji sentence" is valid in 2-valued truth tables: $(p \leftrightarrow q) \vee (p \leftrightarrow r) \vee (q \leftrightarrow r)$.[10] In general, if there is an n-valued characteristic matrix, the following Dugundji sentence is valid just because there are not enough values to make enough distinctions (see Dugundji 1940):

Proposition 7.8 $(p_1 \leftrightarrow p_2) \vee \cdots \vee (p_1 \leftrightarrow p_n) \vee (p_1 \leftrightarrow p_{n+1}) \vee \cdots \vee (p_n \leftrightarrow p_{n+1})$.

All that is needed to make this work is for $p_i \leftrightarrow p_j$ to be designated whenever p_i and p_j are assigned the same value, and for a disjunction to be designated just when at least one disjunct is designated. Dugundji sentences are quite counter-intuitive, in effect saying that once you have enough sentences, two of them must be equivalent.

10 No Chain Property. The Chain Property is $(p \rightarrow q) \vee (q \rightarrow p)$. I am sorry to say that this is quite counter-intuitive. It says that given two possibly very distinct sentences, say $p =$. "The moon is made of green cheese," and $q =$ "The cat is on the mat," one or the two will imply the other. I am feeling qualms here. **RM** satisfies the Chain Property, and so does the classical propositional calculus **TV.** The Chain Property has problems similar to those caused by a Dugundji sentence. It is a serious weakness for **RM**. It is worth pointing out that **EM** does not share this weakness. No chain requirement is put upon the accessibility relation R in its Kripke-style semantics. I have thought about removing the chain requirement from the Kripke semantics for **RM**, but so far, no good.

In Dunn (2018), I took a pragmatist approach to logic and defended the idea that the foundation for logic is to view logics (note the plural) as tools. I said "Once we think of logics as tools, issues become clearer I think as to how to choose among logics. We use the same general considerations as we use for choosing tools." I then went on to give what I called "my amateur list" for choosing tools, including logics:

(1) importance of task(s) the tool is designed to perform,
(2) usability,
(3) direct costs and indirect costs (including environmental costs),
(4) compatibility and integration with other tools,
(5) longevity,
(6) ease of maintenance,
(7) elegance.

I did say that this is likely not a complete list, and also that the items on the list are not completely independent. Thus, for example, ease of maintenance very likely

[10] We do not write the parentheses needed for disjunction to be a binary connective. Think of a long disjunction without parentheses as associated to the left.

lowers indirect costs and also increases longevity, which then again lowers indirect costs. The reader can think of more interdependencies, but the main point of my list is to see that logic might be evaluated using some such list of considerations.

For the fun of it, let us evaluate **RM** versus **R** using this list. I will skip a few items when I do not see how they might apply.

(1) I do not think I have to argue for the importance of logic, particularly as we move into an age when our machines will do much of our reasoning for us.

(2) **RM** is relatively easy to use. **TV** is even easier, and **R** is more difficult.

(3) Clearly, the lower complexity of both **TV** and **RM** make them more environmentally friendly, particularly in the environment of automated theorem proving.

(7) I think that all of **R**, **RM**, and **TV** are elegant in their own ways.

Arieli et al. (2011a, b) have their own evaluation criteria, focusing just on logics for reasoning with inconsistency. I only discovered this list while writing this chapter. I shall list here the criteria as stated in their first paper (they are similar in both papers):

> 1. *Paraconsistency.* The rejection of the principle of explosion, according to which any proposition can be inferred from an inconsistent set of assumptions, is a primary condition for properly handling contradictory data.
>
> 2. *Sufficient expressive power.* Clearly, a logical system is useless unless it can express non-trivial, meaningful assertions. In our framework, a corresponding language how to choose one for a specific application. This should contain at least a negation connective, which is needed for defining paraconsistency, and an implication connective admitting the deduction theorem.
>
> 3. *Faithfulness to classical logic.* As observed by Newton da Costa, one of the founders of paraconsistent reasoning, a useful paraconsistent logic should be faithful to classical logic as much as possible. This implies, in particular, that entailments of a paraconsistent logic should also be valid in classical logic.
>
> 4. *Maximality.* The aspiration to "retain as much of classical logic as possible, while still allowing non-trivial inconsistent theories" is reflected by the property of maximal paraconsistency, according to which any extension of the underlying consequence relation yields a logic that is not paraconsistent anymore

RM meets the first three of these requirements, but not the fourth. Of course, it barely meets the first, but as I have explained, Safety is "safe" even though it is of the form $(A \wedge \backsim A) \rightarrow$ ___. The blank is filled in with something already known to be provable, $(B \vee \backsim B)$. I think that **RM** satisfies the second and the third are clear enough. For the fourth, I offer some explanation.

It is easy to see that all of the logic corresponding to the Sugihara matrices $S_i(+0)$ are paraconsistent, except for S_1, which of course gives the truth tables for the classical propositional calculus **TV**. This means that only the three-valued logic **RM3** is maximally paraconsistent among these. Thus, given a Sugihara matrix $S_i(+0)$ with 4 or more elements, we can falsify $(p \wedge \backsim p) \rightarrow q$ by assigning $+1$ to p and -2 to q. But **RM3** has other undesirable features. For example, to "quote" Arnon Avron (see above), it has a finite characteristic matrix. So when we talk about an "ideal" paraconsistent logic as being maximal, we should not mean simply maximal as a paraconsistent logic, but maximal among paraconsistent logic with other pleasant characteristics. In this sense **RM** seems to fit the bill, especially given Meyer (1968) result that every proper extension of it has a characteristic finite Sugihara matrix. But I wish it did not have the Chain property. However, to quote from a famous song by the Rolling Stones, "You can't always get what you want." Another of my favorite sayings is: "If at first you don't succeed, try, try again." I think I will finish this paper and start to think more about dropping the Chain requirement from the Kripke-style semantics. :)

Acknowledgements Of course, I want to thank Ofer Arieli and Anna Zamansky for their skill and patience in editing this volume, from A to Z, so to speak. I thank the two referees for their very careful and very helpful reports, which greatly improved my exposition. I wish to thank Alasdair Urquhart for telling me I could/should share his observations about the complexity of **RM**. Finally I want to thank Arnon Avron for his many years of good work on **RM** and other non-classical logics, and, of course, for being so nice.

References

Anderson, A., & Belnap, N. D. (1962). The pure calculus of entailment. *The Journal of Symbolic Logic*, *27*, 19–52.

Anderson, A., & Belnap, N. D. (1975). *Entailment: The Logic of Relevance and Necessity* (Vol. 1, p. 542 + xxxii). Princeton: Princeton University Press.

Anderson, A., Belnap, N. D., & Dunn, J. M. (1992). *Entailment: The Logic of Relevance and Necessity* (Vol. 2, pp. 749 + xxvii). Princeton University Press.

Arieli, O., Avron, A., & Zamansky, A. (2011a). Ideal paraconsistent logic. *Studia Logica*, *99*, 31–60.

Arieli, O., Avron, A., & Zamansky, A. (2011b). What is an ideal logic for reasoning with inconsistency? *Proceedings of the Twenty-Second International Joint Conference on Artificial Intelligence*, *2*, 706–711.

Avron, A. (1985). *The Semantics and Proof Theory of Relevance Logics and Nontrivial Theories Containing Contradictions, Ph.D. dissertation*. Tel Aviv University.

Avron, A. (1987). A constructive analysis of RM. *Journal of Symbolic Logic, 52*, 939–951.

Avron, A. (1992). Whither relevance logic. *Journal of Philosophical Logic*, *21*, 243–281.

Avron, A. (1996). The Method of Hypersequents in Proof Theory of Propositional Non-classical Logics. In W. Hodges, M. Hyland, C. Steinhorn, & J. Trusse (Eds.), *Logic: Foundations to Applications* (pp. 1–32). Oxford Science Publications.

Avron, A. (2014). What is relevance logic? *Annals of Pure and Applied Logic*, *165*, 26–48.

Avron, A. (2016). R-Mingle and its Nice Properties. In K. Bimbó (Ed.), *J. Michael Dunn on Information Based Logics* (Vol. 8, pp. 15–44). Outstanding Contributions to Logic. Springer.

Belnap, N. D. (1982). Display logic. *Journal of Philosophical Logic*, *11*, 375–417.

Bimbó, K. (2009). Dual gaggle semantics for entailment. *Notre Dame Journal of Formal Logic*, *50*, 23–41.

Bimbó, K. (2014). *Proof Theory: Sequent Calculi and Related Formalisms*. Boca Raton, FL: CRC Press.

Blok, W. J., & Raftery, J. G. (2004). Fragments of R-mingle. *Studia Logica*, *78*, 59–106.

Church, A. (1951). Kontrolliertes Denken, Untersuchungen zum Logikkalkül und der Logik der Einzelwissenschaften. In Menne-Wilhelmy-Angsil (Ed.) *The Weak Theory of Implication* (pp. 22–37). Munich: Kommissions-Verlag Karl Alber.

Dugundji, J. (1940). Note on a property of matrices for Lewis and Langford's calculi of propositions. *The Journal of Symbolic Logic*, *5*, 150–151.

Dunn, J. M. (1970). Algebraic completeness results for R-mingle and its extensions. *The Journal of Symbolic Logic*, *35*, 1–13.

Dunn, J. M. (1973). A 'Gentzen System' for positive relevant implication. *The Journal of Symbolic Logic*, *38*, 356–357.
Dunn, J. M. (1976a). A Kripke style semantics for R-mingle using a binary accessibility relation. *Studia Logica*, *35*, 163–172.
Dunn, J. M. (1976b). Quantification and RM. *Studia Logica*, *35*, 315–322.
Dunn, J. M. (1976c). A variation on the binary semantics for RM. *Relevance Logic Newsletter*, *1*(56), 67.
Dunn, J. M. (1976d). Intuitive semantics for first degree entailments and coupled trees. *Philosophical Studies*, *29*, 149–168.
Dunn, J. M. (1986). Relevance Logic and Entailment. In D. Gabbay & F. Guenthner (Eds.), *Handbook of Philosophical Logic* (Vol. 3, pp. 117–224). Dordrecht: D. Reidel.
Dunn, J. M. (2000). "Partiality and its Dual, *Partiality and Modality*. In E. Thijsse, F. Lepage, & H. Wansing (Eds.), special issue of *StudiaLogica 66*, 5–40.
Dunn, J. M. (2018). Humans as Rational Toolmaking Animals, translated into Russian by Vasilyi Shangin and published in Modern logic: Its Subject Matter, Foundations and Prospects (Russian). D. Zaitsev (Ed.) (pp. 128–160). Moscow: Forum.
Dunn, J. M., & Restall, G. (2002). Relevance Logic. In D. Gabbay and F. Guenthner (Eds.), *Handbook of Philosophical Logic* (2nd edn, vol. 6, pp. 1–128). Amsterdam: Kluwer.
Fine, K. (1988). Semantics for quantified relevance logic. *Journal of Philosophical Logic*, *17*, 27–59.
Fine, K. (1989). Incompleteness for Quantified Relevance Logics. In J. Norman & R. Sylvan (Eds.), *Directions in Relevant Logic*. Dordrecht: Kluwer.
Grzegorczyk, A. (1964). A philosophically plausible formal interpretation of intuitionistic logic. *Indagationes Mathematicae*, *26*, 596–601.
Hacking, I. (1963). What is strict implication? *The Journal of Symbolic Logic*, *28*, 51–71.
Kripke, S. (1963). Semantical considerations on modal logics. *Acta Philosophica Fennica*, *6*, 83–94.
Kripke, S. (1965). Semantical Analysis of Intuitionistic Logic I. In J. A. Crossley & M. A. Dummett (Eds.), *Formal Systems and Recursive Functions -Proceedings of the Eighth Logic Colloquium Oxford* (pp. 92–130). Amsterdam: North-Holland. July (1963).
Mares, E. D., & Goldblatt, R. (2006). An alternative semantics for quantified relevant logic. *The Journal of Symbolic Logic*, *71*, 163–187.
McCall, S. (1963). *Non-classical Propositional Calculi*, Ph.D. Dissertation, University of Oxford, Oxford University Research Archive https://ora.ox.ac.uk/objects/uuid:6cc1feb8-fe41-4e8c-a725-801077a32ea4.
Meyer, R. K. (1968). A Characteristic Matrix for RM, unpublished manuscript. Published in Anderson and Belnap (1975), 393–420.
Meyer, R. K., & Zane Parks, R. (1972). Independent matrices for the implicational fragment of three-valued logic. *Zeitschrift für Mathmatische Logik und Grundlagen der Mathematik*, *18*, 291–295.
Mints, G. (1971). On some calculi of modal logic (Russian). *Proceedings of the Steklov Institute of Mathematics*, *98*, 97–122.
Mints, G. (1976). Cut-elimination theorem for relevance logics. *Journal of Soviet Mathematics*, *6*, 422–428.
Ohnishi, M., & Matsumoto, K. (1962). A system for strict implication. *Proceedings of the Symposium on the Foundations of Mathematics, Katada, Japan*, *1*, 98–108. Reprinted in *Annals of the Japan Association for Philosophy of Science* 2(4), 183–188 (1964).
Parks, Z. (1972). A note on R-mingle and Sobociński's three-valued logic. *Notre Dame Journal of Formal Logic*, *13*, 227–228.
Pottinger, G. (1983). Uniform, cut-free formulations of T, S4 and S5 (abstract). *The Journal of Symbolic Logic*, *48*, 900.
Priest, G. (1998). What is so bad about contradictions? *The Journal of Philosophy*, *95*, 410–426.
Scroggs, S. J. (1951). Extensions of the Lewis system S5. *The Journal of Symbolic Logic*, *16*, 112–120.

Sobociński, B. (1952). Axiomatization of a partial system of three valued calculus of propositions. *The Journal of Computing Systems*, *1*, 23–55.
Sugihara, T. (1955). Strict implication free from implicational paradoxes. *Memoirs of the Faculty of Liberal Arts, Fukui University, Series*, *1*(4), 55–59.
Thomas, I. (1970). Review of Masao Ohnishi, Kazuo Matsumoto's "A System for Strict Implication". *The Journal of Symbolic Logic*, *35*(2), 326.
Urquhart, A. (1984). The undecidability of entailment and relevant implication. *The Journal of Symbolic Logic*, *49*, 1059–1073.

J. Michael Dunn research has included information-based logics, algebraic logic, proof theory, non-standard logics (especially relevance logic), quantum computation/logic, and logic related to computer science. He has an A.B. in Philosophy, Oberlin College (1963) and a Ph.D. in Philosophy (Logic), University of Pittsburgh (1966). His career started at Wayne State University in 1966, and he retired from Indiana University in 2007 after 38 years there, where he is Emeritus Oscar Ewing Professor of Philosophy, Emeritus Professor of Computer Science and Informatics, and was Founding Dean of the School of Informatics. He is an affiliate of the Info-Metrics Institute at the American University. He has published 8 books and over 100 articles. He has been an Editor of The Journal of Symbolic Logic and Chief Editor of The Journal of Philosophical Logic. He has received many honors, including Fellow of the American Academy of Arts and Sciences.

Chapter 8
The Strict/Tolerant Idea and Bilattices

Melvin Fitting

Abstract Strict/tolerant logic is a formally defined logic that has the same consequence relation as classical logic, though it differs from classical logic at the metaconsequence level. Specifically, it does not satisfy a cut rule. It has been proposed for use in work on theories of truth because it avoids some objectionable features arising from the use of classical logic. Here we are not interested in applications, but in the formal details themselves. We show that a wide range of logics have strict/tolerant counterparts, with the same consequence relations but differing at the metaconsequence level. Among these logics are Kleene's K_3, Priest's LP, and first-degree entailment, FDE. The primary tool we use is the *bilattice*. But it is more than a tool, it seems to be the natural home for this kind of investigation.

Keywords Strict/tolerant · Bilattice · Many-valued logic · Kleene logic · Logic of paradox · First-degree entailment

8.1 Introduction

A natural companion to the question "What is a logic?" (which won't be asked here) is the question "When are logics the same?" It is common to say that sameness for logics means they have the same consequence relations. But then there is the curious example of ST, which stands for *strict/tolerant* for reasons that will become clear later. The idea of holding premises and conclusions of a consequence relation to different standards comes from Malinowski (1990, 2002, 2007), where the standards for premises were weaker. Today it corresponds to what is called TS, for *tolerant/strict*. Holding premises to stronger standards was introduced in Frankowski (2004a, b) and today is called ST. It turns out that the ST consequence relation coincides with that of classical logic, but a good case has been made that ST is not identical with classical logic because the two differ at the metaconsequence level. In fact, Barrio et al.

M. Fitting (✉)
The Graduate School and University Center, City University of New York, New York, NY, USA
e-mail: melvin.fitting@gmail.com

O. Arieli and A. Zamansky (eds.), *Arnon Avron on Semantics and Proof Theory of Non-Classical Logics*, Outstanding Contributions to Logic 21,
https://doi.org/10.1007/978-3-030-71258-7_8

(2021) shows there is a hierarchy of logical pairs, with ST and classical logic at the bottom, where each pair agrees at the consequence level, the metaconsequence level, the metametaconsequence level, and so on up to some arbitrary finite level, and then they differ at the next level. Very curious indeed, and very interesting.

In this paper, we examine a different sort of generalization of the ST phenomenon: wide instead of high. We show there is a family of logic pairs consisting of an ST-like logic and a corresponding classical-like logic, where each pair agrees on consequences but differs on metaconsequences. We do not examine working our way up the meta, meta^2, meta^3, …hierarchy as in Barrio et al. (2021). Instead we complicate the structure of the truth value space itself, of course going to three values and beyond. We set up the basics for study, but we leave the meta-levels to another time or to other people.

The machinery we use comes from bilattice theory, with the original ST/classical example as the simplest case. We sketch the necessary bilattice background, to keep this paper relatively self-contained.

8.2 ST, Classical Logic, and One New Example

Logics can be specified proof theoretically, or semantically. In this paper, we make no use of proof-theoretic methods. The work is entirely semantic.

Many-valued logics are specified by giving a set of truth values, an interpretation for propositional connectives, and a specification of what counts as "true." A bit more precisely, let $\mathcal{T}$ be a non-empty set of truth values, and for each logical connective (in this paper conjunction, disjunction, and negation) assume we have a corresponding operation on $\mathcal{T}$. We will overload the use of the symbols $\wedge$, $\vee$, and $\neg$ to serve as logical connectives and also as operations on $\mathcal{T}$, with context determining which is intended. And finally a non-empty proper subset $\mathcal{D}$ of the truth value space is specified as the *designated truth values*, often with some structural properties imposed.

With respect to a many-valued logic, a valuation v is a mapping from propositional variables to truth values, that is, to $\mathcal{T}$. A valuation extends to all formulas in the usual way, for instance, setting $v(X \wedge Y) = v(X) \wedge v(Y)$, where on the left $\wedge$ is an operation symbol, and on the right $\wedge$ is the corresponding operation on $\mathcal{T}$. A *sequent* is an expression of the form $\Gamma \Rightarrow \Delta$ where Γ and Δ are finite sets of formulas. For a valuation v, we write $v \Vdash \Gamma \Rightarrow \Delta$ provided, if $v(X) \in \mathcal{D}$ for every $X \in \Gamma$ then $v(Y) \in \mathcal{D}$ for some $Y \in \Delta$. More informally, $v \Vdash \Gamma \Rightarrow \Delta$ provided that if every member of Γ is designated under v then some member of Δ also is. A sequent $\Gamma \Rightarrow \Delta$ is *valid* in a many-valued logic provided, for every valuation v in that logic, $v \models \Gamma \Rightarrow \Delta$. We take this notion of validity as determining the consequence relation of the many-valued logic.

Among the best-known three-valued logics are Kleene's strong, K_3, from Kleene (1938, 1950), and Priest's logic of paradox, LP, from Priest (1979) but with truth tables originating in Asenjo (1966). We can take the truth values of both to be 0, $\frac{1}{2}$, and 1. (Other names for these values will also be used from time to time in this paper.)

The intended intuition is that in K_3 the value $\frac{1}{2}$ represents a truth value *gap* while in LP it represents a *glut*. Either way, the truth tables for propositional operators turn out to be the same. Assume we have an ordering so that $0 \leq \frac{1}{2} \leq 1$. Conjunction, $\wedge$, is greatest lower bound (equivalently minimum in this case); disjunction, $\vee$, is least upper bound (or maximum); and negation, $\neg$, is an order reversal so that $\neg 0 = 1$, $\neg 1 = 0$, and $\neg\frac{1}{2} = \frac{1}{2}$. The two logics differ in their choice of designated values. For K_3, the designated value set is $\{1\}$ while for LP it is $\{\frac{1}{2}, 1\}$. We do not go into the motivation for these choices; discussions are available in many places—see Priest (2008), for instance.

The logic known as ST combines aspects of both K_3 and LP through a mixed definition of consequence. Note that since the space of truth values is the same for K_3 and for LP, and the behavior of logical connectives is the same, these two standard logics have the same valuation behavior. $\Gamma \Rightarrow \Delta$ is taken to be valid in ST provided, for every valuation, if every member of Γ is designated in the K_3 sense, then some member of Δ is designated in the LP sense. Now the reason for the name *strict/tolerant* becomes a bit clearer: members of the antecedent Γ are held to stricter standards, only 1 is acceptable, while we are more tolerant with members of Δ accepting both 1 and $\frac{1}{2}$.

Of course classical logic also fits the many-valued paradigm. Truth values are 0 and 1, with $\wedge$ and $\vee$ defined as greatest lower bound and least upper bound, respectively, and negation as order reversal. $\{1\}$ is the set of designated truth values. And $\Gamma \Rightarrow \Delta$ is defined in the expected way: every valuation mapping all members of Γ to a designated value must map some member of Δ to a designated value.

The important connection between ST and classical logic is very simply stated: they have the same consequence relation, see Cobreros (2012), Barrio et al. (2021) among other places.

But it has been argued that they still are not the same logics because they differ at the metainference level. In particular, classical logic validates the cut rule but ST does not, and there are other metainferences on which they differ as well. Current work in Barrio et al. (2021) generalizes this result upward, as we discussed in Sect. 8.1. We will generalize it laterally. We will show there is an abundance of pairs of many-valued logics where one logic is analogous to ST, the other to classical logic, such that both agree on consequence but differ on metaconsequence. Indeed, there are strict/tolerant analogs for strong Kleene logic itself, for the logic of paradox of Priest, and for first-degree entailment. We present one example now, to give an idea of things. It will, perhaps, seem a bit mysterious, but motivations and proofs for our assertions will come later on.

In Fig. 8.1, two lattices are shown. The lattice names will be explained in Sect. 8.7. The truth value names trace back to Ginsberg, d is supposed to represent default, for instance. Here the truth value names play no role other than letting us specify what node we are talking about. Think of both lattices as having an ordering relation, $\leq$, represented graphically as upward (with reflexivity tacitly assumed). For both lattices, $\wedge$ and $\vee$ are interpreted as greatest lower and least upper bound, respectively. Negation is order reversal for both, so $\neg d\top = d\top$ in each, for instance. For the

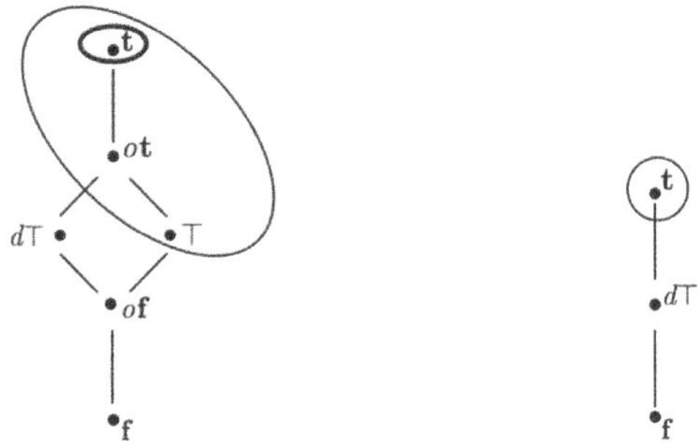

(a) $\mathsf{ST}\langle\mathcal{NINE}, \{\mathbf{t}, o\mathbf{t}, \top\}\rangle$ (b) $\mathsf{C}\langle\mathcal{NINE}, \{\mathbf{t}, o\mathbf{t}, \top\}\rangle$, or Kleene's K_3

Fig. 8.1 A strict/tolerant pair

lattice in Fig. 8.1b, the only designated truth value is **t**, shown circled, thus this is a presentation of strong Kleene logic, K_3. For the lattice in Fig. 8.1a, we introduce both a strict and a tolerant designated set, analogous to what is done with the logic ST. In the present example, the strict set of truth values is $\{\mathbf{t}\}$, shown heavily circled, and the tolerant set is $\{\mathbf{t}, o\mathbf{t}, \top\}$, shown lightly circled. We say $\Gamma \Rightarrow \Delta$ is valid in the resulting strict/tolerant logic provided that for every valuation, if every member of Γ is strictly designated then some member of Δ is tolerantly designated. It will be shown later on that the logics corresponding to these two lattices have the same consequence relation, but differ at the metaconsequence level, and are thus connected in the same way that ST and classical logic are. As we said earlier, we do not analyze higher level differences in this paper.

8.3 ST and $\mathcal{FOUR}$

Our unifying machinery will be bilattices. Before discussing the general machinery we begin with the paradigm example, the Belnap-Dunn system, called $\mathcal{FOUR}$. This was presented in a very influential paper, Belnap (1977). Its truth values were intended to represent sets of ordinary truth values, only true (**t**), only false (**f**), neither ($\bot$), both ($\top$). It has two partial orderings, one on degree of truth, one on degree of information. All of this is shown in Fig. 8.2, in which the information ordering is vertical, and is denoted $\leq_k$. This has become customary in bilattice literature, with *k* standing for *knowledge*, though *i* for *information* would probably be better. The truth ordering is denoted $\leq_t$ and is shown horizontally.

Each of the two orderings gives us the structure of a bounded, distributive lattice. For the truth ordering, greatest lower bound is symbolized using $\wedge$ and least upper bound by $\vee$. A negation operation, denoted $\neg$, is a horizontal symmetry, $\neg\mathbf{t} = \mathbf{f}$, $\neg\mathbf{f} = \mathbf{t}$, $\neg\top = \top$, and $\neg\bot = \bot$. The De Morgan laws hold, so with respect to $\leq_t$ we

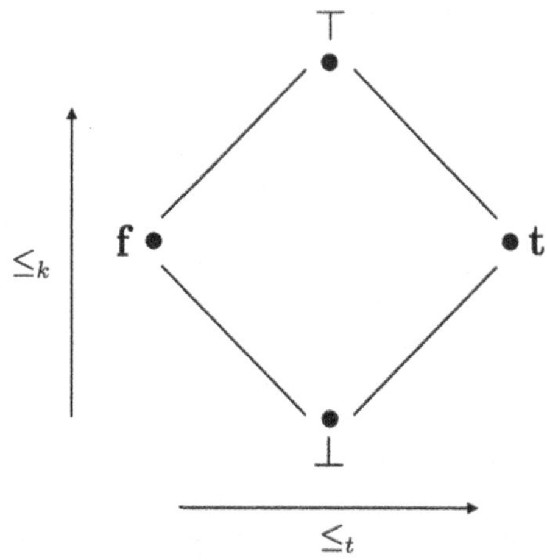

Fig. 8.2 The bilattice $\mathcal{FOUR}$

have a De Morgan algebra. The $\leq_k$ ordering plays an important role, but we postpone discussion until we have introduced the full notion of bilattice, for which $\mathcal{FOUR}$ is the simplest non-trivial example.

In order to turn $\mathcal{FOUR}$ into a many-valued logic, a set of designated truth values must be specified. This is taken to be $\{\mathbf{t}, \top\}$, which one can think of as *at least true*. The values we would naturally think of as consistent are $\mathbf{f}$, $\bot$, $\mathbf{t}$, and the $\leq_t$ ordering, restricted to them, gives us the operations of the strong Kleene logic, K_3. Likewise the set of designated truth values of $\mathcal{FOUR}$, restricted to $\{\mathbf{f}, \bot, \mathbf{t}\}$, gives us $\{\mathbf{t}\}$, appropriate for K_3. Similarly, $\leq_t$ restricted to $\mathbf{f}$, $\top$, $\mathbf{t}$ gives us the operations of LP, and the set of designated truth values of $\mathcal{FOUR}$, similarly restricted, gives us $\{\mathbf{t}, \top\}$, appropriate for LP.

When working with ST we need both three-valued logics K_3 and LP and, although their representation in $\mathcal{FOUR}$ as described above is quite natural, it has the consequence of giving us different carrier sets for the two logics, with one containing $\bot$ and the other $\top$. To avoid this, we do not work with the representation of K_3 just described. Instead we work with the set $\{\mathbf{f}, \top, \mathbf{t}\}$, and we refer to $\{\mathbf{t}, \top\}$ as *tolerantly* designated, and $\{\mathbf{t}\}$ as *strictly* designated. That is, we have one space of truth values, and two versions of designated value. We will do something similar for other bilattices, when we come to them.

Much more can be said about $\mathcal{FOUR}$, but this is enough for the time being. It is better to continue our discussion after the general family of bilattices has been introduced.

8.4 Bilattices

A *bilattice* is an algebraic structure with two lattice orderings. Various conditions can be imposed, connecting the orderings. We start at the simplest level.

A *pre-bilattice* is a structure $\mathcal{B} = \langle \mathcal{B}, \leq_t, \leq_k \rangle$ where each of $\leq_t$ and $\leq_k$ are bounded partial orderings on $\mathcal{B}$. (Notice that we overload $\mathcal{B}$ to stand for both the structure with its orderings, and for its domain. This should cause no confusion since context can sort things out. We do similar things with other structures as well.) Think of the members of domain $\mathcal{B}$ as generalized truth values. The relation $\leq_t$ is intended to order degree of truth in some sense (though it was noted in Shramko and Wansing 2005 that the ordering is really about truth-and-falsity, and that to separate the two something more complex than a bilattice is needed, namely, a *trilattice*. We do not pursue this point here). Meet and join operations with respect to this ordering are denoted $\wedge$ and $\vee$, and the least and greatest elements are denoted **f** and **t**. The other relation, $\leq_k$, is intended to order degree of information, again in some sense. The meet operation with respect to this ordering is denoted $\otimes$ and is called *consensus*; the join operation is denoted $\oplus$ and is called *gullability*, or sometimes *accept all*. The least and greatest elements with respect to this ordering are denoted $\bot$ and $\top$. The Belnap-Dunn structure $\mathcal{FOUR}$ from Fig. 8.2 is the simplest pre-bilattice.

If a pre-bilattice has an operation $\neg$ that reverses $\leq_t$, preserves $\leq_k$, and is an involution, such an operation is simply called *negation*. Formally, the conditions are as follows:

(Neg-1) $a \leq_t b$ implies $\neg b \leq_t \neg a$;
(Neg-2) $a \leq_k b$ implies $\neg a \leq_k \neg b$;
(Neg-3) $\neg\neg a = a$.

Bilattices were introduced by Ginsberg (1988, 1990), who defined a bilattice to be a pre-bilattice with negation (though without using the terminology "pre-bilattice"). $\mathcal{FOUR}$ is the simplest bilattice in Ginsberg's sense. In any such bilattice, $\neg\mathbf{t} = \mathbf{f}$, $\neg\mathbf{f} = \mathbf{t}$, $\neg\top = \top$, $\neg\bot = \bot$. It is not hard to show that we also have De Morgan's laws for the t operations and something akin to them for the k operations.

(NDeM-1) $\neg(a \wedge b) = (\neg a \vee \neg b)$;
(NDeM-2) $\neg(a \vee b) = (\neg a \wedge \neg b)$;
(NDeM-3) $\neg(a \otimes b) = (\neg a \otimes \neg b)$;
(NDeM-4) $\neg(a \oplus b) = (\neg a \oplus \neg b)$.

A pre-bilattice may have a negation-like operation with respect to $\leq_k$ as well. If one exists it is denoted $-$ and is called *conflation*, with the following conditions:

(Con-1) $a \leq_k b$ implies $-b \leq_k -a$;
(Con-2) $a \leq_t b$ implies $-a \leq_t -b$;
(Con-3) $- - a = a$;
(Con-4) $-\neg a = \neg - a$.

Fig. 8.3 The bilattice $\mathcal{DEFAULT}$

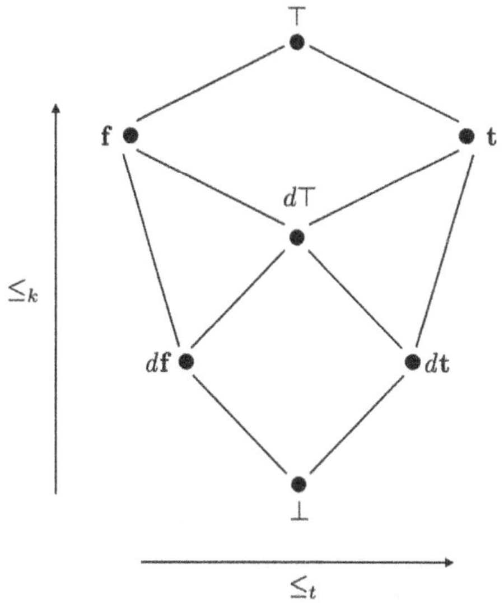

The last condition, that negation and conflation commute, is occasionally not assumed, but will be here. $\mathcal{FOUR}$ is an example of a bilattice with conflation, where $-\top = \bot$, $-\bot = \top$, $-\mathbf{t} = \mathbf{t}$, $-\mathbf{f} = \mathbf{f}$. When conflation is present, we have dual versions of the De Morgan laws given earlier.

(CDeM-1) $-(a \wedge b) = (-a \wedge -b)$;
(CDeM-2) $-(a \vee b) = (-a \vee -b)$;
(CDeM-3) $-(a \otimes b) = (-a \oplus -b)$;
(CDeM-4) $-(a \oplus b) = (-a \otimes -b)$.

Monotonicity conditions for the operations with respect to the ordering defining it are standard, because we have lattice structures. Thus, for instance, $a \leq_t b$ implies $a \wedge c \leq_t b \wedge c$. A bilattice is called *interlaced* if such conditions hold across the two orderings. More precisely, we have interlacing if the following hold:

(Int-1) $a \leq_t b$ implies $a \otimes c \leq_t b \otimes c$;
(Int-2) $a \leq_t b$ implies $a \oplus c \leq_t b \oplus c$;
(Int-3) $a \leq_k b$ implies $a \wedge c \leq_k b \wedge c$;
(Int-4) $a \leq_k b$ implies $a \vee c \leq_k b \vee c$.

In any interlaced bilattice, $\mathbf{f} \wedge \mathbf{t} = \bot$, $\mathbf{f} \vee \mathbf{t} = \top$, $\bot \otimes \top = \mathbf{f}$, and $\bot \oplus \top = \mathbf{t}$. Once again $\mathcal{FOUR}$ is an example, this time of an interlaced bilattice. There are bilattices that are not interlaced. $\mathcal{DEFAULT}$, shown in Fig. 8.3, is an example of one. In it $\mathbf{f} \leq_t d\mathbf{f}$ but $\mathbf{f} \otimes d\top = d\top \not\leq_t d\mathbf{f} = d\mathbf{f} \otimes d\top$. $\mathcal{DEFAULT}$ goes back to Ginsberg (1988), but will play no further role here.

The following plays an important role in Avron (1996) when establishing representation theorems for interlaced bilattices. Representation theorems are discussed in Sect. 8.8.

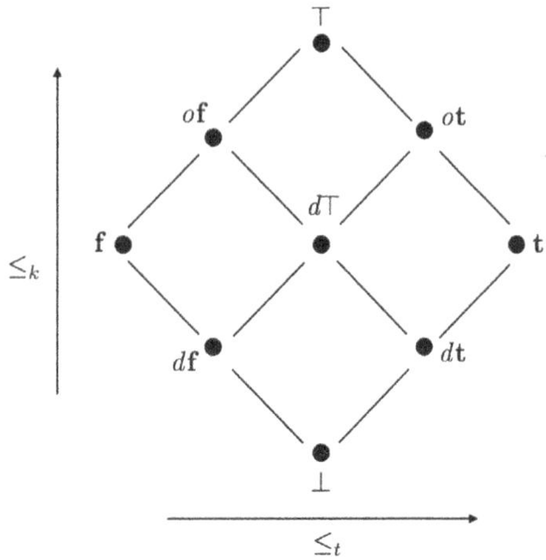

Fig. 8.4 The bilattice $\mathcal{NINE}$

Proposition 8.4.1 *In an interlaced bilattice:*

(1) if $a \leq_k b$ *then* $a \leq_k x \leq_k b$ *if and only if* $a \wedge b \leq_t x \leq_t a \vee b$;
(2) if $a \leq_t b$ *then* $a \leq_t x \leq_t b$ *if and only if* $a \otimes b \leq_k x \leq_k a \oplus b$.

Proof We give the proof of the first, taken from Avron (1996), to give an idea of the uses of interlacing. The second part is similar. Throughout, assume $a \leq_k b$ (it is actually needed in only one part).

Suppose $a \leq_k x \leq_k b$. Using interlacing, $a \vee a \vee b \leq_k x \vee a \vee b \leq_k b \vee a \vee b$, and hence $a \vee b \leq_k x \vee a \vee b \leq_k a \vee b$. Then $x \vee a \vee b = a \vee b$ and so $x \leq_t a \vee b$. By a dual argument, $a \wedge b \leq_t x$, and so $a \wedge b \leq_t x \leq_t a \vee b$.

Now suppose $a \wedge b \leq_t x \leq_t a \vee b$. Using interlacing, $a \otimes (a \wedge b) \leq_t a \otimes x \leq_t a \otimes (a \vee b)$. We have $a \leq_k b$ so by interlacing again, $a = a \wedge a \leq_k a \wedge b$ and hence $a \otimes (a \wedge b) = a$. Similarly $a \otimes (a \vee b) = a$. Then $a \leq_t a \otimes x \leq_t a$, so $a \otimes x = a$, and so $a \leq_k x$. By a dual argument, $x \leq_k b$. ■

A bilattice is *distributive* if all possible distributive laws hold. For instance, not only should $\wedge$ and $\vee$ distribute over each other, as in $a \wedge (b \vee c) = (a \wedge b) \vee (a \wedge c)$, but over $\otimes$ and $\oplus$ as well, for example, $a \wedge (b \otimes c) = (a \wedge b) \otimes (a \wedge c)$. Altogether there are 12 such distributive laws combining $\wedge$, $\vee$, $\otimes$, and $\oplus$.

$\mathcal{FOUR}$ is a distributive bilattice. Figure 8.4 shows a distributive bilattice, $\mathcal{NINE}$, a bit more complex than $\mathcal{FOUR}$. This time the node names come from Arieli and Avron (1998). It is rather easy to show that every distributive bilattice is interlaced. The converse is not true.

In Sect. 8.8, we will discuss bilattice representation theorems, which will help account for where our examples are coming from.

by Proposition 8.6.3, v' maps every logical formula to an exact member of $\mathcal{B}$. We show $v' \not\models \Gamma \Rightarrow \Delta$ in $\mathsf{C}\langle \mathcal{B}, \mathcal{F} \rangle$.

Since $v'(P) \leq_k v(P)$ for each propositional letter then for every logical formula X, $v'(X) \leq_k v(X)$ by Proposition 8.6.4. Since both v and v' map all members of Γ to exact members of $\mathcal{B}$ then, by Proposition 8.5.4, v and v' agree on members of Γ. So v' maps every member of Γ to $\mathcal{D}_s\langle \mathcal{B}, \mathcal{F} \rangle$.

We have a logical formula $Y \in \Delta$ such that $v(Y) \notin \mathcal{D}_t\langle \mathcal{B}, \mathcal{F} \rangle$. We show $v'(Y) \notin \mathcal{D}_s\langle \mathcal{B}, \mathcal{F} \rangle$, which will finish the proof. Well, otherwise $v'(Y)$ would be exact (which it is) and in the prime bifilter $\mathcal{F}$. But $v'(Y) \leq_k v(Y)$ and prime bifilters are upward closed in both bilattice orderings, Proposition 8.6.6, so $v(Y)$ would be in $\mathcal{F}$ (which it is not).

■

It has been vehemently argued whether or not, despite validating the same sequents, classical logic and strict/tolerant logic are the same logic. See Barrio et al. (2021) for a good summary of this, as well as further references to the issue. Their difference is that they do not agree at the metaconsequence level, something that has been generalized upward as we noted at the beginning of this paper. A similar phenomenon applies to the bilattice-based generalizations considered in this paper, and with the same examples.

A metaconsequence is represented by the following general form:

$$\frac{\Gamma_1 \Rightarrow \Delta_1, \ldots, \Gamma_n \Rightarrow \Delta_n}{\Gamma_0 \Rightarrow \Delta_0.}$$

Here the members of Γ_i and Δ_i are taken to be schemata. Validity is understood to mean each instance of such a scheme is valid. Validity for an instance, with respect to a logic, actually has two versions, local and global. The global version is: if each sequent above the line is valid in the logic, so is the sequent below. The local version is: for each valuation, if that valuation validates each sequent above the line then that valuation validates the sequent below. Local is easily seen to imply global. It is the local version that is appropriate here. The particular metaconsequence scheme of interest is the familiar one of cut.

Proposition 8.7.3 *Let $\langle \mathcal{B}, \mathcal{F} \rangle$ be a logical bilattice, where $\mathcal{B}$ is an interlaced bilattice with negation and conflation. The metaconsequence scheme*

$$\frac{\Gamma, A \Rightarrow \Delta \quad \Gamma \Rightarrow \Delta, A}{\Gamma \Rightarrow \Delta}$$

is locally valid in $\mathsf{C}\langle \mathcal{B}, \mathcal{F} \rangle$ but not in $\mathsf{ST}\langle \mathcal{B}, \mathcal{F} \rangle$.

Proof

Local Validity in $\mathsf{C}\langle\mathcal{B}, \mathcal{F}\rangle$: Assume we have a specific instance of the cut scheme, and let v be a mapping from propositional letters to exact members of $\mathcal{B}$. Reasoning in $\mathsf{C}\langle\mathcal{B}, \mathcal{F}\rangle$ we show that if $v \not\models \Gamma \Rightarrow \Delta$ then either $v \not\models \Gamma, A \Rightarrow \Delta$ or $v \not\models \Gamma \Rightarrow \Delta, A$.

Assume $v \not\models \Gamma \Rightarrow \Delta$. Then $v(X) \in \mathcal{F}$ for every $X \in \Gamma$ and $v(Y) \notin \mathcal{F}$ for every $Y \in \Delta$. Either $v(A) \in \mathcal{F}$ or $v(A) \notin \mathcal{F}$. If we have the first, then $v(X) \in \mathcal{F}$ for every X in Γ, A, so $v \not\models \Gamma, A \Rightarrow \Delta$. If we have the second, then $v(Y) \notin \mathcal{F}$ for every $Y \in \Delta, A$, so $v \not\models \Gamma \Rightarrow \Delta, A$.

Local Non-Validity in $\mathsf{ST}\langle\mathcal{B}, \mathcal{F}\rangle$: Let $\Gamma \Rightarrow \Delta$ be any specific sequent that is not valid in $\mathsf{ST}\langle\mathcal{B}, \mathcal{F}\rangle$, and let P be a propositional letter that does not occur in Γ or in Δ. Let us say v is a valuation such that $v \not\models \Gamma \Rightarrow \Delta$ in $\mathsf{ST}\langle\mathcal{B}, \mathcal{F}\rangle$. That is, v maps propositional letters to anticonsistent members of $\mathcal{B}$, maps every member of Γ to $\mathcal{D}_s\langle\mathcal{B}, \mathcal{F}\rangle$, and maps no member of Δ to $\mathcal{D}_t\langle\mathcal{B}, \mathcal{F}\rangle$.

Since P does not occur in Γ or Δ we are free to reassign a value to P without affecting the behavior of v on Γ or Δ. The bilattice value $\top$ must be in $\mathcal{F}$ because $\mathcal{F}$ is non-empty and we have Proposition 8.6.6. Set $v(P) = \top$. Then $v \models \Gamma, P \Rightarrow \Delta$ because v does not map every member of Γ, P to $\mathcal{D}_s\langle\mathcal{B}, \mathcal{F}\rangle$, since $\top$ is anticonsistent but not exact. But also $v \models \Gamma \Rightarrow \Delta, P$ because v maps some member of Δ, P to $\mathcal{D}_t\langle\mathcal{B}, \mathcal{F}\rangle$ since $v(P) = \top$ is anticonsistent and is in $\mathcal{F}$. Thus, v is a counterexample to the local validity, in $\mathsf{ST}\langle\mathcal{B}, \mathcal{F}\rangle$, of the following metainference:

$$\frac{\Gamma, P \Rightarrow \Delta \quad \Gamma \Rightarrow \Delta, P}{\Gamma \Rightarrow \Delta}.$$

■

We conclude this section with a few examples. Starting in Sect. 8.10 we discuss where such examples "really" come from.

Example 8.7.4 In Fig. 8.2, we gave the ur-bilattice, $\mathcal{FOUR}$. For it the exact values are just $\mathbf{f}$ and $\mathbf{t}$, the classical ones, and the anticonsistent ones are these together with $\top$. The only prime bifilter is $\{\mathbf{t}, \top\}$ which, if taken as designated in the set of anticonsistent values, gives us LP. Then $\mathsf{ST}\langle\mathcal{FOUR}, \{\mathbf{t}, \top\}\rangle$ is the usual strict/tolerant logic ST while $\mathsf{C}\langle\mathcal{FOUR}, \{\mathbf{t}, \top\}\rangle$ is just classical logic, and the two theorems above specialize to what are, in effect, the beginnings of the subject.

Example 8.7.5 Figure 8.4 shows a bilattice, $\mathcal{NINE}$, having two prime bifilters. We, quite arbitrarily, choose to work with the smaller one, $\{\mathbf{t}, o\mathbf{t}, \top\}$. The exact members of $\mathcal{NINE}$ are $\{\mathbf{f}, d\top, \mathbf{t}\}$ and the overlap with the prime bilfilter contains just $\mathbf{t}$. Thus, the analog of classical logic from the original ST example turns out to be $\mathsf{K_3}$. There are six anticonsistent values, and these, displayed a bit differently, are shown in Fig. 8.1.

8.8 Bilattice Representation Theorems

Where do bilattices come from? There is an intuitively appealing way of constructing them that is completely representative, in the sense that every bilattice with "reasonable" properties is isomorphic to a bilattice constructed in this way. In bilattice history, this construction dates from Ginsberg (1988), with subsequent extensions by others. In fact, many of the ideas predate bilattices as such, though that was not generally known until later. See Gargov (1999) for an interesting prehistory. In this section, we sketch the ideas, without the proofs, and then add an extension that will be applied to the present investigation in Sect. 8.9.

A central intuition for truth values in a bilattice is that they encode evidence for and evidence against an assertion, treating positive and negative evidence independently. An interesting family of examples is based on groups of experts. Suppose we have one group whose members announce their opinions for something, or don't, and another group similarly announcing opinions against, or keeping silent. The two groups could be distinct, overlap, or be identical. We can identify the opinions in favor with the set of experts declaring for, and similarly for the set of experts against. In this way, a generalized truth value becomes a pair of sets of experts, the set of those for, and the set of those against. We have an increase in knowledge, or more properly information, if additional experts declare their opinions. We have an increase in degree of truth (understood loosely) if additional experts declare in favor while some withdraw from declaring against. This is a good model to have in mind while reading the following, but it is not fully general. The collection of all sets of experts, drawn from some fixed group, is a lattice under the subset ordering, but not all lattices are of this kind, hence the move to general lattice structures, L_1 intuitively representing evidence for, and L_2 intuitively representing evidence against.

Definition 8.8.1 (Bilattice Product) Let $L_1 = \langle L_1, \leq_1\rangle$, and $L_2 = \langle L_2, \leq_2\rangle$ be bounded lattices. Their *bilattice product* is defined as follows:

$$L_1 \odot L_2 = \langle L_1 \times L_2, \leq_t, \leq_k\rangle$$
$$\langle a, b\rangle \leq_k \langle c, d\rangle \text{ iff } a \leq_1 c \text{ and } b \leq_2 d$$
$$\langle a, b\rangle \leq_t \langle c, d\rangle \text{ iff } a \leq_1 c \text{ and } d \leq_2 b.$$

Note the reversal of the $\leq_2$ ordering in the definition of $\leq_t$. The following items are now rather straightforward to check. In stating the results, we assume that 0_1 and 0_2 are the least members of L_1 and L_2, and 1_1 and 1_2 are the greatest. We write $\sqcup_1$ and $\sqcup_2$ for the respective joins, and $\sqcap_1$ and $\sqcap_2$ for the meets. If the two lattices are identical, we omit subscripts. Also recall that a De Morgan algebra is a bounded distributive lattice with a De Morgan involution, written here as an overbar, such that $\overline{a \sqcap b} = \overline{a} \sqcup \overline{b}$ and $\overline{\overline{a}} = a$. (The other De Morgan law follows.) It is often the case in what follows that the distributivity laws of De Morgan algebras are not needed. By a *non-distributive De Morgan algebra* we mean something meeting the conditions for a De Morgan algebra except, possibly, satisfaction of the distributive laws.

(BP-1) $L_1 \odot L_2$ is always a pre-bilattice that is interlaced. In $L_1 \odot L_2$, the extreme elements are $\bot = \langle 0_1, 0_2 \rangle$, $\top = \langle 1_1, 1_2 \rangle$, $\mathbf{f} = \langle 0_1, 1_2 \rangle$, and $\mathbf{t} = \langle 1_1, 0_2 \rangle$. The bilattice operations evaluate to the following:

$$\begin{aligned}
\langle a, b \rangle \wedge \langle c, d \rangle &= \langle a \sqcap_1 c, b \sqcup_2 d \rangle \\
\langle a, b \rangle \vee \langle c, d \rangle &= \langle a \sqcup_1 c, b \sqcap_2 d \rangle \\
\langle a, b \rangle \otimes \langle c, d \rangle &= \langle a \sqcap_1 c, b \sqcap_2 d \rangle \\
\langle a, b \rangle \oplus \langle c, d \rangle &= \langle a \sqcup_1 c, b \sqcup_2 d \rangle.
\end{aligned}$$

(BP-2) If L_1 and L_2 are distributive lattices then $L_1 \odot L_2$ is a distributive bilattice.

(BP-3) If $L_1 = L_2 = L$ then $L \odot L$ is an bilattice with negation, where $\neg\langle a, b \rangle = \langle b, a \rangle$.

(BP-4) If $L_1 = L_2 = L$ is a non-distributive De Morgan algebra then $L \odot L$ is a bilattice with a conflation that commutes with negation, where $-\langle a, b \rangle = \langle \overline{b}, \overline{a} \rangle$.

Combining several of the items above, if L is a De Morgan algebra (which assumes distributivity), then $L \odot L$ is a distributive bilattice with a negation and a conflation that commute.

What is more difficult to establish is that these conditions reverse. For instance, if we have an interlaced bilattice, it is isomorphic to $L_1 \odot L_2$, where L_1 and L_2 are bounded lattices, and L_1 and L_2 are unique up to isomorphism. And so on. Thus we have very general *representation theorems*. These results were proved over time, and various parts can be found in Ginsberg (1988), Fitting (1990, 1991), Avron (1996).

We will not need a detailed proof of these representation theorems but a few basic items from the proof will be of importance to us, since we will be adding one more piece. For an interlaced bilattice $\mathcal{B}$, L_1 can be taken to be $\{x \vee \bot \mid x \in \mathcal{B}\}$ and L_2 to be $\{x \wedge \bot \mid x \in \mathcal{B}\}$, each with the ordering resulting when $\leq_t$ is restricted to L_1 or L_2, respectively. If we have a bilattice with negation the lattices L_1 and L_2 just described are isomorphic and we can simply use L consisting of $\{x \vee \bot \mid x \in \mathcal{B}\}$ with the ordering induced by $\leq_t$. In Proposition 8.8.3, we make use of these pieces of the proof to add one new part to the representation theorem collection.

Suppose $\mathcal{B}$ is a bilattice with negation, $L = \{x \vee \bot \mid x \in \mathcal{B}\}$, and $f : \mathcal{B} \to L$ is defined by $f(x) = x \vee \bot$. This mapping is always many-one. For instance, if $\mathcal{B} = \mathcal{NINE}$ from Fig. 8.4, $f(\top) = f(o\mathbf{t}) = f(\mathbf{t}) = \mathbf{t}$. Even in the paradigm case of $\mathcal{FOUR}$ from Fig. 8.2, $f(\top) = f(\mathbf{t}) = \mathbf{t}$. Thus, each member of the lattice L, generated by the proof of the representation theorem, always has multiple pre-images in the bilattice $\mathcal{B}$ that we are representing. We will show that there are special and unique pre-images of particular significance in our current strict/tolerant investigation. These are simply the *exact* members (provided we have the conflation machinery to define them).

The following lemma provides everything we need for our proof of the central role of exact bilattice members. Using the bilattice representation results above, much of it could be left as an exercise in computation. Instead we give direct proofs, which provide some insights of their own.

Lemma 8.8.2 *Assume $\mathcal{B}$ is an interlaced bilattice with a negation and a conflation. For every $x, y \in \mathcal{B}$:*

(1) $(x \vee \bot) \wedge -(x \vee \bot)$ *is exact;*
(2) $x = (x \vee \bot) \wedge (x \vee \top)$*;*
(3) $(x \wedge y) \vee \bot = (x \vee \bot) \wedge (y \vee \bot)$*;*
(4) $[(x \vee \bot) \wedge -(x \vee \bot)] \vee \bot = x \vee \bot$*;*
(5) if x and y are exact then $x \vee \bot \leq_t y \vee \bot$ if and only if $x \leq_t y$;
(6) if x and y are exact and $x \vee \bot = y \vee \bot$ then $x = y$.

Proof

(1) Exactness is simple.

$$\begin{aligned} -[(x \vee \bot) \wedge -(x \vee \bot)] &= [-(x \vee \bot) \wedge --(x \vee \bot)] \\ &= [-(x \vee \bot) \wedge (x \vee \bot)]. \end{aligned}$$

(2) (This is Corollary 2.8 part 4 in Avron (1996).) Since $\bot \leq_k x \leq_k \top$, using interlacing, $x \vee \bot \leq_k x \vee x \leq_k x \vee \top$, and so $x \vee \bot \leq_k x \leq_k x \vee \top$. Then by Proposition 8.4.1, $(x \vee \bot) \wedge (x \vee \top) \leq_t x \leq_t (x \vee \bot) \vee (x \vee \top)$ so in particular, $(x \vee \bot) \wedge (x \vee \top) \leq_t x$. Also $x \leq_t x \vee \bot$ and $x \leq_t x \vee \top$, so $x \leq_t (x \vee \bot) \wedge (x \vee \top)$.
(3) From $\bot \leq_k x$ by interlacing, $x \vee \bot \leq_k x \vee x = x$. Similarly $y \vee \bot \leq_k y$. Then $\bot \leq_k (x \vee \bot) \wedge (y \vee \bot) \leq_k x \wedge y$. Then by Proposition 8.4.1, $(x \wedge y) \wedge \bot \leq_t (x \vee \bot) \wedge (y \vee \bot) \leq_t (x \wedge y) \vee \bot$, so in particular $(x \vee \bot) \wedge (y \vee \bot) \leq_t (x \wedge y) \vee \bot$. Also $x \wedge y \leq_t x$ so $(x \wedge y) \vee \bot \leq_t (x \vee \bot)$, and similarly $(x \wedge y) \vee \bot \leq_t (y \vee \bot)$. Then $(x \wedge y) \vee \bot \leq_t (x \vee \bot) \wedge (y \vee \bot)$.
(4) Using item (3),

$$\begin{aligned} [(x \vee \bot) \wedge -(x \vee \bot)] \vee \bot &= [(x \vee \bot \vee \bot) \wedge (-(x \vee \bot) \vee \bot)] \\ &= [(x \vee \bot) \wedge (-x \vee \top \vee \bot)] \\ &= [(x \vee \bot) \wedge (-x \vee \mathbf{t})] \\ &= [(x \vee \bot) \wedge \mathbf{t}] \\ &= x \vee \bot. \end{aligned}$$

(5) If $x \leq_t y$ then $x \vee \bot \leq_t y \vee \bot$, using the interlacing conditions. In the other direction, suppose $x \vee \bot \leq_t y \vee \bot$ and both x and y are exact. Then

$$\begin{aligned} x &= (x \vee \bot) \wedge (x \vee \top) \quad \text{part (2)} \\ &= (x \vee \bot) \wedge (-x \vee \top) \quad \text{exactness} \\ &= (x \vee \bot) \wedge -(x \vee \bot) \\ &\leq_t (y \vee \bot) \wedge -(y \vee \bot) \quad \text{interlacing} \\ &= (y \vee \bot) \wedge (-y \vee \top) \\ &= (y \vee \bot) \wedge (y \vee \top) \quad \text{exactness} \\ &= y \quad \text{part (2)}. \end{aligned}$$

(6) This follows from part (5).

■

Proposition 8.8.3 *Suppose L is a non-distributive De Morgan algebra, and $\mathcal{B} = L \odot L$. The set of exact members of $\mathcal{B}$, under the ordering $\leq_t$, is isomorphic to L.*

Proof The proofs of the usual bilattice representation theorems discussed earlier say that $\mathcal{B}$ is isomorphic to $L' \odot L'$ where $L' = \{x \vee \bot \mid x \in \mathcal{B}\}$ with ordering $\leq_t$ restricted to L'. They also say this is unique up to isomorphism, so L and L' are isomorphic. It is enough, then, to show that L' and $\mathcal{E}$ are isomorphic, where $\mathcal{E}$ is the set of exact members of $\mathcal{B}$.

Let $f : \mathcal{E} \to L'$ be defined by $f(x) = x \vee \bot$. We show that f is $1-1$, onto, and an order isomorphism. We begin with onto. An arbitrary member of L' must be $x \vee \bot$ for some $x \in \mathcal{B}$. Let y be $(x \vee \bot) \wedge -(x \vee \bot)$. By Lemma 8.8.2 part 1, y is exact and by part 4, $f(y) = x \vee \bot$. Hence f is onto. It is $1-1$ by Lemma 8.8.2 part 6. Finally, we have an order isomorphism by Lemma 8.8.2 part 5. ■

8.9 Logical De Morgan Algebras

Quite a few common many-valued logics validate De Morgan's laws. An extensive investigation of these can be found in Leitgeb (1999), where applications to the theory of truth were examined. In that paper, being a prime filter was one of the conditions considered for the set of designated truth values. We will take it as central here, and we investigate the resulting family with respect to its relation to bilattices. Actually, since the distributive laws assumed in De Morgan algebras play little role here, we use the more general family of *non-distributive* De Morgan algebras. Everything we say applies, of course, if we also have distributivity.

Definition 8.9.1 (Non-Distributive Logical De Morgan Algebras) Let L be a non-distributive De Morgan algebra (writing $\sqcap$ and $\sqcup$ for meet and join, and overbar for De Morgan complement). A subset D of L is a *prime filter* in L if it meets the following two conditions:

1. $a \sqcap b \in D$ if and only if $a \in D$ and $b \in D$;

2. $a \sqcup b \in D$ if and only if $a \in D$ or $b \in D$.

We call the pair $\langle L, D \rangle$ a non-distributive *logical De Morgan algebra*, thinking of it as a many-valued logic with D as the set of designated truth values.

We will show that each member of the family of logics determined by non-distributive logical De Morgan algebras has a strict/tolerant version. Classical logic is determined by the best-known De Morgan example, and so is part of a large family with strict/tolerant logics.

Using the bilattice construction sketched in Sect. 8.8, if L is a non-distributive De Morgan algebra then $L \odot L$ is an interlaced bilattice with negation and conflation. This can be extended from algebras to logics, as we will now show.

Lemma 8.9.2 *Let $\langle L, D \rangle$ be a non-distributive De Morgan logic. Then $\langle L \odot L, D \times L \rangle$ is a logical bilattice (interlaced, with negation and conflation).*

Proof Given earlier items, all that needs to be shown is that $D \times L$ is a prime bifilter in $L \odot L$, Definition 8.6.5. In the following, $\langle x, y \rangle$ and $\langle z, w \rangle$ are any two members of $L \odot L$. Since we are working with L throughout, membership in L is automatic and can be mentioned or dropped whenever useful. We show one prime bifilter case as sufficiently representative:

$$\begin{aligned}
\langle x, y \rangle \vee \langle z, w \rangle \in D \times L \text{ iff } & \langle x \sqcup z, y \sqcap w \rangle \in D \times L \\
\text{iff } & x \sqcup z \in D \\
\text{iff } & x \in D \text{ or } z \in D \\
\text{iff } & (x \in D \text{ and } y \in L) \text{ or } (z \in D \text{ and } w \in L) \\
\text{iff } & \langle x, y \rangle \in D \times L \text{ or } \langle z, w \rangle \in D \times L.
\end{aligned}$$

■

Proposition 8.9.3 *Let $\langle L, D \rangle$ be a non-distributive logical De Morgan algebra. $\langle L, D \rangle$ is isomorphic to the bilattice-based logic structure $\mathsf{C}(\langle L \odot L, D \times L \rangle)$ from Definition 8.7.1. To state this more precisely, first recall that $\mathsf{C}(\langle L \odot L, D \times L \rangle)$ is the many-valued logic $\langle \mathcal{E}, (D \times L) \cap \mathcal{E} \rangle$, where $\mathcal{E}$ is the set of exact members of $L \odot L$. Then, there is an isomorphism between $\mathcal{E}$ and L that pairs the members of $(D \times L) \cap \mathcal{E}$ with those of D.*

Proof Begin with a non-distributive logical De Morgan algebra $\langle L, D \rangle$, and construct the interlaced bilattice $L \odot L$. Proposition 8.8.3 says the set of exact members of $L \odot L$ is isomorphic to L. We can extract more information from the proof of that proposition. We know the mapping $f(x) = x \vee \bot$ maps the exact members of the bilattice isomorphically to a non-distributive De Morgan algebra $L' = \{x \vee \bot \mid x \in L \odot L\} = \{x \vee \bot \mid x \in L \odot L \text{ and } x \text{ exact}\}$. And further, L' must be isomorphic to L. We now examine the details of the mapping f.

Let $\langle a, b \rangle$ be an arbitrary member of $L \odot L$. Then $f(\langle a, b \rangle) = \langle a, b \rangle \vee \bot = \langle a, b \rangle \vee \langle 0, 0 \rangle = \langle a \sqcup 0, b \sqcap 0 \rangle = \langle a, 0 \rangle$. Now the isomorphism from L' to L is obvious: $\langle a, 0\rangle \mapsto a$. Then further, the mapping from $L \odot L$ to L is, in fact, just $\langle a, b\rangle \mapsto a$, and so it was this that was shown in the proof of Proposition 8.8.3 to be an order-preserving isomorphism when restricted to the exact members of $L \odot L$. So to finish the present proof, we must show this mapping is $1-1$ and onto between $(D \times L) \cap \mathcal{E}$ and D.

We first show the mapping $\langle a, b\rangle \mapsto a$, restricted to $(D \times L) \cap \mathcal{E}$, is onto D. Suppose $a \in D$. Of course $\langle a, \overline{a}\rangle \in D \times L$ and $-\langle a, \overline{a}\rangle = \langle \overline{\overline{a}}, \overline{a}\rangle = \langle a, \overline{a}\rangle$, so $\langle a, \overline{a}\rangle$ is exact. Thus, $\langle a, \overline{a}\rangle \in (D \times L) \cap \mathcal{E}$, and of course $\langle a, \overline{a}\rangle \mapsto a$.

Finally, we show the mapping $\langle a, b\rangle \mapsto a$, restricted to the exact members of $L \odot L$, is $1-1$. To show this it is enough to show that if $\langle a, b\rangle$ and $\langle a, c\rangle$ are both exact, then $\langle a, b\rangle = \langle a, c\rangle$. If $\langle a, b\rangle$ is exact, $\langle a, b\rangle = -\langle a, b\rangle = \langle \overline{b}, \overline{a}\rangle$, so $a = \overline{b}$. Similarly, $a = \overline{c}$, and it follows that $\overline{b} = \overline{c}$ and hence $b = c$. ∎

8.10 Generating Strict/Tolerant Examples

We now have everything we need for the central result of this paper.

Proposition 8.10.1 *For each non-distributive logical De Morgan algebra, there is a strict/tolerant logic having the same consequence relation but differing from it at the metaconsequence level. There is an algorithm for constructing the strict/tolerant logic from the logical De Morgan algebra.*

Proof We present the algorithm and cite the various earlier results proven earlier that establish what we need.

Gen-1 Start with a (non-distributive) logical De Morgan algebra, $\langle L, D\rangle$.

Gen-2 $\langle L \odot L, D \times L\rangle$ is an interlaced logical bilattice with negation and conflation.

Gen-3 Using notation from Definition 8.7.1, $\mathsf{ST}\langle L \odot L, D \times L\rangle$ is a strict/tolerant logic analog and $\mathsf{C}\langle L \odot L, D \times L\rangle$ is a classical logic analog.

Gen-4 By Proposition 8.7.2, $\mathsf{ST}\langle L \odot L, D \times L\rangle$ and $\mathsf{C}\langle L \odot L, D \times L\rangle$ validate the same sequents.

Gen-5 By Proposition 8.7.3, $\mathsf{ST}\langle L \odot L, D \times L\rangle$ and $\mathsf{C}\langle L \odot L, D \times L\rangle$ differ at the metaconsequence level.

Gen-6 Finally, the structure $\mathsf{C}\langle L \odot L, D \times L\rangle$ is isomorphic to the logical De Morgan algebra $\langle L, D\rangle$ with which we began, by Proposition 8.9.3.

∎

Example 8.10.2 Continuing Example 8.7.4. Let L be the lattice $\{0, 1\}$ with the ordering $0 \le 1$, and let D be $\{1\}$. $\langle L, D\rangle$ is not just some logical De Morgan algebra but is that of classical logic, the most basic of all. The bilattice product $L \odot L$ is isomorphic to $\mathcal{FOUR}$ from Fig. 8.2, with $\bot$ corresponding to $\langle 0, 0\rangle$, $\mathbf{f}$ to $\langle 0, 1\rangle$, $\mathbf{t}$ to $\langle 1, 0\rangle$, and $\top$ to $\langle 1, 1\rangle$. The logical bilattice $\langle L \odot L, D \times L\rangle$ is then isomorphic to $\mathcal{FOUR}$ with $\{\mathbf{t}, \top\}$ as designated values. It follows that $\mathsf{C}(\langle L \odot L, D \times L\rangle)$ is classical logic and $\mathsf{ST}(\langle L \odot L, D \times L\rangle)$ is the usual version of strict/tolerant logic, **ST**.

Example 8.10.3 Continuing Example 8.7.5. We start with the Kleene strong three-valued logic, $\{0, \frac{1}{2}, 1\}$, with $\{1\}$ as designated truth value. These give us a (distributive) logical De Morgan algebra, $\mathsf{K}_3 = \langle\{0, \frac{1}{2}, 1\}, \{1\}\rangle$, Kleene's strong three-valued logic. We use this to create the logical bilattice, $\langle \mathsf{K}_3 \odot \mathsf{K}_3, \{1\} \times \mathsf{K}_3\rangle$, which is isomorphic to $\mathcal{NINE}$, from Fig. 8.4. Then, as discussed in Example 8.7.5, this generates the strict/tolerant logic pair from Fig. 8.1.

Example 8.10.4 This time we do something like Example 8.10.3 but modify the work so that we produce a strict/tolerant counterpart of **LP**, the logic of paradox, instead of K_3. Formally, the only difference between **LP** and K_3 is the choice of designated truth values. For **LP**, from $\{0, \frac{1}{2}, 1\}$ we take $\{\frac{1}{2}, 1\}$ as designated, so we have the (again distributive) logical De Morgan algebra $\langle\{0, \frac{1}{2}, 1\}, \{\frac{1}{2}, 1\}\rangle$. The bilattice $\mathcal{NINE}$, from Fig. 8.4, is still the bilattice we must work with (isomorphically). The prime bifilter we now want from $\mathcal{NINE}$ is $\{d\mathbf{t}, \mathbf{t}, d\top, o\mathbf{t}, o\mathbf{f}, \top\}$ (though note that $d\mathbf{t}$ is not anticonsistent) and the intersection of this with the exact members is $\{d\top, \mathbf{t}\}$. The details are much like those of Example 8.10.3 and we wind up with the diagrams shown in Fig. 8.5, which can be compared with the earlier ones. The strictly designated values are $\{d\top, \mathbf{t}\}$ and the tolerantly designated values are $\{\mathbf{t}, d\top, o\mathbf{t}, o\mathbf{f}, \top\}$.

As a simple instance, it is well known that $P, \neg P \Rightarrow Q$ is not valid in **LP**, as the valuation $v(P) = d\top$, $v(Q) = \mathbf{f}$ shows. The same valuation, in the strict/tolerant version, also works as a counterexample.

Example 8.10.5 We start with the bilattice $\mathcal{FOUR}$, shown in Fig. 8.2. As is standard, we take $\{\mathbf{t}, \top\}$ as designated truth values. Using the $\leq_t$ ordering, the resulting logic is the well-known *first-degree entailment*, **FDE**. Thus, we have a logical De Morgan algebra, $\langle\langle\mathcal{FOUR}, \leq_t\rangle, \{\mathbf{t}, \top\}\rangle$, completing **Gen-1** of the construction outlined earlier.

Fig. 8.5 A strict/tolerant counterpart of LP

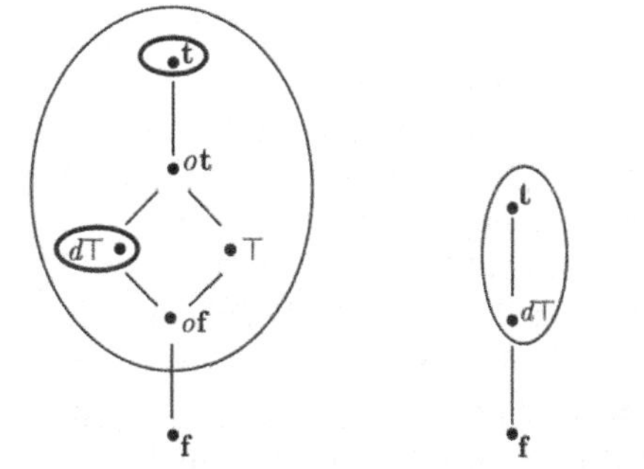

(a) Strict/Tolerant LP (b) Standard LP

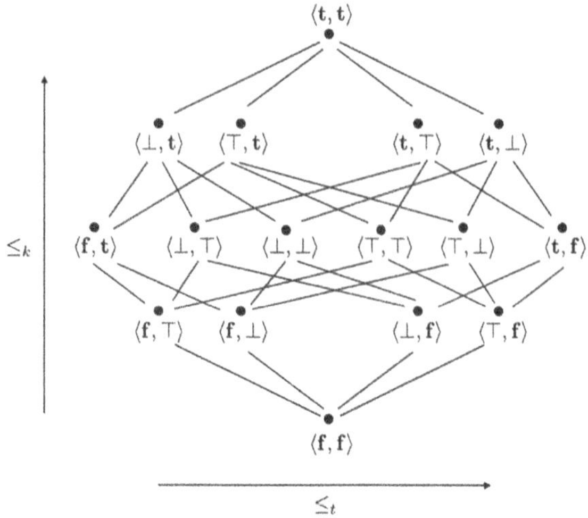

Fig. 8.6 The bilattice $\mathcal{SIXTEEN}$

For **Gen**-2 of the construction, we form the bilattice product $\langle\langle \mathcal{FOUR}, \leq_t\rangle \odot \langle \mathcal{FOUR}, \leq_t\rangle$, which is shown in Fig. 8.6 and given the name $\mathcal{SIXTEEN}$. It may be best to think of the construction simply as formal, without trying to attach intuitive significance to possible meanings for node labels. $\mathcal{SIXTEEN}$ becomes a logical bilattice when we take as designated values $\{\mathbf{t}, \top\} \times \{\mathbf{t}, \top, \mathbf{f}, \bot\}$. That is,
Designated Values: $\langle \mathbf{t}, \mathbf{t}\rangle$, $\langle \mathbf{t}, \top\rangle$, $\langle \mathbf{t}, \mathbf{f}\rangle$, $\langle \mathbf{t}, \bot\rangle$, $\langle \top, \mathbf{t}\rangle$, $\langle \top, \top\rangle$, $\langle \top, \mathbf{f}\rangle$, $\langle \top, \bot\rangle$.

We began with $\mathcal{FOUR}$, using the ordering $\leq_t$, and so our De Morgan operation is the negation of $\mathcal{FOUR}$. Then conflation in $\mathcal{SIXTEEN}$ is: $-\langle a, b\rangle = \langle \neg b, \neg a\rangle$. It is now easy to check that the truth values of $\mathcal{SIXTEEN}$ divide up as shown below. What might be a bit surprising, after the previous examples, is that not everything falls into the exact, consistent, and anticonsistent categories.

Exact Values: $\langle \mathbf{f}, \mathbf{t}\rangle$, $\langle \bot, \bot\rangle$, $\langle \top, \top\rangle$, $\langle \mathbf{t}, \mathbf{f}\rangle$.

Consistent Values: Exact together with $\langle \mathbf{f}, \mathbf{f}\rangle$, $\langle \mathbf{f}, \bot\rangle$, $\langle \mathbf{f}, \top\rangle$, $\langle \bot, \mathbf{f}\rangle$, $\langle \top, \mathbf{f}\rangle$.
Anticonsistent Values: Exact together with $\langle \bot, \mathbf{t}\rangle$, $\langle \top, \mathbf{t}\rangle$, $\langle \mathbf{t}, \bot\rangle$, $\langle \mathbf{t}, \top\rangle$, $\langle \mathbf{t}, \mathbf{t}\rangle$.
None of the above: $\langle \bot, \top\rangle$, $\langle \top, \bot\rangle$.

We now move on to **Gen**-3 of the construction. By Proposition 8.9.3 the classical logic analog, $\mathsf{C}\langle \mathcal{SIXTEEN}, \{\mathbf{t}, \top\} \times \{\mathbf{t}, \top, \mathbf{f}, \bot\}\rangle$, is isomorphic to the bilattice $\mathcal{FOUR}$ under the $\leq_t$ ordering, with $\{\mathbf{t}, \top\}$ designated. As to $\mathsf{ST}\langle \mathcal{SIXTEEN}, \{\mathbf{t}, \top\} \times \{\mathbf{t}, \top, \mathbf{f}, \bot\}\rangle$, it has as members the anticonsistent values from $\mathcal{SIXTEEN}$, with the ordering induced by $\leq_t$. The set of strictly designated values is the intersection of the set of Designated Values for $\mathcal{SIXTEEN}$ with the set of Exact Values, and this is $\{\langle \top, \top\rangle, \langle \mathbf{t}, \mathbf{f}\rangle\}$. Finally, the set of tolerantly designated values is the intersection of the set of Designated Values with the set of Anticonsistent Values, and this is $\{\langle \top, \top\rangle, \langle \mathbf{t}, \mathbf{f}\rangle, \langle \top, \mathbf{t}\rangle, \langle \mathbf{t}, \bot\rangle, \langle \mathbf{t}, \top\rangle, \langle \mathbf{t}, \mathbf{t}\rangle\}$. All this is shown schematically in Fig. 8.7a. The standard formulation of **FDE** is shown as part Fig. 8.7b. Our gen-

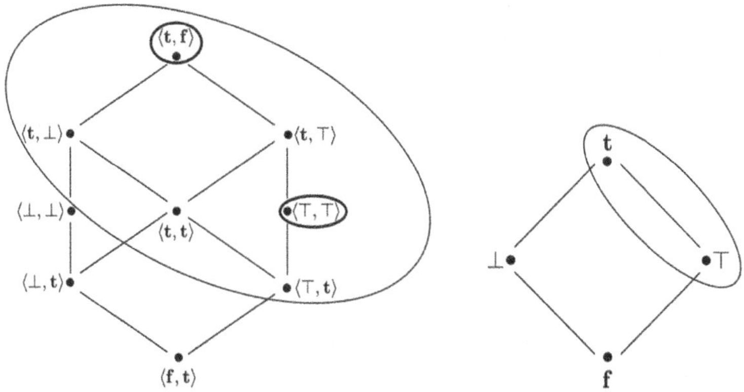

(a) Strict/Tolerant FDE (b) Standard FDE

Fig. 8.7 A strict/tolerant counterpart of FDE

eral results show that these validate the same consequence relation, but differ on the metaconsequence level.

8.11 And More?

The family of what we called logical De Morgan algebras (distributive or not) is mostly made up of examples of purely technical interest. But the fact that K_3, LP, and FDE have strict/tolerant counterparts may have useful consequences, or at least consequences that someone might argue are useful. I leave this to others. But there are some more technical items that I plan to develop further in subsequent work.

Here we looked at K_3, strong Kleene logic. There is also weak Kleene logic. This was generalized to the bilattice context, in Fitting (2006), using what I called "cut down operations." Such operations have been further investigated in Ferguson (2015), and dualized in Szmuc (2018). It is likely that strict/tolerant analogs based on cut down (or track down) operations can be developed, similar to what has been done here.

Analogous to strict/tolerant logic, but with things reversed, there is also tolerant/strict logic, see Frankowski (2004a, b) for background. This is a more complicated family than that of strict/tolerant logic, and will be investigated in a separate paper.

In a private communication, Eduardo Alejandro Barrio raised the question of what is the minimum size of a strict/tolerant counterpart. It may be the case that the algorithm given as proof of Proposition 8.10.1 produces minimal sized counterparts, but perhaps not. This is open.

Finally, Barrio et al. (2021) generalizes the original strict/tolerant phenomenon in an "upward" direction, as we discussed in Section 8.1. Their work examines the structure of consequence, metaconsequence, metametaconsequence, and so on. It is likely that this work also generalizes to the present setting, but it is deferred to a later paper.

References

Arieli, O., & Avron, A. (1996). Reasoning with logical bilattices. *Journal of Logic, Language, and Information*, *5*(1), 25–63.

Arieli, O., & Avron, A. (1998). The value of four values. *Artificial Intelligence*, *102*, 97–141.

Asenjo, F. G. (1966). A calculus of antinomies. *Notre Dame Journal of Formal Logic*, *7*, 103–105.

Avron, A. (1996). The structure of interlaced bilattices. *Mathematical Structures in Computer Science*, *6*(3), 287–299.

Barrio, E. A., Pailos, F., & Szmuc, D. (2021) Hierarchies of Classical and Paraconsistent Logics. submitted for publication.

Belnap, N. D, Jr. (1977). A useful four-valued logic. In J. M. Dunn & G. Epstein (Eds.), *Modern Uses of Multiple-Valued Logic*. D. Reidel.

Cobreros, P., et al. (2012). Tolerant, classical, strict. *Journal of Philosophical Logic*, *41*(2), 347–385.

Ferguson, T. M. (2015). Cut-down operations on bilattices. In *2015 IEEE International Symposium on Multiple-Valued Logic* (pp. 24–29).

Fitting, M. C. (1989). Bilattices and the theory of truth. *Journal of Philosophical Logic*, *18*, 225–256.

Fitting, M. C. (1990). Bilattices in logic programming. In G. Epstein (Ed.), *The Twentieth International Symposium on Multiple-Valued Logic* (pp. 238–246). IEEE.

Fitting, M. C. (1991). Bilattices and the semantics of logic programming. *Journal of Logic Programming*, *11*, 91–116.

Fitting, M. C. (1997). A theory of truth that prefers falsehood. *Journal of Philosophical Logic*, *26*, 477–500.

Fitting, M. C. (2006). Bilattices are nice things. In T. Bolander, V. Hendricks, & S. A. Pederrsen (Eds.), *Self-Reference* (pp. 53–77). Center for the Study of Language and Information.

Frankowski, S. (2004a). Formalization of a plausible inference. *Bulletin of the Section of Logic*, *33*, 41–52.

Frankowski, S. (2004b). P-consequence versus q-consequence. *Bulletin of the Section of Logic*, *33*, 197–207.

Gargov, G. (1999). Knowledge, uncertaity and ignorance in logic: Bilattices and beyond. *Journal of Applied Non-Classical Logics*, *9*(2–3), 195–283.

Ginsberg, M. L. (1988). Multivalued logics: a uniform approach to reasoning in artificial intelligence. *Computational Intelligence*, *4*, 265–316.

Ginsberg, M. L. (1990). Bilattices and modal operators. *Journal of Logic and Computation*, *1*(1), 41–69.

Kleene, S. C. (1938). On notation for ordinal numbers. *Journal of Symbolic Logic*, *3*, 150–155.

Kleene, S. C. (1950). *Introduction to Metamathematics*. Princeton, NJ: D. Van Nostrand.

Leitgeb, H. (1999). Truth and the Liar in De Morgan-valued models. *Notre Dame Journal of Formal Logic*, *40*(4), 496–514.

Malinowski, G. (1990). Q-consequence operation. *Reports on Mathematical Logic*, *24*, 49–59.

Malinowski, Grzegorz. (2002). Referential and Inferential Many-Valuedness. In W. A. Carnielli, M. E. Coniglio, & I. M. L. DOttaviano (Eds.), *Paraconsistency, the Logical Way to the Inconsistent* (Vol. 228, pp. 341–352). Lecture Notes in Pure and Applied Mathematics 228. Marcel Dekker.

Malinowski, G. (2007). That p + q = c(onsequence). *Bulletin of the Section of Logic*, *36*(1/2), 7–19.

Priest, G. (1979). The logic of paradox. *Journal of Philosophical Logic*, *8*(1), 219–241.

Priest, G. (2008). *An Introduction to Non-Classical Logic; From If to Is* (2nd ed.). Cambridge: Cambridge University Press.

Shramko, Y., & Wansing, H. (2005). Some useful 16-valued logics: how a computer network should think. *Journal of Philosophical Logic*, *34*, 121–153.

Shramko, Y., & Wansing, H. (2011). Truth and falsehood: an inquiry into generalized logical values. In *Trends in Logic* (Vol. 36). Springer.

Szmuc, D. (2018). Track-down operations on bilattices. In R. Wille & M. Lukac (Eds.), *Proceedings of the 48th IEEE International Symposium on Multiple-Valued Logic, Los Alamitos, California* (pp. 74–79).

Melvin Fitting was born in Troy, New York, in 1942. His undergraduate degree was from Rensselaer Polytechnic Institute, in mathematics, and his 1968 Ph.D. was supervised by Raymond Smullyan, at Yeshiva University. His dissertation became his first book, Intuitionistic Logic, Model Theory, and Forcing (1969). He has worked in many areas including intensional logic, semantics for logic programming, and theory of truth. A significant part of his work has involved developing tableau systems for non-classical logics, thus generalizing the classical systems of his mentor Smullyan. In 2012, he received the Herbrand Award from the Conference on Automated Deduction, largely for this work. He was the faculty of the City University of New York from 1969 until his retirement in 2013. He worked at the undergraduate Lehman College, and at the City University Graduate Center, where he was in the Departments of Mathematics, Computer Science, and Philosophy. He has authored or co-authored 10 books, the most recent appearing in 2019, as well as numerous research papers. These cover philosophical logic, computability, automated theorem proving, and, with Raymond Smullyan, set theory. In 2019, he received an Honorary Doctorate from the University of Bucharest. He is now an Emeritus Professor, but very much active.

Chapter 9
What Is Negation in a System 2020?

Dov M. Gabbay

Abstract The notion of negation is basic to any formal or informal logical system. When any such system is presented to us, it is presented either as a system without negation or as a system with some form of negation. In both cases, we are supposed to know intuitively whether there is no negation in the system or whether the form of negation presented in the system is indeed as claimed. To be more specific, suppose Robinson Crusoe writes a logical system with Hilbert-type axioms and rules, which includes a unary connective $*A$. He puts the document in a bottle and lets it lose at sea. We find it and take a look. We ask: is the connective "$*$" a negation in the system? Yet the notion of what is negation in a formal system is not clear. When we see a unary connective $*A$, (A a wff) together with some other axioms for some additional connectives, how can we tell whether $*A$ is indeed a form of negation of A? Are there some axioms which the connective "$*$" must satisfy in order to qualify $*$ as a negation?

9.1 Negation in Deductive (Monotonic or Non-monotonic) Systems with Cut

We need to start with a definition of what kind of deductive systems we are going to work with. To choose a definition of a deductive system, we first consider which candidates for known accepted negations we want to address and how these are presented to us. The main candidates for known negation we consider are classical negation, intuitionistic negation, relevance logic negation, linear Logic negation,

---

This paper is an expanded version of the 1986 paper (Gabbay 1986). It is basically a position paper for new research for the year 2020, honouring Arnon Avron. I thank Arnon Avron, Anna Zamansky and Ofer Arieli for reviews and comments.

D. M. Gabbay (✉)
University of Luxembourg, Luxembourg, Luxembourg

King's College London, London, England
e-mail: dov.gabbay@kcl.ac.uk

O. Arieli and A. Zamansky (eds.), *Arnon Avron on Semantics and Proof Theory of Non-Classical Logics*, Outstanding Contributions to Logic 21,
https://doi.org/10.1007/978-3-030-71258-7_9

Łukasiewicz many-valued negation and, last but not least, negation as failure in Logic Programming. Some of these logics are presented as Hilbert systems (such as relevance implication). Some have many representations including Tarski-type consequence systems. The best representation from the point of view of answering the question of "What is negation in a deductive system" is for us to look at Tarski systems based on multi-sets.[1]

Definition 9.1.2 Let L be any propositional logical system and let $\vdash_L$ be its provability/consequence relation. We do not specify how L is presented to us, it can be as a Hilbert style system with axioms and rules, or as a natural deduction system or by semantics, etc. The main point is that we have a faithful Tarski style formulation of the provability/consequence relation of L:

$$A_1, \ldots, A_k \vdash_L B$$

between a finite multi-set Δ containing the formulas A_j, $j = 1, \ldots, k$ and a single B satisfying thc following three conditions:

1. $\Delta \vdash A$ for $A \in \Delta$. (reflexivity).
2. If $\Delta \vdash A$ and $\Delta' \supseteq \Delta$ then $\Delta' \vdash A$. (monotonicity).
3. If $\Delta' \vdash A$ and $\Delta \cup \{A\} \vdash B$ then $\Delta \cup \Delta' \vdash B$. (Transitivity, or cut).

In fact any relation $\vdash$ on wffs satisfying (1), (2) and (3), can be regarded as a monotonic logical system for sets or multi-sets of data. Note the repetition in rule (3) above for the multi-set case.

Note that in the non-monotonic case (Gabbay 1985a), condition (2.) above is replaced by condition (2non) (Gabbay called it "Restricted Monotonicity"):

2non. If $\Delta \vdash A$ and $\Delta \vdash B$ then $\Delta \cup \{A\} \vdash B$

Also note that for resource logics, where multi-sets are used (for example, monotonic affine linear logic), we may also wish to investigate the question of what is negation for a consequence relation with condition (3res) instead of condition (3), where we have:

---

[1] The perceptive reader might ask why is it that we are considering "what is negation" in a deductive consequence system, why not present a consequence system semantically?. The answer is not technical but psychological. When the question was considered in 1986 (see Gabbay 1986), the author had an image of Robinson Crusoe stranded on an island writing a Hilbert System on a sheet of paper, putting it in a bottle, and throwing it into the water. We find it years later and we see a unary connective $*$ in the system and we ask ourselves "Is $*$ a negation in this system?".

Of course, a logic can be presented semantically, but then we can see the intended meaning of the system from the semantics and the challenge is smaller. Consider, for example, classical logic with the connectives $\{\wedge, \vee, \rightarrow\}$ defined semantically via the traditional truth tables for these connectives. We add a unary symbol "$\neg$" giving it the non-deterministic truth table of Arnon Avron (see Olivetti and Terracini 1992), namely:

$$\neg\mathbf{t} = \{\mathbf{f}\} \text{ and } \neg\mathbf{f} = \{\mathbf{t}, \mathbf{f}\}.$$

The consequence relation can be defined semantically.
Question 1.1 Is this $\neg$ (Avron "negation") a negation? (We think it is not a negation).

3res. If $\Gamma \vdash A$ and $\Delta \cup \{A\} \vdash B$ then $\Gamma \cup \Delta \vdash B$.

(3res) does not imply (2) for these logics. But note that (3res) implies (3), but the converse is true only if the consequence relation is monotonic, and it is between "sets" and formulas (not multisets).

The fact that we allow Δ or Γ to be a multi-set presents no technical difficulties.[2]

Our strategy is to give several candidate definitions of what should constitute a negation in a system and test them against our intuitions and against known examples. The examples we look at are as follows:

Example 9.1.3 1. Let us consider the following system in a language with $\neg$ and $\rightarrow$.

(a) $A \rightarrow (B \rightarrow A)$
(b) $[A \rightarrow (B \rightarrow C)] \rightarrow [(A \rightarrow B) \rightarrow (A \rightarrow C)]$
(c) $\neg\neg A \rightarrow A$
(d) $A \rightarrow \neg\neg A$
Rules
(e) Modus Ponens

$$\frac{A, A \rightarrow B}{B}$$

Question 9.1.4 Is $\neg$ a form of negation in this system (i.e. in item 1 of Example 9.1.3)?

2. Let us make life more difficult by adding more axioms to our system. To get the idea of what to add, first we need disjunctions and conjunctions (the system has only $\neg$ and $\rightarrow$). So let us see what can be taken as disjunction.

   In classical logic (with the connectives $\rightarrow$, $\neg$, $\vee$, $\wedge$, and equivalence $\leftrightarrow$) we have:

---

[2] Note that really all we need is to understand, by any precise mathematical-technical means necessary, (proof theoretic, algorithmic, semantic, via translation into another system, via an explicit list/table) the question of when the expression

$$\Delta \vdash A$$

holds.

For Example for the case of relevance implication, in item 2 of Example 9.1.20, we use a translation into a Hilbert system. I do not know at this stage what axiomatic properties to impose on a Tarski consequence relation in order to make it correspond to the relation obtained from the translation in item 2 of Example 9.1.20. More future research is required here.

So Definition 9.1.2 given above is just a very common sample axiomatic definition of a consequence relation. Further note that the author has been claiming for the past 40 years that a logical system should be taken as the declarative set of its theorems as well as an algorithm for demonstrating said theorems. So for example classical logic (perceived as a set of theorems) presented as a Gentzen system is NOT THE SAME LOGIC as classical logic presented via Resolution, which in turn IS NOT THE SAME LOGIC as classical logic presented semantically via Tableaux or truth tables.

$$(a \to b) \to b) \leftrightarrow \neg(a \to b) \vee b \leftrightarrow (a \wedge \neg b) \vee b \leftrightarrow a \vee b.$$

This is in fact a well known definition of $\vee$ in terms of $\to$.

Also let $a \wedge b = \text{def}.\neg(\neg a \vee \neg b) = \neg((\neg a \to \neg b) \to \neg b)$.

Take the following rule:

(f)

$$\frac{\vdash A \to B}{\vdash \neg B \to \neg A}$$

and the further axioms:

(g) $\neg(A \vee B) \leftrightarrow (\neg A \wedge \neg B)$
(h) $\neg(A \wedge B) \leftrightarrow (\neg A \vee \neg B)$
(i) $((A \to \neg\neg B) \to A) \to A$. (This axiom says $(A \to B) \vee A$.)

Question 9.1.5 Is $\neg$ a negation in this system (i.e. in the system of item 2 of Example 9.1.3)?[3]

3. We can ask further: If we also add the axiom

(j) $A \to (\neg A \to B)$.

Question 9.1.6 Does the addition of axiom (j) make $\neg$ a negation in the system (of item 3 of Example 9.1.3)?

(We shall see that answer is no for cases (1) and (2) and yes for case (3).)

It seems from the above Example 9.1.3 that this question does not have an immediate simple answer. Remember that we cannot just write a set of axioms for negation and say that anything satisfying these axioms is a negation. If we write too many axioms we may get only classical negation, and even that is not guaranteed because maybe we do not know how the negation axioms are supposed to interact with other connectives e.g. with $\to$.
Let us look at more examples.

---

[3] Note that axioms (a) and (b), taken together with modus ponens define positive intuitionistic implication. to get positive classical implication we need to add peirce's rule

$$(P)((A \to B) \to A) \to A.$$

Arnon Avron proposed that a better and clearer presentation of the system presented in this item 2 of Example 9.1.3 would be in a language in which disjunction and conjunction are taken as primitive. the system can then be axiomatised by taking some axiomatisation of positive classical logic which has modus ponens as the sole rule of inference, and add to it the axioms (c) and (d), and the rule (f). (Note that if the rule (f) is turned into an axiom, then we get by this a sound and complete system for classical logic.) The other items in Example 9.1.3 can be changed similarly. (that is: in item 1 we take positive intuitionistic logic together with axioms (c) and (d), in item 2 we add (f) and (i), where the latter is taken in a purely positive form, and in item 3 we add (j).

The author prefers the implication based formulation because we need to discuss adding negation as failure to the system.

Example 9.1.7 Consider the system $L3$ below of Wajsberg (1931). It axiomatises the 3 valued logic of Łukasiewicz with $\rightarrow$ and $\neg$.
Axioms:

(W1) $A \rightarrow (B \rightarrow A)$
(W2) $(A \rightarrow B) \rightarrow ((B \rightarrow C) \rightarrow (A \rightarrow C))$
(W3) $(\neg B \rightarrow \neg A) \rightarrow (A \rightarrow B)$
(W4) $((A \rightarrow \neg A) \rightarrow A) \rightarrow A$

The inference rule is modus ponens.

Question 9.1.8 Can one determine on the basis of $\vdash_{L3}$ whether $\neg A$ is a negation in $L3$?

Example 9.1.9 Consider a third system denoted by $LS3$. Its language contains an additional connective $\ominus$ besides $\neg$ and $\rightarrow$. It is obtained from $L3$ by adding the axioms:

($\ominus$1) $\ominus A \rightarrow \neg \ominus A$
($\ominus$2) $\neg \ominus A \rightarrow \ominus A$

Question 9.1.10 1. Is $\neg$ a negation in this system? Is $\ominus$ a negation?
2. If $\neg$ is considered a negation in $L3$, does it have to be considered a negation in the extension $LS3$?

Armed with this stock of examples we now move to a formulation and some possible solutions of our problem.

Problem 9.1.11 *Given a relation $\vdash$ (satisfying (1), (2), (3)) of Definition 9.1.2 and a connective $*A$ in the language of $\vdash$, are there any criteria on the relationship between $\vdash$ and $*$ which will agree with our intuitions regarding the question of when $*$ is to be considered a form of negation? Carnap and Church discussed whether a syntactical characterisation of negation was possible. Carnap thought it was possible and Church thought not. A basic intuition regarding the meaning of $\neg A$ is that* A does not hold *or A is* not wanted *or A is* excluded *or even A is* not confirmed. *Thus if L is a system with a candidate $*A$ for negation, we cannot hope to have A, and its negation $*A$ consistent together (understand "consistent together" intuitively, or maybe "'both provable"). This leads us to our first attempt in answering Problem 9.1.11.*

We must specify a set Θ of unwanted wffs. The wffs of Θ are not allowed to be true (understand "true" intuitively, or maybe "provable"). This is normal and natural for any database. For example we do not want two lecturers to be assigned to the same classroom at the same time. In a formal system L, one can take Θ to be the set containing $\perp$, i.e. falsity *or one can take Θ to be certain conjunctions of atoms, etc.*

*So to get negation into a system we must have a set of unwanted wffs Θ. This set may be different for different negations. The connective (*1) may be a negation because of* $\Theta 1$ *and (*2) may be a negation because of* $\Theta 2$*, and so on.*

We are thus led to the following definition:

Definition 9.1.12 (*Negation as syntactical inconsistency for the monotonic case*) Let $\vdash$ be the provability/consequence relation of a system and $*A$ be a connective. We say that $*$ is a form of negation if there is a fixed non-empty set of wffs Θ^* which is not provably equivalent to the set/multi-set of all wffs, such that for any set or multi-set of wffs Δ and any A the following holds:

$$\Delta \vdash *A \text{ iff } \exists y \in \Theta^*(\Delta, A \vdash y).$$

i.e. A is negated by Δ because A leads to some unwanted y in Θ^*.[4]

Lemma 9.1.13 *Let $*$ be a negation in the logical system $\vdash$, as defined in Definition 9.1.12. Then the set $\{x : \emptyset \vdash^* x\}$ is is non-empty.*

Proof Since $*$ is a negation, let $q \in \Theta^*$, then $\emptyset \vdash *q$, since $q \vdash q$.

The above is a purely syntactic (in terms of $\vdash$) definition. So to check whether $*A$ of an axiom system is a negation, look for a Θ^* and try to prove the above equivalence.

Note that the equivalence must hold for any Δ and A. ■

We may ask ourselves, how do we find a Θ^*? The answer is that if such a $\Theta*$ exists, (i.e. $*$ is a negation according to the above Definition 9.1.12) then it follows from Lemma 9.1.13 that Θ^* can be taken as

$$\Theta^* = \{C | \emptyset \vdash *C\}$$

where $\emptyset$ is the empty set.

Lemma 9.1.14 *Assume $*$ is a negation with a Θ^* according to Definition 9.1.12, then for any Δ and any A, (1) is equivalent to (2):*

1. *$\Delta, A \vdash C$ for some C such that $\emptyset \vdash *C$.*
2. *$\Delta, A \vdash B$ for some $B \in \Theta^*$.*

Proof Let $B \in \Theta^*$ then since $B \vdash B$ we get by Definition 9.1.12 that $\emptyset \vdash *B$. This shows that (2) implies (1).

Assume that for some C such that $\emptyset \vdash *C$, we have $\Delta, A \vdash C$. Since $\emptyset \vdash *C$, we have that for some $B \in \Theta^*$,

---

[4] If there is disjunction $\vee$ in the language, then note that (for technical reasons) Θ will be closed under disjunctions. We need to check what happens when we are dealing with multi-sets. We expect the differences would be technical, the idea of what is negation would be the same.

Furthermore if we are dealing with multi-sets we may need more copies of Δ, I think we might try

$$\Delta \vdash *A \text{ iff for some } y \in \Theta* \text{ and some } k, (\Delta \cup \ldots (k \text{ times}) \ldots \cup \Delta, A \vdash y).$$

This is in the spirit of enhancing the data (via the connective **C**) introduced later in this paper.

Note that we can also negate a set/multi-set Γ, namely $\Delta \vdash *\Gamma$ iff for some $y \in \Theta_*$ and some k; $\Delta\cup$, ...k times, ... $\cup \Delta \cup \Gamma \vdash *y$.

$$C \vdash B.$$

then by cut-res (item (3res in Definition 9.1.2) using $C \vdash B$, we get

$$\Delta, A \vdash B.$$

This proves that (1) implies (2) and we have proved the lemma for the monotonic case. ■

We can thus modify Definition 9.1.12 as follows:

Definition 9.1.15 (*Definition* 9.1.12 *(modified)*) Let $\vdash$ be a monotonic logical system and let $*A$ be a connective. We say that $*$ is a form of negation in $\vdash$ iff for any Δ and any A the following holds.

$\Delta \vdash *A$ iff for some C such that $\emptyset \vdash *C$ we have $\Delta, A \vdash C$.

The above Definition 9.1.15 seems theoretically sound and acceptable. All we have to see now is whether it takes care of all the currently known and agreed upon negations.

We will see later that further modifications are necessary. For this reason, we continue to use Θ^* itself and not $\{C|\emptyset \vdash^* C\}$. Note that Θ^* may contain wffs containing $*$ itself (this means that it is built up also by using the negation connective $*$).

We do not need to exclude this possibility. In fact, for classical logic, we can take Θ^* to be the set $\{q_0 \wedge \neg q_0\}$ for some atom q_0 and we all know that in classical logic $\Delta \vdash \neg A$ iff $\Delta, A \vdash q_0 \wedge \neg q_0$ holds, and so classical negation is a negation. So is an intuitionistic negation because the same equivalence holds.

According to Definition 9.1.12, the $\neg$ defined in Example 9.1.3 axioms (a) to and including axiom (i), i.e. Question 9.1.6 is not a negation. One can see this by taking the following interpretation and verifying that all axioms (a) to and including axiom (i) of Example 9.1.3 are valid. In this interpretation, there are two worlds h and e (heaven for h and earth for e). $\neg A$ is *true* in one if A is *false* in the other. $\rightarrow$ is the usual truth-functional implication. All axioms and rules are valid; i.e. we have

$\vdash A$ iff A is true in e and h under any assignment to the atoms.[5]

---

[5] Additional axioms may be needed for this assertion. If we just add the connective $\neg$ to the language of intuitionistic implication we are simply generating repeatedly/recursively new atoms of the form $\neg A$ for any already generated A, using all the wffs of intuitionistic implication as a basis.

- The axiom $\neg\neg A = A$, says the generating is idempotent.
- The axiom $\neg(A \Rightarrow B) = (A \Rightarrow \neg B)$, if added , takes us in the direction of $\neg$ being negation as failure.
- Adding Peirce's rule, i.e. basing the addition of $\neg$ above on classical implications, takes us to the semantics with e and h.

Let me quote Arnon Avron's comment to me, as follows:

> I can easily see the "only if" part (i.e. soundness). The converse is not obvious. Have you proved it somewhere? If so, you should add a reference. If not, you should give a proof here.

Now we can see that $\neg A$ is not a negation of A, since it just says that A is false in the other world. A $\wedge\neg A$ can be consistent, as A could be true in this world (e.g. e) and false in the other world (e.g. h).

The rule of Definition 9.1.12 for negation does not apply here. If $\neg$ were a negation, then for some Θ, and for all Δ, A we would have:

$$\Delta \vdash \neg A \Leftrightarrow \Delta, A \vdash y, \text{ for some } y \in \Theta.$$

In particular for any $y \in \Theta$ we get $\vdash \neg y$. Let p be atomic then since

$$\neg p \vdash \neg p$$

we get $\neg p, p \vdash y$ for some $y \in \Theta$ and therefore we get that

$\neg p \wedge p \vdash y$ for some $y \in \Theta$, and hence by definition $\vdash \neg(p \wedge \neg p)$.
This means that $p \wedge \neg p$ is false in every model.

Since p is an atom we cannot have the above since $p \wedge \neg p$ is consistent, meaning that it has a model, for example, if p is true at (e) and false at (h) (we can give this assignment since p is atomic), then $p \wedge \neg p$ holds at (e). Thus the $\neg$ above is not a negation according to Definition 9.1.15.

Turning now to Question 9.1.6, we add axiom (j), (of Example 9.1.3), i.e. $A \to (\neg A \to B)$ we get $e = h$ and $\neg$ becomes classical negation. We can take $\Theta = \{(\neg q_0 \wedge q_0)\}$ and derive from the axioms that

$$\vdash \neg A \leftrightarrow (A \to (\neg q_0 \wedge q_0)).$$

In fact, the above additional axiom says simply $\vdash \neg(A \wedge \neg A)$.

Let us check now whether Question 9.1.8, namely whether $\neg$ in the system $L3$ of Example 9.1.7 is indeed a form of negation. This system axiomatises Łukasiewicz 3 valued logic. There are three truth values, 1 (*truth*), $\frac{1}{2}$, and 0 (*falsity*). The truth tables for $\neg$ and $\to$ are as follows:

$$\neg x = 1 - x \text{ and } x \to y = \min(1, 1 + y - x).$$

The idea of the definition for $x \to y$ is that if $x \leq y$ then $x \to y$ is true. (Like $0 \to 1$ in classical logic.) If $x > y$ then $x - y$ is the measure of falsity of $x \to y$ and so

---

This point is very important, since your argument for the claim that $\neg$ becomes classical negation depends (so it seems to me) on the completeness part of the above "iff"!

I believe that you can avoid the above problem if you give a direct, syntactic derivation of $\neg A \vee A$ (i.e. excluded middle) in the system given in item 2 of Example 9.1.3. (This is very easy if you follow my suggestion in Footnote 3) The reason is that it is well known that a complete axiomatization of Classical Logic is obtained by adding to CL+ both (j) and excluded middle. (See P. 27 of our book Avron et al. 2018.)

the value of $x \to y$ is $1 - (x - y)$. $\neg x = 1 - x$ is just the mirror image of the truth value.

Conjunction $x \wedge y$ and disjunction $x \vee y$ have the definition below. They are definable from $\to$ by

$$\begin{aligned} x \vee y &= \text{def.}(x \to y) \to y = \min(x, y). \\ x \wedge y &= \text{def.}\neg(\neg x \vee \neg y) = \max(x, y). \end{aligned}$$

Intuitively, there is no doubt that $\neg x$ is a form of negation in this system because $\neg x = 1 - x$. The farther x is from the truth the nearer $\neg x$ is to the truth.

Remark 9.1.16 The consequence relation for this logic can be defined in two ways, for multi-sets $\Delta = \{A_1, ..., A_n\} \vDash B$:

Option 1. We can write $A_1, \ldots, A_n \vDash_1 B$ in this system to mean that under any assignment: Min (value A_j) $\leq$ val B, and $\vDash B$ to mean that under any assignment val $B = 1$.

Notice that the relation $\vDash_1$, defined semantically above, fulfils the criteria for a logical system. The deduction theorem, however, is not valid for $\vDash$.

The Wajsberg axiom system is complete in the sense that the following holds:

$$A_1, \ldots, A_n \vDash B \text{ iff } \vdash \bigwedge A_j \to B.$$

If we define $A_1, \ldots, A_n \vdash B$ to mean that $\vdash \bigwedge A_j \to B$ it then follows that $\vdash B$ iff val $B = 1$ under all assignment.

Option 2. There is another possibility of deriving /attaching a consequence relation to the axioms of L3. We can let databases be multi-sets and let $A_1, \ldots, A_n \vDash_2 B$ to mean that

$$\text{Max}(0, 1 - \Sigma(1 - \text{Value}(A_i)) \leq \text{Value}(B).$$

We can choose an appropriate **C** for each case.

For multi-sets we can take the formula $\mathbf{C}(x, y) = \text{def.}\neg(x \to \neg y)$, which satisfies the equation:

$$\text{Value}(\mathbf{C}(x, y)) = \max(0, \text{Value}(x) + \text{Value}(y) - 1).$$

This is an enhancement over conjunction $x \wedge y$, which has the value $\min(x, y)$.[6]

[6] Arnon Avron commented as follows:

> What you call $C(x, y)$ is known as the t-norm that underlies Łukasiewicz logic, and is usually denoted by &. (See Hajek's book on fuzzy logics Hájek 2013). Your second Consequence Relation can be characterized as follows:
>
> $$A_1, ..., A_n \vdash B \text{ iff } v(A_1)\&v(A_2)\&...\&v(A_n) \leq v(B)$$
>
> for every valuation v.

Our Definition 9.1.12 of what a negation is should give us that $\neg$ is a negation. Suppose $\neg$ is indeed a negation according to Definition 9.1.12. Then there exists a fixed Θ such that for any Δ and any A of the logic $L3$ we have:

$\Delta \vdash \neg A$ iff $\Delta, A \vdash B$ (for some $B \in \Theta$). Necessarily $\Theta \neq \emptyset$.

Take any $B \in \Theta$ and $\Delta = \emptyset$ then $\vdash \neg B$ iff $B \vdash y$ for some $y \in \Theta$; but since $y = B \in \Theta$ and $B \vdash B$ we get $\vdash \neg B$ for all $B \in \Theta$.

One can verify by looking at the axioms the following lemma:

Lemma 9.1.17 *If* $\vdash A$ *then value* $(A) = 1$ *under all assignments.*

Proof The above is true for the axioms and is preserved under modus ponens and substitution. ■

Thus, we conclude that for any $B \in \Theta$, value $B = 0$ under all assignments.

Now consider an atom q, certainly

$$\neg q \vdash \neg q$$

hence for some $B \in \Theta$,

$$\neg q \wedge q \vdash B$$

hence under all assignments Min (value $\neg q$, value q) $\leq$ value B. In particular for any assignment h with $h(q) = \frac{1}{2}$. This contradicts the previous conclusion that value $B = 0$ always.

We therefore need to improve our Definition 9.1.12 of negation.

Our basic idea in defining negation was that $A \vdash \neg B$ holds if A, B together lead to some undesirable result Θ.

$$\text{i.e. } A, B \vdash \Theta.$$

However, the way the above is written is that A and B are "combined" together via conjunction, i.e. $A \wedge B$. It is quite possible that A, B can be combined together via a different connective, e.g. some connective $\mathbf{C}(A, B)$. Thus $A \vdash \neg B$ holds iff $\mathbf{C}(A, B) \vdash \Theta$. $\mathbf{C}$ is a connective which "brings out" the effect A and B can have together. Of course, $\mathbf{C}(x, y)$ is not an arbitrary connective. It must be monotonic and satisfy some obvious properties. $\mathbf{C}(x, y)$ must say more than just $x \wedge y$ and satisfy the conditions listed in Definition 9.1.18 for it.

Definition 9.1.18 (*Negation as a potential syntactic inconsistency*) Let L be a system with a provability relation $\vdash$ and let $*$ be a unary connective of L. We say $*$ is a form of negation in L iff there exist a non empty set of wffs Θ which is not provably

---

A remark: an option you have not mentioned here is the standard one: $A_1, \ldots, A_n \vdash B$ iff $v(B) = 1$ for every valuation v s.t. $v(A_i) = 1$ for every i. This option is closely related, of course, to Lemma 9.1.17.

equivalent to the set of all wffs, and a binary connective $\mathbf{C}(x, y)$ s.t. the following holds for any D and A:

$$D \vdash^* A \text{ iff } \mathbf{C}(D, A) \vdash y \text{ for some } y \in \Theta.$$

$\mathbf{C}$ must satisfy the following: (*truth* is any provable formula; such formulae exist if $*$ is a negation. See Lemma 9.1.13).

1. $\mathbf{C}(x, y) \vdash x$
2. $\mathbf{C}(x, y) \vdash y$
3. $\mathbf{C}(truth, y) = \mathbf{C}(y, truth) = y$
4. $\frac{x \vdash x'}{\mathbf{C}(x,y) \vdash \mathbf{C}(x',y)}$ $\qquad \frac{y \vdash y'}{\mathbf{C}(x,y) \vdash \mathbf{C}(x,y')}$

where $A = B$ abbreviates $A \vdash B$ and $B \vdash A$.

Think of $\mathbf{C}$ are enhanced conjunction.

Remark 9.1.19 1. We get from the above that (in case that a *falsity* can be defined in the logic, with falsity $\vdash A$, for any A):
$\mathbf{C}(falsity, y) = \mathbf{C}(x, falsity) = falsity$.
2. Definition 9.1.18 was given for D a single formula, if L has conjunction then we can take $\Delta \vdash^* A$ as $\bigwedge \Delta \vdash^* A$. See, however, Option 2 of Remark 9.1.16.

For our negation in the system $L3$, let $\mathbf{C}(x, y) = \neg(x \to \neg y)$, and let $\Theta = \{falsity\} = \{\neg(y_0 \to y_0)\}$. Clearly, by the definition of $\vdash$, $x \vdash \neg y$ iff $\vdash x \to \neg y$ iff value $(x \to \neg y) = 1$ in all assignments, iff value $\neg(x \to \neg y) = 0$ in all assignments, iff $\neg(x \to \neg y) \vdash falsity$.

The truth table for $\mathbf{C}(x, y) = \neg(x \to \neg y)$ is Max (0, value $x+$ value $y - 1$).

As can be seen, since the truth function of $\mathbf{C}(x, y)$ is Max $(0, x + y - 1)$.

We get

1. $\mathbf{C}(x, y) \leq x$
2. $\mathbf{C}(x, y) \leq y$
3. $\mathbf{C}(1, y) = y = \mathbf{C}(y, 1)$
4. (a) $x \leq x' \Rightarrow \mathbf{C}(x, y) \leq \mathbf{C}(x', y)$
   (b) $y \leq y' \Rightarrow \mathbf{C}(x, y) \leq \mathbf{C}(x, y')$

These correspond to the conditions of Definition 9.1.18, and hence $\neg$ in the 3 valued logic is a negation. In fact the above definitions of $\neg$, $\to$ and $\mathbf{C}(x, y)$ as $\neg(x \to \neg y)$ show that $\neg$ is a negation in all Łukasiewicz many valued logics.

Now that we have changed the definition of negation in a formal system, we have to check whether the $\neg$ of Question 9.1.6, i.e. of item 2 of Example 9.1.3 is still not considered a negation. So assume that $\neg$ is a negation in the system of Example 9.1.3, the system with axioms (a) to (i). Then for some Θ and $\mathbf{C}$ the condition of Definition 9.1.18 holds, namely for all D, A

$$D \vdash \neg A \text{ iff } \mathbf{C}(D, A) \vdash B \text{ for some } B \in \Theta.$$

We shall show that

(†) $A \to \neg(x \to x) \vdash \neg A$

using **C** and Θ, and this is impossible because in our two world model (†) says that if A is false in one world, A is false in the other world also. Thus, if we prove (†), then this shows that no **C**, Θ can exist and $\neg$ is not a form of negation.

We now proceed to prove (†):

Since $\mathbf{C}(y, z) \vdash y \wedge z$, we get

$$\mathbf{C}(A \to \neg(x \to x), A) \vdash \neg(x \to x).$$

Hence, by definition of $\neg$,

$$\mathbf{C}(\mathbf{C}(A \to \neg(x \to x), A), x \to x) \vdash B, \text{ for some } B \in \Theta.$$

Since $x \to x$ is *truth* and $C(y, \mathit{truth}) = y$ we get:

$$\mathbf{C}(A \to \neg(x \to x), A) \vdash B, \text{ for some } B \text{ in } \Theta$$

and hence by definition of $\neg$ we get:

(†) $A \to \neg(x \to x) \vdash \neg A$.

Example 9.1.20 (*the system of relevant logic R*) Consider a language with $\to$ only and the following set of axioms and rules, defining the system $R \to$.

Rule. modus ponens

$$\frac{\vdash A, \vdash A \to B}{\vdash B}$$

Axioms.

R1: $A \to A$
R2: $(A \to B) \to ((C \to A) \to (C \to B))$
R3: $(A \to (B \to C)) \to (B \to (A \to C))$
R4: $(A \to (A \to B)) \to (A \to B)$

The above system was introduced by Church (1951) and Moh (1950) respectively. Church called it "weak positive implicational calculus". They proved the following deduction theorem for the system.

1. *Deduction theorem for* $R \to$: If there exists a proof of R from $A_1, \ldots, A_n$ in which all $A_1, \ldots, A_n$ are used in arriving at B then there exists a proof of $A_n \to B$ from $A_1, \ldots, A_{n-1}$ satisfying the same conditions.[7]
   The above calls for the following definition of $\vdash_{R\to}$

[7] Here we need to take the databases Δ as multi-sets. In the Hilbert type formulation this is hidden. When we say that $(A \to B) \to (A \to (A \to B))$ is not a theorem of $R \to$, and use the deduction theorem we get that $\{A \to B, A, A\}$ does not prove B.

2. *Definition of* $\vdash_{R\to}$

$$A_1, \ldots, A_n \vdash_{R\to} B \text{ iff}$$
$$\vdash_{R\to} A_1 \to (A_2 \to \ldots \to (A_n \to B)\ldots)$$

One can see by axiom R3 that the above is independent of the order of $\{A_j\}$. The above system is identical with the implicational relevance logic of Anderson and Belnap. It does not satisfy the conditions of a logical system, but see however, Footnote 2. Negation $\neg$ is introduced into $R \to$ to obtain $R(\to, \neg)$, via the Ackermann negation axioms. These axioms are used to introduce negation not only into $R \to$ but also into all neighbouring systems.

Ackermann axioms for negation

AN1: $(A \to \neg B) \to (B \to \neg A)$
AN2: $(A \to \neg A) \to \neg A$
AN3: $\neg\neg A \to A$

The following can be proved

AN4: $A \to \neg\neg A$
AN5: $(A \to B) \to (\neg B \to \neg A)$

See Anderson and Belnap (1975, pp. 20–21, 107–109) for details. The above definition of negation is indeed negation according to our Definition 9.1.18 of negation. Meyer (1966) has shown that if we add to $R \to$ a symbol f (*falsity*) with the additional axiom

R5: $((A \to f) \to f) \to A$

we get a system equivalent to $R(\to, \neg)$, with $\neg$ via the interpretation

1. $\neg A = \text{def}.A \to f$.
   The following must be proved.
2. $D \vdash \neg A$ iff $D, A \vdash f$
   i.e. $D \vdash A \to f$ iff $D, A \vdash f$

or equivalently by definition of $\vdash_{R\to}$

$$\vdash D \to (A \to f) \text{ iff } \vdash_{R\to} D \to (A \to f), \text{ which is correct.}^8$$

[8] Arnon Avron commented as follows:

> It is not difficult to show that in item 2 above, we can substitute $\neg\neg[(D \to D) \to \neg(A \to A)]$ for f. Since $\neg[(D \to D) \to \neg(A \to A)]$.
>
> This is is provable in $R \to$ for every A, D, you may take Θ as $\{C | \emptyset \vdash \neg C\}$, and there is no need to extend the language in this case.
>
> On the other hand, it is not clear what is $C(x, y)$ in this example, even if you add f! It should satisfy 1–4 of Definition 9.1.18. However, since in Definition 9.1.18 we took "truth" to be *any* provable formula, none of the obvious candidates seems to work.

Remark 9.1.21 Technically, if the system $R(\to, \neg)$ of Example 9.1.20 is formulated with $\neg$ and without f, can we find an f such that $\neg A = A \to f$? In classical logic one can take $f = q_0 \wedge \neg q_0$ or if conjunction is not available, one takes $f = \neg(q_0 \to q_0)$ for some fixed q. We cannot do the same for $R(\to, \neg)$, because if we take $f = \neg(q_0 \to q_0)$ for some fixed atom q_0, we will not have enough axioms on $\to$ to be able to use f as needed. We will have to add axiom R4 for this new $f = \neg(q_0 \to q_0)$ and then show that no new theorems can be proved for any wffs not containing q_0. Thus we see that Definition 9.1.18 is not quite right in the sense that the system considered may be too weak to show that it has a negation. In other words a connective $*$ may indeed be a negation in the system $\vdash$, but $\vdash$ may be too weak to prove the Definition 9.1.18. In fact, a connective $\mathbf{C}(x, y)$ required by Definition 9.1.18 may not be definable in the language of the system, but only in an extension. Intuitively if $*$ is a negation in a conservative extension, then we can and should regard it a negation in the system itself. We are thus led to the following definition:

Definition 9.1.22 (*An improved version of Definition* 9.1.18)

1. Let $L1$ and $L2$ be logical systems such that the language of $L2$ extends the language of $L1$.
   We say $L2$ is a conservative extension of L1 iff the following holds for any Δ, A in the language of $L1$
   $$\Delta \vdash_{L1} A \text{ iff } \Delta \vdash_{L2} A.$$
2. We say that $*$ is a negation in $L1$ iff for some conservative extension $L2$ and some Θ and $\mathbf{C}$ in $L2$ satisfying the conditions of Definition 9.1.18, we have that for any D, A of $L1$ the following holds:
   $$D \vdash^* A \text{ iff } \mathbf{C}(D, A) \vdash_{L1} B, \text{ for some } B \in \Theta.$$

We have now to check whether this new definition of negation turns the connective $\neg$ of Question 9.1.6, i.e. item 2 of Example 9.1.3 axioms (a) to (i) inclusive into a negation. (Recall that we found that $\neg$ is not a negation). The answer is no: $\neg$ is still not a negation. The reason is that it can be proved that (think of the consequence semantically) for any conservative extension of the system in Example 9.1.3 the two world interpretation (with the e world and the h world) is still valid. So the argument for showing that no $\mathbf{C}$ and Θ can make $\neg$ into a negation still goes through.

Example 9.1.23 We now give another example illustrating the need for Definition 9.1.22. Consider the language of classical propositional logic and its consequence relation $\vdash$.

Let $\vdash_1$ be defined as

$$\Delta \vdash_1 A \text{ iff } \Delta \neq \emptyset \text{ and } \Delta \vdash A.$$
$$\vdash_1 \text{ is a consequence relation}$$

However, $\neg$ is not a negation in $\vdash$, according to Definition 9.1.12, since for any non-empty Θ that we choose we would have to have for $B \in \Theta$ that $\emptyset \vdash_1 \neg B$ since certainly $B \vdash_1 B$ contrary to definition of $\vdash_1$. But this is counter-intuitive since certainly $\Theta = \{q \wedge \neg q\}$ should be acceptable.

The example (which was suggested by the referee) is certainly pathological and Definition 9.1.22 handles it nicely. However in our view a more satisfactory solution to this particular problem is to require the following additional property to be fulfilled by a consequence relation.

4. $\Delta \vdash A$ iff $\forall x(\Delta, x \vdash A)$. (Coherence).

We now investigate the possibility that there might be negations for which Θ depends on D. This is quite intuitive since it says that what we do not want, Θ, depends on the data, D, which we have. This is the case for the negation as failure in Logic programming, as shown by Gabbay (Gabbay and Sergot 1986, Sect. 4). Of course, logic programming does not satisfy coherence. In fact, it turns out that we *cannot* have a notion of Θ dependent on D, for a coherent consequence relation. (See Example 9.1.23 above).

Proposition 9.1.24 *Let $\vdash$ be a monotonic logical system with conjunction $\wedge$ and a negation $\neg$ characterised by the following clauses:*

1. *For any D there exists $\Theta(D)$, dependent on D, such that for any A the following hold:*
2. $D \vdash \neg A$ *iff* $\exists y \in \Theta(D)(D, A \vdash y)$
3. $\Delta \vdash A$ *iff* $\forall x(\Delta, x \vdash A)$.

Then there exists an N (independent of D) such that (1) holds, (i.e. $N = \Theta(D)$.)

Proof We prove Proposition 9.1.24 by means of two Lemmas.

Proof of Proposition 9.1.24 **Part 1: Two Lemmas**

Lemma 9.1.25 *Let $\vdash$, $\neg$ and $\Theta(D)$ be as in Proposition 9.1.24. Let $N(D)$ be the set*

$$N(D) = \{y | D \vdash \neg y\}$$

then $\neg$ is a negation satisfying equation (2) of Lemma 9.1.24 with $N(D)$ as a set of unwanted sentences.

Proof Very much as in Lemma 9.1.14, we show that, for any D and A:

- $\exists y \in \Theta(D)(D, A \vdash y)$ iff $\exists z \in N(D)(D, A \vdash z)$

1. Assume $D, A \vdash y$, for some $y \in \Theta(D)$.

By (2) of Proposition 9.1.24 we get that $D \vdash \neg A$ and hence $A \in N(D)$ and therefore there exists a $z \in N(D)$, namely $z = A$ such that $D, A \vdash z$.

2. Assume $D, A \vdash z$, for some $z \in N(D)$.

Since $z \in N(D)$, we therefore have that $D \vdash \neg z$. Hence by (2) of Lemma 9.1.24 again, there exists a $y \in \Theta(D)$ such that $D, z \vdash y$. We now have:

$$D, A \vdash z \text{ and } D, z \vdash y$$

and by the cut rule (3res) we get

$$D, D, A \vdash y.$$

This completes the proof of Lemma 9.1.25.

Note that the proof in part (2) above can be modified to show that $D, A \vdash B$ and $D, A \vdash \neg B$ implies $D, D \vdash \neg A$. ■

Remark 9.1.26 We draw several conclusions from Lemma 9.1.25:

1. First that if $\neg$ is indeed a negation dependent on D (via $\Theta(D)$) then equation (2) of Lemma 9.1.24 is really an uninformative tautology. By Lemma 9.1.24, $\Theta(D)$ can be taken as $N(D) = \{y | D \vdash \neg y\}$ and equation (2) of Lemma 9.1.24 becomes:
$$D \vdash \neg A \text{ iff } \exists y (D \vdash \neg y \text{ and } D, A \vdash y)$$
which is trivially true for $y = A$.
Note that for the case where Θ was fixed (independent of D) we got that $D \vdash \neg A$ iff $\exists y (\vdash \neg y$ and $D, A \vdash y)$ which is more informative.
2. The second conclusion is that Θ is dependent on D in a special way.

   As D gets stronger, Θ increases. This is not intuitive! Why should (a priori) what we do not want increase with the database?

   This property follows since we have:
$$\frac{D' \vdash D, D \vdash \neg A}{D' \vdash \neg A}$$
3. The third conclusion follows from the proof of Lemma 9.1.25 and the assumption (3) of Lemma 9.1.24 .

   We get the following for $\neg$:

   (c1) $\frac{D, A \vdash B; D, A \vdash \neg B}{D, D \vdash \neg A \{A, A\}}$
   Furthermore, since we saw in (2) that $D' \vdash D \Rightarrow N(D') \supseteq N(D)$ we can get that (see Footnote 4, and read $A \wedge B$ as $\{A, B\}$. Thus adding $\wedge$ is always conservative):
   (c2) $\frac{D \vdash \neg A}{D \vdash \neg (A \wedge B)}$

   The reason is that if $D, A \vdash y$, $y \in N(D)$, then certainly $D, B, A \vdash y$ and since $D, B \vdash D$, we have $y \in N(D, B)$ and hence $D \vdash \neg (A \wedge B)$.

We now proceed to use Lemma 9.1.25 to prove Proposition 9.1.24 namely that $\neg$ can be taken to be a negation with a fixed Θ (independent of D). We assumed that the language contains conjunction $\wedge$. $\wedge$ satisfies the three axioms:

$$A \wedge B \vdash A$$
$$A \wedge B \vdash B$$
$$A, B \vdash A \wedge B.$$

We proceed now to the second Lemma:

Lemma 9.1.27 *Let $\vdash$ be a system with negation $\neg$. satisfying the rule:*

1. $\frac{D, A \vdash B;\, D, A \vdash \neg B}{D \vdash \neg A}$
   Then for $N = \{B \wedge C | B \vdash \neg C\}$ we have for any D, A
2. $D \vdash \neg A$ *iff* $\exists y \in N(D, A \vdash y)$.

Proof 1. Assume $D \vdash \neg A$. We are looking for a y such that $y \in N$ and $D, A \vdash y$. Let $y = D \wedge A$. Certainly $D, A \vdash D \wedge A$ and $D \wedge A \in N$ since $D \vdash \neg A$.
2. Assume that for some $y \in N$, we have $D, A \vdash y$. y is then equal to some $B \wedge C$ with $B \vdash \neg C$. Since $D, A \vdash B \wedge C$ we get $D, A \vdash C$. Since $B \vdash \neg C$ we get $D, A \vdash \neg C$ and hence by rule 1, $D \vdash \neg A$.

■

Part 2 of the proof of Proposition 9.1.24: Having proved our two Lemmas (9.1.25 and 9.1.27) we can proceed. Assume the conditions of Proposition 9.1.24 for $\vdash$ and $\neg$ hold. By conclusion (c1) of item 3 of Remark 9.1.26 the conditions of Lemma 9.1.27 hold and hence $\neg$ is a negation with a fixed $\Theta = N$. ■

The above considerations show that there is no hope for a formulation of a negation $\neg$ with a Θ dependent on the database, within the framework of monotonic logics. The assumption that $\wedge$ is available does not restrict generality since $\wedge$ can always be added to the language and Definition 9.1.22 for negation be used. ■

9.2 Calculus of Failure

We give examples from other papers to show that negation as failure is negation in our sense. We give no proofs. It is too complicated for our current paper which is essentially a position paper, see Gabbay (1985b), Gabbay and Horne (2020).

Example 9.2.1 This example illustrates the idea of why we think negation as failure is a proper negation.

Consider a logic program without loops (where every atom either succeeds or fails):

$$\Delta = \{\neg b \Rightarrow a\}.$$

From this program, a succeeds and b fails. Let $\Theta(\Delta) = \{y | y \text{ fails}\}$.

Add the axioms for a new negation symbol $\mathbf{n}_\Delta$ to be

$$\Delta_{\mathbf{n}} = \{y \Rightarrow \mathbf{n}_\Delta | y \in \Theta_\Delta\}.$$

In Δ translate any $\neg x$ as $x \Rightarrow \mathbf{n}_\Delta$. This translation gives a new theory Δ'.

We get $\Delta_1 = \Delta' \cup \Delta_{\mathbf{n}}$ to be $\{(b \Rightarrow \mathbf{n}_\Delta) \Rightarrow a, b \Rightarrow \mathbf{n}_\Delta\}$.

This is an intuitionistic theory for intuitionistic $\Rightarrow$. We have

$$\Delta_1 \vdash x \text{ iff } x \text{ succeeds from } \Delta$$
$$\Delta_1 \vdash (x \Rightarrow \mathbf{n}_\Delta) \text{ iff } x \text{ fails from } \Delta.$$

This was done in my paper Gabbay and Sergot (1986).[9]

Example 9.2.2 When we have loops, we can use answer set programming (Gelfond 2008).

Consider Γ:

$$\Gamma = \{\neg a \Rightarrow b, \neg b \Rightarrow a\}.$$

Answer set 1. $a =$ in, $b =$ out. We get Γ_1.

$$\Gamma_1 = \{b \Rightarrow \mathbf{n}_1, (b \Rightarrow \mathbf{n}_1) \Rightarrow a, (a \Rightarrow \mathbf{n}_1) \Rightarrow b\}.$$

$\Gamma_1 \vdash a$ but $\Gamma_1 \nvdash b$.

Similarly

Answer set 2. $a =$ out, $b =$ in.

We get

$$\Gamma_2 = \{a \Rightarrow \mathbf{n}_2, (b \Rightarrow \mathbf{n}_2) \Rightarrow a, (a \Rightarrow \mathbf{n}_2) \Rightarrow b\}.$$

We get $\Gamma_2 \vdash b$, $\Gamma_2 \nvdash a$.

Example 9.2.3 This covers a general loop. Consider the loop

$$\{\neg a \Rightarrow a\}.$$

---

[9] This is Theorem B on p. 29 of Gabbay and Sergot (1986). It says and we quote:

> Theorem B. Let P be any database. Let L be $L = \{yP(?F)y = 0\}$. Assume that P is such that every goal either succeeds or fails. Then for any G $P(?F)G = 1$ iff $(P, L)(?I)G = 1$.

Note the assumption that every goal either succeeds or fails (i.e. no loops). This is noted on the same page of the paper, we Quote further:

> Theorem B is important. It says that if our mechanical theorem proving is compete (i.e. $P?A = 0$ or $P?A = 1$), then negation as failure is the truly sound classical negation. This holds because it is equal to negation as inconsistency, which is complete. However, in the case that the theorem prover $P?G$ is not complete, e.g. when we have loops, negation as failure may not behave logically.

Answer set programming does not help here, but we can add a historical loop checker. Let us try. We use the notation "$?a = 1$" means $?a =$ success. "$?a = 0$;; means $?a =$ failure.

Part 1: We start the computation with the query $?a = 1$.

$$\neg a \Rightarrow a \quad ?a = 1$$

iff

$$\neg a \Rightarrow a?a = 0$$

iff

$$\neg a \Rightarrow a \quad ?\neg a = 0$$

iff

$$\neg a \Rightarrow a \quad ?a = 1.$$

We loop. So the query $?a = 1$ fails, so $?a = 0$ succeeds.

Part 2: We start the computation with the query $?a = 0$. We implement this by continuing the computation beyond the loop point of Part 1:

$$?\neg a = 1$$

iff

$$?a = 0$$

we get another looping point where we loop again.

This means that if we start with $?a = 0$ then $?a = 0$ also loops, so the query fails an we get answer $a = 1$.

So we get two possibilities.

We thus get no agreement using the loop checker. If we ask $?a = 1$ we loop and therefore we get that a fails and if we start with the query $?a = 0$ we also loop and get that a succeeds.

In case we consider that a fails, we get we get $\{a\}$ for the fail set and we have:

$$\begin{aligned}&\Gamma_1 = \{a \Rightarrow \mathbf{n}_1, (a \Rightarrow \mathbf{n}_1) \Rightarrow a\}\\&\Gamma_1 \vdash a\end{aligned}$$

In case a fails and in case a succeeds we get $\varnothing_\perp$ for the fail set. We use $\perp$

$$\Gamma_2 = \{\perp \Rightarrow \mathbf{n}_2, (a \Rightarrow \mathbf{n}_2) \Rightarrow a\}$$

we have $\Gamma_2 \nvdash a$.

For the sake of comparison, let us re-do Example 9.2.2 using a loop checker and see whether we get the same two possibilities or not. In other words independently of whether we ask $?a = 1$ or if we ask $?a = 0$, we get the same result for a, namely either a succeeds or a fails.[10]

Example 9.2.4 Example 9.2.2 using a loop checker.

Let

$$\Gamma = \{\neg a \Rightarrow b, \neg b \Rightarrow a\}.$$

1. Start with $?a = 1$. We get:

$$?a = 1$$

if

$$?\neg b = 1$$

if

$$?b = 0$$

if

$$\neg a = 0$$

if

$$?a = 1.$$

We loop.
Therefore $?a = 1$ fails, so $?a = 0$ succeeds.
If we continue after the loop we get
if

$$?\neg b = 1$$

if

$$?b = 0$$

We see that we get $?b = 0$ looping if we were starting with $b = 0$.
So in this case $b = 0$ fails, so $b = 1$ succeeds. So the answer success set is $b = 1$ and $a = 0$.
2. Note that if we start with "$?a = 1$" or with "$?b = 0$" we never get the query "$?a = 0$" or the query "$?b = 1$".
3. Let us start and ask $?b = 1$.

$$?b = 1$$

if

$$?\neg a = 1$$

if

[10] In view of the restriction of Theorem B, (the restriction of no loops, see Footnote 9), we want to eliminate loops by a loop checker, and ask will the theorem go through?

$$?a = 0$$

if

$$?\neg b = 0$$

if

$$?b = 1$$

if

$$?\neg a = 1$$

if

$$?a = 0$$

The looping elements are "$?b = 1$" and "$?a = 0$". If the looping elements are failures then the answers are consistent $a = 1$ and $b = 0$ are the successes.

4. We see that (1) and (3) completely agree with the answer set programming answer.

We now want to add negation as failure to full implicational intuitionistic logic.[11] This negation, when added to intuitionistic implication, is very difficult to handle mainly because the meaning of "$\neg$" keeps on changing depending on where $\neg$ appears in the formulas. This is different from answer sets where there may be several options for the meaning of "$\neg$", but once we choose an option, the meaning of $\neg$ gets fixed for all occurrences of "$\neg$".

The next Example 9.2.5 will illustrate the problem.

Example 9.2.5 Let Δ_1 be $\{(1)-(4)\}$

1. $(d \Rightarrow (c \Rightarrow \neg a)) \Rightarrow c$
2. $c \Rightarrow a$
3. $\neg d \Rightarrow x$
4. $\neg x \Rightarrow a$

and let Δ_2 be $\{(1)-(6)\}$, where

5. d
6. c

Note that in the logic of intuitionistic implication, we have, for any X, Y, Z and Δ,

$$\Delta \vdash X \Rightarrow (Y \Rightarrow Z)$$

iff (by definition, or by the deduction theorem)

[11] This means that we take implication with Axioms (a) and (b) of Item 1 of Example 9.1.3, and add the negation as failure symbol and define the computation as in N-Prolog, see Gabbay (1985b). The reader need not reference (Gabbay 1985b), but follow the computation in the examples which follow. It is very intuitive.

$$\Delta \cup \{X\} \cup \{Y\} \vdash Z$$

So if we add $\neg$ to $\Rightarrow$, then for $Z = \neg a$ we get that "$\neg$" says that in our computation, $?a$ is a failure from the database $\Delta \cup \{X, Y\}$.

So looking at clause (1), we can see that "$\neg a$" would need to fail from the database Δ_1 with clauses (5) and (6) added i.e. from Δ_2).

While "$\neg$" in clauses (2) and (3) do not add anything to the database, so $\neg$ needs to fail from Δ_1 above.

Let us now do some specific computations to illustrate the problems involved.

Computation, part 1. We ask "$?a = 1$" from database $\Delta_1 = \{(1), (2), (3), (4)\}$ and use clause (4) first. So

$$\Delta_1 \vdash ?a = 1$$

using (4), if

$$\Delta_1 ?\neg x = 1$$

if

$$\Delta_1 ?x = 0$$

using (3), if

$$\Delta_1 ?\neg d = 0$$

if

$$\Delta_1 ?d = 1.$$

We get that the query $?d = 1$ fails from the database Δ_1, because this database has no clause with head d. Therefore the original query, namely $?a = 1$, fails for the choice of the above initial clause (4) with head a. However, we also have clause (2) with head a and so let us backtrack and ask $?a = 1$ again and this time choose clause (2).

Computation, part 2. Let us backtrack, and ask $\Delta_1 \vdash ?a = 1$ using clause (2).

$$\Delta_1 \vdash ?a = 1$$

using (2), if

$$\Delta_1 ?c = 1$$

using (1), if

$$?(d \Rightarrow (c \Rightarrow \neg a)) = 1$$

and we ask, if

$$\Delta_2 ?\neg a = 1$$

Since

$$\Delta_2 = \Delta_1 \cup \{d, c\}$$

if

$$\Delta_2?a = 0$$

from (4), if

$$\Delta_2?\neg x = 0$$

Computation, part 3. Note that "$\neg x$" is now asked from Δ_2 and not from the original Δ_1! The meaning of "$\neg$" has changed for its same occurrence in "$\neg x$".

We have two options:

Option 1. Continue and ask $\neg x$ from Δ_1.
Option 2. Continue and ask $\neg x$ from Δ_2.
Let us do them in parallel

Option 1	Option 2
from (4) $?\neg x = 0$	from (4), if $?\neg x = 0$
if $?x = 1$	if $?x = 1$
from (3), if $?\neg d = 1$	from (3) if $?\neg d = 1$
if $?d = 0$	if $?d = 0$
Success.	Fail.
Clause (5) not available because we are using Δ_1	Because clause (5) is available.

Question. Which option do we adopt?
Answer. Option 1 is better from the point of view of "what is negation" because we want the meaning of each occurrence of negation to become fixed. We write clause 1 as

1*. $(d \Rightarrow (c \Rightarrow \neg_{\Delta_2} a)) \Rightarrow c$.

and clauses (3) and (4) as

3*. $\neg_{\Delta_1} d \Rightarrow x$
4*. $\neg_{\Delta_1} x \Rightarrow a$.

Option 2 is known in the literature as $\mathcal{N}$ = Prolog (Gabbay 1985b) and its negation as failure was extensively investigated and has complex semantics (Gabbay and Horne 2020; Gelfond 2008).

Example 9.2.6 Let us revisit Example 9.2.5 and be very simple minded about it. We saw in Examples 9.2.1 and 9.2.2. The very simple approach that for a logic program Δ without loops or with semantics where every atom (or literal) x appearing in the program x either succeeds of fails, we can take as Θ_Δ for negation the set of all y which fail.

So let us apply the same procedures to the program Δ_1 of Example 9.2.5.

The clauses are:

1. $(d \Rightarrow (c \Rightarrow \neg a) \Rightarrow c$
2. $c \Rightarrow a$
3. $\neg d \Rightarrow x$

4. $\neg x \Rightarrow a$.

The atoms appearing in this program are $\{a, c, d, x\}$. So let us check first whether every atom either succeeds or fails. We get

a succeeds
c succeeds
x succeeds
d fails.

Here we use the $\mathcal{N}$-Prolog computation, namely Example 9.2.5 computation Parts 1, 2 and Part 3, option 2.

Second, let Θ_1 be $\{d\}$ (as our recipe dictates) and rewrite the program as (with $\mathbf{n}_1$ as negation)

1*. $(d \Rightarrow (c \Rightarrow (a \Rightarrow \mathbf{n}_1))) \Rightarrow c$
2*. $c \Rightarrow a$
3*. $(d \Rightarrow \mathbf{n}_1) \Rightarrow x$
4*. $(x \Rightarrow \mathbf{n}_1) \Rightarrow a$

and the additional clause for d

7*. $d \Rightarrow \mathbf{n}_1$

We now expect the same results of success or failure for $\{a, c, x, d\}$ and of course failure for $\mathbf{n}$.

Let us check.

Case $\mathbf{n}_1$:

$$?\mathbf{n}_1 = 1$$

using (7), if

$$?d = 1$$

fail.

Case a. Using 4*

$$?a = 1$$

$$?x \Rightarrow \mathbf{n}_1$$

add x

$$?\mathbf{n}_1$$

using (7), if

$$?d = 1$$

fail.

We backtrack and use clause 2*.
using 2*

$$?a = 1$$

if

$$?c = 1$$

using 1*,
add d
add c
add a, if

$$?\mathbf{n}_1 = 1$$

using (7)

$$?d$$

Success for $?a = 1$.
Case c. Success. Follows from case a that $?c = 1$.
Case d.

$$?d = 1$$

fails.

The big question we ask is:

Big Question BQ Is this very simple minded approach an indication of a possible a general truth (big theorem) or does it only work sometimes? See Gabbay and Horne (2020).
Answer to BQ: It is an accident, as the next Example 9.2.7 shows. However, there might be a general theorem which is inductive on the structure of nested negations and its proof would certainly be quite complicated.

Example 9.2.7 Part 1: The problem. Consider the following program $\Delta_1 = \{(1)\}$:

1. $((a \wedge (\neg a \wedge \neg b \Rightarrow x)) \Rightarrow x) \Rightarrow z$.
   Let us query

$$\Delta_1 ?z = 1$$

We use clause (1) and ask if $\Delta_2 ?x = 1$ where $\Delta_2 = \Delta_1 \cup \{(2), (3)\}$ where

2. a
3. $\neg a \wedge \neg b \Rightarrow x$.

We continue the computation using clause (3) ad ask a conjunction if $\Delta_2 ?\neg a \wedge b = 1$ which splits to two queries.

if

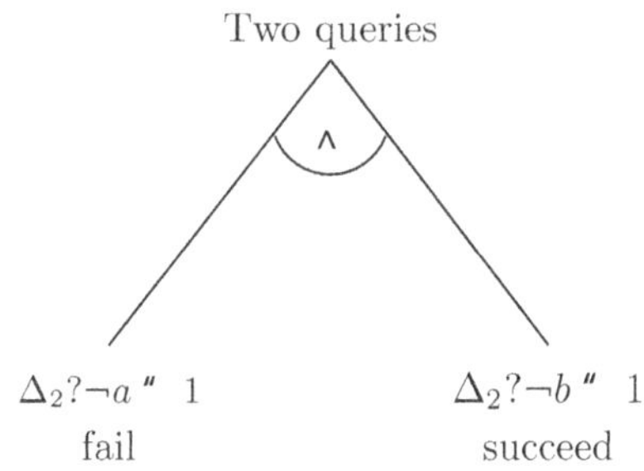

Thus overall we get that $\Delta_1?z$ fails.

Therefore we have that $\Delta_1?z$ fails as well as $\Delta_2?a$, $\Delta_2?b$ and $\Delta_1?x$ all fail because they are not heads of any clauses. So all atoms fail from Δ_1. So

$$\Theta_{\Delta_1} = \{a, b, x, z\}.$$

We therefore translate Δ_1 to Δ_1', namely Δ_1' includes the following clauses

1*. $((a \wedge ((a \Rightarrow \mathbf{n}) \wedge (b \Rightarrow \mathbf{n}) \Rightarrow x)) \Rightarrow x) \Rightarrow z$
4*. $a \Rightarrow \mathbf{n}$
5*. $b \Rightarrow \mathbf{n}$
6*. $x \Rightarrow \mathbf{n}$
7*. $z \Rightarrow \mathbf{n}$.

Let us ask

$$\Delta_1'?z = 1$$

if

$$\Delta_2'?x = 1,$$

where $\Delta_2' = \Delta_1' \cup \{(2*), (3*)\}$, where

2*. a
3*. $((a \Rightarrow \mathbf{n}) \wedge (b \Rightarrow \mathbf{n})) \Rightarrow x$

We continue:
if

$$\Delta_2'?((a \Rightarrow \mathbf{n}) \wedge (b \Rightarrow \mathbf{n})) = 1$$

if

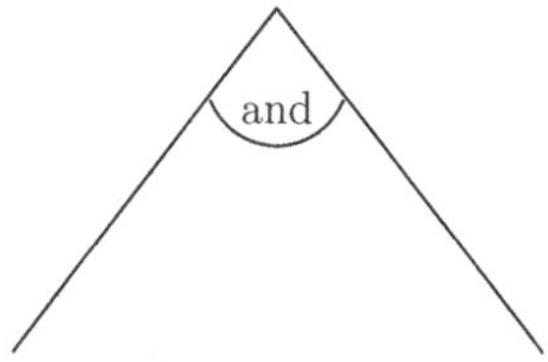

$\Delta_2'?(a \Rightarrow \mathbf{n})a = 1$ Success from (4*) $\qquad$ $\Delta_2?(b \Rightarrow \mathbf{n})b = 1$ Success from (5*)

This result does not match.

$$z \text{ should fail from } \Delta_1.$$

Part 2: The remedy. The remedy is that we need to eliminate the negations inductively on their nestings. Suppose we ask as in Part 1, the query $?\neg z$. The '*neg*" in "$\neg z$" is from Δ_1, but the "$\neg$" in "$\neg a$" and in "$\neg b$" is from Δ_2. So from Δ_2 a succeeds. What fails from Δ_2 are b and x and z. So the proper translation of Δ_2' is

$$\begin{aligned}\Delta_2'' &= \{(1*), (2*), (3*), (5*), *6*), (7*)\} \\ &= \Delta_2' - \{(4*)\}\end{aligned}$$

Let us now follow the computation of Part 1, up to the point where we have $\Delta_2''?a \Rightarrow \mathbf{n}$ This will fail as we want.

We need to define mathematically the induction. If we manage that, then we will get that nested negation as failure is a negation in our sense.

It seems that we need to follow the idea of defining/introducing several negations at once and characterise them together in terms of each other, and they will be negations in our sense. This is a new ball game and is the subject of active research.

9.3 Conclusion and Future Research

Let us summarise what we have learnt in this position paper about the question of what is negation in a system.

1. Assume a logical language with well-formed formulas and a relation $\vdash$ between multisets Δ of wffs and a single wff A of the form

$$\Delta \vdash A.$$

We need not assume any properties of "$\vdash$" nor do we need to know how $\vdash$ is defined.

2. We put forward the basic intuition that a unary connective $*$ in the language, is a negation in $\vdash$ if for every Δ there exists a multiset of wffs $\Theta_*(\Delta)$ such that $\Delta \vdash *A$ iff for some $y \in \Theta_*(\Delta)$

$$\Delta \cup \{A\} \vdash y.$$

3. This intuitive definition works in one form or another also for non-monotonic consequence systems, such as negation as failure (that is why we have that Θ_* depends on Δ). For a monotonic consequence relation, we would expect that Θ_* would be the same for any Δ.

   We did observe, however, that we might introduce a negation $*(\Delta)$ for each Δ and write some axioms connecting all the $*(\Delta)$ negations. We gave some hints in Sect. 2, on how this can be done, for the case of negation as failure added to intuitionistic implication.
4. There are systems $\vdash$ with $*$ where $*$ is not a negation.
5. Most of the well-known systems with $*$, which are considered as a negation, are also negations according to our definition, but not all of them.
6. There are systems such as paraconsistency systems, where the question of whether their negation candidate is indeed a negation is debated in the literature (see Beziau 2020). Our approach might be able to offer a verdict.
7. There is a need for a systematic examination of all candidates for negation in the literature with a view to improve our definition of what is negation and possibly also refute some community misconceptions.

References

Anderson, A. R., & Belnap, N. (1975). *Entailment 1*. Princeton University Press.

Avron, A., Arieli, O., & Zamansky, A. (2018). *Theory of Effective Propositional Paraconsistent Logics* (576 pp.). College Publications.

Beziau, J.-Y. (2020). Are paraconsistent negations negations? https://www.jyb-logic.org/papers12-11/paraconsistent%20negations.pdf.

Church, A. (1951). The weak theory of implications. In *Kontrolliertes Dluken* (Vol. 16, pp. 22–37, 236). J. S. L.

Gabbay, D. (1985a). Theoretical foundations for non-monotonic reasoning in expert systems. In K. Apt (Ed.), *Logics and Models of Concurrent Systems* (pp. 439–459). Springer.

Gabbay, D. M. (1985b). N-prolog: An extension of prolog with hypothetical implication. Part 2. logical foundations, and negation as failure. *The Journal of Logic Programming*, *2*(4), 251–283.

Gabbay, D. (1986). What is negation in a system? In F. R. Drake & J. K. Truss (Eds.), *Logic Colloquium '86* (pp. 95–112). North Holland: Elsevier Science Publishers.

Gabbay, D., & Horne, R. (2020). University of Luxembourg Proof theory of positive linear intuitionistic logic with negation as failure and anti-formulas. Draft paper 637A.

Gabbay, D., & Sergot, M. (1986). Negation as inconsistency. *Journal of Logic Programming*, *1*, 1–35.

Gelfond, M. (2008). Answer sets. In F. van Harmelen, V. Lifschitz, & B. Porter (Eds.), *Handbook of Knowledge Representation* (pp. 285–316). Elsevier.

Hájek, P. (2013). *Metamathematics of Fuzzy Logic* (312 pp.). Trends in Logic. Springer.

Meyer, A. (1966). *Topics in Modal and Many-valued Logics*. Thesis.

Moh, S. K. (1950). The deduction theorems and two new logical systems. *Methods*, *2*, 56–75.

Olivetti, N., & Terracini, L. (1992). N-prolog and equivalence of logic programs. *Journal of Logic, Language and Information*, *1*(4), 253–340.

Wajsberg, M. (1931). Axiomatization of the three valued propositional calculus (in Polish). Comptes rendus des seances de la Societe des Sciences et des Lettres de varsovie, classe Ill (Vol. 24, pp. 126–148). Translation in S. McCall editor *Polish Logic*, 1920–1939 Oxford University Press 1967, pp. 264–284.

Dov M. Gabbay has been working in logic and its applications in the past 50 years. He is one of the world's most active and influential researchers in logic. He has authored over 550 research papers and over 30 research monographs. He has initiated several new and active research areas. He is editor and owner of several international journals, and over 50 handbooks of logic. Recently (past 10 years), he has been working on argumentation and its applications to several areas including Sex Offender Reasoning, Fake News, Security, Medical Reasoning and Talmudic Logic, and Logic and Law. He is a talented prize-winning teacher and has developed the Data-Driven Instruction (DADI) methods for teaching advanced topics to first-year students and the general public. His paper in this book is a product of this method.

Chapter 10
Relevance Domains and the Philosophy of Science

Edwin Mares

Abstract This paper uses Avron's algebraic semantics for the logic RMI to model some ideas in the philosophy of science. Avron's relevant disjunctive structures (RDS) are each partitioned into *relevance domains*. Each relevance domain is a boolean algebra. I employ this semantics to act as a formal framework to represent what Nancy Cartwright calls the "dappled world". On the dappled world hypothesis, local scientific theories each represent restricted aspects and regions of the universe. I use relevance domains to represent the domains of each of these local theories and I provide a formalisation of the salient relationships between so-called fundamental theories and local theories. I also examine ways in which the paraconsistent nature of RMI can be used to deal with inconsistencies within and between theories adopted by scientists. The paper ends with some suggestions about updating RDS given changes in the theories that science adopts.

10.1 Introduction

In a trio of articles in the early 1990s (Avron 1990a, b, 1991), Arnon Avron develops an approach to paraconsistency and relevance that uses a strong relevant logic as its base. This logic, called "RMI" is closely related to Dunn and McCall's system R-Mingle (Anderson and Belnap 1975, §8.15). My interest is not in the complete system RMI, but only in its negation and implication fragment, $RMI_{\rightarrow,\neg}$. Avron shows that $RMI_{\rightarrow,\neg}$ is an extremely interesting logic. He gives it an algebraic semantics. Its models are "relevant disjunction structures". One of the very elegant properties of relevant disjunction structures is that each can be partitioned into a set of boolean algebras. Avron calls these boolean algebras "relevance domains". An application of relevance domains in the philosophy of science is my topic here.

My idea is to use RDS to model what Nancy Cartwright calls the "dappled world". On her view, the most successful scientific theories are not very general theories, but

E. Mares (✉)
School Philosophy, Victoria University, Wellington, New Zealand
e-mail: edwin.mares@vuw.ac.nz

O. Arieli and A. Zamansky (eds.), *Arnon Avron on Semantics and Proof Theory of Non-Classical Logics*, Outstanding Contributions to Logic 21,
https://doi.org/10.1007/978-3-030-71258-7_10

rather specialised ones, which treat only a limited range of phenomena. I model a limited range of phenomena of this sort by a relevance domain – that is, I model it in terms of the propositions about it. In this way, the various specialised (or "local") theories are separated from one another semantically, represented as being in different relevance domains. Within this formal framework, I also formalise two different ways in which more general (or "fundamental") theories are related to local theories. These two relationships correspond to what Cartwright calls "pluralism" and "fundamentalism" in the philosophy of science.

My understanding of relevance domains is epistemological. The fundamentalist thinks that once perfect general theories are found, there will be no need for local theories, and hence in terms of the present framework that there should only be one big relevance domain. Even the pluralist, who thinks that only local theories can be exact, does not necessarily think that domains really match natural boundaries. She may think that dividing the world up into domains is just the best we can do. In contrast to this epistemological view of domains, Avron adopts an ontological interpretation of them. He says that they represent different "levels of reality":

> The idea behind it is not new. Gentzen, for example, divided ... the world of mathematics into three grades, representing three "levels of reality". The elementary theory of numbers has the highest degree or level of reality; set theory has the smallest degree and mathematical analysis occupies the intermediate level. In the theory of types, or in the accumulative von Neumann universe for set theory, we can find indication of a richer hierarchy (Avron 1990a, p. 707).

Another version of the levels of reality view can be found in Plato's *Republic*, in his simile of the divided line. On Plato's view, the concrete world has less reality than mathematical objects, which in turn have less reality than Ideas, such as the idea of the good or the idea of beauty (see Cresswell 2012). More modern views concerning levels of reality can be found in Meinong (in his distinction between entities that exist and those that subsist as well as those that merely have "Aussersein" but not "Sein" (i.e. those that have "Nichtsein")) and early Russell (in his distinction between being and existence Russell 1903, §427).

I employ relevance domains to understand scientific theories, but the use of domains also raises an important problem. Various sorts of inconsistency appear when studying science. But the logic of each individual domain is classical. Classical logic makes valid the principle of explosion – that every proposition is entailed by a contradiction. When using two theories that are about the same domain, but inconsistent with one another, we may encounter local explosions – every proposition in the domain is derivable. In Sect. 10.7, I suggest a method to limit the number and severity of explosions that arise in the simultaneous employment of conflicting theories.

The plan of the paper is as follows. In Sects. 10.2 and 10.3, I introduce relevant disjunction structures and relevance domains. In the following sections, I treat the theories that scientists use as filters on a relevant disjunction structure or on domains within that relevant disjunction structure. In Sect. 10.4, I set out Cartwright's thesis that scientific theories divide the world into domains and suggest that relevance domains can be used formally to represent this thesis. I also formalise the opposing theories of fundamentalism and pluralism, which concern the relation between general theories (theories about an entire relevant disjunction structure) and local theories (theories about individual domains). In Sects. 10.5–10.7, I turn to the use of the logic RMI to deal with inconsistency in science. In the final section before the conclusion, I suggest using a semantic analog to the AGM theory of belief revision to deal with updating and otherwise revising relevant disjunction structures as a way of understanding changes in the views of scientists or scientific communities.

10.2 Relevant Disjunction Structures

At the heart of Avron's semantic theory is his employment of an intensional disjunction operator, $\oplus$. He defends his use of an intensional, rather than extensional, disjunction in a brief discussion of C.I. Lewis's argument for the principle of exposition – that every proposition can be derived from a contradiction. Recall that Lewis's argument starts with the premise $A \wedge \neg A$, then from this he infers by simplification that A and then that $A \vee B$ by the principle of weakening. By simplification again from the premise he infers that $\neg A$ and from $\neg A$ and $A \vee B$ by disjunctive syllogism he derives B. Avron says:

> It follows that no paraconsistent logic can have an operation of disjunction for which both weakening and disjunctive syllogism are always valid. The validity of at least one must be given up.
>
> Which of these two rules should be rejected? It seems obvious to me that if Lewis's argument does not apply to concrete situations it is because nobody will try to infer $A \vee B$ from A unless he sees a connection between A and B. In contrast, applications of disjunctive syllogism are frequent and indispensable. Accordingly, it seems preferable to retain disjunctive syllogism while limiting the validity of weakening (Avron 1990a, p. 170).[1]

The intensional disjunction that Avron adopts does just this: it makes valid every instance of disjunctive syllogism but does not allow weakening except in certain circumstances.[2]

Avron's semantics is built around this intensional disjunction. A *relevant disjunction structure* (RDS) is a quintuple $\mathbf{D} =< D, \leq, ', \oplus, T_D >$ that satisfies the following conditions:

1. $\leq$ is a partial order on D;
2. $'$ is a unary operation and an involution on $< D, \leq >$; that is, for all $a \in D, a'' = a$;
3. $\oplus$ is a binary operation on D; it is associative, commutative, and order preserving on $< D, \leq >$;
4. $a = a \oplus a$;

[1] Avron's relevance relation (Avron 1990a, p. 713ff) is his formalisation of this notion of connection between propositions.

[2] For a similar but more protracted defence of intensional disjunction, see Read (1988).

5. For all $a, b, c \in D$, if $a \leq b \oplus c$, then $b' \leq a' \oplus c$;
6. For all $a, b \in D$, $a \oplus (b' \oplus b)' \leq a$;
7. T_D is a truth set on D.

A *truth set* T_D on D is a subset of D that is closed upward under $\leq$ and satisfies the following semantic entailment condition:

$$\text{(SE) } a \leq b \text{ iff } a \Rightarrow b \in T_D.$$

I think of the elements of RDS as propositions, and call them such throughout this paper. The partial order on propositions is an entailment ordering. $a \leq b$ if and only if a entails b according to the RDS. The operator $\oplus$ is an intensional disjunction, often called "fission" in the relevant logic literature. The operator $'$ is an intensional complement and is used to represent the negation of the language. Postulate 4 (henceforth called 'RDS4') states that $\oplus$ is idempotent. This postulate is essential for the proof that every RDS is partitioned into relevance domains, as is RDS6. RDS5 states that a form of antilogism holds for RDS.

I use the following defined operators on RDS:

$$a \otimes b =_{df} (a' \oplus b')'$$

$$a \Rightarrow b =_{df} a' \oplus b$$

The "fusion" operator, $\otimes$, represents what is usually thought of as an intensional conjunction. It is easily shown that fusion is idempotent, commutative, and order preserving. Avron (1990a, p. 712) thinks that this operator is not a form of conjunction because a proposition $a \otimes b$ can be true while neither a nor b is true. For this reason, Avron also adds another intensional conjunction to his logic. I do not follow him in this, not because of any ideological dispute with Avron, but because my present purposes do not require a conjunction in this sense. The implication operator, $\Rightarrow$, is a relevant implication.

It is useful to know that the standard "contraposition" form of RDS5 also holds:

Lemma 10.1 *If* **D** *is an RDS, then if* $a \leq b$, $b' \leq a'$.

Proof Suppose that $a \leq b$. Then, $a \Rightarrow b \in T_D$. That is, $a' \oplus b \in T_D$. Thus, $b'' \oplus a' \in T_D$, and so $b' \Rightarrow a' \in T_D$. Therefore, $b' \leq a'$. ■

The following lemma is useful, and it also shows in part that fission is a form of disjunction.

Lemma 10.2 *If* $a \leq c$ *and* $b \leq c$, *then* $a \oplus b \leq c$.

Proof Suppose that $a \leq c$ and $b \leq c$. By RDS3 ($\oplus$ is order preserving), $a \oplus b \leq c \oplus c$. By idempotence, $a \oplus b \leq c$. ■

Here are two lemmas relating fusion, fission, and implication that will come in handy later:

Lemma 10.3 $a \leq b \Rightarrow (a \otimes b)$

Proof

1.	$a' \oplus b' \leq a' \oplus b'$	RDS1
2.	$a'' \leq (a' \oplus b')' \oplus b'$	1, RDS5
3.	$a \leq b' \oplus (a \otimes b)$	2, commutativity, RDS2, definition of $\otimes$
4.	$a \leq b \Rightarrow (a \otimes b)$	3, definition of $\Rightarrow$

■

Lemma 10.4 *If* $a \otimes b \leq c$ *then* $a \leq b' \oplus c$.

Proof

1.	$a \otimes b \leq c$	hypothesis
2.	$(a' \oplus b')' \leq c$	1, definition of $\otimes$
3.	$c' \leq a' \oplus b'$	2, lema1 , RDS2
4.	$a'' \leq c'' \oplus b'$	3, RDS5
5.	$a \leq b' \oplus c$	4, RDS2, commutativity

■

An RDS is said to be *residuated* if and only if the following biconditional holds for all propositions a, b, and c:

$$a \leq b \Rightarrow c \text{ if and only if } a \otimes b \leq c$$

Theorem 10.5 *Every RDS is residuated.*

Proof Suppose that $a \otimes b \leq c$. By Lemma 10.4, $a \leq b' \oplus c$. By the definition of $\Rightarrow$, $a \leq b \Rightarrow c$.

Now suppose that $a \leq b \Rightarrow c$. By the definition of $\Rightarrow$, $a \leq b' \oplus c$. By the commutativity of fission, $a \leq c \oplus b$. By RDS5, $c' \leq a' \oplus b'$. By Lemma 10.1, $(a' \oplus b')' \leq c''$. By RDS2 and the definition of fusion, $a \otimes b \leq c$. ■

The notion of a *fusion filter* plays a large role in this paper. It is defined as follows. A set of propositions $\mathcal{F}$ is a fusion filter on an RDS D if and only if (1) $\mathcal{F}$ is closed upwards under $\leq$, that is, if $a \in \mathcal{F}$ and $a \leq b$, then $b \in \mathcal{F}$ and (2) if a and b are both in $\mathcal{F}$, then $a \otimes b$ is also in $\mathcal{F}$.

Lemma 10.6 *If* $< D, T_D >$ *is a RDS, then* T_D *is a fusion filter.*

Proof By the definition of a truth set, T_D is closed upward under $\leq$.

Assume that a and b are in T_D. By Lemma 10.3, $a \leq b \Rightarrow (a \otimes b)$. Since T_D is closed upwards under $\leq$, $b \Rightarrow (a \otimes b) \in T_D$. By the definition of a truth set, $b \leq a \otimes b$, hence $a \otimes b \in T_D$. ■

It is easy, furthermore, to show that every fusion filter is closed under detachment for $\Rightarrow$, that is, if $a \Rightarrow b$ and a are both in a fusion filter $\mathcal{F}$, then b is also in $\mathcal{F}$. Here's a little proof:

1.	$a \Rightarrow b \leq a \Rightarrow b$	RDS1
2.	$(a \Rightarrow b) \otimes a \leq b$	1, lemma 10.5

So, suppose that $a \Rightarrow b$ and a are both in $\mathcal{F}$. $\mathcal{F}$ is closed under fusion so $(a \Rightarrow b) \otimes a \in \mathcal{F}$. By the above argument and the fact that $\mathcal{F}$ is closed upwards under $\leq$, $b \in \mathcal{F}$.

I prove one final lemma before moving on to discuss relevance domains. This lemma connects fusion and implication again in an interesting and (for the present project) important manner.

Lemma 10.7 $(a \Rightarrow c) \otimes (b \Rightarrow d) \leq (a \otimes b) \Rightarrow (c \otimes d)$.

Proof Proof is in the style of a sequent proof, since this proof has a tree structure that is easier to present in this form.

$$\dfrac{\dfrac{\dfrac{(a \Rightarrow c) \otimes a \leq c \quad (b \Rightarrow d) \otimes b \leq d}{(a \Rightarrow c) \otimes a \otimes (b \Rightarrow d) \otimes b \leq c \otimes d}}{((a \Rightarrow c) \otimes (b \Rightarrow d)) \otimes (a \otimes b) \leq c \otimes d}}{(a \Rightarrow c) \otimes (b \Rightarrow d) \leq (a \otimes b) \Rightarrow (c \otimes d)}$$

■

10.3 Relevance Domains

If D is an RDS, for each proposition $a \in D$ there is a *relevance domain*, $|a|$, which is the set of b in D such that $b \oplus b' = a \oplus a'$. Clearly no two distinct relevance domains overlap, and since there is a relevance domain for each proposition, every RDS is partitioned into relevance domains. It so happens that every relevance domain is a boolean algebra, $< |a|, 1_a, 0_a, \oplus_a, \otimes_a, '_a >$, where $1_a = a \oplus a'$, $0_a = a \otimes a'$, $\oplus_a$ is just $\oplus$ restricted to $|a|$, and similarly for the other two operators. In what follows, I drop the subscripts, since the restriction is obvious.

What is especially interesting about the fact that relevance domains are boolean algebras is that within domains $\oplus$ and $\otimes$ act like the join and meet respectively of a lattice. In logical (as opposed to algebraic) terms, they act like extensional disjunction and conjunction when restricted to a domain.

Here is a proof of the algebraic correlate of the logical principle of addition. Assume that $b \in |a|$, so $b' \oplus b = a' \oplus a$.

$$\dfrac{\dfrac{\dfrac{\dfrac{\dfrac{a' \leq a' \oplus a'}{(a' \oplus a')' \leq a''}}{a \otimes a \leq a}}{a \leq a \Rightarrow a}}{a \leq a' \oplus a} \qquad a' \oplus a = b' \oplus b}{\dfrac{a \leq b' \oplus b}{b \leq a' \oplus b}}$$

Substituting a' for a throughout gives us the usual $b \leq a \oplus b$. The principle of simplification – $a \otimes b \leq b$, where $b \in |a|$ – follows from addition in the usual way.

The proof that fusion distributes over fission is rather more involved. Here is a proof of one direction of the principle of distribution:

$$\dfrac{\dfrac{\dfrac{\dfrac{\dfrac{a \otimes b \leq a \otimes b}{b \leq a' \oplus (a \otimes b)} \quad \dfrac{a \otimes c \leq a \otimes c}{c \leq a' \oplus (a \otimes c)}}{b \oplus c \leq (a' \oplus (a \otimes b)) \oplus (a' \oplus (a \otimes c))}}{b \oplus c \leq a' \oplus a' \oplus (a \otimes b) \oplus (a \otimes c)}}{b \oplus c \leq a' \oplus (a \otimes b) \oplus (a \otimes c)}}{a \otimes (b \oplus c) \leq (a \otimes b) \oplus (a \otimes c)}$$

And here is a proof of the converse. Assume that $b, c \in |a|$:

$$\dfrac{\dfrac{a \otimes b \leq a \quad \dfrac{a \otimes b \leq b}{a \otimes b \leq b \oplus c}}{a \otimes b \leq a \otimes (b \oplus c)} \qquad \dfrac{a \otimes b \leq a \quad \dfrac{a \otimes c \leq c}{a \otimes c \leq b \oplus c}}{a \otimes c \leq a \otimes (b \oplus c)}}{(a \otimes b) \oplus (a \otimes c) \leq a \otimes (b \oplus c)}$$

Note that this proof appeals not only to addition and simplification, but to Lemma 10.2 as well.

For each relevance domain, $|a|$, there is a partial order, $\leq_a$, defined as follows. Where $b, c \in |a|$,

$$b \leq_a c \text{ iff } b \otimes c = b$$

A *fusion filter on* $|a|$ is a set of propositions of $|a|$ closed upwards under $\leq_a$ and closed under fusion. Extracting $|a|$ from the RDS in which it resides, fusion filters on $|a|$ are just filters in the standard algebraic sense, since $\oplus$ is just a greatest lower bound on $|a|$. I use fusion filters on domains to represent certain scientific theories in Sect. 10.4 below.

The proposition below shows that when restricted to a relevance domain, $|a|$, the partial order $\leq$ is a subset of $\leq_a$.

Proposition 10.8 *If* $a \leq b$ *and* $|a| = |b|$, *then* $a \leq_a b$.

Proof Suppose that $|a| = |b|$ and that $a \leq b$.

I first show that $a \leq b \otimes a$. By assumption $a \leq b$. Then, since fusion is order preserving, $a \otimes a \leq b \otimes a$. By the idempotence of fusion, $a \leq b \otimes a$.

Now I show that $b \otimes a \leq a$. By idempotence, $b \otimes b \leq b$. By Lemma 10.5 and the definition of $\Rightarrow$, $b \leq b' \oplus b$. Since $|a| = |b|$, $b' \oplus b = a' \oplus a$, and so $b \leq a' \oplus a$. By commutativity, $b \leq a \oplus a'$. By RDS5, $a' \leq b' \oplus a'$. By Lemma 10.1, $(b' \oplus a')' \leq a''$. By the definition of fusion and RDS2, $b \otimes a \leq a$, as required.

Putting this altogether, if $a \leq b$ and $|a| = |b|$, then $a \leq_a b$. ■

Suppose that f is a fusion filter on $|a|$ and $\mathcal{F}$ is a fusion filter on the entire RDS such that every proposition b in f such that $b \neq a \oplus a'$ and $b \neq a \otimes a'$ is also in $\mathcal{F}$. The following lemma entails that $f \subseteq \mathcal{F}$:

Lemma 10.9 *If $\mathcal{F}$ is a fusion filter on D and a is in $\mathcal{F}$, then $a \oplus a' \in \mathcal{F}$.*

Proof Suppose that $a \in \mathcal{F}$.

1. $a \otimes a \leq a$ idempotence
2. $a \leq a \oplus a'$ 1, lemma 10.1, definition of $\otimes$

Since $\mathcal{F}$ is closed upwards under $\leq$, $b \in \mathcal{F}$. ■

Lemma 10.10 *Suppose that $b \in \mathcal{F}$ and $b \leq_a c$. Then $c \in \mathcal{F}$.*

Proof Assume that $b \leq_a c$. Thus, $b, c \in |a|$. By the definition of a relevance domain,

$$b \oplus b' = c \oplus c' = a \oplus a'.$$

First, I show that $(a \otimes b) \Rightarrow b = a \oplus a'$:

1. $a \oplus a' = a \oplus a'$
2. $a' \oplus a' \oplus a = a \oplus a'$ 1, idempotence, associativity
3. $a' \oplus b' \oplus b = a \oplus a'$ 2, $b \oplus b' = a \oplus a'$, associativity
4. $(a \otimes b)' \oplus b = a \oplus a'$ 4, associativity, defintion of $\otimes$
5. $(a \otimes b) \Rightarrow b = a \oplus a'$ 4, definition of $\Rightarrow$

By Lemma 10.9, we know that $a \oplus a' \in \mathcal{F}$. So, by the above, $(a \otimes b) \Rightarrow b \in \mathcal{F}$. Since fusion filters are closed under detachment for $\Rightarrow$, $b \in \mathcal{F}$. ■

10.4 A Dappled World?

I employ RDS to represent a particular view of scientific theories and their relationship to reality. On this picture of science, scientists employ various theories that deal only with local phenomena but also they accept more fundamental theories that contain laws that are supposed to underly the various local laws contained in the other theories. The fundamental theories unify the local theories. This sort of unification is often thought by philosophers of science to be a form of explanation (see, e.g., Kitcher 1981).

Nancy Cartwright, however, has raised serious questions about the nature of fundamental laws and their relationship to local laws. She outlines two views concerning this relationship that she calls "nomological pluralism" and "fundamentalism". She supports pluralism over fundamentalism:

> Metaphysical nomological pluralism is the doctrine that nature is governed in different domains by different systems of laws not necessarily related to each other in any systematic or uniform way; by a patchwork of laws. Nomological pluralism opposes any kind of fundamentalism. I am here concerned with the attempts of physics to gather all phenomena into its own abstract theories. In *How the Laws of Physics Lie* I argued that most situations are brought under physics only by distortion, whereas they can often be described fairly correctly by concepts from more phenomenological laws (Cartwright 1999, p. 31).

Local theories, on Cartwright's terminology, have their own "domains". In what follows, I formalise this view using relevance domains in RDS. But before I get to that, let us look more closely at the dispute between fundamentalism and pluralism, as Cartwright sees it.

In order to understand better the dispute between fundamentalism and nomological pluralism, let's look at an example from Cartwright (1983) and Cartwright (1999). Kepler's laws of planetary motion covered just that range of phenomena: the motions of the (then known) planets in our solar system around the sun. On the fundamentalist view, Newton's laws of gravity and force explained Kepler's laws by being much more general and allowing the derivation of those more specific laws.[3] On Cartwright's pluralism, however, the fundamentalist view is wrong. Kepler's laws are of a different character than the supposedly more general laws. Kepler's laws have no exceptions. On Cartwright's reading, Newton's laws are to be understood as *ceteris paribus* conditionals. Consider Newton's second law of motion, $Force = mass \times acceleration$. My pen sits on my desk in front of me. The earth acts on it in terms of a force of gravity. Yet the pen does not accelerate towards the earth. On the fundamentalist understanding of Newton's theory, the pen may have no *net* acceleration, but it does have a component of acceleration towards the earth that is counteracted by the force exerted on it by the desk. Cartwright interprets components of force, acceleration, and so on, as useful mathematical fictions, but not as being real. She takes Newton's second law to say, in effect, that if ideal circumstances obtain (where there are no counterbalancing forces), $Force = mass \times acceleration$. The fundamentalist, on the other hand, takes the components of force and acceleration to be real entities underlying the phenomena that we perceive.

It is not my aim here to adjudicate between pluralism and fundamentalism. Rather, I suggest how to use the theory of RDS to formalise both theories. I begin with fundamentalism.

[3] In fact, Newton's laws allow the derivation of laws that are *approximately* the same as Kepler's laws. I can adjust my view to claim that Newton's laws allow the derivation of a theory that is approximately the same as Kelper's, but this will require the use of a relation of approximate similarity, and this will add rather a lot of complexity to the formalism. So I just make the simplifying assumption that on the fundamentalist theory local theories are straightforwardly derivable from general ones.

I assume for the purposes of the present section that all the scientific theories I am treating are consistent. In Sect. 10.5 below and those that follow, I treat inconsistent theories as well.

A local theory is a fusion filter on a relevance domain. The domain of a theory is the set of propositions about the things and circumstances to which it applies and does not include any other propositions. A fundamental theory, on the other hand, is a fusion filter on the entire RDS.

According to fundamentalism, a fundamental theory T supersedes a local theory t if and only if the laws of t can be derived from T. It may be that the laws of t do not exhaust all of t, that t contains some propositions that are not laws and hence not derivable from T. This possibility only complicates matters; it does not change the view in any substantial way, and so I will ignore it.

Let us look briefly at an example of what the fundamentalist means when he says that T supersedes t. Suppose that t just contains Galileo's claim that all bodies in free fall fall with the same acceleration (and is also closed upwards under $\leq_a$, where $|a|$ is the domain of t). And let T be Newton's theory of force and gravity. Let F and α be the force of gravity on and acceleration on a body i and m and M be the masses of i and the earth respectively. Then we have:

$$F = m\alpha = \frac{GMm}{r^2}.$$

Dividing through by m we get:

$$\alpha = \frac{GM}{r^2}.$$

So, no matter what mass i has, its acceleration will be GM/r^2. In this example we can see both that the laws of t are straightforwardly contained in T and that T contains more information than t. I generalise this rather straightforward example to represent the entire fundamentalist position on the relationship between general and local theories. I claim that a local theory t is derived from a general one T (in the sense used by scientists and historians of science) if and only if $t \subseteq T$.

Turning to the pluralist position, I have a very different view of the relationship between general and local theories. The pluralist says that fundamental theories generate the laws of local theories in ideal contexts. So, I need to include the notion of an ideal context in my representation. Let $I(T, t)$ pick out a set of sets of propositions (i.e. $I(T, t) \subseteq \wp(D)$). If $C \in I(T, t)$, then C describes an ideal context from the perspectives of T and t. For example, where t is Galileo's theory of terrestrial motion and T is Newton's dynamics, C might describe a context in which a thing is acted upon by only a single external force. I require that all contexts are closed under fusion. That is to say, if $a, b \in C$, for a $C \in I(T, t)$, then $a \otimes b \in C$.

The idea is to formalise the claim that, where c is an ideal circumstance from the perspectives of T and t, that the laws of t are generated by T in C. In order to represent this claim, I first define a fusion operator on sets of propositions in the manner of Fine (1974):

Definition 10.11 (*Set Fusion*) Where X and Y are sets of propositions, $X \otimes Y$ is the set of propositions b for which there is some proposition $a \in Y$ such that $a \Rightarrow b \in X$.

Suppose that a theory T says that under some condition d, the law that $a \Rightarrow b$ holds. In the present framework, I represent this by $d \Rightarrow (a \Rightarrow b)$. Suppose, for example, that d says that there is only force acting on an object i. Then a might say that this force has magnitude m and then b could be 'the force on i is $m \times \alpha$ where α is i's acceleration'.

A theory T generates t in context C is characterised in the formal framework as follows:

$$T \otimes C \supseteq t$$

To say that T is a general theory for t in the pluralist sense is to say that for all C in $I(T, t)$, $T \otimes c \supseteq t$.

The reason that I require C to be closed under fusion is to satisfy the condition of the following theorem:

Theorem 10.12 *If T is a fusion filter on D and C is a set of propositions closed under fusion, then $T \otimes C$ is fusion filter on D.*

Proof Suppose that T is a fusion filter on D and C is a set of propositions closed under fusion.

(1) Assume that b_1 and b_2 are both in $T \otimes C$. Then there are a_1 and a_2 in C such that $(a_1 \Rightarrow b_1)$ and $(a_2 \Rightarrow b_2)$ are both in T. Since T is closed under fusion, $(a_1 \Rightarrow b_1) \otimes (a_2 \Rightarrow b_2) \in T$. By Lemma 10.7, $(a_1 \otimes a_2) \Rightarrow (b_1 \otimes b_2) \in T$. Since c is also closed under fusion, $a_1 \otimes a_2 \in c$. Hence, by the definition of set fusion, $b_1 \otimes b_2 \in T \otimes C$.

(2) Assume that $b \in T \otimes C$ and that $b \leq d$. Then there is some $a \in C$ such that $a \Rightarrow b \in T$. By the definition of $\Rightarrow$, $a' \oplus b \in T$. From the fact that $b \leq d$ and the fact that $\oplus$ is order preserving (RDS3), $a' \oplus b \leq a' \oplus d$. T is closed upwards under $\leq$, so $a' \oplus b \in T$, hence $a \oplus d \in T$. Thus, by the definition of set fusion, $d \in T$. ∎

The treatment of laws using relevant implication does have the virtue that it makes conditions non-monotonic. The conditions on laws, at least on Cartwright's view, are understood as *ceteris paribus* conditions. They imply the consequents of the laws only if certain background conditions obtain or others fail to occur. The addition of new information to the antecedent of a law does not necessarily yield a law in T. But one might worry that the conditional of an RDS is nevertheless too monotonic. Consider a case in which $|a| = |b| = |d|$. If there is a condition $e \in |a|$, then if $d \Rightarrow (a \Rightarrow b) \in T$ then $(d \otimes e) \Rightarrow (a \Rightarrow b) \in T$. But this is rather harmless, at least in the way that I think of conditions like d. These conditions exclude other conditions. For example, d tells us that there is only one force acting on i. Suppose that another condition e tells us that there is more than one force acting on i. Then $e \otimes d$ is inconsistent and quite useless as a condition on a ceteris paribus law.

Note that I am not claiming that the implication of RMI or any other standard relevant logic is perfectly adequate to represent ceteris paribus laws. Rather I do think that a form of counterfactual relevant conditional is right for this task. In particular I believe that the theory of conditionals that André Fuhrmann and I develop in Mares and Fuhrmann (1995) can be used to produce relevant ceteris paribus conditionals of the right sort. One feature that I would build into such conditionals is that they imply the corresponding relevant implications. That is to say, where $\rightsquigarrow$ is the ceteris paribus conditional and $\rightarrow$ is the corresponding relevant implication, $A \rightsquigarrow B \vdash A \rightarrow B$.

Similarly, I am not claiming that relevant implication is by itself adequate to represent laws of science. Laws are, in my opinion, strict relevant implications. But these strict implications are alethic (again, $A \dashv B \vdash A \rightarrow B$). In order to include strict relevant implication and counterfactual relevant implication in the present theory, however, is rather complicated. There are semantical theories for both, but they are worlds-based, not algebraic semantics. From the point of view of the present project, the advantages of replicating these semantics in the present algebraic framework are unclear. So, I use only the implication of RMI here.

Before I leave the topic I need to address an important point that one of the referees raised. This is whether RMI is too weak to represent actual scientific reasoning. It is difficult to answer this question without extensive empirical examination of actual scientific reasoning, and painstaking analysis of whether that reasoning can be recast in terms of RMI. My conjecture is that even when scientists are dealing with global theories, they generally treat only parts of them, which can be understood as representing particular relevance domains. If this is the case, then much of the classical inference that goes on in science can be understood and justified in the present framework.

10.5 The Problems of Inconsistency

In Sect. 10.4, I assume that all the theories that are employed by scientists are consistent. This is a useful assumption but not one that is borne out completely by the history of science. There are at least three types of inconsistency that are important here. The first is what is sometimes called *internal inconsistency*. A theory is internally inconsistent if and only if contradictory propositions can be derived directly from it. The second is what is called *external inconsistency*. This is an inconsistency between two or more theories that are adopted by a scientist or by a scientific community. The third is less often discussed. It is *conceptual inadequacy*. This is an inconsistency that appears when a theory can be shown to be inconsistent with an imagined counterexample that appeals only to uncontroversially possible situations.

Internally inconsistent theories in the natural sciences are quite rare, if they exist at all. Newton's cosmological theory (his physics together with the propositions that space is infinite and that matter is relatively homogeneously distributed throughout space) is one theory that is sometimes said to contain a contradiction (Norton 1999).

There are some uncontroversially inconsistent theories in mathematics, such as Cantor's naïve set theory and Frege's theory of arithmetic. Ross Brady, Richard Sylvan, Graham Priest, Zach Weber, and others have used weak relevant logics to formulate naïve set theories. The logic RMI is too strong for this purpose. In a naïve version of set theory based on RMI, one can prove Curry's paradox. There is a fairly limited range of internal inconsistencies that RMI is capable of treating.

One mathematical theory that is sometimes claimed to be inconsistent but has been used in natural science is Newton's version of the calculus.[4] Vickers (2013, Chap. 6) has recently claimed that Newton's calculus is consistent. I do not want to become involved in this debate.

The second sort of inconsistency, one which plays a greater role in this paper, is that of *external inconsistency*. A theory is said to be externally inconsistent when it contradicts some other theory that is accepted by the scientific community. A standard example of an external inconsistency concerns Neils Bohr's 1913 model of the hydrogen atom. On this model, the sole electron in the atom is in a stable orbit at 0.53×10^{-10}m around the nucleus. But the only view of electrodynamics that was available at the time was the classical theory (with its classical understanding of Maxwell's equations). According to Maxwell's equations, the orbiting electron should radiate energy and loose charge (see Brown and Priest 2015).

The third sort of inconsistency is conceptual inadequacy. Here are two examples. The first is reported by van Fraassen (1989, p. 217). The following three statements seem appropriate to include in a theory of shadows:

1. If X casts a shadow, then there is light falling directly on X;
2. No object can cast a shadow through an opaque object;
3. Every shadow is a shadow of something.

Now consider the case in which a very large building dwarfs a small building. Suppose the sun is in a place such that no light falls directly on the small building. Now consider the shadow on the ground to the side of the small building opposite to the side on which the large building is. There is a region on the ground that is in shadow, but this shadow, by statement 2, cannot be of the large building and by statement 1 cannot be of the small building. By 3, it must be of something, but there are no reasonable candidates for the owner of the shadow.

Van Fraassen states that this theory of shadows is not empirically adequate. That is, it does not satisfy all the empirical evidence. Van Fraassen says:

> This theory is not inconsistent. Taken by itself, it is logically impeccable. But there are phenomena that do not fit the theory – and our little experiment points to a large class of these. So here we have two distinct concepts of inadequacy: *inconsistency* and what I propose to call *empirical inadequacy* (van Fraassen 1989, p. 218).

A theory is empirically inadequate (on Van Fraassen's use of the term) if and only if it does not fit the empirical phenomena. It is true that the theory of shadows given here does not fit with all existing empirical phenomena. But there is more to it than that. What is given as a refutation is not a physical experiment, but a simple thought experiment. This thought experiment is such that even if there were no actual physical circumstances that are as described in the statement of the experiment, it would still

---

[4] The claim stems, perhaps, from Berkeley's criticism of Newton's calculus (see Boyer 1959, pp. 225–226). This interpretation has been adopted by Brown and Priest (2004).

refute the theory. We know that there could be such circumstances and this is enough to show that there are conceptual problems with the theory, even if these problems do not amount to an internal inconsistency. A theory that has a clash with (even merely) physically possible features of the world that are discernible in a conceptual fashion I call *conceptually inadequate*.

A historically more important example of a theory that is inconsistent in application is Aristotle's theory of gravity. On Aristotle's theory, heavy things fall towards the earth with greater acceleration than lighter things. Wanting to refute Aristotle's view, Galileo imagines tying two stones together, a heavier one and a lighter one.[5] The stones are released from a structure taller than the length of the stones tied together. Using Aristotle's theory, we get two very different answers when we try to calculate how quickly the stones accelerate when tied together. First, if we consider the stones separately and then their effect on one another, the lighter stone should slow down the heavier stone (and the heavier stone speed up the lighter one) by pulling on it. If we think of the new object – stones tied together – as a single thing, it should fall faster than the heavier one on its own (Galilei 1933, pp. 59–64) (and see Brown 1993, pp. 1–2). Galileo concludes that the theory is absurd because it gives two conflicting answers to the same question about a given situation. Once again, this a conceptual problem with the theory, but one that falls short of internal inconsistency. It could be that no such situation would ever arise in which this inconsistency emerges. I claim that Aristotle's theory is another case of a conceptual inadequate theory.

I treat conceptual inadequacy, but not empirical inadequacy, as a form of inconsistency, at least in some cases. A conceptually inadequate theory can contradict features that we put in place a priori in the semantical frameworks that we use in order to interpret a class of theories or an area of science. For example, in order to interpret a theory of gravity of a class of physical theories, we might employ a set of possible worlds, in which bodies can vary with regard to their physical properties, including whether or not any of them are tied with rope to any others. In this case we can determine *a priori* that there is an inconsistency between the semantical framework adopted and Aristotle's theory of gravity. I represent this sort of inconsistency in terms of RDS by placing propositions in the truth set which represent the semantical setup which the theory contradicts.

The foregoing discussion suggests that we should not merely think of scientific theories as believed or rejected by the scientific community. Two theories that are inconsistent with one another, for example, should not both be believed, but may both be used. Similarly, as we shall see, a theory that contradicts the belief set becomes trivial. Thus, a more complicated taxonomy of attitudes towards theories is required.

Van Fraassen distinguishes belief in a theory from the *acceptance* of a theory: "acceptance of a theory involves as belief only that it is empirically adequate" (van Fraassen 1980, p. 12). A theory is empirically adequate, moreover, if and only if all of its empirical predictions are true. Van Fraassen characterises his own position, which he calls "constructive empiricism", as holding that science aims only to give

[5] At some places in Galileo's text, the bodies described are stones, and at others they are a musket ball and a cannon ball.

us empirically adequate, or acceptable, theories. I neither agree nor disagree with constructive empiricism in the present essay, but I do think that at times scientists use theories that they do not believe but merely accept in van Fraassen's sense.[6]

In order to characterise theories that are used by scientists but (perhaps because of their conceptual inadequacy or conflict with other theories) are neither believed nor accepted, I also need a notion of a positive attitude towards theories that is weaker than either belief or acceptance. I call it *adoption*. A scientist (or anyone else) can be said to adopt a theory if and only if he or she thinks that the theory is useful in deriving empirically testable predictions. He or she need not believe that every prediction derived from the theory will fit with the empirical evidence, only that some of its results will do so. Thus, I propose a three-tiered understanding of the positive attitudes that scientists may hold to theories. In terms of their extensions, the relationships between the three positive attitudes towards theories are as follows:

$$\text{Belief} \subseteq \text{Acceptance} \subseteq \text{Adoption}$$

I treat a theory that is believed in terms of RDS as a subset of the truth set of the RDS. The truth set is the set of propositions that a scientist or scientific community believes. A theory that is merely accepted is a fusion filter on a domain (if it is a local theory) or on the RDS as a whole (if it is a general theory). A theory that is merely adopted (and neither believed nor accepted) is not represented in terms of RDS. These are discussed in the following two sections.

10.6 RDS and the Limits of Paraconsistency

Although RMI is a paraconsistent logic and some RDS incorporate elements of this paraconsistency, I argue in this section that the straightforward use of RDS to model theories that are inconsistent in the senses discussed above is inadequate as a means to understanding inconsistency in science.

The topic here concerns theories' being applied to one another. Scientists often use one theory to interpret another. They use mathematical theories to interpret physical theories. They use more fundamental theories to interpret less fundamental theories, and so on. I think of this in terms of an algebraic operation on theories. I represent a theory t_1 being applied to another theory t_2 by

$$t_1 \otimes t_2.$$

[6] For example, I have heard some physicists say that they like to use the elegant mathematics of string theory, and that it gives the right results, but that they do not believe it. As my student Tim Irwin pointed out to me, in the social sciences it is a platitude that "all theories are false, but some are useful".

In other words, the application of one theory to another is just the set fusion of the two theories. Note that in RMI, set fusion (like fusion) is commutative.

I start by examining externally inconsistent theories. That is, theories that contradict one another. such theories are not problematic if they are internally constant, and are never applied to one another. Suppose that t_1 contains a proposition a and t_2 contains a'. We shall see that $t_1 \otimes t_2$ contains all of $|a|$. If $|a|$ itself contains a lot of propositions, and there is good reason to apply t_1 to t_2, then this can cause a serious problem.

To show that this problem really exists, I begin by proving the following lemma:

Lemma 10.13 *Suppose that t_1 and t_2 are both fusion filters on $|a|$. Then $t_1 \otimes t_2$ is a fusion filter on $|a|$.*

Proof Suppose first that $b \in t_1 \otimes t_2$ and that $b \leq_a c$. Then $b = b \otimes c$. By the definition of set fusion, there is some proposition $d \in t_2$ such that $d \Rightarrow b \in t_1$. Since $b = b \otimes c$, $d \Rightarrow (b \otimes c) \in t_1$. By the definition of $\Rightarrow$,

$$d' \oplus (b \otimes c) \in t_1.$$

$\oplus$ and $\otimes$, when restricted to $|a|$, are just the join and meet of a distributive lattice (i.e. here, a boolean algebra), hence by the distributive law,

$$(d' \oplus b) \otimes (d' \oplus c) \in t_1.$$

Moreover, $(d' \oplus b) \otimes (d' \oplus c) \leq_a d' \oplus c$. Therefore,

$$d' \oplus c \in t_1,$$

that is,

$$d \Rightarrow c \in t_1$$

and so,

$$c \in t_1 \otimes t_2.$$

It suffices now to show that $t_1 \otimes t_2$ is closed under fusion. Suppose that b and c are both in $t_1 \otimes t_2$. Then, there is some d and e both in t_2 such that both $d \Rightarrow b$ and $e \Rightarrow c$ are in t_1. Since t_1 is closed under fusion,

$$(d \Rightarrow b) \otimes (e \Rightarrow c) \in t_1.$$

By Lemma 10.7,

$$(d \otimes e) \Rightarrow (b \otimes c) \in t_1.$$

t_2 is also closed under fusion, so $d \Rightarrow e \in t_2$. Therefore, $b \otimes c \in t_1 \otimes t_2$. ■

Now note that it follows from $a \leq a \oplus a$ and Lemma 10.4 that $a \otimes a' \leq a$. From this, the following theorem can be proven easily:

Theorem 10.14 *If* $|a| = |b|$, *then* $a \otimes a' \leq b$.

Proof Suppose that $|a| = |b|$. By the definitions of fusion and relevance domains and Lemma 10.1, $b \otimes b' = a \otimes a'$. Since $b \otimes b' \leq b$, then $a \otimes a' \leq b$. ■

Theorem 10.14 together with Proposition 10.8 entails the following corollary:

Corollary 10.15 *If there is some proposition* $a \in t_1$ *such that* $a' \in t_2$ *and both* t_1 *and* t_2 *are fusion filters on* $|a|$, *then* $t_1 \otimes t_2 = |a|$.

This means that if we apply two local theories on the same domain that are inconsistent with one another, the resulting theory will be trivial in the sense that it will contain all the propositions of that domain. This makes RDS problematic for studying the application of local theories that are inconsistent with one another (as in the case of classical dynamics and the old quantum theory).

There is more bad news. Consider a theory that is conceptually inadequate. It seems that a reasonable way of understanding this in terms of RDS is to say that the theory contains a proposition that is the negation of some proposition in the truth set T_D. But this leads to another form of explosion, as is shown by Corollary 10.16.

Corollary 10.16 *If* $a \in T_D$ *and* $|b| = |a|$, *then* $a' \leq b$.

Proof Suppose that $a \in T_D$ and $|b| = |a|$. By Theorem 10.14, $a \otimes a' \leq b$. Since all RDS are residuated, $a \leq a' \Rightarrow b$. But $a \in T_D$, so $a' \Rightarrow b \in T_D$. By the definition of a truth set, $a' \leq b$. ■

Now suppose that t is a local theory on $|a|$ that contradicts T_D. By Proposition 10.8 and Corollary 10.16, t contains every proposition in $|a|$. This shows that RDS are problematic in treating conceptually inadequate theories.

What about internally inconsistent theories? A local theory that is internally inconsistent, as represented on an RDS contains all the propositions in its domain, and hence is quite useless. But a general theory that is internally inconsistent may not be made unusable by that inconsistency. Let's say that T contains both a and a'. Then it contains all of $|a|$. But if $|a|$ is not very large or is not of much interest with regard to the reasons that a scientist is using T, then the contradiction may be ignored.

10.7 A Syntactic Turn

The previous section may have made the situation with regard to the treatment of inconsistency in science seem rather bleak. But things are not that bad. There are now many different logical approaches to inconsistency in science, and they have all had some measure of success. For example, there are the adaptive logics of Batens and his Ghent school (Batens 2017), Carnielli's logics of formal inconsistency (Carnielli and Coniglio 2015), Schotch, Jennings, and Brown's logic of forcing (Brown 1992), the partial models of Da Costa and French (2003), and the chunking and permeating of Brown and Priest (2004, 2015), among others. I have adopted a purely heuristic method of dealing with inconsistencies. I use $RMI_{\rightarrow,\neg}$ to formulate the theories in question, apply them to one another, and then extract the desired propositions from the resulting theory.

I am here giving proof theory an important role to play in treating inconsistent theories, but overall my approach remains semantic. The idea is to treat theories that are inconsistent or inconsistent with one another as syntactic entities (sets of formulas) and use the logic RMI to apply them to one another. Taking theories to be sets of formulas, rather than subsets of an RDS, is that in the logic itself there are relatively few theorems that construct the syntactic counterparts of relevance domains. Relevance domains, for all their virtues in reconstructing ideas in the philosophy of science, cause problems in dealing with inconsistencies. So, I move to proof theory in order to avoid them when dealing with inconsistent theories. The move to proof theory is not the final move in the interpretation of inconsistent theories and their use. After applying theories in the syntactic sense to one another, one must extract the useful results. These results, in order to be integrated into one or one's community's beliefs or acceptances, needs (on the present view) to be interpreted in terms of the RDS of the community, but updating or revising it.

In terms of metaphysics, I treat the RDS as representing the contents of scientific world views, but I treat theories, qua purely syntactic entities, heuristically. That is, I take RDS to represent the actual world, or at least an approximation to it. The RDS are to be updated with new discoveries and to be thus made more adequate as representations of the world. In contrast, theories are treated here as mere means to an end. They help us determine sets of sentences that we want to represent in terms of propositions in RDS. The goal of theorising and testing (in this framework) is always to produce an RDS, as the semantic representation of the world as science sees it.

In order to see how to use the proof theory for RMI in this project, I need first to present it.

The propositional language is made up of a countable set of propositional variables ($p, q, r, \ldots$), a unary negation connective ($\neg$), the binary implication connective ($\rightarrow$), and parentheses. The usual formation rules apply. Fusion, fission, and the biconditional are also useful connectives and are defined as follows:[7]

$$A \circ B =_{df} \neg(A \rightarrow \neg B)$$

$$A + B =_{df} \neg A \rightarrow B$$

$$A \leftrightarrow B =_{df} (A \rightarrow B) \circ (B \rightarrow A)$$

---

[7] In R and stronger systems, $(A \rightarrow B) \circ (B \rightarrow A)$ is equivalent to $(A \rightarrow B) \wedge (B \rightarrow A)$, where $\wedge$ is extensional conjunction.

In order to connect theories in the syntactic sense with their semantics, I assume a valuation function V such that for all propositional variables, p in the language, $V(p) \in D$, and for all formulas A, $V(A)$ is determined in the usual recursive manner:

- $V(\neg A) = V(A)'$;
- $V(A \to B) = V(A) \to V(B)$.

I also define a function V^{-1} from propositions in D to sets of formulas such that $A \in V^{-1}(a)$ if and only if $V(A) = a$. Let X be a subset of D, then $\bigcup_{a \in X} V^{-1}(a)$ is the set of formulas that express propositions in X. Where X is a fusion filter, I will show that $V^{-1}(X)$ is a theory, although the converse does not always hold.

Here is an axiomatisation of $RMI_{\to,\neg}$:

1. $A \to A$
2. $A \to (A \to A)$
3. $(B \to C) \to ((A \to B) \to (A \to C))$
4. $(A \to (B \to C)) \to (B \to (A \to C))$
5. $(A \to (A \to B)) \to (A \to B)$
6. $(A \to \neg B) \to (B \to \neg A)$
7. $\neg\neg A \to A$
8. $(A \to \neg A) \to \neg A$

$RMI_{\to,\neg}$ needs only one rule – modus ponens:

$$\frac{A \to B \qquad A}{B}$$

I use Avron's Gentzen system, $GMRI_{\to,\neg}$ from Avron (1991). This system allows both multiple premises and conclusions.

Axioms:

$$\underbrace{A, A, \ldots, A}_{m \text{ times}} \vdash \underbrace{A, A, \ldots, A}_{n \text{ times}}$$

where A is an atomic formula and $m, n > 0$.

Structural Rules:

Exchange

$$\frac{\Gamma, A, B \vdash \Delta}{\Gamma, B, A \vdash \Delta} \qquad \frac{\Gamma \vdash \Delta, A, B}{\Gamma \vdash \Delta, B, A}$$

Contraction

$$\frac{\Gamma, A, A \vdash \Delta}{\Gamma, A \vdash \Delta} \qquad \frac{\Gamma \vdash \Delta, A, A}{\Gamma \vdash \Delta, A}$$

Meaning Rules:

Implication

$$\frac{\Gamma, A \vdash B, \Delta}{\Gamma \vdash A \to B, \Delta} \qquad \frac{\Gamma, B \vdash \Delta \quad \Gamma^* \vdash A, \Delta^*}{\Gamma, \Gamma^*, A \to B \vdash \Delta, \Delta^*}$$

Negation

$$\frac{\Gamma, A \vdash \Delta}{\Gamma \vdash \neg A, \Delta} \qquad \frac{\Gamma \vdash \Delta, A}{\neg A, \Gamma \vdash \Delta}$$

Derived Rules:

Fusion

$$\frac{\Gamma(A, B) \vdash \Delta}{\Gamma(A \circ B) \vdash \Delta} \qquad \frac{\Gamma \vdash A, \Delta \quad \Gamma^* \vdash B, \Delta^*}{\Gamma, \Gamma' \vdash A \circ B, \Delta, \Delta^*}$$

Fission

$$\frac{\Gamma, A \vdash \Delta \quad \Gamma^*, B \vdash \Delta^*}{\Gamma, \Gamma^*, A + B \vdash \Delta, \Delta^*} \qquad \frac{\Gamma \vdash A, B, \Delta}{\Gamma \vdash A + B, \Delta}$$

I generalise the consequence relation such that, where Γ is a set of sentences (which is either finite or infinite), $\Gamma \vdash A$ if and only if there is a finite subset Γ^* of Γ such that $\Gamma^* \vdash A$. And, where Γ is a set of formulas,

$$Cn(\Gamma) =_{df} \{A : \Gamma \vdash A\}.$$

$Cn(\Gamma)$ is the consequence set of Γ and Cn is a consequence operator. This consequence operator has the following useful Tarskian properties:

- $\Gamma \subseteq Cn(\Gamma)$
- $Cn(Cn(\Gamma)) \subseteq Cn(\Gamma)$
- If $\Gamma \subseteq \Delta$ then $Cn(\Gamma) \subseteq Cn(\Delta)$

An $\mathrm{RMI}_{\rightarrow\neg}$ *theory* (henceforth 'syntactic theory') is a set of formulas Γ such that $\Gamma = Cn(\Gamma)$. The fusion rules (together with the structural rules) entail that every syntactic theory is closed under fusion.

Now I return to the treatment of inconsistent scientific theories. One problem concerns the application of two theories to one another that are inconsistent with one another. As I have said, the natural way in relevant logic to represent the application of theories to one another is by fusion. The syntactic notion of set fusion is very much like the semantic notion. Where T_1 and T_2 are theories,

$$T_1 \circ T_2 =_{df} \{B : \exists A(A \in T_2 \wedge A \rightarrow B \in T_1)\}.$$

It is easy to show that $T_1 \circ T_2$ is a syntactic theory.

There is still an apparent problem here with inconsistent syntactic theories. This problem is closely related to the problem with inconsistencies in semantic theories in an RDS. Let T be a syntactic theory. Let us call the a set of formulas B such that $(B \circ \neg B) \leftrightarrow (A \circ \neg A) \in T$ the A-*domain* of T. The following proofs show that if A and $\neg A$ are both in T, then all the formulas in the A-domain of T are also in T.

I first show that a biconditional entails the corresponding implications:

$$\dfrac{\dfrac{\dfrac{\dfrac{\dfrac{B \vdash B \qquad \dfrac{A \vdash A \qquad \dfrac{B \vdash B \qquad A \vdash B}{A \to B, A \vdash B}}{B \to A, A \to B, A \vdash A}}{A \to B, B \to A, A \to B, A \vdash B}}{A \to B, B \to A, A \vdash B}}{(A \to B) \circ (B \to A), A \vdash B}}{A \leftrightarrow B, A \vdash B}}{A \leftrightarrow B \vdash A \to B}$$

Now I show that $(A \circ \neg A) \to (B \circ \neg B)$ together with $A \circ \neg A$ entails B:

$$\dfrac{\dfrac{\dfrac{\dfrac{B \vdash B, B}{B, \neg B \vdash B}}{B \circ \neg B \vdash B} \qquad \dfrac{\vdots}{A \circ \neg A \vdash A \circ \neg A}}{(A \circ \neg A) \to (B \circ \neg B), A \circ \neg A \vdash B}}{(A \circ \neg A) \to (B \circ \neg B) \vdash (A \circ \neg A) \to B}$$

Let's say that B is in the A-domain of T. This means that $(B \circ \neg B) \leftrightarrow (A \circ \neg A) \in T$. From this and the first derivation, $(A \circ \neg A) \to (B \circ \neg B) \in T$. Then the second derivation shows that $(A \circ \neg A) \to B \in T$. Suppose then that the the contradictory formulas A and $\neg A$ are both in T. T is closed under fusion, so $A \circ \neg A \in T$ and so $B \in T$ as well.

Suppose that T_1 and T_2 are syntactic theories and that B is in the A-domain of T_1. Then $(A \circ \neg A) \to B \in T_1$. It is easy to show that $A \to (\neg A \to B) \in T_1$. Now suppose that $A \in T_1$ and $\neg A \in T_2$. Then $B \in T_2$. This shows that if T_1 and T_2 are inconsistent with one another with regard to A, every formula in the A-domain of T_1 is in $T_1 \circ T_2$.

The saving feature of the move to proof theory, though, is that in order to giving syntactic representations of actual scientific theories, we rarely have cause to add formulas that entail substantive statements of the form $(A \circ \neg A) \leftrightarrow (B \circ \neg B)$, especially where A and B have intuitively different content. This means that domains in theories are for the most part relatively small and harmless from a practical standpoint.

I represent both local and general theories syntactically by theories of the logic. The difference between them is that local theories might be formulated in more restricted languages. Let T be a theory that is supposed to be about a domain $|a|$. Then the language of T will include only formulas in $V^{-1}(|a|)$. Note that just because T is a theory and is formulated over $|a|$, the set of propositions represented by T may not be a fusion filter over $|a|$. It is for this reason that I turn to theories in the syntactic sense. For consider two theorics T_1 and T_2 that are inconsistent with one another. There is some formula $A \in T_1$ such that $\neg A \in T_2$. $T_1 \circ T_2$ need not contain all of $V^{-1}(|V(A)|)$. It does contain all of the formulas B such that it is provable in the logic that $(A \circ \neg A) \leftrightarrow (B \circ \neg B)$, but, as I have said, in general there are relatively few such formulas B that are substantively different in content from A or $\neg A$.

After applying one syntactic theory to another, scientists need to extract those results in which they are interested. The collection of those results can then be given an interpretation in terms of fusion filter on a domain $|a|$ of D (or perhaps on D as a whole), if this collection is consistent from the perspective of $|a|$ (and does not contradict T_D).

In some cases, however, the inconsistency between T_1 and T_2 will make straightforward application of them to one another useless. To use again an example mention briefly earlier, if T_1 contains a value n for a particular parameter α and T_2 contains a different value m (and the logic is extended to include as a theorem, $\alpha = n \rightarrow (\alpha = m \rightarrow n = m)$), then $T_1 \circ T_2$ contains $n = m$, which is bad enough, but if the logic also contains the laws of subtraction and addition, then $T_1 \circ T_2$ will contain $p = q$ for all numbers p and q. What needs to be done in such circumstances is to "chunk" T_1, T_2, or both of them (Brown and Priest 2004, 2015). This means that one or both of the theories need to broken down into subtheories before application can usefully take place such that the application of a salient subtheory of T_1 to a subtheory of T_2 produces a useful result.

The movement back and forth between the consideration of theories syntactically and semantics, together with heuristics governing the choice of results to extract from the application of theories to one another allows the productive use of theories that are adopted but neither believed nor accepted.[8]

10.8 Towards a Theory of RDS Revision

One of the virtues of having a algebraic structure like an RDS in which theories are interpreted is that this structure can be updated or otherwise revised in response to the belief or acceptance of new theories. Revising an RDS, **D** is to revise T_D. The integration of a new belief, on the view that I have presented here, amounts to adding a new proposition to T_D. But to add any new proposition to T_D is to add a new implication to T_D (since every proposition in RMI entails an implication). Thus, adding a new proposition to T_D alters the partial order on D and the whole RDS is changed.

Formulating an update (the addition of a proposition to T_D) is quite easy. If we start with an RDS, $\mathbf{D} =<< D, \oplus, \leq, ' >, T_D >$, we can construct a new RDS, $\mathbf{D}^* =<< D^*, \oplus^*, \leq^*, ' >, T_{D^*} >$, where T_{D^*} contains the new belief, a. I begin by taking the smallest fusion filter extending T_D and containing a. Let's call this filter, T^*. Then, for each $b \in D$, I define $[b]$ as the set of propositions c such that $b \Rightarrow c$ and $c \Rightarrow b$ are both in T^*. Now we can define the revised RDS:

- $D^* = \{[a] : a \in D\}$;
- $\leq^* = \{< [a], [b] >: a \Rightarrow b \in T^*\}$;

[8] As one referee helpfully pointed out, in real life theories quantifiers are used. I would be interesting to see what effect the addition of quantifiers has to the theory of relevance domains. My idea is this: extend RDS using Halmos's theory of polyadic algebras (Halmos 1962). *I think* relevance domains would remain and would look like little classical polyadic algebras. But I do not have a proof of this yet.

- $[a]' = [a']$;
- $[a] \oplus^* [b] = [a \oplus b]$;
- $T_{D^*} = \{[a] : a \in T^*\}$.

It is easy (but rather tedious) to show that $\mathbf{D}^*$ is an RDS.

Updates of this kind need to be done with care. Adding a proposition to T_D can make accepted theories inconsistent with T_D, which we have seen can cause serious difficulties. Perhaps adding an AGM-style entrenchment relation on accepted theories would help one judge whether adding new beliefs is worthwhile.

Defining belief contraction on an RDS can be much more difficult than defining updates. Suppose that new scientific theories that community wants to accept requires that there be two distinct propositions that are conflated on the existing semantics. Two statements, in this case, say the same thing on the existing RDS, but the new theories require that they be distinguished from one another.[9] In such cases, the old RDS has to be replaced with a new one that contains more propositions in its carrier set.

Let's say that the proposition a needs to be replaced by b and c. Then we first construct an RDS-like structure in which $\leq$ is weakened to a reflexive and transitive relation, $\precsim$. The definition of a relevance domain, and other definitions, use $a \approx b$ (i.e. $d \precsim e \wedge e \precsim d$) instead of $a = b$. Now T_D can be contracted, using some form of AGM-style contraction, to remove either $b \Rightarrow c$ or $c \Rightarrow b$ and then a new RDS can be constructed using the definitions given above for updating RDS.

A revision operator could be defined in terms of sequences of contractions and updates. An idealising assumption that we could add is to restrict T_D to be consistent. If that is so, then we could ban all updates that result in an inconsistent truth set and treat revision in terms of the so-called Levi identity according to which the revision of T_D by a is the same as the contraction of T_D by a' and the update of the result by a.

10.9 Summing Up

In this paper I have suggested that relevant disjunction structures (RDS) can be used formally to represent the dappled world – a world in which precise scientific theories characterise distinct domains of the world. Each domain – a relevance domain in the RDS used as a framework – is a boolean algebra and each local theory is a filter on one such algebra. General scientific theories cover all domains. I give a formal characterisation of the two positions that Nancy Cartwright outlines with regard to general and local theories. Foundationalists think that local theories are straightforwardly derivable from general theories. I represent this relation by a simple

[9] The opposite has occurred several times in science. For example, whereas Newton's theory distinguishes conceptually between gravitational and inertial mass, Einstein's theory takes them to be the same. One could imagine, however, a future physical theory that once again distinguishes between them.

subset relation. Pluarlists, on the other hand, thinks that general theories contain only ceteris paribus laws, which I characterise using the fusion of general theories and ideal circumstances.

I then turn to problems concerning the use of theories that either contain inconsistencies or the application of theories to one another that are inconsistent with each other. It is shown that a straightforward semantic approach is inadequate. The fact that RDS are partitioned into boolean algebras entails that various forms of explosion are possible when dealing with inconsistencies. Instead, I suggest that inconsistent theories be dealt with syntactically in order to derive the desired results and that when these results are obtained the results (not the theories as a whole) be integrated somehow into RDS.

References

Anderson, A., & Belnap, N. D. (1975). *Entailment: Logic of Relevance and Necessity* (Vol. I). Princeton: Princeton University Press.

Avron, A. (1990). Relevance and paraconsistency - a new approach. *Journal of Symbolic Logic, 55*, 707–732.

Avron, A. (1990). Relevance and paraconsistency - a new approach, part II: The formal systems. *Notre Dame Journal of Formal Logic, 31*, 169–202.

Avron, A. (1991). Relevance and paraconsistency - a new approach, part III: Cut-free gentzen-type systems. *Notre Dame Journal of Formal Logic, 32*, 147–160.

Batens, D. (2017). Scientific problem solving: Why inconsistency is no big deal. *Humana Mente, 10*, 149–177.

Boyer, C. B. (1959). *The History of the Calculus and its Conceptual Development*. New York: Dover.

Brown, B. (1992). The old quantum theory: A paraconsistent approach. *PSA: Proceedings of the Biennial Meeting of the Philosophy of Science Association, 1992*, 397–411.

Brown, B., & Priest, G. (2004). Chunk and permeate: An inconsistent inference strategy. Part I: The infinitesimal calculus. *Journal of Philosophical Logic, 33*, 379–388.

Brown, B., & Priest, G. (2015). Chunk and permeate II: Bohr's hydrogen atom. *European Journal for the Philosophy of Science, 5*, 297–314.

Brown, J. R. (1993). *The Laboratory of the Mind: Thought Experiments in the Natural Sciences*. London: Routledge.

Carnielli, W., & Coniglio, M. E. (2015). *Paraconsistent Logic: Consistency, Contradiction, and Negation*. Dordrecht: Springer.

Cartwright, N. (1983). *How the Laws of Physics Lie*. Oxford: Oxford University Press.

Cartwright, N. (1999). *The Dappled World: A Study of the Boundaries of Science*. Cambridge: Cambridge University Press.

Da Costa, N. C. A., & French, S. (2003). *Science and Partial Truth: A Unitary Approach to Models and Scientific Reasoning*. Oxford: Oxford University Press.

Cresswell, M. J. (2012). Mathematical entities in the divided line. *The Review of Metaphysics, 66*, 89–104.

Fine, K. (1974). Models for entailment. *Journal of Philosophical Logic, 3*, 347–372.

Galilei, G. (1933). *Dialogues Concerning Two New Sciences*. New York: McGraw-Hill.

Halmos, P. (1962). *Algebraic Logic*. New York: Chelsea.

Kitcher, P. (1981). Explanatory unification. *Philosophy of Science, 48*, 507–531.

Mares, E., & Fuhrmann, A. (1995). A relevant theory of conditionals. *Journal of Philosophical Logic*, *24*, 645–665.

Norton, J. D. (1999). The cosmological woes of Newtonian gravitational theory. In H. Goenner, J. Renn, J. Ritter, & T. Sauer (Eds.), *The Expanding Worlds of General Relativity* (pp. 271–323). Basil: Birkhauser.

Read, S. (1988). *Relevant Logic: A Philosophical Interpretation of Inference*. Oxford: Basil Blackwell.

Russell, B. (1903). *Principles of Mathematics*. New York: W.W. Norton and Company.

van Fraassen, B. C. (1980). *The Scientific Image*. Oxford: Oxford University Press.

van Fraassen, B. C. (1989). *Laws and Symmetry*. Oxford: Oxford University Press.

Vickers, P. (2013). *Understanding Inconsistent Science*. Oxford: Oxford University Press.

Edwin Mares is a Professor in the Philosophy Programme of Victoria University of Wellington, New Zealand. He has written extensively on logic and formal epistemology, including three books: *Relevant Logic: A Philosophical Interpretation* (2004), *A Priori* (2011), and (with Stuart Brock) *Realism and Anti-Realism* (2007).

Chapter 11
Consequence Relations with Real Truth Values

Daniele Mundici

Abstract Syntax and semantics in Łukasiewicz infinite-valued sentential logic Ł are harmonized by revising the Bolzano-Tarski paradigm of "semantic consequence," according to which, θ follows from Θ iff every valuation v that satisfies all formulas in Θ also satisfies θ. For θ to be a consequence of Θ, we also require that any infinitesimal perturbation of v that preserves the truth of all formulas of Θ also preserves the truth of θ. An elementary characterization of Łukasiewicz implication shows that the Łukasiewicz axiom $((X \to Y) \to Y) \to ((Y \to X) \to X)$ guarantees the continuity and the piecewise linearity of the implication operation $\to$, an appropriate fault-tolerance property of any logic of [0, 1]-valued observables. The directional derivability of the functions coded by all $\psi \in \Theta$ and by θ then provides a quantitative formulation of our refinement of Bolzano-Tarski consequence, which turns out to coincide with the time-honored syntactic Ł-consequence.

Keywords Łukasiewicz logic · Łukasiewicz calculus · Wajsberg algebra · MV-algebra · Łukasiewicz axioms · Łukasiewicz implication · [0, 1]-valued logic · [0, 1]-valued observable · Modus Ponens · Consequentia Mirabilis · Differential semantics · Stable semantics

2010 Mathematics Subject Classification Primary: 06D35 · Secondary: 03B50

11.1 Introduction

In his paper (Avron 2015), A. Avron investigates a general notion of implication that does not assume the availability of any proof system and thus does not depend on the notion of "use" of a formula in a given proof—a notion typically occurring in systems of relevance logic. Avron's generalized implication, called "semi-implication,"

D. Mundici (✉)
Department of Mathematics and Computer Science "Ulisse Dini", University of Florence, Viale Morgagni 67/A, I-50134 Florence, Italy
e-mail: daniele.mundici@unifi.it

O. Arieli and A. Zamansky (eds.), *Arnon Avron on Semantics and Proof Theory of Non-Classical Logics*, Outstanding Contributions to Logic 21,
https://doi.org/10.1007/978-3-030-71258-7_11

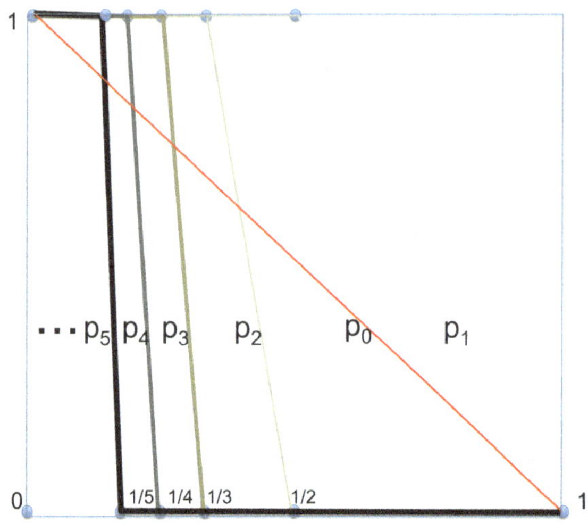

Fig. 11.1 The functions $p_n\colon [0,1] \to [0,1]$, $n = 0, 1, 2, \ldots$, where $p_0(x) = 1 - x$, and $p_n(x) = \min(1, \max(0, -n(n+1)x + n + 1))$ for every $n = 1, 2, 3, \ldots$. Each p_n is continuous, piecewise linear, and each linear piece of p_n agrees with a polynomial with integer coefficients. For short, p_n is a one-variable *McNaughton function*

hinges on a weak form of the classical-intuitionistic deduction theorem, called the "relevant deduction property" (RDP). A binary connective $\to$ of a logic L is a *semi-implication* for L if it has the RDP and there are formulas ϕ, ψ in L such that $\phi \to \psi$ is provable in L but $\psi \to \phi$ is not. It is shown that a finitary logic L has a semi-implication iff L has a strongly sound and complete Hilbert-type system which is an extension by axiom schemas of the standard Hilbert-type system for the implicational fragment of relevance logic R. Minimal logics with RDP are characterized in Avron (2015).

The notions of "(always sentential) logic," "implication," "consequence," "proof" stem from the Polish tradition (Tarski 1936; Rasiowa 1974; Wójcicki 1988) and its developments. In particular, the Bolzano-Tarski paradigm of semantic consequence $\models_L$, (see Tarski 1936, footnote on p. 417), states that for any formula θ and set Θ of formulas in a logic L,

$$\Theta \models_L \theta \text{ iff every model of } \Theta \text{ is also a model of } \theta. \tag{11.1}$$

While the notion of a (tarskian) model is perfectly clear in first-order logic, for sentential logics we may reformulate (11.1) as follows:

$$\Theta \models_L \theta \text{ iff } v(\theta) = 1 \text{ for every valuation } v \text{ such that } v(\phi) = 1 \text{ for all } \phi \in \Theta. \tag{11.2}$$

In the present paper, the methodological approach of Avron (2015) to implication and consequence in relevance logics is taken as a template to our approach to implication and consequence in [0, 1]-valued Łukasiewicz logic Ł. The logic Ł, of course, falls outside the scope of Avron (2015). Suffice to say that Łukasiewicz implication is not an implication in the sense of Avron (2015), if only because the usual deduction theorem fails in Ł, (Mundici 2011, Corollary 1.9).

As an illustration of the inadequacy of definition (11.2) in Ł, for each $n = 0, 1, 2, \ldots$ let p_n be the function in Fig. 11.1. Since p_n is continuous and piecewise linear with integer coefficients, McNaughton's representation theorem (Cignoli et al. 2000, Corollary 3.2.8) yields a formula ψ_n coding p_n in Ł. In particular, the function $p_0(x) = 1 - x$ is coded by the formula $\neg X$. Let $\Psi = \{\psi_1, \psi_2, \ldots\}$. According to formulation (11.2) of the Bolzano-Tarski paradigm, $\neg X$ is a semantic consequence of Ψ but is not a semantic consequence of any finite subset of Ψ. Thus, if one sticks to this formulation, Ł fails to be a finitary logic, despite its syntactic consequence relation is finitary, (Cignoli et al. 2000, §4; Mundici 2011, Definition 1.8, Corollary 1.9).

To overcome this incompleteness phenomenon, definition (11.2) must be tailored to Łukasiewicz logic. To this purpose, we first observe that valuations in the traditional sense amount to taking quotients by *maximal* ideals (in the sense of Lemma 11.3.5), or maximal implicative filters (defined in Cignoli et al. 2000, 4.2.6 as the dual counterparts of maximal ideals). Building on Mundici (2015), in Definition 11.4.1 and Construction 11.4.2, we will construct a semantics for Ł in terms of quotients by *prime* ideals (or their dually defined prime implicative filters). By Proposition 11.4.3, the resulting "prime" valuations have a geometric counterpart, hinging on the differentiability properties of the McNaughton function $\hat{\phi}$ coded by any formula ϕ in Łukasiewicz logic.[1]

Specifically, let Θ be a set of formulas and θ a formula. Using the directional derivability of $\hat{\psi}$, $(\psi \in \Theta)$ and $\hat{\theta}$ as $[0, 1]$-valued functions defined on the valuation space $[0, 1]^{\kappa}$, we say that θ is a *stable consequence* of Θ if whenever an infinitesimal perturbation dv preserves the truth of all formulas ψ in Θ, (in the sense that $(v + dv)(\psi) = 1$), then dv also preserves the truth of θ.

Turning to our example, for each $i = 1, 2, \ldots$, the McNaughton function p_i has the constant value 1 on an open right neighborhood $\mathcal{N}_i \ni 0$. Since $\bigcap_j \mathcal{N}_j = \{0\}$ and p_0 attains value 1 only at 0, then $\neg X$ a consequence of Ψ according to (11.2), but *not* a stable consequence of Ψ, because

$$0 = \frac{\partial p_1}{\partial x^+}(0) = \frac{\partial p_2}{\partial x^+}(0) = \frac{\partial p_3}{\partial x^+}(0) \ldots, \text{ but } \frac{\partial p_0}{\partial x^+}(0) = -1.$$

Remarkably, $\neg X$ also fails to be a syntactic consequence of Ψ, i.e., $\neg X$ cannot be obtained from finitely many applications of Modus Ponens to formulas in Ψ and substitution instances of the Łukasiewicz axioms.

Theorem 11.4.4 shows that in Łukasiewicz logic stable consequence coincides with syntactic consequence. Our valuations as quotients by prime ideals (equivalently, quotients by prime implicative filters) have a *quantitative* content, given by the directional derivatives of the McNaughton functions coded by formulas in Ł. The usual valuations in (11.2) are just valuations of order 0.

---

[1] McNaughton functions stand to Łukasiewicz logic as boolean functions stand to boolean propositional logic.

As a warm-up to consequence relations in general [0, 1]-valued logics, we conclude this section with the following logic-free result, essentially pertaining to college mathematics.

Theorem 11.1.1 *Suppose* $\to$ *is a continuous* $[0, 1]$*-valued function defined on the unit real square* $[0, 1]^2$*, having the following properties:*

$x \to (y \to z) = y \to (x \to z)$, *and*
$x \to y = 1$ *iff* $x \leq y$.

(i) Then upon setting $\neg x = x \to 0$, the algebra $W = ([0, 1], 1, \neg, \to)$ satisfies the following equations:

$$1 \to x = x$$
$$(x \to y) \to ((y \to z) \to (x \to z)) = 1$$
$$((x \to y) \to y) = ((y \to x) \to x)$$
$$(\neg x \to \neg y) \to (y \to x) = 1.$$

(ii) As a consequence, there is a unique one-one order-preserving bijection $\phi\colon [0, 1] \to [0, 1]$ such that for all $x, y \in [0, 1]$

$$x \to y = \phi^{-1}(\min(1, 1 - \phi(x) + \phi(y))). \tag{11.3}$$

Proof (i) has a tedious but straightforward proof. See the present author's paper: https://doi.org/10.1017/jsl.2020.74. Then (ii) follows from (i) as an exercise in first year calculus, using the continuity and monotonicity properties of $\to$. An alternative proof of (ii) independent of (i) can be found in Baczyński and Balasubramaniam (2008, p. 65, Theorem 2.4.20, and references therein) using t-norm theory and other less elementary tools. □

11.2 Syntax and Semantics of Sentential Logics

A reformulation of Theorem 11.1.1 will be given in Theorem 11.3.2. To fit Theorem 11.1.1 into the framework of non-classical logics, let us briefly consider the hendiadys syntax/semantics in the time-honored (Polish style) approach to a logic L.

Syntax. One is given an unlimited supply of sentential variables $X_1, X_2, \ldots$ and a set of connectives and constant symbols. The set FORM_n of formulas $\psi(X_1, \ldots, X_n)$ is then defined by induction on the number of connectives in a formula. A certain (usually, Turing computable) set of formulas is called the set of *syntactic L-tautologies*. Next, for Θ a set of formulas and θ a formula, one says that θ is a *syntactic L-consequence* of Θ, in symbols, $\Theta \vdash_L \theta$, if θ is obtainable from the syntactic L-tautologies and the formulas of Θ by some specific algorithmic manipulation, typically ensuring that $\vdash_L$ is closed under substitutions. A main nontrivial feature of syntax is the "unique readability" of each formula ϕ, in the sense that ϕ is uniquely decomposable into its immediate subformulas. This allows one to define and argue by induction on the number of connectives of ϕ.

Semantics. One is given a *matrix*, i.e., an algebra $\mathcal{M}$ over a universe whose elements are called "truth values," containing a subset $\mathcal{D}$ of truth values designated to stand for "true." For simplicity, let us assume that $\mathcal{D}$ is a singleton. It is, of course, assumed that formulas have connectives and constant symbols in correspondence with the operations and the distinguished elements in $\mathcal{M}$. The semantics of the logic $L = L_{\mathcal{M}}$ begins with the definition of a "(truth) valuation" over n variables $X_1, \ldots, X_n$, i.e., an arbitrary map $v\colon \{X_1, \ldots, X_n\} \to \mathcal{M}$. The "absolute freeness" of FORM_n, together with the unique readability of each formula, ensures that v uniquely extends to a homomorphism, also denoted v, from FORM_n into $\mathcal{M}$. This is the "truth-functionality" property of the logic L that distinguishes logic from probability, where syntax hardly has any role. A *semantic tautology* $\tau(X_1, \ldots, X_n)$ of L is a formula whose value is "true" for all homomorphisms $v\colon \mathsf{FORM}_n \to \mathcal{M}$. The semantics of L culminates with the definition of semantic L-consequence, usually following formulation (11.2).

Completeness. One now hopes to prove that semantic and syntactic L-tautologies coincide. A more challenging, no less important task is to prove that syntactic and semantic L-consequence coincide, in agreement with the perception of L as a *calculus ratiocinator-cum-semantica*.

Motivation. Given the mushrooming plethora of logics on the market, it helps if some motivation is given for the newcomer L.

Example 11.2.1 Let us consider boolean logic L_{boole}.
Syntax. A formula ϕ is a syntactic tautology of L_{boole} if can be derived by substitution and Modus Ponens from the following axioms (called the *basic syntactic tautologies*):

$$\boxed{\begin{array}{l} A \to (B \to A) \\ (A \to B) \to ((B \to C) \to (A \to C)) \\ ((A \to B) \to B) \to ((B \to A) \to A) \\ (\neg B \to \neg A) \to (A \to B) \\ (\neg A \to A) \to A \ \ (\textit{Consequentia Mirabilis}). \end{array}}$$

A formula θ is a syntactic consequence of a set Θ of formulas iff θ is derivable from the syntactic tautologies and Θ by a finite number of applications of Modus Ponens.

Semantics. The algebra $\mathcal{M} = \mathcal{M}_{\text{boole}}$ in boolean logic is the two-element set $\{0, 1\}$ equipped with the operation $\neg x = 1 - x$ and the natural lattice operations $\wedge, \vee$. Implication turns out to be the derived operation $x \to y = \neg x \vee y$. The set of operations $\{\neg, \to\}$, as well as $\{\neg, \vee\}$, can express any function from $\mathcal{M}^m$ to $\mathcal{M}$: this is the "functional completeness" of boolean logic based on these operations. So the actual choice of a functionally complete basis of connectives is essentially a matter of aesthetics. That's why, e.g., in the theory of boolean algebras, implication plays second fiddle. By the deduction theorem, "semantic consequence" is definable in terms of "semantic tautology." The resulting notion of consequence agrees with the Bolzano-Tarski paradigm (11.2).

Completeness. The completeness theorem of boolean logic states that syntactic consequence coincides with semantic consequence. In particular, syntactic and semantic tautologies coincide.

Motivation. Boolean logic provides a calculus ratiocinator-cum-semantica complying with the "bivalence principle," according to which every sentence/assessment is either true or false, but not both. There are no other "truth values." This is the quite exceptional status of statements in complete theories in first-order logic. The power of boolean formulas to code yes-no observables is so strong that relatively few basic observations on the state, position, and scanned symbol in a configuration of a nondeterministic Turing machine $\mathcal{T}$ over an input x are efficiently coded by Cook's celebrated formula, $\phi = \phi_{\mathcal{T},x,t,}$ in such a way that for any input x and deadline t, ϕ states

$$\text{“}\mathcal{T} \text{ on input } x \text{ accepts } x \text{ in } t \text{ steps.”} \tag{11.4}$$

More precisely, some homomorphism v satisfies $v(\phi) = 1$ iff (11.4) is actually observed. Thus, the tautology problem of boolean sentential logic is coNP-complete.

11.3 Dropping the "Consequentia Mirabilis" Axiom

Syntax. The set FORM_n of n-variable formulas coincides with the set of boolean n-variable formulas. By dropping the last item in Example 11.2.1, we are left with the following list of basic tautologies of a new (almost centennial) logic Ł, known as *infinite-valued Łukasiewicz sentential logic* (Borkowski and Łukasiewicz 1970, p. 144):

$$\boxed{\begin{array}{l} A \to (B \to A) \\ (A \to B) \to ((B \to C) \to (A \to C)) \\ ((A \to B) \to B) \to ((B \to A) \to A) \\ (\neg B \to \neg A) \to (A \to B). \end{array}}$$

The additional axiom

$$((A \to B) \to (B \to A)) \to (B \to A)$$

occurring in Łukasiewicz's original list was proved to follow from these four, by Chang and Meredith, (see Cignoli et al. 2000, p. 102 for bibliographical details). The definition of a *syntactic Ł-tautology* τ is the same as for syntactic tautologies in boolean logic: τ must be obtainable from the above basic syntactic tautologies via substitution and Modus Ponens. This (nontrivially) makes the tautology problem in Ł coNP-complete, just as its boolean fragment, (Cignoli et al. 2000, Theorem 9.3.4). Given a set Θ of formulas and a formula θ, *syntactic Ł-consequence* $\vdash_Ł$ *is defined (as in boolean logic) by*

> $\Theta \vdash_Ł \theta$ iff θ is obtainable from a subset of Θ and the syntactic tautologies of Ł by finitely many applications of Modus Ponens.

Semantics: Tautologies. The semantics of Ł is more delicate than the semantics of boolean logic. We start from the matrix $\mathcal{M}_Ł$ whose universe of truth values is the unit real interval $[0, 1]$, equipped with the operations $x \to y = \min(1, 1 - x + y)$ and $\neg x = 1 - x = x \to 0$. Upon restriction to $\{0, 1\}$ we recover boolean negation and implication. Then, as in the case of boolean logic, the *semantic Ł-tautologies* are those formulas $\phi(X_1, \ldots, X_n)$ which are evaluated to 1 by every homomorphism $v\colon \mathsf{FORM}_n \to \mathcal{M}_Ł$.

Completeness (first part): the Rose-Rosser theorem for tautologies. The following nontrivial result shows that the semantical and the syntactical notion of a tautology agree in Łukasiewicz logic.

Theorem 11.3.1 (Rose-Rosser completeness theorem for tautologies, Rose and Rosser 1958; Cignoli et al. 2000, §4) *semantic Ł-tautologies coincide with syntactic Ł-tautologies.*

Henceforth, we will use the term "tautology" for both syntactic and semantic Ł-tautologies.

Semantics: Consequence. The construction of a semantic Ł-consequence relation beyond the paradigm (11.2) will take the rest of this paper. We first provide some motivation.

Motivation Most observables and random variables in everyday life (as well as in physics) are real-valued. For every bounded observable $\mathcal{O}$, by taking the distance from the maximum and the minimum value as the unit of measure, $\mathcal{O}$ becomes $[0, 1]$-valued and adimensional, like angular amplitude. Thus, no measurement unit is needed. Needless to say, the change of the max/min bounds results in a different observable. (Fahrenheit temperature differs from Celsius temperature.)

Continuity is assumed at the outset, to ensure that the inevitable imprecision of our assessments of the basic observables does not have fatal effects on the assessment of composite (derived) observables. Continuity in physics has the same function, allowing physical laws to be expressed by formulas, notwithstanding that: (i) the result of every measurement is a real number *together with an error estimate*, and (ii) already units of measurement are imprecisely defined: think of the sequence of more and more precise "definitions" of the fundamental physical unit named "meter."

Our study of [0, 1]-observables is not aimed at discovering new physical laws. Rather, generalizing what boolean logic does for {0, 1}-valued observables/statements, we will give a *semantic* method to draw consequences concerning [0, 1]-observables from premises concerning [0, 1]-observables. This will turn out to coincide with the *algorithmic-syntactic* method based on Modus Ponens.

Variables in [0, 1]-logic transform into a truth value $y \in [0, 1]$ our assessments of the bounded *basic* observables of the problem we are considering. As for the coding of nondeterministic computations, the choice of the basic variables is a matter of convenience. Truth functionality then allows one to code by formulas certain *composite* [0, 1]-valued observables. A brute force counting shows that functional completeness fails, whence the choice of the operations in $\mathcal{M}$ is no longer a matter of aesthetics.

While addition and multiplication have a basic role in algebra and analysis, in logic the central status of "consequence" is mostly reflected by an "implication" operation $\rightarrow_{\mathcal{M}}\colon [0, 1]^2 \rightarrow [0, 1]$. If formulas are to code [0, 1]-observables (just as formulas in boolean logic code {0, 1}-observables), then $\rightarrow_{\mathcal{M}}$ must be *continuous*. Since the continuity assumption merely takes care of the fault-tolerance property of observables, in order to capture the bare minimum essentials of implication, let us put forward the following two conditions:

Order Property $x \rightarrow_{\mathcal{M}} y = 1$ iff $x \leq y$,
in accordance with the natural order structure of [0, 1].

Exchange Property $x \rightarrow_{\mathcal{M}} (y \rightarrow_{\mathcal{M}} z)$ iff $y \rightarrow_{\mathcal{M}} (x \rightarrow_{\mathcal{M}} z)$,
making the order of appearance of x and y irrelevant for the conclusion z.

The following restatement of Theorem 11.1.1 shows that the four equations therein are an algebraic counterpart of the basic tautologies of Łukasiewicz logic Ł.

Theorem 11.3.2 *Suppose the continuous function* $\rightarrow\colon [0, 1]^2 \rightarrow [0, 1]$ *has the order and the exchange properties. Let* $\neg x = x \rightarrow 0$. *Then there is a unique isomorphism of the algebra* $W = ([0, 1], 1, \neg, \rightarrow)$ *onto the* standard Wajsberg algebra, $W_{Ł} = ([0, 1], 1, \neg_{xŁ}, \rightarrow_{Ł})$, *where* $\neg_{Ł} x = 1 - x$ *and* $x \rightarrow_{Ł} y = \min(1, 1 - x + y)$ *is* Łukasiewicz implication.

We refer to Borkowski and Łukasiewicz (1970, pp.129–130) for historical information on Łukasiewicz implication $\rightarrow_{Ł}$. Wajsberg algebras are the algebras satisfying the four equations in Theorem 11.1.1(i). Equivalently, they are the Lindenbaum algebras of Łukasiewicz logic based on the implication connective, (Cignoli et al. 2000, §4.4).

Theorem 11.3.2 sheds new light on the meaning of the Łukasiewicz basic tautologies listed at the outset of this section, notably the intriguing tautology

$$((A \rightarrow B) \rightarrow B) \rightarrow ((B \rightarrow A) \rightarrow A).$$

As a matter of fact, given a [0, 1]-valued function $\to$ on $[0, 1]^2$ having the order and the exchange property, *failure of this tautology entails failure of the continuity property of the implication operation* $\to$.

The main properties of Wajsberg algebras needed in the sequel are summarized in the following result.

Lemma 11.3.3 *Let* $\text{card}(\mathcal{X})$ *denote the cardinality of a set* $\mathcal{X} \neq \emptyset$ *of variables. The set* $\mathsf{V}(\mathsf{FORM}(\mathcal{X}))$ *of valuations of* $\mathsf{FORM}(\mathcal{X})$ *into the* standard Wajsberg algebra $W_Ł$ *can be identified with its homeomorphic copy* $\mathsf{V}(\mathcal{X})$*, consisting of all valuations of the variables in* $\mathcal{X}$*. In symbols,*

$$\mathsf{V}(\mathsf{FORM}(\mathcal{X})) = \mathsf{V}(\mathcal{X}) = [0, 1]^{\text{card}(\mathcal{X})}\,. \tag{11.5}$$

In this way, $\mathsf{V}(\mathsf{FORM}(\mathcal{X}))$ *inherits the connected compact Hausdorff topology of the Tychonoff cube* $[0, 1]^{\text{card}(\mathcal{X})}$*. In view of* (11.5)*, we have the following:*

(i) Every formula $\phi \in \mathsf{FORM}(\mathcal{X})$ *belongs to* FORM_n *for some* $n = 1, 2, \ldots$ *and codes the piecewise linear continuous function* $\hat{\phi}\colon\ [0, 1]^n \to [0, 1]$ *with integer coefficients by the stipulation*

$$\hat{\phi}(v) = v(\phi), \quad v \in \mathsf{V}(\mathsf{FORM}_n). \tag{11.6}$$

As ϕ *ranges over* FORM_n*,* $\hat{\phi}$ *ranges over the set of McNaughton functions on* $[0, 1]^n$.

(ii) For every $\psi \in \mathsf{FORM}_n$*, point* $v \in [0, 1]^n$*, and nonzero vector* $u \in \mathbb{R}^n$ *such that the segment* $[v, v + \epsilon u]$ *is contained in* $[0, 1]^n$ *for some* $\epsilon > 0$*, the directional derivative* $\partial\hat{\psi}(v)/\partial u$ *exists and varies continuously with* u*, once* v *is kept fixed.*

(iii) Let $\mathcal{W}_{\text{card}(\mathcal{X})}$ *denote the free Wajsberg algebra on* $\text{card}(\mathcal{X})$ *many generators. Up to variable renaming,* $\mathcal{W}_{\text{card}(\mathcal{X})}$ *consists of all functions* $f\colon\ [0, 1]^n \subseteq [0, 1]^{\text{card}(\mathcal{X})} \to [0, 1]$, $(n = 1, 2, \ldots,$ *with* n *finite and* $\leq \text{card}(\mathcal{X}))$ *obtained from the coordinate functions* $\pi_i(z_1, \ldots, z_n) = z_i$ *by pointwise application of the operations of negation* $\neg_Ł x = 1 - x$ *and Łukasiewicz implication* $x \to_Ł y = \min(1, 1 - x + y)$*. Thus,* $\mathcal{W}_{\text{card}(\mathcal{X})}$ *coincides with the set of McNaughton functions* $f(Y_1, \ldots, Y_r)$ *with* $\{Y_1, \ldots, Y_r\} \subseteq \mathcal{X}$.

Proof (i) A routine exercise in Łukasiewicz logic (Mundici 2011, 1.2, 1.5, 4.1). (ii) Immediate from (i). (iii) follows combining Chang's completeness theorem (Chang 1959; Cignoli et al. 2000, 3.1.4, 3.6.7) with McNaughton's representation theorem (McNaughton 1951; Cignoli et al. 2000, Theorem 9.1.5). □

Remark 11.3.4 By (11.5) the valuation space of n-variable formulas in Łukasiewicz logic is naturally endowed with the topological-differential structure of the cube $[0, 1]^n$. This allows us to infinitesimally perturb a valuation $v \mapsto v + dv$ and see the effect of this perturbation on any formula $\psi = \psi(X_1, \ldots, X_n)$. As proved in the next section, a complete quantitative account of this effect is given by the value $v(\psi)$ *together with* suitable directional derivatives at v of the McNaughton function $\hat{\psi}$ coded by ψ. Given a set Θ of formulas, the differential analysis of Θ and ψ under perturbations $v \mapsto v + dv$ provides a refinement of the Bolzano-Tarski semantic consequence which turns out to coincide with syntactic consequence.

The following elementary lemma rephrases in algebraic terms for Ł the standard formulation (11.2) of Bolzano-Tarski semantic consequence, only depending on the evaluation of Θ and ψ at v.

Lemma 11.3.5 *For Θ a set of formulas and θ a formula, let $\mathcal{W}$ be a free Wajsberg algebra containing $\hat{\theta}$ as well as the McNaughton function $\hat{\psi}$ for each $\psi \in \Theta$. (See Lemma 11.3.3(i) for this notation.) Then the following conditions are equivalent:*

(1) For every maximal implicative filter $\mathfrak{f}$ of $\mathcal{W}$, if $\hat{\psi} \in \mathfrak{f}$ for all $\psi \in \Theta$ then $\hat{\theta} \in \mathfrak{f}$.
(2) Every maximal ideal $\mathfrak{m}$ of $\mathcal{W}$ with $\hat{\psi}/\mathfrak{m} = 1$ for all $\psi \in \Theta$ satisfies $\hat{\theta}/\mathfrak{m} = 1$.
(3) θ is a Bolzano-Tarski semantic consequence of Θ in Ł in the sense of (11.2).

Proof Routine. See, e.g., Mundici (2011, Theorem 4.16), which uses the equivalent language of MV-algebras and their ideals.[2] □

11.4 The Differential Semantics of Łukasiewicz Infinite-Valued Logic

In any ring R with unit 1_R, one has two equivalent definitions of the ideal $I(S)$ generated by a subset S of R:

Internal $I(S) = \{r_1x_1s_1 + \cdots + r_kx_ks_k \mid k \in \mathbb{N};\; r_i, s_i \in R;\; x_i \in S\}$.
External $t \in I(S)$ iff t belongs to every ideal of R containing S.

The external definition of $t \in I(S)$ has the following equivalent reformulation:

$t/J = 0$ for every ideal J of R such that $x/J = 0$ for all $x \in S$.

Every ideal of R is the intersection of the irreducible ideals containing it. Thus

$t \in I(S)$ iff any irreducible ideal J that gives value 0 each $x \in S$ (taking quotients by J), also gives value 0 to t.

Modulo the inessential transformation $x \mapsto 1_R - x$, this definition of $I(S)$ is reminiscent the notion of "θ belongs to the set $D(\Theta)$ of semantic consequences of a set

[2] For implicative filters of Wajsberg algebras, we refer to Cignoli et al. (2000, §4.2). By an "ideal" of a Wajsberg algebra we mean an ideal of its corresponding MV-algebra, as defined in Cignoli et al. (2000, §1.2).

Θ of formulas" arising from definition (11.2). On the other hand, the internal definition parallels the traditional method of deriving θ by suitable syntactic-algorithmic manipulations on $\Theta \cup$ the tautologies.

In boolean rings, maximal and irreducible ideals coincide. Further, every maximal quotient of a boolean ring B is just the two-element boolean ring $\{0, 1\}$, whence the quotient operation amounts to assigning either value 0 or 1 to elements of B—as befits the simplest possible non-constant evaluation of the elements of B.

Differently from maximal ideals, prime ideals of the algebras of Łukasiewicz logic have the irreducibility property, (Cignoli et al. 2000, Corollary 1.2.14). A natural notion of semantic consequence will be obtained in this section by realizing "valuations" as prime quotients of these algebras: Indeed all these quotients have a quantitative description in terms of directional derivatives.

Our construction is inspired by the following quotation from C.C. Chang, (Chang 1998, pp. 5–6):

> My failure to prove the completeness in Chang (1958) using MV-algebras was a disappointment to me at that time. I tried that year and even after I left Cornell, but with no success. My mistake was in trying to pound the thing out by sticking to maximal ideals. [...] But a lucky break occurred when Dana Scott realized, with far-reaching insight, that there is a notion of prime ideals in MV-algebras (a notion I had not considered until then).

Mimicking what Chang did to obtain his celebrated proof in Chang (1959) that the equational class of MV-algebras is generated by the standard MV-algebra, let us first replace maximal ideals by prime ideals in Lemma 11.3.5(1).

Definition 11.4.1 For Θ, a set of formulas and θ a formula, let $\mathcal{W}$ be a free Wajsberg algebra containing $\hat{\theta}$ as well as the McNaughton function $\hat{\psi}$ for each $\psi \in \Theta$. We say that θ is a *stable consequence* of Θ if for every prime ideal $\mathfrak{p}$ of $\mathcal{W}$ such that $\hat{\psi}/\mathfrak{p} = 1$ (i.e., $\neg\hat{\psi} \in \mathfrak{p}$) for all $\psi \in \Theta$, we also have $\hat{\theta}/\mathfrak{p} = 1$ (i.e., $\neg\hat{\theta} \in \mathfrak{p}$).

Stated otherwise, for every *prime* implicative filter $\mathfrak{f}$ of $\mathcal{W}$ such that $\hat{\psi} \in \mathfrak{f}$ for all $\psi \in \Theta$, we also have $\hat{\theta} \in \mathfrak{f}$.

While, by Lemma 11.3.5, taking the quotient of $f \in \mathcal{W}$ by a maximal ideal $\mathfrak{m}$ amounts to evaluating f at the only point $x_{\mathfrak{m}} \in \bigcap\{g^{-1}(0) \mid g \in \mathfrak{m}\}$, taking quotients by prime non-maximal ideals *prima facie* does not have a similarly appealing quantitative counterpart. However, as will be seen in Proposition 11.4.3, every prime non-maximal valuation u actually computes suitable directional derivatives of f at $x_{\mathfrak{m}}$. This quantitative and differential geometric content of u makes the quotient operation $\phi \mapsto \hat{\phi}/\mathfrak{p}_u$ a genuine semantic notion, no less significant than the usual pointwise valuation $\phi \mapsto \hat{\phi}/\mathfrak{m}$.

Construction 11.4.2 (Mundici 2015) For $n = 1, 2, \ldots$ and $0 \leq t \leq n$ let $u = (u_0, u_1, \ldots, u_t)$ be a $(t+1)$-tuple of elements of $\mathbb{R}^n$ where $u_1, \ldots, u_t$ are linearly independent vectors, and $u_0 \in [0, 1]^n$. For each $m = 1, 2, \ldots$, denoting by $\mathrm{conv}(Z)$ the convex hull of $Z \subseteq \mathbb{R}^n$, let

$$T_{u,m} = \mathrm{conv}(u_0, u_0 + u_1/m,\ u_0 + u_1/m + u_2/m^2, \ldots, u_0 + u_1/m + \cdots + u_t/m^t).$$

Any such u is said to be a *differential (or "prime") valuation of order t, in $\mathbb{R}^n$* if there is an integer $k > 0$ such that for all $m \geq k$ the n-cube $[0, 1]^n$ contains $T_{u,m}$. When this is the case, the subset $\mathfrak{p}_u$ of the free n-generator Wajsberg algebra $\mathcal{W}_n$ is defined by

$$\mathfrak{p}_u = \{f \in \mathcal{W}_n \mid f^{-1}(0) \supseteq T_{u,m} \text{ for some } m = 1, 2, \ldots\}.$$

Then $\mathfrak{p}_u$ is a prime ideal of $\mathcal{W}_n$.

For any formula $\psi(X_1, \ldots, X_n)$ and differential valuation $u = (u_0, u_1, \ldots, u_t)$ in $\mathbb{R}^n$, we then make the following stipulation:

u satisfies ψ means $\neg\hat{\psi} \in \mathfrak{p}_u$ (equivalently, $\hat{\psi}/\mathfrak{p}_u = 1$).

Further, for any $\Theta \subseteq \mathsf{FORM}_n$ and $\theta \in \mathsf{FORM}_n$, we write

$\Theta \models_\partial \theta$ if θ is satisfied by every differential valuation that satisfies each $\psi \in \Theta$.

By (11.5) and Lemma 11.3.5, valuations in the usual pointwise sense coincide with differential valuations of order 0. Readers familiar with ring theory will recognize the order of a differential valuation as (the counterpart for Wajsberg algebras of) the Krull depth of the prime ideal corresponding to it.

In view of Definition 11.4.1, the deep connection between the geometric and the algebraic content of differential (= prime) valuations is the subject matter of the following result.

Proposition 11.4.3 *Let* $u = (u_0, u_1, \ldots, u_t)$ *be a differential valuation in* $\mathbb{R}^n$*. Then* $\mathfrak{p}_{(u_0)} \supseteq \mathfrak{p}_{(u_0,u_1)} \supseteq \cdots \supseteq \mathfrak{p}_{(u_0,u_1,\ldots,u_{t-1})} \supseteq \mathfrak{p}_{(u_0,u_1,\ldots,u_t)}$*. Further:*

$\mathfrak{p}_{(u_0)}$ *coincides with the maximal ideal of* $\mathcal{W}_n$ *given by all functions of* $\mathcal{W}_n$ *that vanish at* u_0*, (*$\mathcal{W}_n$ *being the free n-generator Wajsberg algebra).*

$\mathfrak{p}_{(u_0,u_1)}$ *coincides with the prime ideal of* $\mathcal{W}_n$ *given by all functions vanishing over some interval of the form* $\mathrm{conv}(u_0, u_0 + u_1/m)$ *for some integer* $m > 0$*. Equivalently, by Lemma 11.3.3(ii),* $f(u_0) = 0$ *and* $\partial f(u_0)/\partial u_1 = 0$.

$\mathfrak{p}_{(u_0,u_1,u_2)}$ *coincides with the prime ideal of* $\mathcal{W}_n$ *given by those* f *such that for some integer* $m > 0$, f *vanishes on the segment* $\mathrm{conv}(u_0, u_0 + u_1/m)$*, and* $\partial f(y)/\partial u_2 = 0$ *for every* y *in the relative interior of the segment* $\mathrm{conv}(u_0, u_0 + u_1/m)$.

Inductively, $\mathfrak{p}_{(u_0,u_1,\ldots,u_t)}$ *equals the prime ideal of* $\mathcal{W}_n$ *consisting of all* f *such that for some integer* $m > 0$, f *vanishes on the* $(t-1)$*-simplex S defined by*

$$S = \mathrm{conv}\left(u_0, u_0 + u_1/m, u_0 + u_1/m + u_2/m^2, \ldots, u_0 + u_1/m + \cdots + u_{t-1}/m^{t-1}\right),$$

and $\partial f(y)/\partial u_t = 0$ *for every* y *in the relative interior of S.*

Proof See Mundici (2015). □

Completeness (second part): syntactic Ł-consequence = stable consequence. The following completeness theorem highlights the finitary character of the stable consequence relation $\models_\partial$ for Łukasiewicz logic.

Theorem 11.4.4 *Fix $n = 1, 2, \ldots,$ $\Theta \subseteq \mathsf{FORM}_n$ and $\theta \in \mathsf{FORM}_n$. Then $\Theta \models_\partial \theta$ iff θ is a syntactic Ł-consequence of Θ iff θ is a stable consequence of Θ. It follows that $\Theta \models_\partial \theta$ iff $\{\theta_1, \ldots, \theta_k\} \models_\partial \theta$ for some finite subset $\{\theta_1, \ldots, \theta_k\}$ of Θ.*

Proof The proof rests on the following two main facts:

Irreducibility Every ideal of a Wajsberg algebra coincides with the intersection of all prime ideals containing it. This follows from the Subdirect Representation Theorem, (Cignoli et al. 2000, Corollary 1.2.14).

Differential geometric interpretation Ideals of the form $\mathfrak{p}_u$ exhaust all possible prime ideals of the free Wajsberg algebra $\mathcal{W}_n$, (Busaniche and Mundici 2007, Corollary 2.18).

See Mundici (2015) for details. □

The generalization of stable consequence to sets $\Theta \subseteq \mathsf{FORM}(\mathcal{X})$ for arbitrary sets $\mathcal{X}$ of variables is straightforward, (Mundici 2015).

The case of semisimple algebras. A Wajsberg algebra is said to be *semisimple* if the intersection of its maximal ideals is the zero ideal. By Cignoli et al. (2000, Theorem 4.6.6), the Bolzano-Tarski semantic consequences (11.2) of a set $\Theta \subseteq \mathsf{FORM}(\mathcal{X})$ coincide with the stable consequences of Θ iff the Lindenbaum algebra of Θ is semisimple. Since free Wajsberg algebras are semisimple, in the particular case when Θ is the empty set, its stable consequences coincide with semantic tautologies. Thus *Theorem* 11.4.4 *extends Theorem* 11.3.1.

By Cignoli et al. (2000, Theorem 6.3.2), boolean algebras (i.e., idempotent MV-algebras) are hyperarchimedean MV-algebras, whence they are semisimple. So *in the fragment of Łukasiewicz logic given by boolean logic, stable semantics coincides with Bolzano-Tarski semantics.* We also have the following.

Corollary 11.4.5 *For any finite set Θ of formulas, the set of stable consequences of Θ coincides with the set of Bolzano-Tarski consequences of Θ.*

Proof By Hay's theorem (Hay 1963; Mundici 2011, Theorem 3.4.9), the Lindenbaum algebra of Θ is semisimple. □

11.5 Concluding Remarks

As remarked by one of the referees of this paper, the usual algebraic notion associated to consequence in Łukasiewicz logic is in terms of implicative filters. Accordingly, most results on implication and consequence in this paper are stated in the framework of Wajsberg algebras.

This said, just as boolean algebras provide the standard algebraic counterpart of classical propositional logic rather than idempotent Wajsberg algebras, similarly MV-algebras and their ideals are a key tool in the study of Łukasiewicz logic. Thus, in Proposition 11.4.3, building on the analogy between the internal and the external definition of ideals in rings, we have highlighted the role of irreducibility and the differential properties of prime ideals in MV-algebras. As an extra bonus of Elliott's classification, (Effros 1981; Elliott 1976), the associative-commutative structure of MV-algebras provides a direct application of the natural deductive machinery of Łukasiewicz logic to the algorithmic theory of approximately finite-dimensional (AF) C*-algebras whose Murray-von Neumann order of projections is a lattice. See Mundici (1986), Mundici (2018), and Mundici and Panti (1993).

Historically, after the syntactic proof of the completeness theorem for tautologies by Rose and Rosser (1958), Chang was able to give a neat algebraic proof of Theorem 11.3.1 focusing on prime, rather than maximal ideals of MV-algebras. The former, and not the latter, satisfy the subdirect representation theorem ensuring their irreducibility. In the same way, the completeness theorem for stable consequence rests on the wealth of geometric properties of prime ideals in MV-algebras, rather than on the pointwise valuations provided by maximal quotients. Indeed, the Bolzano-Tarski paradigm (11.2) of semantic consequence based on pointwise valuations is insensitive of the stability properties of truth under infinitesimal perturbations in the valuation spaces $[0, 1]^{\kappa}$ of Łukasiewicz logic Ł.

Refining paradigm (11.2), stable consequence stipulates that for a formula θ to be an Ł-consequence of Θ one must also guarantee that the truth of θ be preserved under any infinitesimal perturbation that preserves the truth of each formula ψ in Θ. Semantically, stability under perturbations amounts to the vanishing of all directional derivatives of the piecewise linear functions $\hat{\theta}$ and $\hat{\psi}$ associated to the prime ideals $\mathfrak{p}_u$ introduced in Proposition 11.4.3.

In the same way, as the dynamics of a material point depends on its initial position *and* its initial speed, the deductive closure of a set Θ of formulas in Łukasiewicz logic depends on the pointwise valuations satisfying Θ *and* on the differential valuations of order 1,2, ... satisfying every formula in Θ.

Acknowledgements The author is grateful to both referees of this paper for their competent reading and valuable suggestions for improvement.

References

Avron, A. (2015). Semi-implication: a chapter in universal logic. In A. Koslow & A. Buchsbaum (Eds.), *The Road to Universal Logic: Festschrift for 50th Birthday of Jean-Yves Béziau Volume I* (pp. 59–72). Switzerland: Birkhäuser, Springer International Publishing.

Baczyński, M., & Balasubramaniam, J. (2008). *Fuzzy Implications*. Berlin: Springer.

Borkowski, L., & Łukasiewicz, J. (Ed.). (1970). *Selected Works*. Studies in Logic and the Foundations of Mathematics. Warsaw: NorthHolland, Amsterdam and Polish Scientific Publishers.

Busaniche, M., & Mundici, D. (2007). Geometry of Robinson consistency in Łukasiewicz logic. *Annals of Pure and Applied Logic*, *147*, 1–22.

Chang, C. C. (1958). Algebraic analysis of many-valued logics. *Transactions of the American Mathematical Society*, *88*, 467–490.

Chang, C. C. (1959). A new proof of the completeness of the Łukasiewicz axioms. *Transactions of the American Mathematical Society*, *93*, 74–90.

Chang, C. C. (1998). The writing of the MV-algebras. *Studia Logica*, *61*, 3–6.

Cignoli, R., D'Ottaviano, I. M. L., & Mundici, D. (2000). *Algebraic Foundations of Many-Valued Reasoning* (Vol. 7). Trends in Logic. Dordrecht: Kluwer Academic Publishers. Reprinted, Springer Science & Business Media (2013).

Effros, E. G. (1981). *Dimensions and C*-Algebras* (Vol. 46)., CBMS Regional Conference Series in Mathematics Providence, R.I.: American Mathematical Society.

Elliott, G. A. (1976). On the classification of inductive limits of sequences of semisimple finite-dimensional algebras. *Journal of Algebra*, *38*, 29–44.

Hay, L. S. (1963). Axiomatization of the infinite-valued predicate calculus. *Journal of Symbolic Logic*, *28*, 77–86.

McNaughton, R. (1951). A theorem about infinite-valued sentential logic. *Journal of Symbolic Logic*, *16*, 1–13.

Mundici, D. (1986). Interpretation of AF C*-algebras in Łukasiewicz sentential calculus. *Journal of Functional Analysis*, *65*, 15–63.

Mundici, D. (2011). *Advanced Łukasiewicz Calculus and MV-Algebras* (Vol. 35)., Trends in Logic Berlin: Springer.

Mundici, D. (2015). The differential semantics of Łukasiewicz syntactic consequence, Chapter 7. In F. Montagna (Ed.), *Petr Hájek on Mathematical Fuzzy Logic* (Vol. 6, pp. 143–157). Outstanding Contributions. Switzerland: Springer International Publishing. https://doi.org/10.1007/978-3-319-06233-4.

Mundici, D. (2018). Word problems in Elliott monoids. *Advances in Mathematics*, *335*, 343–371.

Mundici, D., & Panti, G. (1993). Extending addition in Elliott's local semigroup. *Journal of Functional Analysis*, *117*, 461–471.

Rasiowa, H. (1974). *An Algebraic Approach to Non-Classical Logics*. Warszawa: PWN - Polish Scientific Publishers.

Rose, A., & Rosser, J. B. (1958). Fragments of many-valued sentential calculus. *Transactions of the American Mathematical Society*, *87*, 1–53.

Tarski, A. (1936). On the concept of logical consequence, Chapter XVI in Tarski (1956).

Tarski, A. (1956). *Logic, Semantics, Metamathematics*. Oxford: Clarendon Press. Reprinted: Hackett, Indianapolis, 1983.

Wójcicki, R. (1988). *Theory of Logical Calculi: Basic Theory of Consequence Operations, Synthese Library* (Vol. 199). Dordrecht: Kluwer.

Daniele Mundici studied Physics at the University of Modena, Italy. After working for 11 years as a junior and high school teacher, he became an Assistant Professor at the University of Florence. From 1987 to 2002, he was a Professor in Computer Science at the University of Milan. In 2002, he became a Professor in Mathematical Logic at the University of Florence. He is a member of the International Academy of Philosophy of Science, AIPS, Bruxelles, and corresponding member of the National Academy of Exact, Physical and Natural Sciences, ANCEFN, Buenos Aires. Former

President of the Kurt Gödel Society, KGS, Vienna, and of the Associazione Italiana di Logica e Applicazioni, AILA. His research work includes many-valued logic, polyhedra, probability, and AF C*-algebras. He is a managing editor of various journals, including the Journal of Algebra and its Applications, and Studia Logica.

Chapter 12
Geometric Rules in Infinitary Logic

Sara Negri

Abstract Large portions of mathematics such as algebra and geometry can be formalized using first-order axiomatizations. In many cases it is even possible to use a very well-behaved class of first-order axioms, namely, what are called *coherent* or *geometric* implications. Such class of axioms can be translated to inference rules that can be added to a sequent calculus while preserving its structural properties. In this work, this fundamental result is extended to their infinitary generalizations as extensions of sequent calculi for both classical and intuitionistic infinitary logic. As an application, a simple proof of the infinitary Barr's theorem without the axioms of choice is shown.

Keywords Geometric axioms · Axioms-as-rules · Infinitary logic · G3 calculi · Barr's theorem

12.1 Introduction

Large portions of mathematics such as algebra and geometry can be formalized using first-order axiomatizations. In some cases it is even possible to use just a very well-behaved class of first-order axioms, namely, what are called *coherent*[1] implications. A *coherent implication* (also known in the literature as a "geometric implication", a "geometric axiom", a "geometric sentence", a "coherent axiom", a "basic geometric sequent", or a "coherent formula") is a first-order sentence that is the universal closure of an implication $D_1 \supset D_2$, where both D_1 and D_2 are positive formulas, i.e., formulas built up from atoms using conjunction, disjunction, and existential

[1] We adopt here the use of *coherent* to replace the use of "geometric", reserving the latter for the infinitary version.

This article is an extension of an homonymous unpublished manuscript from 2014 with the late Roy Dyckhoff. The work has been partially supported by the Academy of Finland research project no. 1308664.

S. Negri (✉)
Department of Mathematics, University of Genoa, Genoa, Italy
e-mail: sara.negri@unige.it

O. Arieli and A. Zamansky (eds.), *Arnon Avron on Semantics and Proof Theory of Non-Classical Logics*, Outstanding Contributions to Logic 21,
https://doi.org/10.1007/978-3-030-71258-7_12

quantification. Coherent theories are important for many reasons; to start with, there is a methodological rationale for focusing on coherent theories.

Gentzen's systems of deduction, sequent calculus, and natural deduction, have been recognized as an answer to Hilbert's 24th problem (as emphasized, among others, in Negri and von Plato 2011, 2019). They provide the basis for a general theory of proof methods in mathematics that overcomes the limitations of Hilbert-style axiomatic systems.

Natural deduction and sequent calculus give a transparent analysis of the structure of proofs that works to perfection for pure logic, but once they are augmented with axioms for mathematical theories, much of their strong structural properties, such as eliminability of cut, is lost. Transformation of axioms into rules of inference of a suitable form, however, is by now an established method to regain such properties (see, e.g., Negri and von Plato 1998, 2001, 2011, 2019). Coherent theories are very well suited to this methodology, in fact, they can be translated to inference rules in a natural fashion: In the context of a sequent calculus such as **G3c** (Negri and von Plato 2001; Troelstra and Schwichtenberg 2000), special coherent implications as axioms can be converted directly (Negri 2003) to inference rules without affecting the admissibility of the structural rules.[2] Convertibility of coherent axioms into rules thus has the great advantage of importing to mathematical theories many of the results that typically hold only for calculi for pure logic, i.e., unextended calculi.

Coherent theories are rather ubiquitous. As emphasized by Johnstone (2002a, b), Negri and von Plato (2011), there are many examples of coherent theories in mathematics: all algebraic theories, such as group theory and ring theory, all essentially algebraic theories, such as category theory (Freyd 1972), the theory of fields, the theory of local rings, lattice theory (Skolem 1920), projective and affine geometry (Bezem and Hendriks 2008; Skolem 1920; Negri and von Plato 2011),[3] the theory of separably closed local rings (also known as "strictly Henselian local rings") (Wraith 1979).

Occurrence of coherent theories is not limited to mathematics: special coherent implications $\forall \mathbf{x}.\, C \supset D$ generalize the Horn clauses from logic programming, where D is required to be an atomic formula; in fact, they generalize the "clauses" of disjunctive logic programs (Minker 1994), where D is allowed to be a disjunction of atoms. In the context of modal and non-classical logics, coherent implications are used to characterize semantically, through properties of accessibility relations in Kripke frames, a wealth of systems (Negri 2005; Dyckhoff and Negri 2012; Negri 2014).

Last but not least, every first-order theory has a conservative coherent extension. Starting with a modification of Skolem's argument from 1920 for his "normal form" theorem, various approaches to the result have been presented and discussed in detail

---

[2] Coherent theories can be fruitfully internalized also in the context of natural deduction (Simpson 1994) and in the formalization of algebraic reasoning through the method of *dynamical algebra* (Coste et al. 2001; Yengui 2015).

[3] Indirectly, through a suitable treatment of negation, also Euclidean geometry is covered by coherent theories, cf. Avigad et al. (2009).

in Dyckhoff and Negri (2015) together with applications to intermediate and modal logics. A serious shortcoming of the standard "Morleyisation" technique is the fact that coherent implications are generated using conjunctive or disjunctive normal form that typically destroys the structure of formulas, whereas a "conservative" algorithm has been presented in Dyckhoff and Negri (2015).

Many fundamental notions in mathematics, however, escape first-order logic. Examples are arithmetical induction, certain axioms in algebra such as the axioms of torsion abelian groups or of Archimedean ordered fields, or in the theory of connected graphs, as well as in the modeling of epistemic social notions such as *common knowledge*. The insufficiency of first-order logic in expressing fundamental notions and constructions in mathematics led Arnon Avron to the problem of finding the right logical framework for the formalization and mechanization of mathematics. In Avron (2003), he gave evidence that *ancestral logic*, first-order logic extended with transitive closure of binary relations, is a viable answer. A proof-theoretic study of ancestral logic is offered in Cohen and Avron (2014) via a suitable extension of the sequent system $\mathbf{LK}_=$ with an induction-like rule for a transitive closure operator. A more detailed analysis of ancestral logic is given in Cohen and Avron (2019).

A different, but related route toward the goal of a logical framework for formalizing mathematics, is taken in this work. All the examples mentioned above can be axiomatized by means of *geometric axioms*, which are extensions of coherent axioms that allow infinitary disjunctions. We therefore take, as a ground logic, infinitary logic,[4] for which we provide both a classical and an intuitionistic G3-style[5] sequent system.

It was shown in Negri (2003) how to extend the standard classical cut-free first-order sequent calculus **G3c** with rules that capture the meaning of coherent axioms without losing admissibility of structural rules such as *Contraction* and *Cut*. The results and proofs therein, however, are only for finitary languages and calculi, and the question arises of what happens if we allow, more generally, rules to capture the meaning of geometric formulas, i.e., those using arbitrary (rather than just finite) disjunctions. We shall stick to countable disjunctions, arising in axioms such as $\forall x. \bigvee_{n>0} nx = 0$. The appropriate inference rule for this axiom would appear to be the infinitary rule

$$\frac{\{nx = 0, \Gamma \Rightarrow \Delta \mid n > 0\}}{\Gamma \Rightarrow \Delta}$$

(this has countably many premisses—one for each $n > 0$—and without loss of generality has x instantiable by any term in the conclusion). With the addition of this rule, derivations are infinitely branching trees, and proofs of standard cut-admissibility results from, e.g., Troelstra and Schwichtenberg (2000), Negri and von Plato (2011) no longer apply.

---

[4] For a useful background on infinitary logic see Sundholm (1983).

[5] By G3-style, we mean a calculus with a formulation of the rules obtained upon the model of the calculus **G3c** and that enjoys the same structural properties (Negri and von Plato 2001).

There is however a rich theory of cut admissibility for infinitary logics, of which Feferman (1968) and Takeuti (1987, Ch. 4) offer useful early surveys; we mention especially Problem 22.21 of the latter, based on López-Escobar Lopez-Escobar (1965). The main cut-admissibility result in Feferman (1968) (theorem 3.3), however, uses a rule (S) that identifies sequents modulo multiplicity of formulas—a form of contraction—and eliminates multicut rather than cut.[6] On the other hand, that of López-Escobar and Takeuti use a Schütte-style argument, and we prefer to offer a more traditional argument.[7]

We study the version of infinitary logic with finite sequents rather than the one with countably infinite sequents. If sequents are finite, the analysis of a succedent infinitary disjunction $\bigvee_{n>0} A_n$ needs a rule such as (for $k > 0$)

$$\frac{\Gamma \Rightarrow A_k, \bigvee_{n>0} A_n, \Delta}{\Gamma \Rightarrow \bigvee_{n>0} A_n, \Delta} R\bigvee_k$$

rather than (if sequents can be countably infinite)

$$\frac{\Gamma \Rightarrow A_1, A_2, \ldots, A_n, \ldots, \Delta}{\Gamma \Rightarrow \bigvee_{n>0} A_n, \Delta} R\bigvee$$

with the premiss having an infinite succedent. Observe however that using infinitary sequents would not dispense from the need of rules with infinitely many premisses for $L\bigvee$ and $R\bigwedge$.

We prove the structural results for geometric extensions of both the classical and the intuitionistic calculus (see Sections 4 and 5 below). The problems of root-first proof-search in analytic calculi with infinite branching are not considered, other than to remark that once cut admissibility is established and, by other means, finite model property results are established, then finitization techniques illustrated by Jäger et al. (2007) for the logic of common knowledge may be useful in constraining the search.

Perhaps the most important property of coherent theories to the constructively minded is that coherent implications I form sequents that give a Glivenko class (Orevkov 1968). In this case, the result (Negri 2003; Dyckhoff and Negri 2017),

---

[6] Warning: the notation for infinitary disjunction in Feferman (1968) is Σ (where we use $\bigvee$), and that for (finitary) existential quantification is $\bigvee$ (where we use $\exists$).

[7] Tait (2006) presents a cut-elimination result for infinitary classical logic with quantification defined in terms of infinitary conjunction and disjunction; the calculus uses single-sided sequents with signed formulas and adheres to the sequents-as-sets paradigm. Lopez-Escobar (1965) is interested in interpolation for $L_{\omega_1\omega}$, which is proved through a cut-free sequent calculus. The calculus has sequents-as-sets, infinitary sequents as well as infinitary rules. Also Takeuti (1987) has infinitary sequents and infinitary rules, but sequents are sequences rather than sets; there is an explicit weakening rule, and contraction is in-built in the rules. Rathjen (2016) has both a classical and an intuitionistic calculus $L_{\omega_1\omega}$ without invertible rules nor admissible contraction. The intuitionistic calculus is obtained through the usual single-succedent projection from the classical one, which takes to a perhaps less direct proof of Barr's theorem; the full form of Barr's theorem, with the Axiom of Choice, is investigated.

known as the first-order Barr Theorem,[8] states that if each $I_i : 0 \leq i \leq n$ is a coherent implication and the sequent $I_1, \ldots, I_n \Rightarrow I_0$ is classically provable, then it is intuitionistically provable. By these results, the proof-theoretic study of coherent theories gives a general twist to the problem of extracting the constructive content of mathematical proofs because a classical proof is already an intuitionistic one, without the need of being modified.

As an application of the availability of a cut-free system with uniform behavior for classical and intuitionistic background logic, we obtain a very simple proof of Barr's theorem that extends to the infinitary case the approach presented in Negri (2003): If a finitary or infinitary geometric implication is derivable in the classical sequent system for geometric theories $\mathbf{G3c}_\omega\mathbf{T}$, it is derivable also in its intuitionistic counterpart $\mathbf{G3i}_\omega\mathbf{T}$.

12.2 Syntax

"Countable" means finite or countably infinite. Atomic formulas are indicated by the letters P, Q, ... as usual; there may be countably many predicate and function symbols, and equality. *Formulas* A are built up using the extension of the usual first-order syntax with countable disjunctions $\bigvee_{n>0} A_n$ and conjunctions $\bigwedge_{n>0} A_n$; quantifiers are as usual, subject to the constraint that no formula may have infinitely many free variables. *Sentences* are closed formulas.

By induction on the definition, each formula A has a countable ordinal $d(A)$ as its *depth* (the successor of the supremum of the depths of its immediate subformulas). For example, $\bot$ and atoms P have depth 1, since they have no immediate subformulas and the supremum of an empty family of ordinals is 0. It follows that, if A' is a proper subformula of A, then $d(A') < d(A)$.

Sequents $\Gamma \Rightarrow \Delta$ have a finite multiset of formulas on each side. The inference rules for $\bigvee$ are thus:

$$\frac{\{\Gamma, A_n \Rightarrow \Delta \mid n > 0\}}{\Gamma, \bigvee_{n>0} A_n \Rightarrow \Delta}\ L\bigvee \qquad \frac{\Gamma \Rightarrow \Delta, \bigvee_{n>0} A_n, A_k}{\Gamma \Rightarrow \Delta, \bigvee_{n>0} A_n}\ R\bigvee_k.$$

Observe that $L\bigvee$ has countably many premisses, one for each $n > 0$. The rules for $\bigwedge$ are dual to the above ones.

Derivations built using these rules are thus (in general) infinite trees, with countable branching but where (as may be proved by induction on the definition of derivation) each branch has finite length. The *leaves* of the trees are those where the two sides have an atomic formula in common, and also instances of $L\bigvee$ where the disjunction is empty, i.e. is $\bot$. To make this precise, we give a formal definition of

[8] The general form of Barr's theorem (Barr 1974; Wraith 1980; Rathjen 2016) is higher order and includes the axiom of choice.

the notion of *derivation* $\mathcal{D}$ and the associated notions of its *height* $ht(\mathcal{D})$ and its *end-sequent*.

1. Any sequent $\Gamma \Rightarrow \Delta$, where some atomic formula occurs in both Γ and Δ, is a derivation, of *height* 0 and with *end-sequent* $\Gamma \Rightarrow \Delta$.
2. If each $\mathcal{D}_n$ is a derivation, of height α_n, with end-sequent $\Gamma_n \Rightarrow \Delta_n$ and

$$\frac{\ldots \quad \Gamma_n \Rightarrow \Delta_n \quad \ldots}{\Gamma \Rightarrow \Delta} R$$

   is an inference (i.e., an instance of a rule), then

$$\frac{\ldots \quad \dfrac{\mathrm{d}_n}{\Gamma_n \Rightarrow \Delta_n} \quad \ldots}{\Gamma \Rightarrow \Delta} R$$

   is a derivation, of *height* the countable ordinal $sup_n(\alpha_n) + 1$ and with *end-sequent* $\Gamma \Rightarrow \Delta$.

Thus, each derivation has a countable ordinal *height* (the successor of the supremum of the heights of its immediate subderivations). Thus, if Γ and Δ have an atomic formula in common, then $\Gamma \Rightarrow \Delta$ has a derivation $\mathcal{D}$ of height $ht(\mathcal{D}) = 0$. The sequent $\bot, \Gamma \Rightarrow \Delta$ (regarded as a zero-premiss rule) has a derivation of height 1. Observe that the definitions of depth and height differ from those in Feferman (1968): we use the successor of a supremum rather than the supremum of the successors: note that $sup_{n>0}(n + 1) = \omega \neq \omega + 1 = (sup_{n>0}(n)) + 1$. It follows that, if $\mathcal{D}'$ is a sub-derivation of $\mathcal{D}$, then $ht(\mathcal{D}') < ht(\mathcal{D})$.

For classical infinitary logic it is possible to use a minimal language with only infinitary disjunction, negation, and the existential quantifier as primitive. Infinitary conjunction can be defined in such a way that the usual De Morgan laws hold (cf. Proposition 12.4.7 below). However, it will be useful for our purposes to consider a calculus where all the connectives and quantifiers, even if interdefinable, are given as primitive and negation is defined as $\sim A \equiv A \supset \bot$; this is not just useful but even necessary since our purpose is to extract the constructive content of classical proofs and many of the interdefinabilities do not hold in intuitionistic logic.

12.3 Geometric Implications

By a *geometric implication* we mean a sentence G of the form $\forall \mathbf{x}.\, C \supset D$ where the quantifier (over a finite list $\mathbf{x}$ of variables) binds all free variables of $C \supset D$, the antecedent C is a finite conjunction of atoms $P_1 \ldots P_k$ and the succedent D is a finite or countably infinite disjunction $\bigvee E_n$ of existentially quantified finite conjunctions of atoms Q_{ni}, i.e., each $E_n = \exists \mathbf{y}_n (Q_{n1} \wedge \cdots \wedge Q_{nm_n})$. The restrictions

Table 12.1 The calculus $\mathbf{G3}c_\omega$

Initial sequents:

$P, \Gamma \Rightarrow \Delta, P$

Logical rules:

$$\frac{}{\bot, \Gamma \Rightarrow \Delta} L\bot$$

$$\frac{A, B, \Gamma \Rightarrow \Delta}{A\&B, \Gamma \Rightarrow \Delta} L\& \qquad \frac{\Gamma \Rightarrow \Delta, A \quad \Gamma \Rightarrow \Delta, B}{\Gamma \Rightarrow \Delta, A\&B} R\&$$

$$\frac{A_k, \bigwedge_{n>0} A_n, \Gamma \Rightarrow \Delta}{\bigwedge_{n>0} A_n, \Gamma \Rightarrow \Delta} L\bigwedge_k \qquad \frac{\{\Gamma \Rightarrow \Delta, A_n \mid n > 0\}}{\Gamma \Rightarrow \Delta, \bigwedge_{n>0} A_n} R\bigwedge$$

$$\frac{A, \Gamma \Rightarrow \Delta \quad B, \Gamma \Rightarrow \Delta}{A \vee B, \Gamma \Rightarrow \Delta} L\vee \qquad \frac{\Gamma \Rightarrow \Delta, A, B}{\Gamma \Rightarrow \Delta, A \vee B} R\vee$$

$$\frac{\{\Gamma, A_n \Rightarrow \Delta \mid n > 0\}}{\Gamma, \bigvee_{n>0} A_n \Rightarrow \Delta} L\bigvee \qquad \frac{\Gamma \Rightarrow \Delta, \bigvee_{n>0} A_n, A_k}{\Gamma \Rightarrow \Delta, \bigvee_{n>0} A_n} R\bigvee_k$$

$$\frac{\Gamma \Rightarrow \Delta, A \quad B, \Gamma \Rightarrow \Delta}{A \supset B, \Gamma \Rightarrow \Delta} L\supset \qquad \frac{A, \Gamma \Rightarrow \Delta, B}{\Gamma \Rightarrow \Delta, A \supset B} R\supset$$

$$\frac{A(t/x), \forall x A, \Gamma \Rightarrow \Delta}{\forall x A, \Gamma \Rightarrow \Delta} L\forall \qquad \frac{\Gamma \Rightarrow \Delta, A(y/x)}{\Gamma \Rightarrow \Delta, \forall x A} R\forall\ (y\ \textit{fresh})$$

$$\frac{A(y/x), \Gamma \Rightarrow \Delta}{\exists x A, \Gamma \Rightarrow \Delta} L\exists\ (y\ \textit{fresh}) \qquad \frac{\Gamma \Rightarrow \Delta, \exists x A, A(t/x)}{\Gamma \Rightarrow \Delta, \exists x A} R\exists$$

on the language already adopted ensure that, even if D is an infinite disjunction, it only has finitely many free variables.

Such a sentence G determines a (finitary or infinitary) *geometric rule* where the name L_G indicates that it is a *left rule*, determined by the geometric sentence G:

$$\frac{\ldots \qquad Q_{n1}(\mathbf{x}, \mathbf{y}_n), \ldots, Q_{nm_n}(\mathbf{x}, \mathbf{y}_n), P_1(\mathbf{x}), \ldots, P_k(\mathbf{x}), \Gamma \Rightarrow \Delta \qquad \ldots}{P_1(\mathbf{x}), \ldots, P_k(\mathbf{x}), \Gamma \Rightarrow \Delta} L_G$$

with one premiss for each of the countably many disjuncts E_n of D. The variables in $\mathbf{y}_n$ are chosen to be *fresh*, i.e., are not in the conclusion; and without loss of generality they are all distinct. The list $\mathbf{y}_n$ of variables may vary as n varies, and maybe no finite list suffices for all the countably many cases. The variables $\mathbf{x}$ (finite in number) may be instantiated with arbitrary terms. Henceforth we shall normally omit mention of the variables. We need also a further condition for admissibility of contraction to hold:

Closure condition: *Given a system with geometric rules, if it has a rule with an instance of form*

$$\frac{\ldots \quad Q_{n1}(\mathbf{x},\mathbf{y}_n),\ldots,Q_{nm_n}(\mathbf{x},\mathbf{y}_n),P_1(\mathbf{x}),\ldots,P_{k-2}(\mathbf{x}),P(\mathbf{x}),P(\mathbf{x}),\Gamma\Rightarrow\Delta \quad \ldots}{P_1(\mathbf{x}),\ldots,P_{k-2}(\mathbf{x}),P(\mathbf{x}),P(\mathbf{x}),\Gamma\Rightarrow\Delta}$$

then also the rule

$$\frac{\ldots \quad Q_{n1}(\mathbf{x},\mathbf{y}_n),\ldots,Q_{nm_n}(\mathbf{x},\mathbf{y}_n),P_1(\mathbf{x}),\ldots,P_{k-2}(\mathbf{x}),P(\mathbf{x}),\Gamma\Rightarrow\Delta \quad \ldots}{P_1(\mathbf{x}),\ldots,P_{k-2}(\mathbf{x}),P(\mathbf{x})\Gamma\Rightarrow\Delta}$$

has to be included in the system.

As was the case for the finitary case treated in Negri (2003), also in the infinitary case the condition is unproblematic, since each atomic formula contains only a finite number of variables and therefore so are the instances; it follows that the number of rules to be added to a given system with geometric rules is finite.

Theorem 12.3.1 *If we add to the basic system for $L_{\omega_1\omega}$ a finite or infinite family of geometric rules L_G, then we can prove all of the geometric sentences G from which they were determined.*

Proof It suffices to deal with a single such sentence and its associated rule (or, rather, since we must satisfy the closure condition, rules); let it be the sentence G given above. Here is a derivation (with the variables omitted):

$$\cfrac{\ldots \quad \cfrac{\cfrac{\cfrac{\ldots \quad Q_{j1},\ldots,Q_{jm_j},P_1,\ldots,P_k\Rightarrow\bigvee E_n,Q_{jl}\quad\ldots}{Q_{j1},\ldots,Q_{jm_j},P_1,\ldots,P_k\Rightarrow\bigvee E_n,Q_{j1}\wedge\cdots\wedge Q_{jm_j}}\,R\wedge^{m_j-1}}{Q_{j1},\ldots,Q_{jm_j},P_1,\ldots,P_k\Rightarrow\bigvee E_n,E_j}\,R\exists}{Q_{j1},\ldots,Q_{jm_j},P_1,\ldots,P_k\Rightarrow\bigvee E_n}\,R\bigvee_j \quad \ldots}{\cfrac{\cfrac{\cfrac{P_1,\ldots,P_k\Rightarrow\bigvee E_n}{P_1\wedge\cdots\wedge P_k\Rightarrow\bigvee E_n}\,L\wedge^{k-1}}{\Rightarrow P_1\wedge\cdots\wedge P_k\supset\bigvee E_n}\,R\supset}{\Rightarrow\forall\mathbf{x}.\,P_1\wedge\cdots\wedge P_k\supset\bigvee E_n}\,R\forall^*}\,L_G$$

QED

Note that, in contrast to the case with finitary first-order logic, the depth may be infinite; for example, there are as many $R\forall$ steps as the finite number of variables in $\mathbf{x}$; then there is one $R\supset$ step, then $k-1$ steps of $L\wedge$, then one step of L_G; but this has infinitely many premises, and it is not hard to ensure that, for each $j>0$, E_j has (once its bound variables are stripped off) at least j conjuncts. This could of course be fixed by having generalized versions of $R\wedge$ and $R\exists$.

In the following, we shall denote with $\mathbf{G3c}^*_\omega$ any extension of $\mathbf{G3c}_\omega$ with a finite or infinite family of such rules L_G.

Before proceeding with the structural properties of extensions of $\mathbf{G3c}_\omega$ by geometric rules, we give some examples of geometric axioms and their corresponding rules.

12.3.1 Examples of Geometric Axioms and Rules

The axiom of **torsion abelian groups**, $\forall x. \bigvee_{n>1}(nx = 0)$, becomes the rule

$$\frac{\ldots \quad nx = 0, \Gamma \Rightarrow \Delta \quad \ldots}{\Gamma \Rightarrow \Delta} R_{Tor}$$

The axiom of **Archimedean ordered fields**, $\forall x. \bigvee_{n\geq 1}(x < n)$, becomes the rule

$$\frac{\ldots \quad x < n, \Gamma \Rightarrow \Delta \quad \ldots}{\Gamma \Rightarrow \Delta} R_{Arc}$$

The axiom of **connected graphs**,

$$\forall xy.x = y \vee \bigvee_{n\geq 1} \exists z_0 \ldots \exists z_n (x = z_0 \;\&\; y = z_n \;\&\; z_0 R z_1 \;\&\; \ldots \;\&\; z_{n-1} R z_n)$$

becomes the rule

$$\frac{x = y, \Gamma \Rightarrow \Delta \quad xRy, \Gamma \Rightarrow \Delta \quad \ldots \quad x = z_0, y = z_n, z_0 R z_1, \ldots, z_{n-1} R z_n, \Gamma \Rightarrow \Delta \quad \ldots}{\Gamma \Rightarrow \Delta} R_{Conn}$$

Finally, we mention an example that pertains to another domain of application, that of epistemic logic. For what follows, we refer to Marti and Studer (2018) for the basic background and a survey of proof systems. Common knowledge among a set of agents $\{a_1, \ldots, a_k\}$ is defined starting from the *Everybody knows* operator

$$\mathcal{E}(A) \equiv \mathcal{K}_{a_1}(A) \wedge \cdots \wedge \mathcal{K}_{a_k}(A),$$

i.e., the conjunction ranging over all the agents of the individual knowledge operators $\mathcal{K}_{a_i}$, and taking the infinitary conjunction of all the n-ary iterations of $\mathcal{E}$

$$CA \equiv \bigwedge_n \mathcal{E}^n(A).$$

We thus have the following reading (with the group of agents made implicit):

A is common knowledge if everybody knows A, everybody knows that everybody knows A, etc.

In Kripkean terms, each $\mathcal{K}_{a_i}$ is a modality with its own accessibility relation R_{a_i}; it is then not difficult to verify that the accessibility relation $R_{\mathcal{E}}$ associated to $\mathcal{E}$ is the *union* of all the R_{a_i}, i.e.,

$$x R_{\mathcal{E}} y \equiv x R_{a_1} y \vee \cdots \vee x R_{a_k} y$$

and that the accessibility relation for C is the **transitive closure** of $R_{\mathcal{E}}$

$$xR_{\mathcal{C}}y \equiv (\exists n \in \mathbb{N}^{+})R^n_{\mathcal{E}}$$

where $R^n_{\mathcal{E}} \equiv (\exists y_0, y_1 \dots y_{n-1}, y_n \in W)(x = y_0 \,\&\, y = y_n \,\&\, y_0 R_{\mathcal{E}} y_1 \;\&\; \dots \;\&\; y_{n-1} R_{\mathcal{E}} y_n)$, so the above is, in infinitary logic,

$$xR_{\mathcal{C}}y \equiv xR^1_{\mathcal{E}}y \vee \dots \vee xR^n_{\mathcal{E}}y \dots .$$

The truth condition for the common knowledge operator $\mathcal{C}$ is

$$x \Vdash \mathcal{C}A \text{ iff for all } y,\ xR_{\mathcal{C}}y \text{ implies } y \Vdash A.$$

The definition of transitive closure $R_{\mathcal{C}}$ gives the geometric rules

$$\frac{xR^1_{\mathcal{E}}y\Gamma \Rightarrow \Delta \quad \dots \quad xR^n_{\mathcal{E}}y, \Gamma \Rightarrow \Delta \dots}{xR_{\mathcal{C}}y, \Gamma \Rightarrow \Delta}\; T^{\omega}$$

$$\frac{xR_{\mathcal{C}}y, \Gamma \Rightarrow \Delta}{xR^n_{\mathcal{E}}y, \Gamma \Rightarrow \Delta}\; Inc .$$

We remark that also on the modal side, common knowledge is captured by an ω rule; the single-sided sequent calculus of Jäger et al. (2007) the rule is

$$\frac{\dots \mathcal{E}^n A, \Gamma \dots \text{ for alln}}{\mathcal{C}A, \Gamma}\; \omega\mathcal{C} .$$

12.4 Structural Properties

As a first step, we prove

Proposition 12.4.1 *For every formula A, and for every Γ, Δ, the sequent $A, \Gamma \Rightarrow \Delta, A$ is derivable in* $\mathbf{G3c}_{\omega}$.

Proof By (transfinite) induction on the depth of the formula. If $A = \bigvee_{n>0} A_n$, we have

$$\cfrac{\dots \quad \cfrac{\cfrac{\dots}{A_i, \Gamma \Rightarrow \Delta, \bigvee_{n>0} A_n, A_i}\; Ind.Hyp.}{A_i, \Gamma \Rightarrow \Delta, \bigvee_{n>0} A_n}\; R\bigvee_i \quad \dots}{\bigvee_{n>0} A_n, \Gamma \Rightarrow \Delta, \bigvee_{n>0} A_n}\; L\bigvee$$

from which the result follows by transfinite induction, each A_i being of lesser depth than A.

If $A = \bigwedge_{n>0} A_n$ the proof is similar, but with the left and right rules in opposite order. For the finitary connectives and the quantifiers the proof is as for **G3c** (Negri and von Plato 2001). **QED**

In the following we shall use *hp* as an abbreviation of *height preserving*. A property (such as invertibility, admissibility) is qualified as height preserving if it maintains the height of the original derivation.

Lemma 12.4.2 (hp-substitution) *Given a derivation of* $\Gamma \Rightarrow \Delta$ *in* $\mathbf{G3c}^*_\omega$, *with* x *a free variable in* Γ, Δ, t *a term free for* x *in* Γ, Δ *and not containing any of the variables of the geometric rules in the derivation, we can find a derivation of* $\Gamma(t/x) \Rightarrow \Delta(t/x)$ *in* $\mathbf{G3c}^*_\omega$ *with the same height.*

Proof Similar to the proof of hp-substitution in Negri (2003). **QED**

Proposition 12.4.3 (hp-weakening) *The rules of left and right weakening are hp-admissible* $\mathbf{G3c}^*_\omega$.

Proof By a straightforward (transfinite) induction on the height of the derivation of the premiss of each rule, with the usual proviso on variables: if the last rule is a rule with a condition on a variable and the weakening formula contains the same variable, the fresh variable in the rule is first renamed before applying the inductive hypothesis. **QED**

Proposition 12.4.4 (hp-invertibility) *All the rules of* $\mathbf{G3c}^*_\omega$ *are hp-invertible.*

Proof By (transfinite) induction on the height of the derivation of the conclusion of each rule. We consider the case of $L\bigvee$, i.e., a sequent $\Gamma, \bigvee_{n>0} A_n \Rightarrow \Delta$ with derivation height α. If α is zero, then it is either an initial sequent, and thus each $\Gamma, A_n \Rightarrow \Delta$ is also an initial sequent, thus definable with height zero. Else observe that by definition of height of a derivation α has to be a successor ordinal, i.e., $\alpha = \beta + 1$. If α is 1 and the sequent is an instance of $L\bot$, we have a rule with an empty set of assumptions and there is nothing to prove. Let us consider the last (proper) rule and distinguish the case in which $\bigvee_{n>0} A_n$ is a side formula and the case in which it is the principal formula. In the former case the last rule can have one or denumerably many premisses (if the last rule is $L\bigvee$ with a principal formula other than $\bigvee_{n>0} A_n$). The last rule has the form

$$\frac{\{\bigvee_{n>0} A_n, \Gamma_m \Rightarrow \Delta \mid m > 0\}}{\bigvee_{n>0} A_n, \Gamma \Rightarrow \Delta}\ L\bigvee$$

and by definition of height we have, for each m, $\vdash_{\gamma_m} \bigvee_{n>0} A_n, \Gamma_m \Rightarrow \Delta$[9] where for each m, $\gamma_m < \beta$. By inductive hypothesis we have, for each m and n, $\vdash_{\gamma_m} A_n, \Gamma_m \Rightarrow \Delta$ and therefore, by applying the rule, $\vdash_\alpha A_n, \Gamma \Rightarrow \Delta$.

If instead $\bigvee_{n>0} A_n$ is principal, the last rule in the derivation has the form

[9] As usual, we denote by $\vdash_\gamma$ derivability with height bounded by γ.

$$\frac{\{A_n, \Gamma \Rightarrow \Delta \mid m > 0\}}{\bigvee_{n>0} A_n, \Gamma \Rightarrow \Delta} L\bigvee$$

and we have $\vdash_{\gamma_n} A_n, \Gamma \Rightarrow \Delta$ with $\gamma_n < \beta < \alpha$ and thus *a fortiori* $\vdash_\alpha A_n, \Gamma \Rightarrow \Delta$.

The proofs of hp-invertibility of $R\bigwedge$ are similar. For the other infinitary rules, it follows from Proposition 12.4.3. For the finitary rules, the proof is standard. **QED**

Next, we show that the structural rules of left and right contraction are hp-admissible.

Proposition 12.4.5 (hp-contraction) *The rules of left and right contraction*

$$\frac{A, A, \Gamma \Rightarrow \Delta}{A, \Gamma \Rightarrow \Delta} LC \qquad \frac{\Gamma \Rightarrow \Delta, A, A}{\Gamma \Rightarrow \Delta, A} RC$$

are hp-admissible in $\mathbf{G3c}^*_\omega$.

Proof By a simultaneous (transfinite) induction for the left and right contraction rule.

Consider the left rule. If it is an initial sequent, then the conclusion is also an initial sequent and has the same derivation height.

If the contraction formula is $\bigvee_{n>0} A_n$ and it is not principal in the last rule, say $L\bigvee$, we have

$$\frac{\{\bigvee_{n>0} A_n, \bigvee_{n>0} A_n, \Gamma_m \Rightarrow \Delta \mid m > 0\}}{\bigvee_{n>0} A_n, \bigvee_{n>0} A_n, \Gamma \Rightarrow \Delta} L\bigvee$$

we apply the induction hypothesis to each of its premisses and obtain for each $m > 0$ the sequent $\bigvee_{n>0} A_n, \Gamma_m \Rightarrow \Delta$, all with at most the same height as the corresponding premisses. A step of $L\bigvee$ gives the desired conclusion. The cases with other rules are dealt with in a similar way.

If $\bigvee_{n>0} A_n$ is instead principal in the last rule, we have

$$\frac{\{\bigvee_{n>0} A_n, A_n, \Gamma \Rightarrow \Delta \mid n > 0\}}{\bigvee_{n>0} A_n, \bigvee_{n>0} A_n, \Gamma \Rightarrow \Delta} L\bigvee.$$

By hp-invertibility of $L\bigvee$ we obtain derivations, of height at most the height of the premisses, of the sequents $A_n, A_n, \Gamma \Rightarrow \Delta$, and by induction hypothesis of $A_n, \Gamma \Rightarrow \Delta$. A step of $L\bigvee$ gives $\bigvee_{n>0} A_n, \Gamma \Rightarrow \Delta$ with the required bound.

If the last rule is $L\supset$ we have

$$\frac{\Gamma, A \supset B \Rightarrow \Delta, A \quad B, \Gamma, A \supset B \Rightarrow \Delta}{\Gamma, A \supset B, A \supset B \Rightarrow \Delta} L\supset.$$

By application of hp-invertibility of $L\supset$ we obtain derivations of $\Gamma \Rightarrow \Delta, A, A$ and of $B, B, \Gamma \Rightarrow \Delta$, both of smaller height than the conclusion. Application of the induction hypothesis (for LC and RC) gives derivations (of the same height of the latter) of $\Gamma \Rightarrow \Delta, A$ and of $B, \Gamma \Rightarrow \Delta$, and a step of $L\supset$ gives the desired conclusion.

All the other cases are dealt with in a similar or simpler way (i.e., $R\bigvee$ and $R\exists$ do not require use of hp-invertibility and induction hypothesis is applied directly to the premiss of the rule).

The cases in which one or both of the contraction formulas are principal in a (finitary or infinitary) geometric rule are dealt with as usual (using the closure condition in the case of both formulas principal) and the presence of a possibly infinite number of premisses doesn't change the structure of the inductive argument. **QED**

Lemma 12.4.6 *The rules for negation, defined as* $\sim A \equiv A \supset \bot$,

$$\frac{\Gamma \Rightarrow \Delta, A}{\sim A, \Gamma \Rightarrow \Delta}\, L\sim \qquad \frac{A, \Gamma \Rightarrow \Delta}{\Gamma \Rightarrow \Delta, \sim A}\, R\sim$$

are admissible in $\mathbf{G3c}_\omega$.

Proof Immediate using $L\bot$, admissibility of RW, and the implication rules. **QED**

Next we check the De Morgan laws, lest the infinitary nature of disjunctions forces not only infinitary branching but prevents a uniform bound on the length of branches:

Proposition 12.4.7 *The following sequents (the "infinitary De Morgan laws"), with* $\bigwedge$ *defined as the dual of* $\bigvee$, *i.e.,*

$$\bigwedge_{n>0} A_n \equiv \sim \bigvee_{n>0} \sim A_n \ (Def_\bigwedge)$$

are derivable in $\mathbf{G3c}_\omega \setminus \{L\bigwedge, R\bigwedge\}$:

1. $\sim\bigwedge_{n>0} A_n \Rightarrow \bigvee_{n>0} \sim A_n$
2. $\bigvee_{n>0} \sim A_n \Rightarrow \sim\bigwedge_{n>0} A_n$
3. $\sim\bigvee_{n>0} A_n \Rightarrow \bigwedge_{n>0} \sim A_n$
4. $\bigwedge_{n>0} \sim A_n \Rightarrow \sim\bigvee_{n>0} A_n$

Proof 1. $\sim\bigwedge_{n>0} A_n \Rightarrow \bigvee_{n>0} \sim A_n$ follows thus:

$$\cfrac{\cfrac{\cfrac{\bigvee_{n>0} \sim A_n \Rightarrow \bigvee_{n>0} \sim A_n}{\Rightarrow \sim\bigvee_{n>0} \sim A_n, \bigvee_{n>0} \sim A_n}\, R\sim}{\Rightarrow \bigwedge_{n>0} A_n, \bigvee_{n>0} \sim A_n}\, Def_\bigwedge}{\sim\bigwedge_{n>0} A_n \Rightarrow \bigvee_{n>0} \sim A_n}\, L\sim$$

2. $\bigvee_{n>0} \sim A_n \Rightarrow \sim\bigwedge_{n>0} A_n$ is similar;

3.

$$\cfrac{\cdots \quad \cfrac{\sim\sim A_n \Rightarrow A_n}{\sim\sim A_n \Rightarrow \bigvee_{n>0} A_n} R\bigvee \quad \cdots}{\cfrac{\bigvee_{n>0} \sim\sim A_n \Rightarrow \bigvee_{n>0} A_n}{\cfrac{\Rightarrow \bigvee_{n>0} A_n, \sim\bigvee_{n>0} \sim\sim A_n}{\cfrac{\Rightarrow \bigvee_{n>0} A_n, \bigwedge_{n>0} \sim A_n}{\sim\bigvee_{n>0} A_n \Rightarrow \bigwedge_{n>0} \sim A_n} L\sim} Def_\bigwedge} R\sim} L\bigvee$$

4. $\bigwedge_{n>0} \sim A_n \Rightarrow \sim\bigvee_{n>0} A_n$ is similar.
QED

It is easy to verify that in the language we have chosen, with both infinitary conjunction and disjunction as primitive, the De Morgan laws are derivable, with derivations using both pairs of rules.

Admissibility of

$$\frac{\Gamma \Rightarrow \Delta, A \quad A, \Gamma' \Rightarrow \Delta'}{\Gamma, \Gamma' \Rightarrow \Delta, \Delta'} Cut$$

(in which A is the *cut formula*) for such a system (using finite *sets* rather than finite or infinite multisets) is shown by a Gentzen-style argument in Feferman (1968). We give a detailed proof below, allowing also the use of rules determined by geometric implications. The notion of "height of a derivation" treats instances of *Cut* just like any binary rule and allows also for the rules L_G determined by geometric implications.

12.4.1 Cut Admissibility

For the above results (about geometric rules) to be useful, it remains to show the admissibility of *Cut* for the calculus thus modified. For **G3c**, the proof of cut-elimination eliminates a topmost cut by induction on the complexity of the cut formula and subinduction on the sum of the heights of the derivations of the premisses of cuts. To adapt the proof to the infinitary case, where heights are given by ordinals, we shall employ the standard notion of (natural or Hessenberg) addition $\alpha\#\beta$ for countable ordinals α and β (cf. e.g., 10.1.2B in Troelstra and Schwichtenberg 2000 for the definition). We recall that # is commutative and that if $\alpha < \alpha'$ then $\alpha\#\beta < \alpha'\#\beta$.

The *rank* $\pi(I)$ of an instance I of *Cut* with cut-free premisses $\mathcal{D}$ and $\mathcal{D}'$ is the pair comprising the depth $d(A)$ of the cut formula and the natural sum $h(\mathcal{D})\#h(\mathcal{D}')$ of the heights of the premisses. We will call the second component the *total height* of the cut. Pairs are ordered lexicographically.

Ordinals are well ordered, so we can reason by (transfinite) induction; since we actually do it for pairs, we call this *transfinite lexicographic induction*. It can be converted to ordinary transfinite induction by turning pairs into ordinals, e.g., the pair (δ, σ) can be converted to $\delta \cdot \epsilon_0 + \sigma$, where ϵ_0 has the useful property of being greater than any possible value of σ; but pairs are conceptually clearer.

Lemma 12.4.8 *In*

$$\frac{\Gamma \Rightarrow \Delta, A \quad A, \Gamma' \Rightarrow \Delta'}{\Gamma, \Gamma' \Rightarrow \Delta, \Delta'} \, Cut$$

if the premisses have cut-free derivations in $\mathbf{G3c}^{*}_{\omega}$, *then so has the conclusion.*

Proof By transfinite lexicographic induction on the rank of instances of *Cut* and case analysis. We first show (a) the reduction steps for cuts with cut formula principal in both premisses, i.e., *principal cuts*. Then we show (b) how non-principal cuts are reduced by permutation, maintaining the cut formula but reducing the sum of heights. Different strategies are allowed, and we give the details only of the permutations of cuts into the first premiss; permutations into the second premiss are covered generically. We also omit treatment of the cases involving $\bigwedge$ since they are duals of those of $\bigvee$.

(a) 1. If the cut formula is principal in each premiss for instances of initial sequents, then the conclusion is already an initial sequent, so the cut can be eliminated.
2. If the cut formula $\bigvee_{n>0} A_n$ is principal in each premiss, then we consider the cut

$$\frac{\dfrac{\Gamma \Rightarrow \Delta, \bigvee_{n>0} A_n, A_k}{\Gamma \Rightarrow \Delta, \bigvee_{n>0} A_n} R\bigvee_k \quad \dfrac{\ldots \quad A_n, \Gamma' \Rightarrow \Delta' \quad \ldots}{\bigvee_{n>0} A_n, \Gamma' \Rightarrow \Delta'} L\bigvee}{\Gamma, \Gamma' \Rightarrow \Delta, \Delta'} \, Cut$$

which we transform into

$$\frac{\dfrac{\Gamma \Rightarrow \Delta, \bigvee_{n>0} A_n, A_k \quad \dfrac{\ldots \quad A_n, \Gamma' \Rightarrow \Delta' \quad \ldots}{\bigvee_{n>0} A_n, \Gamma' \Rightarrow \Delta'} L\bigvee}{\Gamma, \Gamma' \Rightarrow \Delta, \Delta', A_k} Cut \quad A_k, \Gamma' \Rightarrow \Delta'}{\Gamma, \Gamma', \Gamma' \Rightarrow \Delta, \Delta', \Delta'} \, Cut$$

followed by contractions to reduce two copies of the finite multisets Γ' and Δ' to one. By the induction hypothesis, we can construct a cut-free derivation of the conclusion of the first cut, since the second component of the rank has been reduced by 1 (and the first, i.e. $d(\bigvee_{n>0} A_n)$, is unchanged). We can do the same for the second cut, since the depth of the first component has been reduced (by 1 or more) from $d(\bigvee_{n>0} A_n)$ to $d(A_n)$. The contractions are admissible by a result above.
3. Principal cuts with formulas with binary connectives and quantifiers as outermost logical constant are reduced as in the standard proof for **G3c** (cf. Negri and von Plato 2001).

(b) 1. If the first premiss is an instance of an initial sequent with the atom P principal and P is the cut formula, then the conclusion may be obtained by *Weakening* from the second premiss, regardless of the rule used in the second premiss, as in

$$\frac{\Gamma, P \Rightarrow P, \Delta \quad P, \Gamma' \Rightarrow \Delta'.}{\Gamma, P, \Gamma' \Rightarrow \Delta, \Delta'} \; Cut$$

2. Similarly to case b.1, if the second premiss is an instance of an initial sequent with the atom P principal and P is the cut formula, then the conclusion may be obtained by *Weakening* from the first premiss, regardless of the rule used in the first premiss.
3. If the first premiss is an instance of an initial sequent with the atom P principal but P is not the cut formula, then the conclusion is already an instance of an initial sequent, regardless of the rule used in the second premiss, as in

$$\frac{\Gamma, P \Rightarrow P, \Delta, C \quad C, \Gamma' \Rightarrow \Delta'.}{\Gamma, P, \Gamma' \Rightarrow P, \Delta, \Delta'} \; Cut$$

4. Similarly to case b.3, if the second premiss is an instance of an instance of an initial sequent with the atom P principal but P is not the cut formula, then the conclusion is already an initial sequent, regardless of the rule used in the first premiss.

If the cut formula C is not principal in the left premiss, we reason by cases on the last rule used to derive it.

5. It is $R\bigvee_k$, we have

$$\frac{\dfrac{\Gamma \Rightarrow \Delta, \bigvee_{n>0} A_n, A_k, C}{\Gamma \Rightarrow \Delta, \bigvee_{n>0} A_n, C} \; R\bigvee_k \quad C, \Gamma' \Rightarrow \Delta'}{\Gamma, \Gamma' \Rightarrow \Delta, \bigvee_{n>0} A_n, \Delta'} \; Cut$$

can be transformed to

$$\frac{\dfrac{\Gamma \Rightarrow \Delta, \bigvee_{n>0} A_n, A_k, C \quad C, \Gamma' \Rightarrow \Delta'}{\Gamma, \Gamma' \Rightarrow \Delta, \bigvee_{n>0} A_n, A_k, \Delta'} \; Cut}{\Gamma, \Gamma' \Rightarrow \Delta, \bigvee_{n>0} A_n, \Delta'} \; R\bigvee_k$$

the cut is "permuted upwards", with unchanged cut formula C and reduced total height. All the other cases of non-principal cuts with finitary rules are treated in a similar way.

6. If the last rule is $L\bigvee$, we have

$$\frac{\dfrac{\ldots \quad A_n, \Gamma \Rightarrow \Delta, C \quad \ldots}{\Gamma, \bigvee_{n>0} A_n \Rightarrow \Delta, C} \; L\bigvee \quad C, \Gamma' \Rightarrow \Delta'.}{\Gamma, \bigvee_{n>0} A_n, \Gamma' \Rightarrow \Delta, \Delta'} \; Cut$$

This is transformed to

$$\dfrac{\cdots \quad \dfrac{A_n, \Gamma \Rightarrow \Delta, C \quad C, \Gamma' \Rightarrow \Delta'}{\Gamma, A_n, \Gamma' \Rightarrow \Delta, \Delta'}\, Cut \quad \cdots}{\Gamma, \bigvee_{n>0} A_n, \Gamma' \Rightarrow \Delta, \Delta'}\, L\bigvee$$

with countably many cuts, each of lower rank (since the cut formula C is unchanged but the total height has, in each case, been reduced).

7. If the last rule is a geometric rule, we have

$$\dfrac{\dfrac{\cdots \quad P_1(\mathbf{t}), \ldots, P_n(\mathbf{t}), Q_{k1}(\mathbf{t}, \mathbf{y}), \ldots, Q_{km_k}(\mathbf{t}, \mathbf{y}), \Gamma \Rightarrow \Delta, C \quad \cdots}{\Gamma, P_1(\mathbf{t}), \ldots, P_n(\mathbf{t}) \Rightarrow \Delta, C}\, L_G \quad C, \Gamma' \Rightarrow \Delta'}{\Gamma, P_1(\mathbf{t}), \ldots, P_n(\mathbf{t}), \Gamma' \Rightarrow \Delta, \Delta'}\, Cut$$

(with $\mathbf{y}$ fresh) that can be transformed to

$$\dfrac{\cdots \quad \dfrac{\Gamma, P_1(\mathbf{t}), \ldots, P_n(\mathbf{t}), Q_{k1}(\mathbf{t}, \mathbf{y}), \ldots, Q_{km_k}(\mathbf{t}, \mathbf{y}) \Rightarrow \Delta, C \qquad C, \Gamma' \Rightarrow \Delta'}{\Gamma, P_1(\mathbf{t}), \ldots, P_n(\mathbf{t}), Q_{k1}(\mathbf{t}, \mathbf{y}), \ldots, Q_{km_k}(\mathbf{t}, \mathbf{y}), \Gamma' \Rightarrow \Delta, \Delta'}\, Cut \quad \cdots}{\Gamma, P_1(\mathbf{t}), \ldots, P_n(\mathbf{t}), \Gamma' \Rightarrow \Delta, \Delta'}\, L_G$$

with countably many cuts of lower rank (since the cut formula C is unchanged but the sum of heights has, in each case, been reduced). A similar treatment applies to any additional rule required for the closure condition.

8. If the cut formula C is not principal in the second premiss, and that premiss is not an initial sequent, then a permutation into the second premiss is applicable, as in (for example)

$$\dfrac{\Gamma \Rightarrow \Delta, C \quad \dfrac{A, C, \Gamma' \Rightarrow \Delta', B}{C, \Gamma' \Rightarrow \Delta', A \supset B}\, R\supset}{\Gamma, \Gamma' \Rightarrow \Delta, \Delta', A \supset B}\, Cut$$

which is transformed to

$$\dfrac{\dfrac{\Gamma \Rightarrow \Delta, C \quad A, C, \Gamma' \Rightarrow \Delta', B.}{A, \Gamma, \Gamma' \Rightarrow \Delta, \Delta', B}\, Cut}{\Gamma, \Gamma' \Rightarrow \Delta, \Delta', A \supset B}\, R\supset$$

We include as a special case that where the second premiss is (in effect) derived by $L\bot$, the cut's conclusion is then similarly derivable, and similarly for some instances of L_G. Other cases produce a cut on the same cut formula but with reduced sum of heights.

To complete the inductive argument, we have to be convinced that in each case we have reduced the rank of the cut. There are two cases of interest: a.2 and b.6. The first of these generates a single cut (of reduced total height) as premiss to a further cut on a reduced cut formula, and then some contractions. The second generates countably many cuts on the same cut formula but reduces in each case the total height. **QED**

Theorem 12.4.9 *The* Cut *rule is admissible in* $\mathbf{G3c}^*_\omega$.

Proof It remains to show that an arbitrary derivation using instances (possibly infinite in number) of the *Cut* rule can be transformed to a cut-free derivation. Since this number may be infinite, we argue by transfinite induction on the height of the derivation. Consider a derivation $\mathcal{D}$; if it does not end in a cut, but with a step by the rule R, then, by inductive hypothesis, each premiss (which has height less than $ht(\mathcal{D})$) can be transformed to a cut-free derivation (with conclusion unchanged), and thus so, by adding an R-step, can $\mathcal{D}$. Otherwise, if $\mathcal{D}$ ends with a cut, the derivations of its premisses both have height less than $ht(\mathcal{D})$; by inductive hypothesis, each can be transformed to a cut-free derivation (with conclusion unchanged). We now use the Lemma to obtain a cut-free derivation of the conclusion of $\mathcal{D}$. **QED**

Observe that by the above we obtain cut elimination, not just cut admissibility, because the proof is a syntactic transformation resulting in a cut-free proof.

More precise accounting can give us bounds on the height of the cut-free derivation thus constructed; for details see the methods used in Feferman (1968). The present methods suffice for our purpose.

12.5 An Intuitionistic Infinitary Calculus

The table of rules of the intuitionistic infinitary calculus is given below. As a justification of the rules, observe that it is obtained from the classical calculus by

1. Imposing to the right rule of infinitary conjunction the same restrictions that are needed for rule $R\forall$;
2. Repeating the principal formula in the left premiss to $L\supset$ to make the rule invertible;
3. Admitting the weakening context only in the conclusion (but not in the premiss) of $R\supset$.

We observe that the restriction on $R\bigwedge$ is the reason why in the multi-succedent calculus we cannot have the rules of binary conjunction as special cases of the infinitary one. The reason why the rule for infinitary conjunction in the intuitionistic multi-succedent calculus should have the same restriction as $R\forall$, i.e., no weakening context in the premiss but only in the conclusion, is best seen semantically. By taking the Lindenbaum algebra that corresponds to the logic, a liberal use of a weakening context in $R\bigwedge$ would amount to imposing that join distributes over arbitrary meet; however, complete Heyting algebras have distributivity (only) for meet over arbitrary join, but not for join over arbitrary meets. Consider the complete Heyting algebra of the topology generated by open intervals of the real line, where join is the operation

Table 12.2 The calculus $\mathbf{G3}i_\omega$

Initial sequents:

$P, \Gamma \Rightarrow \Delta, P$

Logical rules:

$$\frac{A, B, \Gamma \Rightarrow \Delta}{A\&B, \Gamma \Rightarrow \Delta}\ L\& \qquad \frac{\Gamma \Rightarrow \Delta, A \quad \Gamma \Rightarrow \Delta, B}{\Gamma \Rightarrow \Delta, A\&B}\ R\&$$

$$\frac{A_k, \bigwedge_{n>0} A_n, \Gamma \Rightarrow \Delta}{\bigwedge_{n>0} A_n, \Gamma \Rightarrow \Delta}\ L\bigwedge_k \qquad \frac{\{\Gamma \Rightarrow A_n \mid n > 0\}}{\Gamma \Rightarrow \Delta, \bigwedge_{n>0} A_n}\ R\bigwedge$$

$$\frac{A, \Gamma \Rightarrow \Delta \quad B, \Gamma \Rightarrow \Delta}{A \vee B, \Gamma \Rightarrow \Delta}\ L\vee \qquad \frac{\Gamma \Rightarrow \Delta, A, B}{\Gamma \Rightarrow \Delta, A \vee B}\ R\vee$$

$$\frac{\{\Gamma, A_n \Rightarrow \Delta \mid n > 0\}}{\Gamma, \bigvee_{n>0} A_n \Rightarrow \Delta}\ L\bigvee \qquad \frac{\Gamma \Rightarrow \Delta, \bigvee_{n>0} A_n, A_k}{\Gamma \Rightarrow \Delta, \bigvee_{n>0} A_n}\ R\bigvee_k$$

$$\frac{A \supset B, \Gamma \Rightarrow \Delta, A \quad B, \Gamma \Rightarrow \Delta}{A \supset B, \Gamma \Rightarrow \Delta}\ L\supset \qquad \frac{A, \Gamma \Rightarrow B}{\Gamma \Rightarrow \Delta, A \supset B}\ R\supset$$

$$\frac{}{\bot, \Gamma \Rightarrow \Delta}\ L\bot$$

$$\frac{\Gamma, A(y/x) \Rightarrow \Delta}{\Gamma, \exists x A \Rightarrow \Delta}\ L\exists \qquad \frac{\Gamma \Rightarrow \Delta, \exists x A, A(t/x)}{\Gamma \Rightarrow \Delta, \exists x A}\ R\exists$$

$$\frac{\forall x A, A(t/x), \Gamma \Rightarrow \Delta}{\Gamma, \forall x A \Rightarrow \Delta}\ L\forall \qquad \frac{\Gamma \Rightarrow A(y/x)}{\Gamma \Rightarrow \Delta, \forall x A}\ R\forall$$

of union and meet is the interior of intersection. Consider the interval $(-1, 0)$ and the family of intervals $(-1/n, 1)$, where n ranges over the positive integers. For all n, we have $(-1, 0) \vee (-1/n, 1) = (-1, 1)$, so $\bigwedge_n((-1, 0) \vee (-1/n, 1)) = (-1, 1)$. Instead, $\bigwedge_n(-1/n, 1) = (0, 1)$, so $(-1, 0) \vee \bigwedge_n(-1/n, 1) = (-1, 1) - \{0\}$.[10]

The following are provable by an easy adaptation of the proof for $\mathbf{G3c}^*_\omega$. We shall thus limit ourselves to considering in detail only the cases that differ significantly from those of $\mathbf{G3c}^*_\omega$.

Lemma 12.5.1 *Given a derivation of $\Gamma \Rightarrow \Delta$ in $\mathbf{G3i}^*_\omega$, with x a free variable in Γ, Δ, t a term free for x in Γ, Δ and not containing any of the variables of the geometric rules in the derivation, we can find a derivation of $\Gamma(t/x) \Rightarrow \Delta(t/x)$ in $\mathbf{G3i}^*_\omega$ with the same height.*

Proof Similar to the proof for $\mathbf{G3c}^*_\omega$. **QED**

Proposition 12.5.2 *The rules of left and right weakening are hp-admissible in $\mathbf{G3i}^*_\omega$.*

[10] A similar argument to discard the formulation of $R\bigwedge$ without context restriction is presented in Nadel (1978).

Proof Since there are no restrictions on the antecedent in $\mathbf{G3i}_\omega$ rules, the proof of admissibility of left weakening is identical to the proof for $\mathbf{G3c}^*_\omega$. For right weakening, we consider the inductive step in the case in which the last rule is a rule with context restriction ($R\bigwedge$, $R\forall$). In such cases, we cannot apply the inductive hypothesis to the premiss(es) of the rule since this would result to an extra formula in the context that would thus violate the restrictions. Instead, we just apply the rule with a new context weakened on the right with the desired formula with a proviso on eigenvariables: if the weakening formula contains some of them, rename the eigenvariables by hp-substitution (Lemma 12.5.1) to avoid a clash. **QED**

Proposition 12.5.3 *All the rules of* $\mathbf{G3i}_\omega$ *except* $R\bigwedge$, $R\supset$ *and* $R\forall$ *are hp-invertible in* $\mathbf{G3i}^*_\omega$.

Proof Observe that all the rules without context restrictions and the invertibility of which is not an instance of weakening are identical to the rules for $\mathbf{G3c}^*_\omega$, so the proof proceed as for $\mathbf{G3c}^*_\omega$. **QED**

Proposition 12.5.4 *The rules of left and right contraction*

$$\frac{A, A, \Gamma \Rightarrow \Delta}{A, \Gamma \Rightarrow \Delta}\ LC \qquad \frac{\Gamma \Rightarrow \Delta, A, A}{\Gamma \Rightarrow \Delta, A}\ RC$$

are hp-admissible in $\mathbf{G3i}^*_\omega$.

Proof The proof has the same structure as the proof for the classical calculus, by a simultaneous (transfinite) induction for the left and right contraction rule, so we need to consider only the new cases arising from the modified rules ($R\bigwedge$, $R\supset$, $R\forall$).

For (LC), if the last rule is one of the above, the contraction formula cannot be principal, so we can proceed by applying the inductive hypothesis and then the rule.

For (RC), if the last rule is one of the above, we distinguish two cases: either the two occurrences of the contraction formula are both in the context of the rule, or one is in the context and another one is principal in the rule. In the former case, consider the premiss of the rule and apply it with a modified weakening context in which only one copy of the contraction formula is retained. In the latter, use a weakening context without the formula. **QED**

Proposition 12.5.5 *The cut rule is admissible in* $\mathbf{G3i}^*_\omega$.

Proof The proof uses the same induction as the one for the classical calculus. We shall therefore limit ourselves to considering the new cases arising from the modified rules, i.e., $R\bigwedge$, $R\supset$, and $R\forall$.

(a) We start with principal cuts. Again, since the cases for cut formula that is an implication or a universal formula are identical to those for the finitary intuitionistic multi-succedent calculus **G3im** (detailed in Negri and von Plato 2001, theorem 5.3.6), it is enough to consider the case of an infinitary conjunction. A cut with $\bigwedge_{n>0} A_n$ principal in both premisses has the form

$$\dfrac{\dfrac{\{\Gamma \Rightarrow A_n \mid n > 0\}}{\Gamma \Rightarrow \Delta, \bigwedge_{n>0} A_n} R\bigwedge \quad \dfrac{A_k, \bigwedge_{n>0} A_n, \Gamma' \Rightarrow \Delta'}{\bigwedge_{n>0} A_n, \Gamma' \Rightarrow \Delta'} L\bigwedge}{\Gamma, \Gamma' \Rightarrow \Delta, \Delta'} Cut ,$$

This is converted as follows

$$\dfrac{\dfrac{\Gamma \Rightarrow A_k \quad \dfrac{\Gamma \Rightarrow \Delta, \bigwedge_{n>0} A_n \quad A_k, \bigwedge_{n>0} A_n, \Gamma' \Rightarrow \Delta'}{\Gamma, \Gamma', A_k \Rightarrow \Delta, \Delta'} Cut_1}{\Gamma^2, \Gamma' \Rightarrow \Delta, \Delta'} Cut_2}{\Gamma, \Gamma' \Rightarrow \Delta, \Delta'} L/R\text{-}C^* ,$$

where Cut_1 has reduced height, Cut_2 has reduced depth, and $L/R\text{-}C^*$ denotes possible multiple steps of left and right contraction.

(b) We then consider the case of a non-principal cut in which (at least) one of the two premisses is a rule with context restriction. Among the various cases we need to consider only the case in which one of the rules is (1) $R\bigwedge$ or (2) a $\bigvee$ rule and a rule with context restriction, since the other cases are treated as for $\mathbf{G3c}^*_\omega$ or for **G3im**. We have the following cases for the left and right premiss of cut, respectively:

1.1 $R\bigwedge$ and initial sequent. There are two subcases, i.e., cut on the principal formula of the initial sequent or on some other formula. In the first subcase we have

$$\dfrac{\dfrac{\{\Gamma \Rightarrow A_n \mid n > 0\}}{\Gamma \Rightarrow \Delta, P, \bigwedge_{n>0} A_n} R\bigwedge \quad P, \Gamma' \Rightarrow \Delta', P.}{\Gamma, \Gamma', \Rightarrow \Delta, \Delta', \bigwedge_{n>0} A_n} Cut$$

By using a new context for $R\bigwedge$ and steps of LW, the above is converted into the derivation where the cut disappears

$$\dfrac{\dfrac{\{\Gamma \Rightarrow A_n \mid n > 0\}}{\Gamma \Rightarrow \Delta, \Delta', P, \bigwedge_{n>0} A_n} R\bigwedge}{\Gamma, \Gamma' \Rightarrow \Delta, \Delta', P, \bigwedge_{n>0} A_n} LW^* .$$

In case the cut formula is in the context of the initial sequent, also the conclusion of cut is an initial sequent, so the derivation is replaced by the initial sequent and the cut disappears.

1.2 $R\bigwedge$ and a conclusion of $L\bot$. There are again two cases, depending on whether the cut formula is $\bot$ or some other formula. In both cases we proceed as above.

1.3 Initial sequent and $R\bigwedge$. Similar to 1.1.

1.4 Conclusion of $L\bot$ and $R\bigwedge$. Similar to 1.3.

1.5 $R\bigwedge$ and generic rule. We first distinguish two cases: either the cut formula is the principal formula of $R\bigwedge$ or it is a formula in the context. The latter case is easily dealt with by applying $R\bigwedge$ with a new weakening context without the cut formula and using admissible weakening steps. In the former, in the case of a rule with one premiss, we have

$$\cfrac{\cfrac{\{\Gamma \Rightarrow A_n \mid n>0\}}{\Gamma \Rightarrow \Delta, \bigwedge_{n>0} A_n}\, R\bigwedge \qquad \cfrac{\bigwedge_{n>0} A_n, \Gamma' \Rightarrow \Delta'}{\bigwedge_{n>0} A_n, \Gamma'' \Rightarrow \Delta''}\, Rule}{\Gamma, \Gamma'' \Rightarrow \Delta, \Delta''}\, Cut \quad .$$

The expected permutation is as follows, where the new cut has reduced height:

$$\cfrac{\cfrac{\Gamma \Rightarrow \Delta, \bigwedge_{n>0} A_n \qquad \bigwedge_{n>0} A_n, \Gamma' \Rightarrow \Delta'}{\Gamma, \Gamma' \Rightarrow \Delta, \Delta'}\, Cut}{\Gamma, \Gamma'' \Rightarrow \Delta, \Delta''}\, Rule \quad .$$

However such a permutation is blocked if *Rule* is a rule with context restriction. So we need to examine such cases more closely because a different conversion is needed. We have four sub-subcases.

1.5.1 $R\bigwedge$ and $R\bigwedge$. We have the derivation

$$\cfrac{\cfrac{\{\Gamma \Rightarrow A_n \mid n>0\}}{\Gamma \Rightarrow \Delta, \bigwedge_{n>0} A_n}\, R\bigwedge \qquad \cfrac{\{\bigwedge_{n>0} A_n, \Gamma' \Rightarrow B_m \mid n>0\}}{\bigwedge_{n>0} A_n, \Gamma' \Rightarrow \Delta', \bigwedge_{m>0} B_m}\, R\bigwedge}{\Gamma, \Gamma' \Rightarrow \Delta, \Delta', \bigwedge_{m>0} B_m}\, Cut$$

which is converted into one with infinitely many cuts, all of lesser height

$$\cfrac{\cfrac{\cfrac{\{\Gamma \Rightarrow A_n \mid n>0\}}{\Gamma \Rightarrow \bigwedge_{n>0} A_n}\, R\bigwedge \qquad \{\bigwedge_{n>0} A_n, \Gamma' \Rightarrow B_m \mid n>0\}}{\{\Gamma, \Gamma' \Rightarrow B_m \mid n>0\}}\, Cut}{\Gamma, \Gamma' \Rightarrow \Delta, \Delta', \bigwedge_{m>0} B_m}\, R\bigwedge$$

1.5.2 $R\bigwedge$ and $L\supset$. We have the derivation

$$\cfrac{\cfrac{\{\Gamma \Rightarrow A_n \mid n>0\}}{\Gamma, \Rightarrow \Delta, \bigwedge_{n>0} A_n}\, R\bigwedge \qquad \cfrac{\bigwedge_{n>0} A_n, B\supset C, \Gamma' \Rightarrow B \quad \bigwedge_{n>0} A_n, C, \Gamma' \Rightarrow \Delta'}{\bigwedge_{n>0} A_n, B\supset C, \Gamma' \Rightarrow \Delta'}\, L\supset}{\Gamma, \Gamma', B\supset C \Rightarrow \Delta, \Delta'}\, Cut$$

and the conversion of the cut to two cuts of lesser height

$$\cfrac{\cfrac{\cfrac{\cfrac{\{\Gamma \Rightarrow A_n \mid n>0\}}{\Gamma, \Rightarrow \bigwedge_{n>0} A_n}\, R\bigwedge \quad \bigwedge_{n>0} A_n, B\supset C, \Gamma' \Rightarrow B}{\Gamma, \Gamma', B\supset C \Rightarrow B}\, Cut \qquad \cfrac{\cfrac{\{\Gamma \Rightarrow A_n \mid n>0\}}{\Gamma \Rightarrow \bigwedge_{n>0} A_n}\, R\bigwedge \quad \bigwedge_{n>0} A_n, C, \Gamma' \Rightarrow \Delta'}{\Gamma, \Gamma', C \Rightarrow \Delta'}\, Cut}{\Gamma, \Gamma', B\supset C \Rightarrow \Delta'}\, L\supset}{\Gamma, \Gamma', B\supset C \Rightarrow \Delta, \Delta'}\, RW$$

1.5.3 $R\bigwedge$ and $R\supset$. By now, we have learnt the rationale of such reductions. As in the previous case, $R\bigwedge$ is re-applied with an empty right context, which makes the permutation with the rule with context restriction possible, with a reduction in the cut height.

1.5.4 $R\bigwedge$ and $R\forall$. Similar.

1.6 Finitary rule and $R\bigwedge$. We consider the case of a generic one-premiss rule. The derivation then has the form

$$\dfrac{\dfrac{\Gamma\Rightarrow\Delta}{\Gamma'\Rightarrow\Delta',C}\,Rule \qquad \dfrac{\{C,\Gamma''\Rightarrow A_n \mid n>0\}}{C,\Gamma''\Rightarrow\Delta'',\bigwedge_{n>0}A_n}\,R\bigwedge}{\Gamma',\Gamma''\Rightarrow\Delta',\Delta'',\bigwedge_{n>0}A_n}\,Cut$$

We first consider the case of a finitary rule without context restriction. If C is not the principal formula of *Rule*, the cut is simply permuted to its premiss, with reduced height. Then *Rule* is applied. This is easily generalized to the case of a rule with two premisses, with two cuts instead of one, both of reduced height.

If C is principal, there are two cases to consider, namely, $R\vee$ and $R\exists$. For the former we have

$$\dfrac{\dfrac{\Gamma\Rightarrow\Delta,B,C}{\Gamma\Rightarrow\Delta,B\vee C}\,R\vee \qquad \dfrac{\{B\vee C,\Gamma''\Rightarrow A_n \mid n>0\}}{B\vee C,\Gamma''\Rightarrow\Delta'',\bigwedge_{n>0}A_n}\,R\bigwedge}{\Gamma',\Gamma''\Rightarrow\Delta',\Delta'',\bigwedge_{n>0}A_n}\,Cut$$

which is converted as follows, where we use steps of height-preserving invertibility and convert the cut into two cuts both of reduced rank

$$\dfrac{\dfrac{\dfrac{\Gamma\Rightarrow\Delta,B,C \qquad \dfrac{B\vee C,\Gamma''\Rightarrow\Delta'',\bigwedge_{n>0}A_n}{B,\Gamma''\Rightarrow\Delta'',\bigwedge_{n>0}A_n}\,hp\text{-}inv}{\Gamma,\Gamma''\Rightarrow\Delta,\Delta'',C,\bigwedge_{n>0}A_n}\,Cut \qquad \dfrac{B\vee C,\Gamma''\Rightarrow\Delta'',\bigwedge_{n>0}A_n}{C,\Gamma''\Rightarrow\Delta'',\bigwedge_{n>0}A_n}\,hp\text{-}inv}{\Gamma',\Gamma''^2\Rightarrow\Delta',\Delta''^2,\bigwedge_{n>0}A_n^2}\,Cut}{\Gamma',\Gamma''\Rightarrow\Delta',\Delta'',\bigwedge_{n>0}A_n}\,L\text{-}RC^*$$

If *Rule* is a rule with two premisses, the remaining case is $R\wedge$, which is similar to that of $R\vee$.

1.7 Infinitary rule and $R\bigwedge$. We consider the case in which the infinitary rule is $R\bigwedge$ and the cut formula being the principal formula of $R\bigwedge$ since the case of cut formula in the weakening context is treated by just applying $R\bigwedge$ with a modified weakening context. We have

$$\dfrac{\dfrac{\{\Gamma\Rightarrow A_n \mid n>0\}}{\Gamma\Rightarrow\Delta,\bigwedge_{n>0}A_n}\,R\bigwedge \qquad \dfrac{\{\bigwedge_{n>0}A_n,\Gamma'\Rightarrow B_m \mid m>0\}}{\bigwedge_{n>0}A_n,\Gamma'\Rightarrow\Delta',\bigwedge_{m>0}B_m}\,R\bigwedge}{\Gamma,\Gamma'\Rightarrow\Delta,\Delta',\bigwedge_{m>0}B_m}\,Cut$$

and the conversion

$$\dfrac{\cdots \quad \dfrac{\dfrac{\{\Gamma \Rightarrow A_n \mid n>0\}}{\Gamma \Rightarrow \bigwedge_{n>0} A_n}\, R\bigwedge \qquad \bigwedge_{n>0} A_n, \Gamma' \Rightarrow B_m}{\Gamma, \Gamma' \Rightarrow B_m}\, Cut \quad \cdots}{\Gamma, \Gamma' \Rightarrow \Delta, \Delta', \bigwedge_{n>0} B_n}\, R\bigwedge$$

with denumerable many cuts of lower height.

2.1 $L\bigvee$ and rule with context restriction. We have a typical case of the form, for $R\supset$,

$$\dfrac{\dfrac{\{\Gamma, A_n \Rightarrow \Delta, C \mid n>0\}}{\Gamma, \bigvee_{n>0} A_n \Rightarrow \Delta, C}\, L\bigvee \qquad \dfrac{C, \Gamma', A \Rightarrow B}{C, \Gamma' \Rightarrow \Delta', A \supset B}\, R\supset}{\Gamma, \Gamma', \bigvee_{n>0} A_n \Rightarrow \Delta, \Delta', A \supset B}\, Cut$$

this converted to infinitely many cuts of smaller height as follows

$$\dfrac{\cdots \quad \dfrac{\Gamma, A_n \Rightarrow \Delta, C \quad C, \Gamma' \Rightarrow \Delta', A \supset B}{\Gamma, \Gamma', A_n \Rightarrow \Delta, \Delta', A \supset B}\, Cut \quad \cdots}{\Gamma, \Gamma', \bigvee_{n>0} A_n \Rightarrow \Delta, \Delta', A \supset B}\, L\bigvee$$

2.2 $R\bigvee$ and rule with context restriction. If the cut formula is a side formula in the left premiss, the cut is permuted to the left premiss, and then $R\bigvee$ is applied. If the cut formula is $\bigvee_{n>0} A_n$, a typical case, exemplified with $R\supset$, has the form

$$\dfrac{\dfrac{\Gamma \Rightarrow \Delta, \bigvee_{n>0} A_n, A_k}{\Gamma \Rightarrow \Delta, \bigvee_{n>0} A_n}\, R\bigvee \qquad \dfrac{\bigvee_{n>0} A_n, A, \Gamma' \Rightarrow B}{\bigvee_{n>0} A_n, \Gamma' \Rightarrow \Delta', A \supset B}\, R\supset}{\Gamma, \Gamma' \Rightarrow \Delta, \Delta', A \supset B}\, Cut$$

and is converted to two cuts, both of lower rank

$$\dfrac{\dfrac{\dfrac{\Gamma \Rightarrow \Delta, \bigvee_{n>0} A_n, A_k \quad \bigvee_{n>0} A_n, \Gamma' \Rightarrow \Delta', A \supset B}{\Gamma, \Gamma' \Rightarrow \Delta, \Delta', A \supset B, A_k}\, Cut \qquad \dfrac{\bigvee_{n>0} A_n, \Gamma' \Rightarrow \Delta', A \supset B}{A_k, \Gamma' \Rightarrow \Delta', A \supset B}\, hp\text{-}inv.}{\Gamma, \Gamma'^2 \Rightarrow \Delta, \Delta'^2, A \supset B^2}\, Cut}{\Gamma, \Gamma' \Rightarrow \Delta, \Delta', A \supset B}\, L\text{-}RC^*$$

2.3 Rule with context restriction and $L\bigvee$, with cut formula $\bigvee_{n>0} A_n$. The cut formula is necessarily in the weakening context of the rule with context restriction and we have

$$\dfrac{\dfrac{\Gamma \Rightarrow \Delta}{\Gamma' \Rightarrow \Delta', \bigvee_{n>0} A_n}\ Rule \quad \dfrac{\{\Gamma'', A_n \Rightarrow \Delta'' \mid n > 0\}}{\Gamma'', \bigvee_{n>0} A_n \Rightarrow \Delta''}\ L\bigvee}{\Gamma', \Gamma'' \Rightarrow \Delta', \Delta''}\ Cut \quad .$$

The conclusion of cut is obtained, without cut, by applying *Rule* with a new weakening context without $\bigvee_{n>0} A_n$ and using admissibility of left weakening.

2.4 Rule with context restriction and $L\bigvee$, with cut formula the principal formula of the rule with context restriction. We have

$$\dfrac{\dfrac{\Gamma, A \Rightarrow B}{\Gamma \Rightarrow \Delta, A \supset B}\ R\supset \quad \dfrac{\{A \supset B, A_n, \Gamma' \Rightarrow \Delta' \mid n > 0\}}{A \supset B, \bigvee_{n>0} A_n, \Gamma \Rightarrow \Delta'}\ L\bigvee}{\Gamma, \Gamma', \bigvee_{n>0} A_n \Rightarrow \Delta, \Delta'}\ Cut \quad .$$

This is transformed into denumerably many cuts of reduced height as follows

$$\dfrac{\dots \quad \dfrac{\Gamma \Rightarrow \Delta, A \supset B \quad A \supset B, A_n, \Gamma' \Rightarrow \Delta'}{\Gamma, \Gamma', A_n \Rightarrow \Delta, \Delta'}\ Cut \quad \dots}{\Gamma, \Gamma', \bigvee_{n>0} A_n \Rightarrow \Delta, \Delta'}\ L\bigvee \quad .$$

QED

12.5.1 *A Proof of the Infinitary Barr Theorem*

Barr's theorem[11] is a fundamental result in geometric logic: it guarantees that for geometric theories classical derivability of geometric implications entails their intuitionistic derivability. The result has its origin, through appropriate completeness results, in the theory of sheaf models, with the following formulation (cf. e.g., Mac Lane and Moerdijk 1994, Ch. 9, Theorem 2):

Theorem 12.5.6 *For every Grothendieck topos $\mathcal{E}$ there exists a complete Boolean algebra* **B** *and a surjective geometric morphism $Sh(\mathbf{B}) \longrightarrow \mathcal{E}$.*

The most general form of Barr's theorem (Barr 1974; Wraith 1980; Rathjen 2016) is higher-order and includes the axiom of choice, and stated as eliminating not just classical reasoning but also the axiom of choice[12].

If one is interested solely in derivability in geometric logic (finitary or infinitary, but without the axiom of choice), Barr's theorem can be regarded as identifying

[11] While this result is usually attributed to Barr (1974), it was implicit in works such as Grothendieck's Tohoku paper on homological algebra (Grothendieck 1957) or Joyal's letter to Grothendieck on model structure for simplicial sheaves (Joyal 1984).

[12] That such formulations of Barr's theorem should be taken with caution is demonstrated in Rathjen (2016) where *internal* vs. *external* addition of the axiom of choice is considered and it is shown that the latter preserves conservativity whereas the former does not.

a Glivenko class, i.e., a class of sequents for which classical derivability entails intuitionistic derivability[13] and a proof entirely internal to proof theory, without any detour through completeness with respect to topos-theoretic models, obtains.

Consider now a classical theory axiomatized by finitary or infinitary geometric implications. Extending the conversion into rules of Negri (2003) to cover the case of infinitary disjunctions and using the results detailed above, we transform the classical theory into a contraction- and cut-free sequent calculus, denoted by $\mathbf{G3c}_\omega\mathbf{T}$. We shall denote by $\mathbf{G3i}_\omega\mathbf{T}$ the corresponding intuitionistic extension of $\mathbf{G3i}_\omega$. The following holds:

Theorem 12.5.7 *If a finitary or infinitary geometric implication is derivable in* $\mathbf{G3c}_\omega\mathbf{T}$, *it is derivable in* $\mathbf{G3i}_\omega\mathbf{T}$.

Proof *Almost nothing to prove.* Any derivation in $\mathbf{G3c}_\omega\mathbf{T}$ uses only rules that follow the (infinitary) geometric rule scheme and logical rules. Observe that geometric implications contain no $\supset$, $\bigwedge$ or $\forall$ in the scope of $\vee$, which means that no instance of the rules that violate the intuitionistic restrictions is used, so the derivation directly gives (through the addition, where needed, of the missing implications in steps of $L\supset$) a derivation in $\mathbf{G3i}_\omega\mathbf{T}$ of the same conclusion. **QED**

A proof of Barr's theorem for *finitary* geometric theories was given in Negri (2003) through a cut-free presentation of finitary geometric theories and the choice of an appropriate sequent calculus that, in effect, trivializes the result. By the results above, the same trivialization works for infinitary logic: a classical proof *already* is an intuitionistic proof.

12.6 Concluding Remarks and Further Work

This article stems from an unpublished note written in 2014 with Roy Dyckhoff and circulated at that time among colleagues. It was recently presented in seminars in Verona, Florence and Helsinki in the Autumn of 2018 and in a conference in Tübingen in March 2019. I am grateful to the organizers and the audience of those events for valuable feedback, in particular Michele Abrusci, Ingo Blechschmidt, Giulio Fellin, Per Martin-Löf, Paolo Maffezioli, Dale Miller, Pierluigi Minari, Jan von Plato, Gabriel Sandu, Peter Schuster, Göran Sundholm, Daniel Wessel. I am also grateful to the two referees for their insightful suggestions.

There are several lines of extension of the results presented here. To start with, it is well known that using the Axiom of Choice any set can be well ordered and thus, in effect, turned into an ordinal. One can then use the same transfinite induction

[13] Mints (2017) attributes the standard proof of the first-order version to Orevkov (1968), although it does not appear therein. The result can nevertheless be reduced to one of the Glivenko classes (Orevkov 1968) provided one uses, for intuitionistic logic, a multi-succedent calculus with invertible rules, as in Negri (2016).

used here to generalize our results to an infinitary language with $\bigwedge$ and $\bigvee$ indexed by an arbitrary set rather than by the natural numbers. Second, the original Barr's theorem includes the axiom of choice. It would be interesting to see how the methods presented here fare in comparison, e.g., to the treatment of Rathjen (2016). We plan to constructivize the cut-elimination proof with a proof that replaces induction on ordinals with induction on well-founded trees.

The infinitary proof theory investigated in this paper is that of infinite width. However, as pointed out by one of the referees, there is another form of infinitary proof theory, of infinite height. These are called *proofs by infinite-descent*: they are allowed to have infinite height branches, as long as they admit some infinite-descent condition (Brotherston and Simpson 2011). Such systems were recently developed for Avron's favorite ancestral logic in Cohen and Rowe (2018). It would be interesting to investigate the infinite-descent style proof theory of theories of geometric implications and compare it with the present approach.

Last but not least, our motivation in this work was to provide a framework for reasoning both classically and intuitionistically with geometric theories. There are however more general motivations that make proof analysis in infinitary logic of independent interest. There are theories for which a complete, sound, finitary proof system cannot be achieved, so that an infinitary proof system is developed to obtain completeness, e.g., for PA. In this sense, a potentially interesting area of application of the present methodology is infinitary modal logic. Existing results on the proof theory for infinitary modal logic can be found in Minari (2016). Particularly promising would be to use labeled calculi to obtain cut elimination.

References

Avigad, J., Dean, E., & Mumma, J. (2009). A formal system for Euclid's Elements. *Review of Symbolic Logic, 2*(4), 700–768.

Avron, A. (2003). Transitive closure and the mechanization of mathematics. In *Thirty-five years of automating mathematics* (Vol. 28, pp. 149–171). Applied Logic Series. Dordrecht: Kluwer Academic Publishers.

Barr, M. (1974). Toposes without points. *Journal of Pure and Applied Algebra, 5*(3), 265–280.

Bezem, M., & Hendriks, D. (2008). On the mechanization of the proof of Hessenberg's theorem in coherent logic. *Journal of Automated Reasoning, 40*(1), 61–85.

Brotherston, J., & Simpson, A. (2011). Sequent calculi for induction and infinite descent. *Journal of Logic and Computation, 21*(6), 1177–1216.

Cohen, L., & Avron, A. (2014). Ancestral logic: a proof theoretical study. In *Logic, Language, Information, and Computation* (Vol. 8652, pp. 137–151). Lecture Notes in Computer Science. Heidelberg: Springer.

Cohen, L., & Avron, A. (2019). The middle ground-ancestral logic. *Synthese, 196*(7), 2671–2693.

Cohen, L., & Rowe, R. N. S. (2018). Uniform inductive reasoning in transitive closure logic via infinite descent. *Computer science logic 2018* (Vol. 119). LIPIcs Leibniz International Proceedings in Informatics. Art. No. 17, 16. Schloss Dagstuhl. Leibniz-Zent. Inform., Wadern.

Coste, M., Lombardi, H., & Roy, M.-F. (2001). Dynamical method in algebra: Effective Nullstellensätze. *Annals of Pure and Applied Logic, 111*(3), 203–256.

Dyckhoff, R., & Negri, S. (2012). Proof analysis in intermediate logics. *Archive for Mathematical Logic, 51*(1–2), 71–92.
Dyckhoff, R., & Negri, S. (2015). Geometrisation of first-order logic. *Bulletin of Symbolic Logic, 21*(2), 123–163.
Dyckhoff, R., & Negri, S. (2017). Commentary on Grigori Mints' "Classical and intuitionistic geometric logic". *The IfCoLog Journal of Logics and their Applications, 4*(4), 1235–1239.
Feferman, S. (1968). Lectures on Proof Theory. In *Proceedings of the Summer School in Logic (Leeds, 1967)*, (pp. 1–107). Berlin: Springer.
Freyd, P. (1972). Aspects of topoi. *Bulletin of the Australian Mathematical Society, 7*(1), 1–76.
Grothendieck, A. (1957). Sur quelques points d'algèbre homologique. *Tôhoku Mathematical Journal, 2*(9), 119–221.
Jäger, G., Kretz, M., & Studer, T. (2007). Cut-free common knowledge. *Journal of Applied Physics, 5*(4), 681–689.
Johnstone, P. T. (2002a). *Sketches of an Elephant: A Topos Theory Compendium. Vol. 1* (Vol. 43). Oxford Logic Guides. New York: The Clarendon Press Oxford University Press.
Johnstone, P. T. (2002b). *Sketches of an Elephant: A Topos Theory Compendium. Vol. 2* (Vol. 44). Oxford Logic Guides. Oxford: The Clarendon Press Oxford University Press.
Joyal, A. (1984). Lettre d'André Joyal à Alexandre Grothendieck. In G. Maltsiniotis (Ed.). Available from: https://webusers.imj-prg.fr/~georges.maltsiniotis/ps/lettreJoyal.pdf.
Lopez-Escobar, E. G. K. (1965). An interpolation theorem for denumerably long formulas. *Fundamenta Mathematicae, 57*, 253–272.
Mac Lane, S., & Moerdijk, I. (1994). *Sheaves in Geometry and Logic. A First Introduction to Topos Theory*. Universitext. New York: Springer.
Marti, M., & Studer, T. (2018). The proof theory of common knowledge. In *Jaakko Hintikka on knowledge and game-theoretical semantics* (Vol. 12, pp. 433–455). Outst. Contrib. Log. Cham: Springer.
Minari, P. (2016). Some remarks on the proof-theory and the semantics of infinitary modal logic. *Advances in Proof Theory* (Vol. 28, pp. 291–318). Progress Computer Science Applied Logic. Cham: Birkhäuser/Springer.
Minker, J. (1994). Overview of disjunctive logic programming. *Annals of Mathematics and Artificial Intelligence, 12*(1–2), 1–24. Disjunctive logic programming (San Diego, CA, 1991).
Mints, G. (2017). Classical and intuitionistic geometric logic. *Proceedings of the International Conference—Philosophy, Mathematics, Linguistics: Aspects of Interaction, 2012* (Vol. 70, pp. 117–125). Studies in Logic (London). London: College Publications. Reprinted in IfCoLog Journal of Logics and their Applications, vol. 4, pp. 1227–1234, 2017.
Nadel, M. E. (1978). Infinitary intuitionistic logic from a classical point of view. *Annals of Mathematical Logic, 14*, 159–191.
Negri, S. (2003). Contraction-free sequent calculi for geometric theories with an application to Barr's theorem. *Archive for Mathematical Logic, 42*(4), 389–401.
Negri, S. (2005). Proof analysis in modal logic. *Journal of Philosophical Logic, 34*(5–6), 507–544.
Negri, S. (2014). Proofs and countermodels in non-classical logics. *Logica Universalis, 8*(1), 25–60.
Negri, S. (2016). Glivenko sequent classes in the light of structural proof theory. *Archive for Mathematical Logic, 55*(3–4), 461–473.
Negri, S., & von Plato, J. (1998). Cut elimination in the presence of axioms. *Bulletin of Symbolic Logic, 4*(4), 418–435.
Negri, S., & von Plato, J. (2001). *Structural Proof Theory*. Cambridge: Cambridge University Press. Appendix C by Aarne Ranta.
Negri, S., & von Plato, J. (2011). *Proof Analysis. A Contribution to Hilbert's Last Problem*. Cambridge: Cambridge University Press.
Negri, S., & von Plato, J. (2019). From mathematical axioms to mathematical rules of proof: recent developments in proof analysis. *Philosophical Transactions of the Royal Society A, 377*(2140), 20180037.

Orevkov, V. P. (1968). Glivenko's sequence classes. In V. P. Orevkov (Ed.) *Logical and logico-mathematical calculi. Part 1* (pp. 131–154). Leningrad.
Rathjen, M. (2016). Remarks on Barr's theorem. Proofs in geometric theories. In D. Probst & P. Schuster (Eds.), *Concepts of Proof in Mathematics, Philosophy, and Computer Science* (Vol. 6, pp. 347–374). Ontos Mathematical Logic. Berlin: Walter de Gruyter.
Simpson, A. (1994). The Proof Theory and Semantics of Intuitionistic Modal Logic. PhD thesis.
Skolem, T. (1920). Logisch-kombinatorische Untersuchungen über die Erfüllbarkeit oder Beweisbarkeit mathematischer Sätze nebst einem Theorem über dichte Mengen. Videnskapsselskapets skrifter, 1. Mat.-naturv. klasse *4*, 1–36.
Sundholm, G. (1983). *Proof Theory, a survey of the omega-rule*. PhD thesis.
Tait, W. (2006). *Normal derivability in classical logic*, *72*, 204–236.
Takeuti, G. (1987). *Proof Theory* (2nd ed., Vol. 81). Studies in Logic and the Foundations of Mathematics. Amsterdam: North-Holland Publishing Co.
Troelstra, A. S., & Schwichtenberg, H. (2000). *Basic Proof Theory* (2nd ed.). Cambridge: Cambridge University Press.
Wraith, G. (1979). Generic Galois theory of local rings. In M. P. Fourman, C. J. Mulvey, & D. S. Scott (Eds.), *Applications of Sheaves. Proceedings of the Research Symposium on Applications of Sheaf Theory to Logic, Algebra and Analysis, Durham, July 9–21, 1977* (pp. 739–767). Berlin: Springer.
Wraith, G. C. (1980). Intuitionistic algebra: some recent developments in topos theory. *Proceedings of the International Congress of Mathematicians (Helsinki, 1978)* (pp. 331–337). Helsinki. Acad. Sci. Fennica.
Yengui, I. (2015). *Constructive Commutative Algebra. Projective Modules over Polynomial Rings and Dynamical Gröbner Bases* (Vol. 2138). Lecture Notes in Mathematics. Cham: Springer.

Sara Negri is a Professor of Mathematical Logic at the University of Genova. She has been Professor of Theoretical Philosophy at the University of Helsinki since 2015 as well as a Docent of Logic since 1998. After a Ph.D. in Mathematics in 1996 at the University of Padova and research visits at the University of Amsterdam and Chalmers, she has been a Research Associate at the Imperial College in London, a Humboldt Fellow in Munich, and a Visiting Scientist at the Mittag-Leffler Institute in Stockholm and at the Hausdorff Institute for Mathematics in Bonn. Her research interests range from mathematical logic and philosophy of mathematics to proof theory and its applications to modal and non-classical logics.

Chapter 13
Connexive Variants of Modal Logics Over FDE

Sergei Odintsov, Daniel Skurt, and Heinrich Wansing

Abstract Various connexive **FDE**-based modal logics are studied. Some of these logics contain a conditional that is both connexive and strict, thereby highlighting that strictness and connexivity of a conditional do not exclude each other. In particular, the connexive modal logics $\mathbf{cBK}^-$, **cKN4**, $\mathbf{scBK}^-$, **scKN4**, **cMBL**, and **scMBL** are introduced semantically by means of classes of Kripke models. The logics $\mathbf{cBK}^-$ and **cKN4** are connexive variants of the **FDE**-based modal logics $\mathbf{BK}^-$ and **KN4** with a weak and a strong implication, respectively. The system **cMBL** is a connexive variant of the modal bilattice logic **MBL**. The latter is a modal extension of Arieli and Avron's logic of logical bilattices and is characterized by a class of Kripke models with a four-valued accessibility relation. In the systems $\mathbf{scBK}^-$, **scKN4**, and **scMBL**, the conditional is both connexive and strict. Sound and complete tableau calculi for all these logics are presented and used to show that the entailment relations of the systems under consideration are decidable for finite premise set. Moreover, the logics $\mathbf{cBK}^-$ and **cMBL** are shown to be algebraizable. The algebraizability of **cMBL** is derived from proving **cMBL** to be definitionally equivalent to **MBL**. All connexive modal logics studied in this paper are decidable, paraconsistent, and inconsistent but non-trivial logics.

S. Odintsov
Sobolev Institute of Mathematics, Novosibirsk, Russia
e-mail: odintsov@math.nsc.ru

D. Skurt · H. Wansing (✉)
Department of Philosophy I, Ruhr University Bochum, Bochum, Germany
e-mail: heinrich.wansing@rub.de

D. Skurt
e-mail: daniel.skurt@rub.de

O. Arieli and A. Zamansky (eds.), *Arnon Avron on Semantics and Proof Theory of Non-Classical Logics*, Outstanding Contributions to Logic 21,
https://doi.org/10.1007/978-3-030-71258-7_13

13.1 Introduction

In a recent paper titled "Strictness and Connexivity" (Iacona 2019), Andrea Iacona argues that natural language indicative conditionals are adequately formalized as strict implications. According to Iacona, this "strict conditional view" is in conflict with the characteristic principles of connexive logic,[1] namely Aristotle's and Boethius' theses, principles that express a certain contra-classical understanding of implication, $\supset$, and negation, $\sim$:

AT $\sim(\sim A \supset A)$
AT′ $\sim(A \supset \sim A)$
BT $(A \supset B) \supset \sim(A \supset \sim B)$
BT′ $(A \supset \sim B) \supset \sim(A \supset B)$.

He claims that the strict conditional view invalidates **AT** and **AT′** for constantly true propositions A, and that it invalidates **BT** and **BT′** for constantly false propositions A. In the opinion of Iacona, the strict conditional view validates only restricted versions of Aristotle's and Boethius' theses, namely (notation adjusted):

RAT If it is possible that $\sim A$, then $\sim(\sim A \supset A)$
RAT′ If it is possible that A, then $\sim(A \supset \sim A)$
RBT If it is possible that A, then $(A \supset B) \supset \sim(A \supset \sim B)$
RBT′ If it is possible that A, then $(A \supset \sim B) \supset \sim(A \supset B)$.

Moreover, Iacona (2019, p. 8) holds that since **AT**, **AT′**, **BT**, **BT′** "are plausible only insofar as they entail **RAT**, **RAT′**, **RBT**, **RBT′**, the strict conditional view is as plausible as any connexivist theory of conditionals."

According to Iacona, there is thus a substantial contrast between strict and connexive conditionals and a reason to choose between them. In the present paper, we present various ways of adding a connexive conditional to Belnap and Dunn's basic four-valued paraconsistent logic **FDE** (first-degree entailment logic). Three of the five conditionals we will consider are *both* strict and connexive. Whereas Iacona maintains that the strictness of a conditional provides reason for rejecting the characteristic principles of connexive logic, we show that there exists no categorical difference between the strictness and the connexivity of a conditional. This calls for clarification, and we will therefore discuss the notion of strictness and comment on the relationship between strictness and connexivity before we will introduce connexive variants of modal logics over **FDE**.

[1] For overviews of connexive logic see, Wansing (2021), McCall (2012).

When is a conditional strict?
C.I. Lewis attempted to define a conditional that adequately formalizes indicative conditionals in natural language.[2] In his Lewis et al. (1918, p. 124) we can find the following definition (notation slightly adjusted):

> The relation of strict implication can be defined in terms of negation, possibility, and product:
>
> 11.02 $p \prec q :\equiv {\sim}\Diamond(p \wedge {\sim}q)$
>
> Thus "p implies q" or "p strictly implies q" is to mean "It is false that it is possible that p should be true and q false" or "The statement 'p is true and q false' is not self-consistent." When q is deducible from p, to say "p is true and q is false" is to assert, implicitly, a contradiction.

Since in Lewis et al. (1918) the underlying non-modal logic is classical propositional logic, if we assume the interdefinability of $\Box$ (necessity) and $\Diamond$ by means of classical negation, instead of the above definition we can use the following definition:

(Def. $\prec$) $\quad p \prec q :\equiv \Box(p \supset q)$

where $\supset$ is classical, Boolean implication. If we want to define the notion of a strict conditional more generally, we could then assume a conditional, $\supset$, and a necessity operator, $\Box$, as given and use (Def. $\prec$). There are, however, subtleties we have to consider. In the theory of definitions, it is required that definiens and definiendum are interreplaceable in all linguistic contexts without thereby effecting a change of denotation. In logics in which provable equivalence is a congruence relation, an axiom of the form $(p \prec q) \equiv \Box(p \supset q)$ guarantees that $(p \prec q)$ and $\Box(p \supset q)$ are mutually interreplaceable salva veritate, but we will also consider logics in which provable equivalence *fails to be* a congruence relation. In particular, in **FDE** and its extensions, where a clear distinction is drawn between (support of) truth conditions and (support of) falsity conditions, in addition to interreplaceability salva veritate, interreplaceability salva falsitate is a natural requirement, and interreplaceability salva veritate need not always imply interreplaceability salva falsitate, or vice versa. Another complication for defining a notion of a strict conditional comes with the possibility of using (support of) truth conditions that are plausible as (support of) truth conditions of $\Box(p \supset q)$ in order to semantically define $p \prec q$ with $\prec$ as *primitive*,

[2] When Iacona introduces the strict conditional view as a view of indicative conditionals, this seems to presuppose that we are dealing with natural language conditionals. However, one can, and Iacona does, also consider formal languages containing a conditional that is meant to represent a natural language indicative conditional. Also, note that the claim that indicative and subjunctive conditionals are distinct is contentious, see Priest (2018). Iacona (2019, p. 2) explains that his focus on indicative conditionals "is not intended to suggest that counterfactuals differ in some important respect. On the contrary, most of what will be said about conditionals can be extended, mutatis mutandis, to counterfactuals." We will not take a stance on this issue here. Moreover, when we are just interested in defining the strictness and the connexivity of a conditional, we need not define the notion of a conditional but may take it as given. The notion of an implication is often introduced by requiring that it is a binary connective satisfying the Deduction Theorem, see, for example, Avron et al. (2018), Wansing and Odintsov (2016). Note also that we will use the terms "conditional" and 'implication" as synonymous.

i.e., in order to semantically define $p \rightfishtail q$ without having $\Box$ and $\supset$ available in the language.

Iacona (2019) assumes a specific format of truth conditions for defining the notion of a strict implication (the "strict conditional view of indicative conditionals"), namely, restricted universal quantification over possible worlds, and he defines the truth conditions of a strict conditional $p \rightfishtail q$ (notation and terminology adjusted) as follows:

> $p \rightfishtail q$ is true in a possible world w if and only if for every world w' accessible from w, it holds that p is false in w' or q is true in w'.

Different notions of strict implication are then obtained by imposing conditions on the accessibility relation between possible worlds. This seems to be a widely shared understanding of strict conditionals, and for our purposes we need not further elaborate the notion of a strict implication. We shall consider three support of truth conditions for conditionals in terms of restricted universal quantification over worlds (or rather states), and these conditionals are therefore strict in Iacona's sense.

Strictness and connexivity
According to Iacona, the strictness and the connexivity of an implication are in conflict insofar as the strict conditional view invalidates the characteristic principles of connexive logic. As we shall emphasize, with the standard conception of connexivity, the one adopted by Iacona, there is no reason to oppose strictness and connexivity as properties of conditionals.[3] A conditional can be both strict and connexive, and this observation is not new. The implication in the connexive logic **C** and its modal extension **CK** from Wansing (2005), for example, is both strict and connexive. Therefore, what we are considering in this paper could also be presented from the perspective of **CK**, starting with the replacement of the constructive implication of **CK** by the non-strict Boolean implication. Although Iacona grants that his understanding of connexivity validates **AT**, **AT**′, **BT**, and **BT**′, he doubts, however, that these principles are valid in general. According to him, **AT**, **AT**′, **BT**, and **BT**′ are "falsified by vacuously true conditionals, that is, conditionals with necessary consequents or impossible antecedents" (Iacona 2019, p. 8).

This is not the place to present a detailed discussion and criticism of Iacona (2019), but the difference between Iacona's conception of connexive logic and the understanding of connexive logic assumed in the present paper becomes clear also from his claim that (notation adjusted) it "seems that connexivists face a dilemma: either they deny that a conditional of the form $p \supset (p \vee q)$, or $q \supset (p \vee q)$, is true, which is quite implausible, or they deny ... **AT**." Moreover, according to Iacona connexivists face another dilemma, namely, that (notation adjusted) "either they deny that a conditional of the form $(p \wedge q) \supset p$ or $(p \wedge q) \supset q$ is true, which is quite implausible, or they deny **BT**." There are, however, logics that validate these

[3] Usually, it also required that a connexive implication, $\rightarrow$, is non-symmetric, i.e., that $(A \rightarrow B) \rightarrow (B \rightarrow A)$ fails to be a theorem.

principles ("disjunctive addition" and "conjunctive simplification") as well as **AT**, **AT′**, **BT**, and **BT′**.[4]

In his concluding remarks Iacona writes that

> If what we want to preserve is restricted connexivity, rather than full connexivity, then there is no need to abandon classical logic, for conditionals can adequately be formalized as strict conditionals by using the expressive resources of classical modal logic. Even if we were not satisfied with such formalization and wanted to follow a different route, it would still be a route that does not lead to connexive logic.

However, since the strictness of an implication does not prevent it from being connexive, even if it is granted that conditionals are adequately formalized as strict conditionals, this does not entail that one does not end up with a connexive implication in a system of connexive logic after all. In any case, connexive logic rests on the challenge to define formal systems that *unrestictedly* validate **AT**, **AT′**, **BT**, **BT′**. As we will see, connexive modal logics over **FDE** may well be strict.

The paper is structured as follows. In Sect. 13.2, the connexive modal logics **cBK**$^-$, **cKN4**, **scBK**$^-$, **scKN4**, **cMBL**, and **scMBL** are introduced semantically by means of Kripke models that use support of truth as well as support of falsity clauses. The logics **cBK**$^-$ and **cKN4** are connexive variants of the **FDE**-based modal logics **BK**$^-$ and **KN4** with a weak and a strong implication, respectively, see Odintsov and Wansing (2017), Drobyshevich and Wansing (2020). The system **cMBL** is a connexive variant of the modal bilattice logic **MBL**, and in the systems **scBK**$^-$, **scKN4**, and **scMBL**, the conditional is both connexive and strict. Section 13.3 is devoted to the presentation of tableau calculi for our connexive variants of **FDE**-based modal logics, and in Sect. 13.4, these calculi are shown to be sound and complete with respect to their Kripke semantics. In Sect. 13.5 we highlight some properties of **cBK**$^-$, **cKN4**, **scBK**$^-$, **scKN4**, **cMBL**, and **scMBL**. Next, in Sect. 13.6, we study the algebraization problem for the connexive logics **cBK**$^-$, **cKN4**, and **cMBL**. It is shown that **cBK**$^-$ and **cMBL** are algebraizable and that **cKN4** is algebraizable if its global semantical consequence relation is compact. Finally, Sect. 13.7 concludes the paper with a few summarizing remarks.

---

[4] Note that also for Richard Routley (Routley et al. 1982), the failure of conjunctive simplification is characteristic of connexive logic. For Routley, connexive logic and relevance logic more or less coincide. If one shares the containment view of valid implication according to which in a valid implication $A \rightarrow B$, the content of B must be included in the content of A, then disjunctive addition fails, and if connexive logic is motivated by the idea of negation as cancelation, then conjunctive simplification cannot hold in full generality, in particular, $(A \wedge {\sim}A) \supset A$ and $(A \wedge {\sim}A) \supset {\sim}A$ fail to be valid (cf. also Wansing and Skurt 2018).

13.2 Semantics for Connexive FDE-Based Modal Logics

13.2.1 Semantics for KFDE

We begin by defining the (support of) truth and the (support of) falsity conditions for **KFDE**, a simple **FDE**-based modal logic without a primitive, detaching conditional, introduced by Graham Priest (2008). In order to define the semantics, we will make use of a metalogical language which is a two-sorted first-order language containing:

- all formulas of **KFDE** as the first sort of individual variables,
- a non-empty denumerable set V of information state variables as the second sort of variables,
- the classical connectives $⩕$, $⩖$, $\neg$, $\twoheadrightarrow$,
- the classical quantifiers $\forall$ and $\exists$,
- the binary predicate symbols $\Vdash^+$, $\Vdash^-$, and R.

The metalanguage is then defined as follows:

state variables:	$w \in V$
object language formula variables:	A
atomic formulas of the metalanguage:	α
formulas of the metalanguage:	φ

$\alpha ::= w \Vdash^+ A \mid w \Vdash^- A \mid wRw$

$\varphi ::= \alpha \mid \neg\varphi \mid \varphi ⩕ \varphi \mid \varphi ⩖ \varphi \mid \varphi \twoheadrightarrow \varphi \mid \forall\varphi \mid \exists\varphi$

Bi-implication, $\twoheadleftarrow\!\!\!\twoheadrightarrow$, is defined as usual.

The object language $\mathcal{L}_{\mathbf{KFDE}} = \{\vee, \wedge, \sim, \Box, \Diamond\}$ is then based on a non-empty countable set of atomic propositions Prop. We denote by $\mathsf{Form}(\mathbf{KFDE})$ the set of formulas defined as usual, formulas by A, B, C, etc., and sets of formulas by Γ, Δ, Σ, etc.

A **KFDE**-model is a tuple $\mathcal{M} = \langle W, R, v^+, v^- \rangle$, where W is a non-empty set of information states (possible worlds), $R \subseteq W^2$ is an accessibility relation on W, and v^+ and v^- are functions $v^+, v^- : \mathsf{Prop} \to 2^W$. We now define verification (or support of truth) and falsification (or support of falsity) relations $\Vdash^+$ and $\Vdash^-$ between worlds and formulas in a model $\mathcal{M}$ as follows[5]:

[5] Since the metalanguage is classical, all classical equivalences hold in the metalanguage.

$$
\begin{array}{lll}
w \Vdash^+ p & \Leftrightarrow & w \in v^+(p); \\
w \Vdash^- p & \Leftrightarrow & w \in v^-(p); \\
w \Vdash^+ A \wedge B & \Leftrightarrow & (w \Vdash^+ A \mathrel{\&} w \Vdash^+ B); \\
w \Vdash^- A \wedge B & \Leftrightarrow & (w \Vdash^- A \text{ or } w \Vdash^- B); \\
w \Vdash^+ A \vee B & \Leftrightarrow & (w \Vdash^+ A \text{ or } w \Vdash^+ B); \\
w \Vdash^- A \vee B & \Leftrightarrow & (w \Vdash^- A \mathrel{\&} w \Vdash^- B); \\
w \Vdash^+ {\sim} A & \Leftrightarrow & w \Vdash^- A; \\
w \Vdash^- {\sim} A & \Leftrightarrow & w \Vdash^+ A; \\
w \Vdash^+ \Box A & \Leftrightarrow & \forall u (wRu \Rightarrow u \Vdash^+ A); \\
w \Vdash^- \Box A & \Leftrightarrow & \exists u (wRu \mathrel{\&} u \Vdash^- A); \\
w \Vdash^+ \Diamond A & \Leftrightarrow & \exists u (wRu \mathrel{\&} u \Vdash^+ A); \\
w \Vdash^- \Diamond A & \Leftrightarrow & \forall u (wRu \Rightarrow u \Vdash^- A).
\end{array}
$$

We say that a formula A is true at world w in a **KFDE**-model $\mathcal{M} = \langle W, R, v^+, v^- \rangle$ iff $w \Vdash^+ A$. We say that a formula A is A true in $\mathcal{M}$, $\mathcal{M}, \Vdash^+ A$, iff A is true at every world $w \in W$. A formula A is **KFDE**-valid, $\models_{\mathbf{KFDE}} A$, iff A is true in every **KFDE**-model. Finally, a set of formulas Γ entails a formula A, $\Gamma \models_{\mathbf{KFDE}} A$, iff for all **KFDE**-models $\mathcal{M}$ and worlds w, if $w \Vdash^+ B$, for all $B \in \Gamma$, then $w \Vdash^+ A$.

13.2.2 Connexive Extensions of KFDE

We will now successively enrich the language of **KFDE** by four different connexive implications $\rightarrow_c$, $\Rightarrow_c$, $\Box\!\!\rightarrow$, and $\Box\!\!\Rightarrow$. They are defined by the following support of truth and support of falsity conditions:

$$
\begin{array}{lll}
w \Vdash^+ A \rightarrow_c B & \Leftrightarrow & (w \Vdash^+ A \Rightarrow w \Vdash^+ B); \\
w \Vdash^- A \rightarrow_c B & \Leftrightarrow & (w \Vdash^+ A \Rightarrow w \Vdash^- B);
\end{array}
$$

$$
\begin{array}{lll}
w \Vdash^+ A \Rightarrow_c B & \Leftrightarrow & ((w \Vdash^+ A \Rightarrow w \Vdash^+ B) \mathrel{\&} (w \Vdash^- B \Rightarrow w \Vdash^- A)); \\
w \Vdash^- A \Rightarrow_c B & \Leftrightarrow & (w \Vdash^+ A \Rightarrow w \Vdash^- B);
\end{array}
$$

$$
\begin{array}{lll}
w \Vdash^+ A \Box\!\!\rightarrow B & \Leftrightarrow & \forall u (wRu \Rightarrow (u \Vdash^+ A \Rightarrow u \Vdash^+ B)); \\
w \Vdash^- A \Box\!\!\rightarrow B & \Leftrightarrow & \forall u (wRu \Rightarrow (u \Vdash^+ A \Rightarrow u \Vdash^- B));
\end{array}
$$

$$
\begin{array}{lll}
w \Vdash^+ A \Box\!\!\Rightarrow B & \Leftrightarrow & \forall u (wRu \Rightarrow (((u \Vdash^+ A \Rightarrow u \Vdash^+ B) \mathrel{\&} (u \Vdash^- B \Rightarrow u \Vdash^- A)))); \\
w \Vdash^- A \Box\!\!\Rightarrow B & \Leftrightarrow & \forall u (wRu \Rightarrow (u \Vdash^+ A \Rightarrow u \Vdash^- B)).
\end{array}
$$

Note that the verification conditions for $\rightarrow_c$ and $\Rightarrow_c$ are exactly those of the weak and strong implication of **BK**$^-$ and **KN4**, whereas the verification conditions of $\Box\!\!\rightarrow$ and $\Box\!\!\Rightarrow$ can be interpreted a giving rise to a strict conditional. The falsification conditions of the respective conditionals follow the idea that *connexivity is about the falsity conditions for the conditional*, cf. Omori and Wansing (2019). We will call the resulting systems **cBK**$^-$, **cKN4**, **scBK**$^-$, and **scKN4**. Moreover, note that the implication in the mentioned systems **C** and **CK** from Wansing (2005) is also both connexive and strict as it is a connexive variant of intuitionistic implication. In these systems support of truth conditions analogous to those for $\Box\!\!\rightarrow$ in **scBK**$^-$ are stated with respect to the preorder of Kripke models for intuitionistic logic, whereas the modal operators $\Box$ and $\Diamond$ are defined with respect to another binary relation on the set of states. Finally, the system **cBK**$^-$ can also be viewed as a modal extension of the material connexive logic **MC** from Wansing (2021). Note that Kamide (2019) and Omori (2019) are related to the system **MC** with modality as well.

Observation 13.1 It is easy to observe that all the introduced implications together with strong negation satisfy **AT**, **AT**$'$, **BT**, and **BT**$'$ (with $\supset$ being replaced by the connexive conditional in question).

Proof Since the proofs are straightforward, we will leave them to the reader. $\Box$

13.2.3 Semantics for cMBL *and* scMBL

The modal bilattice logic **MBL** is a modal extension of Arieli and Avron's prominent logic of logical bilattices, (Arieli and Avron 1996). The study of **MBL** was motivated in Rivieccio et al. (2015) by obtaining a modal extension of the four-valued logic **FDE** characterized by possible worlds models with a four-valued accessibility relation between possible worlds. In what follows we will enrich the language of **MBL** by two different connexive implications, $\rightarrow_c$ and $\boxplus\!\!\rightarrow$, where the former is the connective introduced above and the latter is a connexive strict implication based on the necessity operator $\boxplus$.

The languages $\mathcal{L}_{\mathbf{cMBL}} = \{\vee, \wedge, \otimes, \oplus, \rightarrow_c, \sim, \boxplus, \bot, \top, \mathsf{b}, \mathsf{n}\}$ and $\mathcal{L}_{\mathbf{scMBL}} = \{\vee, \wedge, \otimes, \oplus, \boxplus\!\!\rightarrow, \sim, \boxplus, \bot, \top, \mathsf{b}, \mathsf{n}\}$ are based, as above, on a non-empty countable set of atomic propositions Prop. Again, if L is a logic, we denote by $\mathsf{Form}(\mathsf{L})$ the set of formulas of the language of L defined as usual, formulas by A, B, C, etc., and sets of formulas by Γ, Δ, Σ, etc.

We define **cMBL**- and **scMBL**-models as tuples $\mathcal{M} = \langle W, R_+, R_-, v^+, v^- \rangle$, where $R_+, R_- \subseteq W \times W$ are accessibility relations on W, and the rest is analogously defined as above. For the connectives and constants not considered so far, we have the following verification and falsification conditions:

$$
\begin{array}{lll}
w \Vdash^+ A \otimes B & \Leftrightarrow & (w \Vdash^+ A \,\&\, w \Vdash^+ B); \\
w \Vdash^- A \otimes B & \Leftrightarrow & (w \Vdash^- A \,\&\, w \Vdash^- B); \\
w \Vdash^+ A \oplus B & \Leftrightarrow & (w \Vdash^+ A \text{ or } w \Vdash^+ B); \\
w \Vdash^- A \oplus B & \Leftrightarrow & (w \Vdash^- A \text{ or } w \Vdash^- B); \\
w \Vdash^+ \top & \& & \neg(w \Vdash^- \top); \\
\neg(w \Vdash^+ \bot) & \& & w \Vdash^- \bot; \\
w \Vdash^+ \mathsf{b} & \& & w \Vdash^- \mathsf{b}; \\
\neg(w \Vdash^+ \mathsf{n}) & \& & \neg(w \Vdash^- \mathsf{n}); \\
w \Vdash^+ \boxplus A & \Leftrightarrow & \forall u(wR_+u \Rightarrow u \Vdash^+ A) \,\&\, \forall u(wR_-u \Rightarrow \neg(u \Vdash^- A)); \\
w \Vdash^- \boxplus A & \Leftrightarrow & \exists u(wR_+u \,\&\, u \Vdash^- A); \\
w \Vdash^+ A \boxplus\!\!\rightarrow B & \Leftrightarrow & \forall u(wR_+u \Rightarrow (u \Vdash^+ A \Rightarrow u \Vdash^+ B)) \,\&\, \forall u(wR_-u \Rightarrow \neg(u \Vdash^+ A \Rightarrow u \Vdash^- B)); \\
w \Vdash^- A \boxplus\!\!\rightarrow B & \Leftrightarrow & \forall u(wR_+u \Rightarrow (u \Vdash^+ A \Rightarrow u \Vdash^- B));
\end{array}
$$

Truth at a world, truth in a model, validity, and entailment are defined in analogy to the definitions of these notions in Sect. 13.2.1.

Observation 13.2 Together with strong negation, $\boxplus\!\!\rightarrow$ satisfies **AT** and **AT**′, but neither **BT** nor **BT**′.

Proof Left for the reader. □

13.3 Tableau Calculi for Connexive FDE-Based Modal Logics

Sound and complete tableau calculi for the connexive extensions of **KFDE** under consideration can easily be obtained by modifying Priest's tableau calculus for **KFDE**, cf. Priest (2008), and the tableau calculi for $\mathbf{BK}^{\Box-}$ and **KN4** from Odintsov and Wansing (2017). Hence, in the following, we will only present the rules for the connexive implications.

We assume some familiarity with the tableau method as applied to modal extensions of **FDE** in Priest (2008), Odintsov and Wansing (2010, 2017). In the tableau for the systems considered below, tableau lines are of the form $A, +i$; $A, -i$; iRj; iR^+j; or iR^-j where A is an object language formula of the connexive extension of **FDE**-based modal logic in question, i and j are natural numbers representing information states or worlds, $+$ indicates verification ($\Vdash^+$), $-$ indicates failure of verification ($\nVdash^+$), and R, R^+ and R^- represent the respective accessibility relations. Tableau for a single conclusion derivability statement $\Gamma \vdash A$ start with a line of the form $B, +0$ for every premise B from the finite premise set Γ and a line of the form $A, -0$. Then tableau rules are applied to tableau lines leading to more complex tableau. A *branch of a tableau closes* iff it contains a pair of lines $C, +i$ and $C, -i$. A *tableau closes* iff all of its branches close. If a tableau (tableau branch) is not closed, it is called *open*. A *tableau branch is said to be complete* iff no more rules can be applied to expand it. A *tableau is said to be complete* iff each of its branches is complete.

13.3.1 Tableau Calculi for cBK$^-$, cKN4, scBK$^-$, *and* scKN4

The tableau rules for $\rightarrow_c$, $\Rightarrow_c$, $\Box\!\!\rightarrow$ and $\Box\!\!\Rightarrow$ make use of the "+"-rules for weak and strong implication of $\mathbf{BK}^{\Box-}$ and **KN4**, cf. Odintsov and Wansing (2017), whereas the "–"-rules have to be adjusted accordingly:

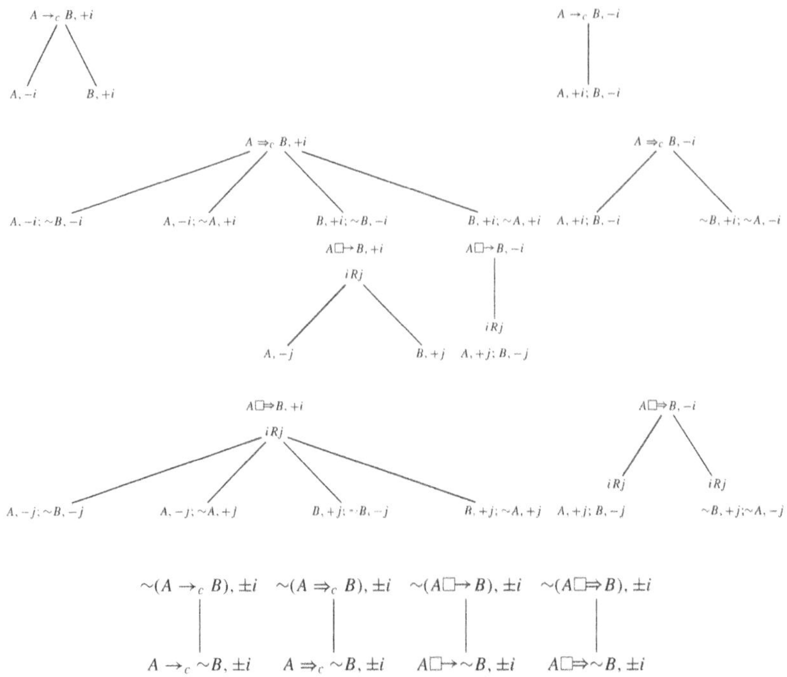

13.3.2 *Tableau Calculi for* cMBL *and* scMBL

The tableau rules for **cMBL** and **scMBL** can be obtained by adding rules for $\rightarrow_c$ and ⊞→ to the tableau rules of **MBL** and by eliminating the rules for $\rightarrow$, cf. Odintsov and Wansing (2017). Since the tableau rules for $\rightarrow_c$ are exactly as above, we will only present the rules for ⊞→. However, since the definition of ⊞ is non-standard and since the semantics for ⊞ is required for the definition of ⊞→, we will present those rules as well.

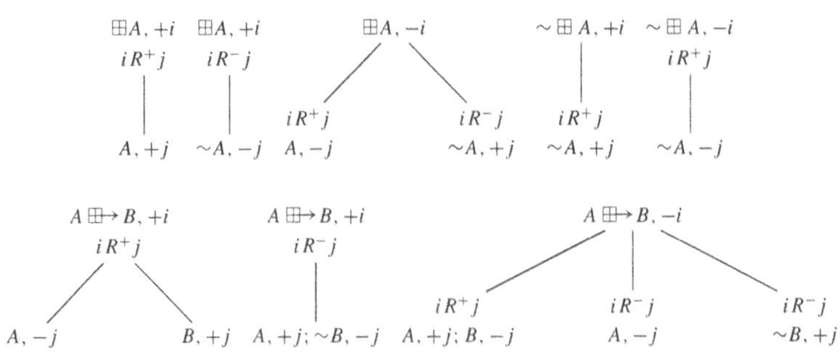

$$\sim(A \boxplus\!\!\rightarrow B), \pm i$$
$$|$$
$$A \boxplus\!\!\rightarrow \sim B, \pm i$$

13.4 Soundness and Completeness

13.4.1 cBK⁻, cKN4, scBK⁻, *and* scKN4

For the soundness and completeness of **cBK**$^-$, **cKN4**, **scBK**$^-$, and **scKN4** we will follow the structure of the proofs given in Odintsov and Wansing (2010), Odintsov and Wansing (2017), and Priest (2008).

Definition 13.3 Let $\mathcal{M} = \langle W, R, v^+, v^- \rangle$ be any L-model, with $\mathsf{L} \in \{$**cBK**$^-$, **cKN4**, **scBK**$^-$, **scKN4**$\}$ and $\mathfrak{b}$ be a tableau branch. The model $\mathcal{M}$ is said to be faithful to $\mathfrak{b}$ iff there exists a function f from the set of natural numbers into W such that

(1) for every line $A, +i$ on $\mathfrak{b}$, $f(i) \Vdash^+ A$.
(2) for every line $A, -i$ on $\mathfrak{b}$, $\neg(f(i) \Vdash^+ A)$.
(3) for every line iRj on $\mathfrak{b}$, $f(i)Rf(j)$.

The function f is said to show that $\mathcal{M}$ is faithful to the branch $\mathfrak{b}$.

Lemma 13.4 *Let* $\mathcal{M} = \langle W, R, v^+, v^- \rangle$ *be any* L*-model, with* $\mathsf{L} \in$ {**cBK**$^-$, **cKN4**, **scBK**$^-$, **scKN4**} *and* $\mathfrak{b}$ *be a tableau branch. If* $\mathcal{M}$ *is faithful to* $\mathfrak{b}$ *and a tableau rule is applied to* $\mathfrak{b}$*, then the application produces at least one extension* $\mathfrak{b}'$ *of* $\mathfrak{b}$*, such that* $\mathcal{M}$ *is faithful to* $\mathfrak{b}'$.

Proof By induction on the construction of the tableau and inspection of the tableau rules. The cases for $\wedge$, $\vee$, $\sim$, $\Box$, $\Diamond$, as well as the cases for $A \rightarrow_c B, \pm i$ and $A \Rightarrow_c B, \pm i$ have been treated in the corresponding lemmata in Odintsov and Wansing (2010), Odintsov and Wansing (2017), and Priest (2008).

We will now suppose the function f shows that $\mathcal{M}$ is faithful to a branch containing the following lines in the respective tableau:

$\sim(A \to_c B), +i$:	Then, $f(i) \Vdash^- A \to_c B \Leftrightarrow (f(i) \Vdash^+ A \twoheadrightarrow f(i) \Vdash^- B) \Leftrightarrow f(i) \Vdash^+ A \to_c \sim B$.
$\sim(A \to_c B), -i$:	Then, $\neg(f(i) \Vdash^+ \sim(A \to_c B)) \Leftrightarrow \neg(f(i) \Vdash^- A \to_c B) \Leftrightarrow \neg(f(i) \Vdash^+ A \twoheadrightarrow f(i) \Vdash^- B) \Leftrightarrow \neg(f(i) \Vdash^+ A \to_c \sim B)$.
$\sim(A \Rightarrow_c B), +i$:	Then, $f(i) \Vdash^- A \Rightarrow_c B \Leftrightarrow (f(i) \Vdash^+ A \twoheadrightarrow f(i) \Vdash^- B) \Leftrightarrow f(i) \Vdash^+ A \Rightarrow_c \sim B$.
$\sim(A \Rightarrow_c B), -i$:	Then, $\neg(f(i) \Vdash^+ \sim(A \Rightarrow_c B)) \Leftrightarrow \neg(f(i) \Vdash^- A \Rightarrow_c B) \Leftrightarrow \neg(f(i) \Vdash^+ A \twoheadrightarrow f(i) \Vdash^- B) \Leftrightarrow \neg(f(i) \Vdash^+ A \Rightarrow_c \sim B)$.
$A \Box\!\!\rightarrow B, +i$:	Then, with $f(i) R f(j)$ we have, $f(i) \Vdash^+ A \Box\!\!\rightarrow B \Leftrightarrow (f(j) \Vdash^+ A \twoheadrightarrow f(j) \Vdash^+ B) \Leftrightarrow (\neg(f(j) \Vdash^+ A) \Cup f(j) \Vdash^+ B)$.
$A \Box\!\!\rightarrow B, -i$:	Then, there is a j such that $f(i) R f(j)$ and we have, $\neg(f(i) \Vdash^+ A \Box\!\!\rightarrow B) \Leftrightarrow \neg(f(j) \Vdash^+ A \twoheadrightarrow f(j) \Vdash^+ B) \Leftrightarrow ((f(j) \Vdash^+ A) \Cap \neg f(j) \Vdash^+ B)$.
$\sim(A \Box\!\!\rightarrow B), +i$:	Then, with $f(i) R f(j)$ we have, $f(i) \Vdash^- A \Box\!\!\rightarrow B \Leftrightarrow (f(j) \Vdash^+ A \twoheadrightarrow f(j) \Vdash^- B) \Leftrightarrow f(i) \Vdash^+ A \Box\!\!\rightarrow \sim B$.
$\sim(A \Box\!\!\rightarrow B), -i$:	Then, there is a j such that $f(i) R f(j)$ and we have, $\neg(f(i) \Vdash^+ \sim(A \Box\!\!\rightarrow B)) \Leftrightarrow \neg(f(j) \Vdash^- A \Box\!\!\rightarrow B) \Leftrightarrow \neg(f(j) \Vdash^+ A \twoheadrightarrow f(i) \Vdash^- B) \Leftrightarrow$
$A \Box\!\!\Rightarrow B, +i$:	Then, with $f(i) R f(j)$ we have, $f(i) \Vdash^+ A \Box\!\!\Rightarrow B \Leftrightarrow ((f(j) \Vdash^+ A \twoheadrightarrow f(j) \Vdash^+ B) \Cap (f(j) \Vdash^- B \twoheadrightarrow f(j) \Vdash^- A)) \Leftrightarrow ((f(j) \Vdash^+ A \twoheadrightarrow f(j) \Vdash^+ B) \Cap (f(j) \Vdash^+ \sim B \twoheadrightarrow f(j) \Vdash^+ \sim A)) \Leftrightarrow (\neg(f(j) \Vdash^+ A) \Cup f(j) \Vdash^+ B) \Cap (\neg(f(j) \Vdash^+ \sim B) \Cup f(j) \Vdash^+ \sim A) \Leftrightarrow (\neg(f(j) \Vdash^+ A) \Cap \neg(f(j) \Vdash^+ \sim B)) \Cup (\neg(f(j) \Vdash^+ A) \Cap (f(j) \Vdash^+ \sim A)) \Cup ((f(j) \Vdash^+ B) \Cap \neg(f(j) \Vdash^+ \sim B)) \Cup ((f(j) \Vdash^+ B) \Cap (f(j) \Vdash^+ \sim A))$.
$A \Box\!\!\Rightarrow B, -i$:	Then, there is a j such that $f(i) R f(j)$ and we have, $\neg(f(i) \Vdash^+ A \Box\!\!\Rightarrow B) \Leftrightarrow \neg((f(j) \Vdash^+ A \twoheadrightarrow f(j) \Vdash^+ B) \Cap (f(j) \Vdash^- B \twoheadrightarrow f(j) \Vdash^- A)) \Leftrightarrow \neg(f(j) \Vdash^+ A \twoheadrightarrow f(j) \Vdash^+ B) \Cup \neg(f(j) \Vdash^- B \twoheadrightarrow f(j) \Vdash^- A)) \Leftrightarrow (f(j) \Vdash^+ A) \Cap \neg(f(j) \Vdash^+ B)) \Cup (f(j) \Vdash^- B) \Cap \neg(f(j) \Vdash^- A)) \Leftrightarrow (f(j) \Vdash^+ A) \Cap \neg(f(j) \Vdash^+ B)) \Cup (f(j) \Vdash^+ \sim B) \Cap \neg(f(j) \Vdash^+ \sim A))$.
$\sim(A \Box\!\!\Rightarrow B), +i$:	Then, with $f(i) R f(j)$ we have, $f(i) \Vdash^- A \Box\!\!\Rightarrow B \Leftrightarrow (f(j) \Vdash^+ A \twoheadrightarrow f(j) \Vdash^- B) \Leftrightarrow f(i) \Vdash^+ A \Box\!\!\Rightarrow \sim B$.
$\sim(A \Box\!\!\Rightarrow B), -i$:	Then, there is a j such that $f(i) R f(j)$ and we have, $\neg(f(i) \Vdash^+ \sim(A \Box\!\!\Rightarrow B)) \Leftrightarrow \neg(f(j) \Vdash^- A \Box\!\!\Rightarrow B) \Leftrightarrow \neg(f(j) \Vdash^+ A \twoheadrightarrow f(j) \Vdash^- B) \Leftrightarrow \neg(f(i) \Vdash^+ A \Box\!\!\Rightarrow \sim B)$.

□

Definition 13.5 Let $\mathfrak{b}$ be a complete open tableau branch. Then the structure $\mathcal{M}_{\mathfrak{b}} = \langle W_{\mathfrak{b}}, R_{\mathfrak{b}}, v^+_{\mathfrak{b}}, v^-_{\mathfrak{b}} \rangle$ induced by $\mathfrak{b}$ is defined as follows:

(1) $W_{\mathfrak{b}} := \{w_i \mid i \text{ occurs on } \mathfrak{b}\}$,
(2) $w_i R_{\mathfrak{b}} w_j$ iff iRj occurs $\mathfrak{b}$,
(3) $w_i \in v^+_{\mathfrak{b}}(p)$ iff $p, +i$ occurs on $\mathfrak{b}$,
(4) $w_i \in v^-_{\mathfrak{b}}(p)$ iff $\sim p, +i$ occurs on $\mathfrak{b}$.

Lemma 13.6 *Suppose $\mathfrak{b}$ is a complete open tableau branch, and let $\mathcal{M}_{\mathfrak{b}} = \langle W_{\mathfrak{b}}, R_{\mathfrak{b}}, v^+_{\mathfrak{b}}, v^-_{\mathfrak{b}} \rangle$ be the model induced by $\mathfrak{b}$. Then,*

- *If $A, +i$ occurs on $\mathfrak{b}$, then $w_i \Vdash^+ A$.*
- *If $A, -i$ occurs on $\mathfrak{b}$, then $w_i \nVdash^+ A$.*
- *If $\sim A, +i$ occurs on $\mathfrak{b}$, then $w_i \Vdash^- A$.*
- *If $\sim A, -i$ occurs on $\mathfrak{b}$, then $w_i \nVdash^- A$.*

From the lemmata above it follows by familiar reasoning, cf. Odintsov and Wansing (2010), Odintsov and Wansing (2017), and Priest (2008) that the respective tableau calculi are sound and complete for **cBK**$^-$, **cKN4**, **scBK**$^-$, and **scKN4**.

Theorem 13.1 *Let $\Gamma \cup \{A\}$ be a finite set of L-formulas, with $\mathsf{L} \in \{$**cBK**$^-$, **cKN4**, **scBK**$^-$, **scKN4**$\}$. Then $\Gamma \models_{\mathsf{L}} A$ iff $\Gamma \vdash_{\mathsf{L}} A$ in the respective tableau calculus.*

13.4.2 cMBL *and* scMBL

For the soundness and completeness of **cMBL** and **scMBL** we will follow, with some modification, the structure of the proofs given in Sect. 13.4.1.

Definition 13.7 Let $\mathcal{M} = \langle W, R^+, R^-, v^+, v^- \rangle$ be any L-model, with $\mathsf{L} \in \{$**cMBL**, **scMBL**$\}$ and $\mathfrak{b}$ be a tableau branch. The model $\mathcal{M}$ is said to be faithful to $\mathfrak{b}$ iff there exists a function f from the set of natural numbers into W such that:

(1) for every line $A, +i$ on $\mathfrak{b}$, $f(i) \Vdash^+ A$.
(2) for every line $A, -i$ on $\mathfrak{b}$, $\neg(f(i) \Vdash^+ A)$.
(3) for every line iR^+j on $\mathfrak{b}$, $f(i)R^+f(j)$.
(4) for every line iR^-j on $\mathfrak{b}$, $f(i)R^-f(j)$.

The function f is said to show that $\mathcal{M}$ is faithful to the branch $\mathfrak{b}$.

Lemma 13.8 *Let* $\mathcal{M} = \langle W, R^+, R^-, v^+, v^- \rangle$ *be any* L*-model, with* $\mathsf{L} \in \{$**cMBL**, **scMBL**$\}$ *and* $\mathfrak{b}$ *be a tableau branch. If* $\mathcal{M}$ *is faithful to* $\mathfrak{b}$ *and a tableau rule is applied to* $\mathfrak{b}$*, then the application produces at least one extension* $\mathfrak{b}'$ *of* $\mathfrak{b}$*, such that* $\mathcal{M}$ *is faithful to* $\mathfrak{b}'$.

Proof By induction on the construction of the tableau and inspection of the tableau rules. We will proof the lemma only for the cases that involve $\boxplus\!\!\to$, since the proofs for the other connectives have been given in Odintsov and Wansing (2017) and Sect. 13.3.1.

We will now suppose the function f shows that $\mathcal{M}$ is faithful to a branch containing the following lines in the respective tableau:

$A \boxplus\!\!\to B, +i$:	Then, with $f(i)R^+f(j)$ we have, $f(j) \Vdash^+ A \twoheadrightarrow f(j) \Vdash^+ B$, which is equivalent to $\neg(f(j) \Vdash^+ A) \Cup f(j) \Vdash^+ B$. And with $f(i)R^-f(j)$ we have $\neg(f(j) \Vdash^+ A \twoheadrightarrow f(j) \Vdash^- B)$, which is equivalent to $f(j) \Vdash^+ A \Cap \neg(f(j) \Vdash^- B)$.
$A \boxplus\!\!\to B, -i$:	Then, there is a j such that $f(i)R^+f(j)$ and we have, $\neg(f(j) \Vdash^+ A \twoheadrightarrow f(j) \Vdash^+ B)$, which is equivalent to $f(j) \Vdash^+ A \Cap \neg(f(j) \Vdash^+ B)$. Or, there is a j such that $f(i)R^-f(j)$ and we have, $f(j) \Vdash^+ A \twoheadrightarrow f(j) \Vdash^- B$, which is equivalent to $\neg(f(j) \Vdash^+ A) \Cup f(j) \Vdash^- B$.
$\sim(A \boxplus\!\!\to B), +i$:	Then, with $f(i)R^+f(j)$ we have, $f(i) \Vdash^- A \boxplus\!\!\to B \Leftrightarrow (f(j) \Vdash^+ A \twoheadrightarrow f(j) \Vdash^- B) \Leftrightarrow f(i) \Vdash^+ A \boxplus\!\!\to \sim B$.
$\sim(A \boxplus\!\!\to B), -i$:	Then, there is a j such that $f(i)R^+f(j)$ and we have, $\neg(f(i) \Vdash^+ \sim(A \boxplus\!\!\to B)) \Leftrightarrow \neg(f(j) \Vdash^- A \boxplus\!\!\to B) \Leftrightarrow \neg(f(j) \Vdash^+ A \twoheadrightarrow f(j) \Vdash^- B) \Leftrightarrow \neg(f(i) \Vdash^+ A \boxplus\!\!\to \sim B)$. □

Definition 13.9 Let $\mathfrak{b}$ be a complete open tableau branch. Then the structure $\mathcal{M}_\mathfrak{b} = \langle W_\mathfrak{b}, R^+_\mathfrak{b}, R^-_\mathfrak{b}, v^+_\mathfrak{b}, v^-_\mathfrak{b} \rangle$ induced by $\mathfrak{b}$ is defined as follows:

(1) $W_\mathfrak{b} := \{w_i \mid i \text{ occurs on } \mathfrak{b}\}$,
(2) $w_i R^+_\mathfrak{b} w_j$ iff iR^+j occurs $\mathfrak{b}$,
(3) $w_i R^-_\mathfrak{b} w_j$ iff iR^-j occurs $\mathfrak{b}$,

(4) $w_i \in v_{\mathfrak{b}}^{+}(p)$ iff $p, +i$ occurs on $\mathfrak{b}$,
(5) $w_i \in v_{\mathfrak{b}}^{-}(p)$ iff $\sim p, +i$ occurs on $\mathfrak{b}$.

Lemma 13.10 *Suppose $\mathfrak{b}$ is a complete open tableau branch, and let $\mathcal{M}_{\mathfrak{b}} = \langle W_{\mathfrak{b}}, R_{\mathfrak{b}}^{+}, R_{\mathfrak{b}}^{-}, v_{\mathfrak{b}}^{+}, v_{\mathfrak{b}}^{-} \rangle$ be the model induced by $\mathfrak{b}$. Then,*

- *If $A, +i$ occurs on $\mathfrak{b}$, then $w_i \Vdash^{+} A$.*
- *If $A, -i$ occurs on $\mathfrak{b}$, then $w_i \nVdash^{+} A$.*
- *If $\sim A, +i$ occurs on $\mathfrak{b}$, then $w_i \Vdash^{-} A$.*
- *If $\sim A, -i$ occurs on $\mathfrak{b}$, then $w_i \nVdash^{-} A$.*

From the lemmata above it follows by familiar reasoning, cf. Odintsov and Wansing (2010), Odintsov and Wansing (2017), and Priest (2008) that the respective tableau calculi are sound and complete for **cMBL** and **scMBL**.

Theorem 13.2 *Let $\Gamma \cup \{A\}$ be a finite set of* L*-formulas, with* L $\in$ {**cMBL**, **scMBL**}. *Then $\Gamma \models_{\mathsf{L}} A$ iff $\Gamma \vdash_{\mathsf{L}} A$ in the respective tableau calculus.*

Note that it is possible to consider infinite premise sets. This requires a modification in the proof of the soundness lemma, cf. Priest (2008, p. 285).

13.5 Some Properties of cBK^{-}, cKN4, scBK^{-}, scKN4, cMBL, and scMBL

Connexive logics are contra-classical insofar as they fail to be subsystems of classical logic, and connexive systems may have unusual properties. The connexive logic **CC1** of Angell and McCall (1966), for example, invalidates conjunctive simplification, $(A \wedge B) \rightarrow A$ and $(A \wedge B) \rightarrow B$. The logics **cBK**$^{-}$, **cKN4**, **scBK**$^{-}$, **scKN4**, **cMBL**, and **scMBL** do validate conjunctive simplification, but they are all *inconsistent* but non-trivial logics. The following schematic formulas are valid in the indicated systems containing the displayed conditionals:

$$
\begin{array}{ll}
\sim(A \rightarrow_c A) \rightarrow_c (A \rightarrow_c A) & \sim(\sim(A \rightarrow_c A) \rightarrow_c (A \rightarrow_c A)) \\
(A \wedge \sim A) \rightarrow_c A, & \sim((A \wedge \sim A) \rightarrow_c A), \\
(A \wedge \sim A) \Rightarrow_c A, & \sim((A \wedge \sim A) \Rightarrow_c A) \\
(A \wedge \sim A) \Box\!\!\rightarrow A, & \sim((A \wedge \sim A) \Box\!\!\rightarrow A), \\
(A \wedge \sim A) \Box\!\!\Rightarrow A, & \sim((A \wedge \sim A) \Box\!\!\Rightarrow A), \\
(A \wedge \sim A) \boxplus\!\!\rightarrow A, & \sim((A \wedge \sim A) \boxplus\!\!\rightarrow A).
\end{array}
$$

Note that Arieli and Avron's logic of logic bilattices **BL**$\supset$ on which **cMBL** is based is already negation-inconsistent, as pointed out in Omori and Wansing (2018). This is due to the fact that the information join, $\oplus$, combines the standard truth conditions of disjunction with the standard falsity conditions of conjunction, so that, both $(A \supset A) \oplus \sim(A \supset A)$ and $\sim((A \supset A) \oplus \sim(A \supset A))$ are valid, where $\supset$ is the weak implication of **BL**$\supset$.

All logics under consideration are, however, decidable.

Corollary 13.11 *For finite premise sets, the entailment relations of the systems* **cBK**$^-$, **cKN4**, **scBK**$^-$, **scKN4**, **cMBL**, *and* **scMBL** *are decidable.*

Proof Each tableau rule for these systems, except of the rules for the propositional constants $\top$, $\bot$, n, b (given in Odintsov and Wansing 2017), either outputs formulas with a smaller number of connectives than present in the formula from the input or with formulas that decompose into formulas with a smaller number of connectives. We restrict the use of the tableau rules for the propositional constants to applications by which a branch is closed. As a result, the tableau construction terminates. We can make this more precise by introducing the following complexity measure, $c(A)$, for formulas:

$$\begin{array}{lll}
c(p) & = c({\sim}p) \; = \; 0 & \text{for atomic propositions } p \\
c(\sharp) & = 0 & \text{for } \sharp \in \{\top, \bot, \mathsf{n}, \mathsf{b}\} \\
c(\sharp A) & = c(A) + 1 & \text{for } \sharp \in \{{\sim}{\sim}, \Box, \Diamond\} \\
c(\sharp A) & = c(A) + 2 & \text{for } \sharp \in \{{\sim}\Box, {\sim}\Diamond, \boxplus, {\sim}\boxplus\} \\
c(A \sharp B) & = c(A) + c(B) + 1 & \text{for } \sharp \in \{\wedge, \vee, \rightarrow_c, \Box\!\!\rightarrow, \oplus, \otimes\} \\
c(A \sharp B) & = c(A) + c(B) + 2 & \text{for } \sharp \in \{\Rightarrow_c, \Box\!\!\Rightarrow, \boxplus\!\!\rightarrow\} \\
c({\sim}(A \sharp B)) & = c(A) + c(B) + 2 & \text{for } \sharp \in \{\boxplus\!\!\rightarrow\} \\
c({\sim}(A \sharp B)) & = c(A) + c(B) + 3 & \text{for } \sharp \in \{\rightarrow_c, \Rightarrow_c, \Box\!\!\rightarrow, \Box\!\!\Rightarrow\} \\
c({\sim}(A \sharp B)) & = c(A) + c(B) + 4 & \text{for } \sharp \in \{\wedge, \vee, \oplus, \otimes\}
\end{array}$$

□

Another peculiarity of the systems **scBK**$^-$, **scKN4**, and **scMBL** is due to the strictness of their implications. The strict and connexive implications in the mentioned logics **C** and **CK** are obtained by laying down suitable support of falsity conditions for the otherwise intuitionistic implication. Since intuitionistic implication is defined with respect to a preorder, the reflexivity of that relation guarantees that *modus ponens* not only is a valid inference rule but that it also holds relativized to a given state. The strict and connexive implications of **scBK**$^-$, **scKN4**, and **scMBL** are defined with respect to an arbitrary relation so that it is not guaranteed, for example, that if $w \Vdash^+ A \Box\!\!\rightarrow B$ and $w \Vdash^+ A$, then $w \Vdash^+ B$. Finally, being extensions of **FDE**, the systems under consideration are paraconsistent logics.

13.6 On Algebraizability of the Connexive Logics cBK$^-$, cKN4, and cMBL

In this section we deal with a certain more complicated property of logics, namely, the existence of an equivalent algebraic semantics. More precisely, we study whether the logics **cBK**$^-$, **cKN4**, and **cMBL** are algebraizable in the sense of Blok and Pigozzi (1989). Roughly speaking, the fact that some logic L is algebraizable means that its consequence relation $\vdash_{\mathsf{L}}$ is mutually interpretable with the equational consequence relation defined over one or another class of algebras. The algebraizability of L allows to apply a variety of strong methods of universal algebra for studying this logic. Blok and Pigozzi (1989) discovered also some intrinsic characterizations of algebraizable logics that give us the possibility to conclude that logic L is algebraizable without explicit construction of its equivalent algebraic semantics. We will apply one of such characterizations and pass now to precise definitions.

The algebraizability theory was developed for logics defined via deductive systems. According to Łoś and Suszko (1958) a logic L may be defined via a deductive system if and only if $\vdash_{\mathsf{L}}$ is a compact and structural Tarskian consequence relation, i.e., $\vdash_{\mathsf{L}}$ satisfies the following five properties for all sets Γ, Δ of L-formulas and all L-formulas A, B:

(1) *Reflexivity.* $A \in \Gamma$ implies $\Gamma \vdash_{\mathsf{L}} A$;
(2) *Monotonicity.* $\Gamma \vdash_{\mathsf{L}} A$ and $\Gamma \subseteq \Delta$ imply $\Delta \vdash_{\mathsf{L}} A$;
(3) *Transitivity.* $\Gamma \vdash_{\mathsf{L}} A$ and $\Delta \vdash_{\mathsf{L}} B$ for every $B \in \Gamma$ imply $\Delta \vdash_{\mathsf{L}} A$;
(4) *Compactness.* $\Gamma \vdash_{\mathsf{L}} A$ implies $\Gamma' \vdash_{\mathsf{L}} A$ for some finite $\Gamma' \subseteq \Gamma$;
(5) *Structurality.* $\Gamma \vdash_{\mathsf{L}} A$ implies $\sigma(\Gamma) \vdash_{\mathsf{L}} \sigma A$ for every substitution σ.

Recall that by a substitution we mean a mapping σ from Prop to the set of all L-formulas, which can be extended homomorphically to the set of all L-formulas, $\sigma(\Gamma) = \{\sigma A \mid A \in \Gamma\}$.

For a set $\Theta(p, q) = \{\theta_i(p, q) \mid 0 \leq i \leq n\}$ of L-formulas and for L-formulas A and B, we consider $\Gamma \vdash_{\mathsf{L}} \Theta(A, B)$ as a tuple of sequents

$$\Gamma \vdash_{\mathsf{L}} \theta_0(A, B), \ \ldots\ , \ \Gamma \vdash_{\mathsf{L}} \theta_n(A, B).$$

On the other hand, $\Gamma, \Theta(A, B) \vdash_{\mathsf{L}} C$ is an abbreviation for $\Gamma, \theta_0(A, B), \ \ldots\ , \theta_n(A, B) \vdash_{\mathsf{L}} C$. In what follows we will write $A \Theta B$ instead of $\Theta(A, B)$.

A formal expression $A \approx B$, where A and B are L-formulas, is called an L-*equation*. For a finite set $\delta \approx \epsilon = \{\delta_j \approx \epsilon_j \mid 0 \leq j \leq m\}$ of L-equations, $\delta \,\Theta\, \epsilon$ is an abbreviation for

$$\{\theta_i(\delta_j, \epsilon_j) \mid 0 \leq i \leq n,\ 0 \leq j \leq m\}.$$

Theorem 13.3 (Theorem 4.7 of Blok and Pigozzi 1989) *Let* L *be a logic such that* $\vdash_{\mathsf{L}}$ *is a compact and structural Tarskian consequence relation. Logic* L *is algebraizable if and only if there exists a finite set* Δ *of* L*-formulas in two variables and a finite set* $\delta \approx \epsilon$ *of* L*-equations in a single variable such that the following conditions (i)–(v) hold for all* L*-formulas* A, B, C:

(i) $\vdash_{\mathsf{L}} A \,\Delta\, A$;
(ii) $A \,\Delta\, B \vdash_{\mathsf{L}} B \,\Delta\, A$;
(iii) $A \,\Delta\, B,\ B \,\Delta\, C \vdash_{\mathsf{L}} A \,\Delta\, C$;
(iv) $A_0 \,\Delta\, B_0, \ldots A_{n-1} \,\Delta\, B_{n-1} \vdash_{\mathsf{L}} \sharp(A_0, \ldots, A_{n-1}) \,\Delta\, \sharp(B_0, \ldots, B_{n-1})$ *for every primitive connective* $\sharp$ *of arity n and for all* L*-formulas* $A_0, \ldots, A_{n-1}, B_0, \ldots, B_{n-1}$;

(v) $A \dashv\vdash_{\mathsf{L}} \delta(A) \Delta \epsilon(A)$.

Up to this moment we considered local semantical consequence relations $\models_{\mathsf{L}}$. Recall that each of the logics L considered in the paper was defined via a class of its models $\mathcal{M}od_{\mathsf{L}}$, and the relation $\Gamma \models_{\mathsf{L}} A$ holds iff for every $\mathcal{M} \in \mathcal{M}od_{\mathsf{L}}$ and every world w of $\mathcal{M}$, if $w \Vdash^{+} B$ for all $B \in \Gamma$, then $w \Vdash^{+} A$. Such local semantical consequence relations corresponds to the inference from premises and tautologies of logic L with the help of *modus ponens* only. The consequence relation of a deductive system uses all rules of the system and it corresponds to global semantical consequence, which can be defined as follows. Let L be a logic defined via a class of models $\mathcal{M}od_{\mathsf{L}}$. An L-formula *is true in* $\mathcal{M} \in \mathcal{M}od_{\mathsf{L}}$, symbolically $\mathcal{M} \models A$, if $w \Vdash^{+} A$ for every world w of $\mathcal{M}$. We associate with L a global consequence relation $\models^{*}_{\mathsf{L}}$ as follows. For a set $\Gamma \cup \{A\}$ of L-formulas, the relation $\Gamma \models^{*}_{\mathsf{L}} A$ holds if for every $\mathcal{M} \in \mathcal{M}od_{\mathsf{L}}$, we have $\mathcal{M} \models A$ whenever $\mathcal{M} \models B$ for all $B \in \Gamma$.

Using the definition of validity of formulas in models we can directly check the following statement.

Proposition 13.12 *Let* $\mathsf{L} \in \{\mathbf{cBK}^{-}, \mathbf{cKN4}, \mathbf{cMBL}\}$. *Then* $\models^{*}_{\mathsf{L}}$ *is a structural Tarskian consequence relation.*

Neither the definition of validity nor the weak completeness theorems of Sect. 13.4 dealing with finite sets of premises can help, however, to prove that the relations $\models^{*}_{\mathsf{L}}$ are compact. Fortunately, as we will see, a minor modification of calculus $\mathbf{HBK}^{\Box -}$ from Drobyshevich and Wansing (2020) produces a Hilbert-style calculus strongly complete w.r.t. $\models^{*}_{\mathbf{cBK}^{-}}$. This completeness result implies that the relation $\models^{*}_{\mathbf{cBK}^{-}}$ is compact. Using Theorem 13.3 we will prove that $\models^{*}_{\mathbf{cBK}^{-}}$ is algebraizable. For the relation $\models^{*}_{\mathbf{cKN4}}$ we prove a conditional statement: compactness implies algebraizability. Finally, we will see that the relation $\models^{*}_{\mathbf{cMBL}}$ is algebraizable due to the fact that **cMBL** is definitionally equivalent to **MBL**.

We continue with the Hilbert-style calculus $\mathbf{HcBK}^{-}$, which has the following axioms:

(1) axioms of the positive fragment of classical propositional logic in the language $\{\vee, \wedge, \rightarrow_c\}$;
(2) axioms of strong negation $\sim$:
$\sim(A \wedge B) \leftrightarrow_c (\sim A \vee \sim B)$; $\sim(A \rightarrow_c B) \leftrightarrow_c (A \rightarrow_c \sim B)$;
$\sim(A \vee B) \leftrightarrow_c (\sim A \wedge \sim B)$; $\sim\sim A \leftrightarrow_c A$;
(3) modal axioms:

$\Box 1)$ $\Box(A \rightarrow_c B) \rightarrow_c (\Box A \rightarrow_c \Box B)$; $\Box 2)$ $\Box(\sim A \rightarrow_c \sim B) \rightarrow_c (\sim\Box A \rightarrow_c \sim\Box B)$;
$\Box 3)$ $\sim\Box(A \wedge B) \rightarrow_c (\sim\Box A \vee \sim\Box B)$; $\Box 4)$ $\Box(A \vee \sim B) \Rightarrow_c (\Box A \vee \sim\Box B)$;
$\Diamond 1)$ $\Diamond A \Leftrightarrow_c \sim\Box\sim A$.

Here and in what follows, working with $\mathbf{cBK}^{-}$ and **cMBL**, we use the following abbreviations:

$$A \Rightarrow_c B := (A \rightarrow_c B) \wedge (\sim B \rightarrow_c \sim A),$$

$$A \leftrightarrow_c B := (A \rightarrow_c B) \wedge (B \rightarrow_c A), \quad A \Leftrightarrow_c B := (A \Rightarrow_c B) \wedge (B \Rightarrow_c A).$$

The inference rules of **cBK**$^-$ include *modus ponens* for $\rightarrow_c$ and *necessitation* for $\Box$:

$$(MP^{\rightarrow_c})\ \frac{A \quad A \rightarrow_c B}{B}; \quad (N^{\Box})\frac{A}{\Box A}.$$

The relation $\Gamma \vdash^*_{\mathbf{cBK}^-} A$ holds if A can be obtained from elements of Γ and **HcBK**$^-$-axioms with the help of $MP^{\rightarrow_c}$ and $N^{\Box}$. The set of **HcBK**$^-$-theorems is defined as $Th(\mathbf{HcBK}^-) = \{A \mid \varnothing \vdash^*_{\mathbf{cBK}^-} A\}$. Finally, the relation $\Gamma \vdash_{\mathbf{cBK}^-} A$ holds if A can be obtained from elements of $\Gamma \cup Th(\mathbf{HcBK}^-)$ with the help of $MP^{\rightarrow_c}$ only. In what follows we omit the lower index in the expressions "$\vdash_{\mathbf{cBK}^-}$" and "$\vdash^*_{\mathbf{cBK}^-}$" if it does not lead to confusion.

A set Γ of **cBK**$^-$-formulas is said to be a *prime* **cBK**$^-$*-theory* if it is non-trivial (different from the set of all **cBK**$^-$-formulas), closed under $\vdash$, and satisfies the *disjunction property*: $A \vee B \in \Gamma$ implies $A \in \Gamma$ or $B \in \Gamma$.

Standardly, one can prove

Lemma 13.13 (Extension lemma) *For any set Γ of* **cBK**$^-$*-formulas and* **cBK**$^-$*-formula A, if $\Gamma \nvdash A$, then there is a prime* **cBK**$^-$*-theory such that $\Gamma \subseteq \Gamma'$ and $\Gamma' \nvdash A$.*

The *canonical* **cBK**$^-$*-model* is a tuple $\mathcal{M}^c = \langle W_c, R_c, v_c^+, v_c^- \rangle$, where 1) W_c is the set of all prime **cBK**$^-$-theories; 2) for prime **cBK**$^-$-theories Γ and Δ, the relation $\Gamma R_c \Delta$ holds if for every **cBK**$^-$-formula A:

$$(\Box A \in \Gamma \text{ implies } A \in \Delta) \text{ and } ({\sim} A \in \Delta \text{ implies } {\sim}\Box A \in \Gamma);$$

3) $v_c^+(p) = \{\Gamma \in W_c \mid p \in \Gamma\}$ and $v_c^-(p) = \{\Gamma \in W_c \mid {\sim} p \in \Gamma\}$.

Lemma 13.14 (Canonical model lemma) *For every prime* **cBK**$^-$*-theory Γ and* **cBK**$^-$*-formula A, we have*

$$\mathcal{M}^c, \Gamma \Vdash^+ A \text{ iff } A \in \Gamma; \quad \mathcal{M}^c, \Gamma \Vdash^- A \text{ iff } {\sim} A \in \Gamma.$$

Proof This statement can be proved by a natural modification of the proof of Drobyshevich and Wansing (2020, Lemma 3.4). There are two main differences between **cBK**$^-$ and **BK**$^{\Box-}$ of Drobyshevich and Wansing (2020). First, the language of **BK**$^{\Box-}$ lacks $\Diamond$, but axiom $\Diamond$1) of **HcBK**$^-$ allows to reduce the case of the possibility operator to that of the necessity operator. Second, the falsification of $\rightarrow_c$ is defined in a different way as compared to the implication $\rightarrow$ of **BK**$^{\Box-}$. The case of falsification (support of falsity) for connexive implication can be treated, however, as follows:

$$\Gamma \Vdash^- A \rightarrow_c B \text{ iff } (\Gamma \Vdash^+ A \twoheadrightarrow \Gamma \Vdash^- B) \text{ iff (ind. hypothesis) } (A \notin \Gamma \Cup {\sim} B \in \Gamma) \text{ iff}$$

$$\text{iff } A \rightarrow_c {\sim} B \in \Gamma \text{ iff } {\sim}(A \rightarrow_c B) \in \Gamma.$$

□

Theorem 13.4 (Strong completeness for global consequence) *For any set* Γ *of* $\mathbf{cBK}^-$*-formulas and any* $\mathbf{cBK}^-$*-formula* A*,*

$$\Gamma \vdash^*_{\mathbf{cBK}^-} A \textit{ iff } \Gamma \models^*_{\mathbf{cBK}^-} A.$$

Proof The soundness part is omitted as usual.

Assume that $\Gamma \nvdash^*_{\mathbf{cBK}^-} A$. Let $\Gamma' = \{B \mid \Gamma \vdash^*_{\mathbf{cBK}^-} B\}$. Obviously, $\Gamma' \nvdash^*_{\mathbf{cBK}^-} A$, moreover, $\Gamma' \nvdash_{\mathbf{cBK}^-} A$. By the Extension lemma, there is $\Gamma'' \in W_c$ such that $\Gamma' \subseteq \Gamma''$ and $\Gamma'' \nvdash_{\mathbf{cBK}^-} A$. From the Canonical model lemma, we have

$$\Gamma'' \Vdash^+ B \text{ for all } B \in \Gamma' \text{ and } \Gamma'' \nVdash^+ A.$$

Let us consider the submodel of $\mathcal{M}^c$ generated by Γ'', i.e., the model $\mathcal{M}^c_{\Gamma''} = \langle W_{\Gamma''}, R_1, v_1^+, v_1^- \rangle$, where $W_{\Gamma''}$ consists of Γ'' and all those worlds from W_c which can be reached from Γ'' in a finite number of steps via R_c. The accessibility relation R_1 and valuations v_1^+, v_1^- are induced by R_c, v_c^+, and v_c^- in a natural way. Standardly one can prove that for every $\Delta \in W_{\Gamma''}$ and every $\mathbf{cBK}^-$-formula B,

$$\mathcal{M}^c, \Delta \Vdash^+ B \text{ iff } \mathcal{M}^c_{\Gamma''}, \Delta \Vdash^+ B. \tag{13.1}$$

If $\Delta \in W_{\Gamma''}$ and it can be reached from Γ'' in n steps, i.e., there are $\Delta_1, \ldots, \Delta_{n-1}$ such that

$$\Gamma'' R_c \Delta_1 R_c \Delta_2 R_c \ldots R_c \Delta_{n-1} R_c \Delta,$$

then $\Delta \Vdash^+ B$ whenever $\Gamma'' \Vdash^+ \Box^n B$.[6] Since Γ' is closed under rule $N^\Box$, we have

$$\mathcal{M}^c_{\Gamma''} \models B \text{ for all } B \in \Gamma'.$$

At the same time from (13.1) we have $\mathcal{M}^c_{\Gamma''}, \Gamma'' \nVdash^+ A$, i.e., $\mathcal{M}^c_{\Gamma''} \not\models A$. This concludes the proof. □

Corollary 13.15 *The relation* $\models^*_{\mathbf{cBK}^-}$ *is compact.*

This corollary together with Proposition 13.12 imply that Theorem 13.3 can be applied to $\models^*_{\mathbf{cBK}^-}$.

Theorem 13.5 *The relation* $\models^*_{\mathbf{cBK}^-}$ *is algebraizable with the equivalence formula* $p \Leftrightarrow_c q$ *and defining equation* $p \rightarrow_c p = p$.

Proof It follows from the definition of validity of formulas that for every model $\mathcal{M}$ and $\mathbf{cBK}^-$-formulas A and B, we have $\mathcal{M} \models A \Leftrightarrow_c B$ if and only for every world w of $\mathcal{M}$,

$$w \Vdash^+ A \text{ iff } w \Vdash^+ B, \quad w \Vdash^- A \text{ iff } w \Vdash^- B.$$

---

[6] As usual, $\Box^0 B = B$ and $\Box^{n+1} B = \Box\Box^n B$.

Using this remark, we can easily check the items (i)–(iv) of Theorem 13.3 for $\models^*_{\mathbf{cBK}^-}$ and formula $p \Leftrightarrow_c q$. It remains to check that for every model $\mathcal{M}$ and $\mathbf{cBK}^-$-formula A,

$$\mathcal{M} \models A \text{ iff } \mathcal{M} \models (A \rightarrow_c A) \Leftrightarrow_c A.$$

Assume that A is true in all worlds of $\mathcal{M}$ and prove that the formulas $A \rightarrow_c A$ and A are true and false in the same worlds of $\mathcal{M}$. Since $A \rightarrow_c A$ is true at every world of every model, we have to check only that $A \rightarrow_c A$ and A are false in the same worlds of $\mathcal{M}$. Let w be a world of $\mathcal{M}$. We have

$$w \Vdash^- A \rightarrow_c A \text{ iff } (w \Vdash^+ A \twoheadrightarrow w \Vdash^- A) \text{ iff } w \models^- A.$$

The second equivalence is due to our assumption that $\mathcal{M} \models A$. Now we assume that $A \rightarrow_c A$ and A are true and false in the same worlds of $\mathcal{M}$. From $\mathcal{M} \models A \rightarrow_c A$ we obtain $\mathcal{M} \models A$.

Theorem 13.3 implies now that $\models^*_{\mathbf{cBK}^-}$ is algebraizable with the equivalence formula $p \Leftrightarrow_c q$ and defining equation $p \rightarrow_c p = p$. □

Modifying the proof of this theorem, we obtain

Corollary 13.16 *If the relation $\models^*_{\mathbf{cKN4}}$ is compact, then it is algebraizable with the equivalence formula $p \Leftrightarrow_c q$ and the defining equation $p \Rightarrow_c p = p$.*

Proof As above we check the items (i)–(iv) of Theorem 13.3 for $\models^*_{\mathbf{cKN4}}$ and formula $p \Leftrightarrow_c q$. It remains to check that for every model $\mathcal{M}$ and **cKN4**-formula A,

$$\mathcal{M} \models A \text{ iff } \mathcal{M} \models (A \Rightarrow_c A) \Leftrightarrow_c A.$$

Since the falsification conditions for $\Rightarrow_c$ and $\rightarrow_c$ are the same, this last check can be done in essentially the same way as in the previous proof. Thus, the compactness of $\models^*_{\mathbf{cBK}^-}$ implies its algebraizability. □

Theorem 13.6 *The relation $\models^*_{\mathbf{cMBL}}$ is algebraizable.*

Proof Let us consider the formula $\odot(p, q) = (p \wedge \mathsf{b}) \vee (\sim q \wedge \mathsf{n})$ from Odintsov et al. (2019). It has one remarkable property. For every model $\mathcal{M}$ and one of its worlds w, we have the following equivalences:

$$w \Vdash^+ \odot(p, q) \text{ iff } w \Vdash^+ p; \quad w \Vdash^- \odot(p, q) \text{ iff } w \Vdash^+ q.$$

In particular, the formula $\odot(A \rightarrow_c B, A \wedge \sim B)$ has the following verification and falsification conditions:

$$w \Vdash^+ \odot(A \rightarrow_c B, A \wedge \sim B) \text{ iff } (w \Vdash^+ A \twoheadrightarrow w \Vdash^+ B);$$
$$w \Vdash^- \odot(A \rightarrow_c B, A \wedge \sim B) \text{ iff } (w \Vdash^+ A \mathbin{\&} w \Vdash^- B).$$

Thus, $\odot(A \rightarrow_c B, A \wedge \sim B)$ has exactly the same verification and falsification conditions as $A \rightarrow B$, where $\rightarrow$ is the conditional connective of the original logic

MBL. Let us define a structural translation ρ from the set of all **MBL**-formulas to the set of all **cMBL**-formulas so that it preserves all propositional variables and constants, commutes with all connectives except for $\to$, and for $\to$ we put $\rho(A \to B) = \odot(\rho A \to_c \rho B, \rho A \wedge {\sim}\rho B)$. It is routine to check that for every set Γ of **MBL**-formulas and **MBL**-formula A, we have

$$\Gamma \models^*_{\mathbf{MBL}} A \text{ iff } \rho(\Gamma) \models^*_{\mathbf{cMBL}} \rho A,$$

where $\rho(\Gamma) = \{\rho B \mid B \in \Gamma\}$. Now we define a structural translation θ, which acts in the inverse direction, from the set of all **cMBL**-formulas to the set of all **MBL**-formulas, and preserves all propositional variables and constants, commutes with all connectives except for $\to_c$, and $\theta(A \to_c B) = \odot(\theta A \to \theta B, \theta A \to {\sim}\theta B)$. Again we can see that the verification and falsification conditions for $\theta(p \to_c q)$ coincide with those for $p \to_c q$. This allows one to prove that for every set Γ of **cMBL**-formulas and **cMBL**-formula A, we have

$$\Gamma \models^*_{\mathbf{cMBL}} A \text{ iff } \theta(\Gamma) \models^*_{\mathbf{MBL}} \theta A.$$

Naturally, $\theta(\Gamma) = \{\theta B \mid B \in \Gamma\}$. Now we calculate the results of double translations:

$$\rho\theta(A \to_c B) = \odot(\odot(\rho\theta A \to_c \rho\theta B, \rho\theta A \wedge {\sim}\rho\theta B), \odot(\rho\theta A \to_c {\sim}\rho\theta B, \rho\theta A \wedge {\sim\sim}\rho\theta B)),$$

$$\theta\rho(A \to B) = \odot(\odot(\theta\rho A \to \theta\rho B, \theta\rho A \to {\sim}\theta\rho B), \theta\rho A \wedge {\sim}\theta\rho B),$$

from which we have

$$w \Vdash^+ \rho\theta(A \to_c B) \text{ iff } w \Vdash^+ \odot(\rho\theta A \to_c \rho\theta B, \rho\theta A \wedge {\sim}\rho\theta B) \text{ iff } w \Vdash^+ \rho\theta A \to_c \rho\theta B,$$

$$w \Vdash^- \rho\theta(A \to_c B) \text{ iff } w \Vdash^+ \odot(\rho\theta A \to_c {\sim}\rho\theta B, \rho\theta A \wedge {\sim\sim}\rho\theta B) \text{ iff } w \Vdash^- \rho\theta A \to_c \rho\theta B,$$

and similarly

$$w \Vdash^+ \theta\rho(A \to B) \text{ iff } w \Vdash^+ \odot(\theta\rho A \to \theta\rho B, \theta\rho A \to {\sim}\theta\rho B) \text{ iff } w \Vdash^+ \theta\rho A \to \theta\rho B,$$

$$w \Vdash^- \theta\rho(A \to B) \text{ iff } w \Vdash^+ \theta\rho A \wedge {\sim}\theta\rho B \text{ iff } w \Vdash^- \theta\rho A \to \theta\rho B.$$

Using these relations, by induction on the complexity of formulas we infer that for every **cMBL**-formula A and every **MBL**-formula B,

$$\varnothing \models^*_{\mathbf{cMBL}} A \Leftrightarrow_c \rho\theta A, \quad \varnothing \models^*_{\mathbf{MBL}} B \Leftrightarrow \theta\rho B.$$

It was noticed in Odintsov and Wansing (2017) that the connective $\Leftrightarrow$ determines Tarski's congruence on the algebra of **MBL**-formulas (the greatest congruence compatible with all $\models^*_{\mathbf{MBL}}$-theories). Similarly, one can prove that $\Leftrightarrow_c$ determines Tarski's congruence on the algebra of **cMBL**-formulas.

We have thus proved that up to Tarski's congruence there are mutually inverse structural translations between **MBL** and **cMBL**, i.e., the logics **MBL** and **cMBL** are definitionally equivalent in the sense of Gyuris (1999). According to Rivieccio et al. (2015), logic **MBL** is algebraizable. Applying the results of Gyuris (1999), we conclude that **cMBL** is algebraizable too. □

13.7 Conclusion

In this paper we introduced various **FDE**-based modal logics through classes of Kripke models with suitable support of falsity conditions. In some of these logics their implication connective is both connexive and strict, which is a combination of properties that have been considered in the context of modeling natural language conditionals. We presented sound and complete tableau calculi for all logics under consideration. Moreover, we proved that the logics $\mathbf{cBK}^-$ and **cMBL** are algebraizable, but an equivalent algebraic semantics for these logics was not presented. In the future, we plan to develop and study the algebraic semantics for these logics. Also, we will try to axiomatize the rest of the logics considered in the paper and we intend to study the algebraization problem for them. Other directions for further investigations include the study of stronger modal logics by imposing the familiar frame conditions for obtaining normal modal logics stronger than the smallest normal modal logic **K** based on classical logic, and the study of first-order extensions of $\mathbf{cBK}^-$, **cKN4**, $\mathbf{scBK}^-$, **scKN4**, **cMBL**, and **scMBL**.

Acknowledgements The work reported in this paper has been carried out as part of the research project **FDE**-based modal logics, supported by the Deutsche Forschungsgemeinschaft, DFG, grant WA 936/13-1, and the Russian Foundation for Basic Research, RFBR, grant No. 18-501-12019. We gratefully acknowledge this support.

References

Arieli, O., & Avron, A. (1996). Reasoning with logical bilattices. *Journal of Logic, Language and Information*, *5*(1), 25–63.

Avron, A., Arieli, O., & Zamansky, A. (2018). *Theory of Effective Propositional Paraconsistent Logics*. London: College Publications.

Blok, W. J., & Pigozzi, D. (1989). *Algebraizable Logics* (Vol. 396). Memoirs of the American Mathematical Society. AMS.

Drobyshevich, S., & Wansing, H. (2020). Proof systems for various FDE-based modal logics. In *Review of Symbolic Logic*, *13*(4), 720–747.

Gyuris, V. (1999). *Variations of Algebraizability*, PhD thesis, The University of Illinois at Chicago.

Iacona, A. (2019). Strictness and Connexivity. In *Inquiry. An Interdisciplinary Journal of Philosophy*, published online October 20.
Kamide, N. (2019). Bi-Classical Connexive Logic and its Modal Extension: Cut-elimination, completeness and duality. *Logic and Logical Philosophy*, *28*(3), 481–511.
Lewis, C. I. (1918). *A Survey of Symbolic Logic*. London: Forgotten Books. Reprinted 2015.
Łoś, J., & Suszko, R. (1958). Remarks on Sentential logic. *Proceedings of the Koninklijke Nederlandse Akademie van Wetenschappen, Series A*, *61*, 177–183.
McCall, S. (1966). Connexive implication. *Journal of Symbolic Logic*, *31*, 415–433.
McCall, S. (2012). A history of connexivity. In D. M. Gabbay, et al. (Eds.), *Handbook of the History of Logic. Volume 11. Logic: A History of its Central Concepts* (pp. 415–449). Amsterdam: Elsevier.
Odintsov, S. P., & Wansing, H. (2010). Modal logics with Belnapian truth values. *Journal of Applied Non-Classical Logics*, *20*(3), 279–301.
Odintsov, S. P., & Wansing, H. (2017). Disentangling FDE-based Paraconsistent Modal Logics. *Studia Logica*, *105*(6), 1221–1254.
Odintsov, S. P., Skurt, D., & Wansing, H. (2019). On definability of connectives and modal logics over FDE. *Logic and Logical Philosophy*, *28*(4), 631–659.
Omori, H. (2019). Towards a bridge over two approaches in connexive logic. *Logic And Logical Philosophy*, *28*(3), 553–566.
Omori, H., & Wansing, H. (2019). Connexive logics: an overview and current trends. *Logic and Logical Philosophy*, *28*(4), 371–387.
Omori, H., & Wansing, H. (2018). On contra-classical variants of Nelson logic N4 and its classical extension. *Review of Symbolic Logic*, *11*(4), 805–820.
Priest, G. (2008). *An Introduction to Non-Classical Logics: From If to Is* (2nd ed.). Cambridge: Cambridge University Press.
Priest, G. (2018). Some new thoughts on conditionals. *Topoi*, *37*, 369–377.
Rivieccio, U., Jung, A., & Jansana, R. (2017). Four-valued modal logic: Kripke semantics and duality. *Journal of Logic and Computation*, *27*(1), 155–199. Published online 15 June 2015.
Routley, R., Meyer, R. K., Plumwood, V., & Brady, R. T. (1982). *Relevant Logics and their Rivals* (Vol. 1). Atascadero: Ridgeview Publishing Company.
Wansing, H., et al. (2005). Connexive Modal Logic. In R. Schmidt (Ed.), *Advances in Modal Logic* (Vol. 5, pp. 367–383). King's College Publications.
Wansing, H. (2021). Connexive Logic. In E. N. Zalta (Ed.), *The Stanford Encyclopedia of Philosophy* (Spring 2021 ed.). https://plato.stanford.edu/archives/spr2016/entries/logic-connexive/.
Wansing, H., & Odintsov, S. P. (2016). On the methodology of paraconsistent logic. In H. Andreas & P. Verde (Eds.), *Logical Studies of Paraconsistent Reasoning in Science and Mathematics* (pp. 175–204). Springer.
Wansing, H., & Skurt, D. (2018). Negation as cancellation, connexive logic, and qLPm. *Australasian Journal of Logic*, *15*(2), 476–488.

Sergei Odintsov is a leading Researcher at Sobolev Institute of Mathematics (Novosibirsk, Russia). He also teaches at the Chair of Algebra and Mathematical Logic of Novosibirsk State University (Associate Professor since 2015 and Professor since 2020). He took his M.A. (1984) in Mathematics at Novosibirsk State University and his Ph.D. (candidate of sciences, 1987) and Habilitation (Doctor of Sciences, 2007) in Algebra, Number Theory and Mathematical Logic at Sobolev Institute of Mathematics. He is the author of Constructive Negations and Paraconsistency (Springer 2008) and of numerous articles. He has been working mainly on algorithm theory and non-classical logic. Moreover, he is a member of a number of editorial boards of logic and philosophy journals.

Daniel Skurt is a Postdoctoral Researcher working with Hitoshi Omori on Dialetheism and Paraconsistent Logics at the Ruhr University Bochum, Germany. In 2017, he took his Ph.D. in Logic at the Ruhr University Bochum, in 2012 his M.A. in Logic at Leipzig University, and in 2007 his

Dipl.-Ing. in Electrical Engineering at the Technical University of Chemnitz. His research interests include many-valued logics, modal logics, and paraconsistent logics.

Heinrich Wansing is a Professor of Logic and Epistemology at the Ruhr University Bochum (Germany). Before that he was a Professor of Philosophy of Science and Logic at TU Dresden (1999-2010). He took his M.A. and his Ph.D. in philosophy at the Free University of Berlin and his Habilitation in Logic and Analytical Philosophy at the University of Leipzig. He is the author of The Logic of Information Structures (Springer 1993), Displaying Modal Logic (Kluwer 1998), Truth and Falsehood. An Inquiry into Generalized Logical Values (with Y. Shramko, Springer 2011), Proof Theory of N4-related Paraconsistent Logics (with N. Kamide, College Publications 2015), and numerous articles. He has been working mainly on philosophical logic. Moreover, he is the Editor-in-Chief of the book series Trends in Logic (Springer), a Managing Editor of the journal Studia Logica, and a member of a number of other editorial boards of logic and philosophy journals.

Chapter 14
Comments on the Papers

Arnon Avron

Abstract This final chapter includes some comments of mine about the papers in this volume and their connections with my work. I am very grateful to all the authors of these papers for their nice contributions!

14.1 Bimbo

Kata is of course one of the world's leading researchers on substructural logics in general, and relevance logics in particular. This paper too is a contribution to the study and understanding of substructural logics. As their very name indicates, such logics are defined using proof systems. However, no proof-theoretical method is used in Kata's paper. Instead, the paper heavily uses *semantic* methods. This fits well my recent tendency (noted upon also in my remarks about the contribution of Colacito, Galatos, and Metcalfe) to rely more on semantic methods in the study of proof systems than on pure proof-theoretical ones. However, the semantic method that I have usually employed in my research on substructural logics and paraconsistent logics were mainly algebraic, or non-deterministic generalizations of algebraic methods. In addition, I have also used Kripke-style semantics when modal or intermediate logics were involved. What I have never used until not long ago was the semantic framework for relevance logics which has been developed by Routley and Meyer. This framework is based on the use of *ternary* relations among worlds (or "setups", or "situations"), and for long I have found it too complicated for me to work with. My opinion about its value changed several years ago, when the use of such semantics was the only way I could find for proving that the relevance logic **R** has what I take

A. Avron (✉)
School of Computer Science, Tel Aviv University, Tel Aviv, Israel
e-mail: aa@cs.tau.ac.il

O. Arieli and A. Zamansky (eds.), *Arnon Avron on Semantics and Proof Theory of Non-Classical Logics*, Outstanding Contributions to Logic 21,
https://doi.org/10.1007/978-3-030-71258-7_14

as the most basic relevance criterion. (This fact is announced on Page 457 of Avron et al. (2018), but its proof has not been published yet.)

Specifically, in this paper Kata uses the relational approach to the study of certain substructural logic which have the standard two modalities (or "exponentials"). But its real goal is to contribute to the understanding of the standard connectives, their general role(s), and the connections between them. Particularly interesting for me is the light it shed on the role of the negation connective—a connective in which I always have had a special interest. This statement might look strange, since deliberately, none of the systems studied here include negation. But seeing what can be done *without* some connective is also essential for understanding the role of that connective. Thus the presence of negation is usually crucial for establishing duality between disjunction and conjunction, as well as between necessity and possibility. Nevertheless, in this paper those dualities are handled and used without negation. The way this is done in the case of disjunction and conjunction is particularly interesting: instead of studying *internal* relations between them within one system, the paper shows *external* duality between different systems, one having disjunction but not conjunction, while in the other it is the other way around.

14.2 Caleiro, Marcelino

Developing and applying the semantic framework of non-deterministic matrices (Nmatrices) is one of the main research programs on which I have been working for the last twenty years. (Actually, at the beginning I thought that I had invented it, but now I know that I had not been the first to come across this idea.) In my work on it I have discovered that what is perhaps its greatest advantage is that not only it makes it possible to provide useful (and frequently finite) semantics to logics for which the use of ordinary matrices fails—in many cases this can be done in a *modular* way. In other words: there were many cases in which I (together with my coauthors) were able to provide simultaneously, and in a compositional way, non-deterministic semantic to every logic in a given family of logics. This mainly happened when all the logics in some family we considered are obtained by adding various axioms to one basic logic. In such a case the process was as follows: using intuitive considerations concerning the basic logic in the family, we have found a corresponding Nmatrix **M** for which it is sound and complete. Then with each axiom used in the family, we have tried to associate a certain way of refining **M**. Usually an Nmatrix which is sound and complete for a logic **L** in the family was obtained by combining the refining methods associated with each of its extra axioms. In case of an axiom for which this did not work, we tried to replace **M** by some rexpansion (Avron and Zohar 2019) of it for which the modular approach works also for the problematic axiom. This method was very successful, but it was applied on a case by case basis. I felt that there should be a general theory behind this, that (among other things) would enable one to know in advance what axioms will be easy to handle, and which will be problematic. Unfortunately, I have never had the opportunity to try to develop such a theory. Therefore I was very pleased to read the contribution of Carlos and Sérgio, and discover that this is precisely what they are doing in it. Using a generalization of Nmatrices called PN-matrices (which has been introduced by Agata Ciabattoni, Lara Spendier, Ori Lahav, and Anna Zamansky), they describe some general, but still useful results, and then a special case which is still general enough to cover all the cases that have been worked out in the literature (so far). This is a very important (and very useful) step in the research on PNmatrics and their applications. I also enjoyed seeing the application given in the paper of the more general result to intuitionistic propositional logic. I do hope that the authors will continue this research in the directions mentioned in the paper (including what they justly call "mandatory topic" in the description of Example 4.3.) I hope also that future papers will clarify the connections between the analytic multiple-conclusion calculi developed in Sect. 14.5 to ordinary Gentzen-style calculi.

14.3 Carnielli and Bueno-Soler

The main subject of this paper is probability logics which are based on LFIs (Logics of Formal Inconsistency) rather than on classical logic. This is a rather interesting and promising application of LFIs. I have never worked on probability logics myself. Nevertheless, this paper is strongly related to subjects on which I did work:

- Paraconsistency is of course an area to which many of my works have been devoted. Among them a great part was devoted to the study of LFIs. The latter is a central family of paraconsistent logics, which have been introduced by Carnielli and Marcos as a generalization of da Costa's systems $\mathbf{C_n}$ $(1 \leq n \leq \omega)$. Providing semantics in a modular way to the main logics in this family was the second, particularly successful application of my non-deterministic semantic framework, and the first one that was employing more than two truth-values. (My framework is actually a special, particularly simple and convenient, case of the possible translation semantics of Carnielli and Marcos.) As mentioned in the paper, the use of logics like **Cie** is much easier than that of da Costa's system $\mathbf{C_1}$, because in contrast to **Cie** (which has a characteristic three-valued Nmatrix), the latter has no finite characteristic Nmatrix. Therefore I agree with the authors' choice to work with **Cie** rather than with $\mathbf{C_1}$. (It should be noted, though, that the addition of the consistency-propagation axioms of $\mathbf{C_1}$ would retain the availability of a characteristic three-valued Nmatrix.)
- My approach to paraconsistent logic is pragmatic. It is not based on the belief that there are contradictions in the world, but on the fact that often there are contradictions in our beliefs about it— and we should be able to cope with them. One main source of contradictions in our beliefs is when we get information from different sources. This state provides the main motivation of using the famous Dunn-Belnap four-valued matrix $\mathcal{FOUR}$. (This matrix, which is mentioned in this paper, has also been a central subject of my research.) It is interesting to

note that like the authors, I too have tried to handle cases for which $\mathcal{FOUR}$ itself cannot be used. Thus Avron and Konikowska (2012) uses a four-valued *Nmatrix* for dealing with cases in which sources provide information on non-atomic formulas (and not only on atomic ones, as in Belnap's model). On the other hand, the present paper provides probability-based tools for handling cases in which sources defer in their reliability. I think it is very important to try to combine the two works into one unified framework.

- Another goal to which I have tried to contribute is combining paraconsistency with fuzziness. (See Sect. 14.5.) This means using all numbers in [0,1] as truth-values. Something similar is done here. The connections, if any, should be an interesting (and potentially rather useful) future research direction.

14.4 Colacito, Galatos, and Metcalfe

The interest for me of the contribution of Alumendo, Nikolaos, and George is twofold.

First of all, it generalizes in an interesting way (and by this put in context) a theorem of mine about one of my favorite systems: the semi-relevant logic **RM**. I was surprised to see that the proof they give of their generalization is easier than both of my direct proofs of the original theorem, which were rather complex. It is true that my theorem is an easy consequence of the cut-elimination theorem for my Gentzen-type system for **RM** (which uses hypersequents). However, as pointed out in this paper, the original proof of the latter theorem was very complicated. It should be pointed out that a significantly simpler (though still far from easy) proof can now be found in my book with O. Arieli and A. Zamansky on paraconsistent logics. The new proof uses a *semantic* method, rather than the very involved syntactic method of my original proof.

This brings me to the second, and more important, aspect of this paper which is particularly interesting for me. For several years now I am usually trusting semantic methods more than proof-theoretical ones, since the latter are much more difficult to verify. And indeed, this paper too uses semantic methods for proving proof-theoretical properties of systems. However, in contrast to what I usually do, it relies on algebraic methods, and the use of general algebraic structures. I have of course used such structures myself in many of my papers (and even introduced new classes of them). However, in most cases this was for me only a step on the way to finding particular cases of the general semantics that are concrete and effective on one hand, but still suffice for completeness on the other. Then what I really used was the more concrete semantics. Here, in contrast, the general algebraic semantics itself is directly applied. The paper demonstrates the power of this approach. I find it amazing how successful it is!

One final remark: I would be very happy to see similar methods applied in the investigation of my logic **RMI** which is employed in Edwin Mares' paper. However, one cannot conservatively add to it the propositional constants 1 and 0 with their usual properties. Hence it seems that a (significant?) change should be made in the techniques used in this paper in order to cope with **RMI**.

14.5 Coniglio, Esteva, Gispert, and Godo

This paper is a continuation of an ongoing research project of its authors (together with other collaborators) on paraconsistent fuzzy logics. The main problem that such a project faces is that the standard basic fuzzy logics are based on preserving *absolute* truth. More precisely: their prime intended semantics is many-valued semantics in which the truth-values are all the real numbers in the interval [0,1], but just one of them (in this case 1, representing absolute truth) is designated. As shown in Avron et al. (2018), such logics cannot be paraconsistent. Accordingly, alternative, less strict, semantics for fuzziness is needed, The most natural approach to this task is to base the consequence relation on preserving some less-than-absolute degree of truth, i.e., instead of taking the set of designated value to be the singleton {1}, we allow it to be some *interval* of the form (t,1] or [t,1], where $0 < t < 1$. This is indeed the approach of this paper, and it has also been my approach in my own works on the same problem, like in Sect. 14.5 of Avron and Zohar (2019).

There are other points of similarity between the content of this paper and the related part of Avron and Zohar (2019). Thus both works concentrate on conservative extensions of Gödel-Dummett logic; in the systems considered in both, this logic is enriched with Łukasiewich negation, and this new connective is taken as the official negation (rather than that of Gödel, which is still available); in both the truth-value 1/2 plays a special role, and a truth-value t leads to paraconsistent logic iff $t \leq 1/2$; and in both a potentially infinite family of logics is shown to actually have a small finite cardinality. On the other hand, there are big differences as well between what is done in these two papers. First, the present paper investigates also extensions of the finite Gödel logics—something that (Avron and Zohar 2019) does not. More important is the fact that the method used in Avron and Zohar (2019) is based on the use of Nmatrices and operations on them that are peculiar to the full class of Nmatrices. In contrast, the approach in the present paper is very algebraic, and uses algebraic techniques and theory for which there are currently no counterparts for Nmatrices. As a result of these different approaches, the families of logics which are studied in the two papers are not the same. Nevertheless, I feel that there should be close connections between them, and between the results of Avron and Zohar (2019) and some of the results of this paper. It might be interesting to find those connections. Another important challenge for the study of Nmatrices is to try to generalize to them the algebraic concepts and theory that are used so efficiently here.

Finally, I would like to note that I like the concept of a saturated paraconsistent logic, which is introduced here as a generalization of the concept of Ofer, Anna, and me of an "ideal" paraconsistent logic. (The examples given in the paper of saturated paraconsistent logics that are not "ideal" are also nice.) I believe that this concept might prove to be rather useful in future investigations of paraconsistency.

14.6 Dunn

Mike and me know each other for many years now. At the beginning of his contribution, he says that he is not sure when we first met. Naturally (as is usually the case in a meeting between a well-known scientist and someone who is at the first stages of his career), I do remember. We first met in 1993, when I was on sabbatical in Stanford, and Mike was kind enough to invite me to be his guest in Bloomington. This was, by the way, also the first (and the last) time in my life in which I met the inside of a long limousine... (To my great surprise, Mike sent one to pick me from the airport.) Of course we were having some correspondence before that, first when I tried to find a place for doing my post-doc, and then when I wrote him about the new linear logic of Girard and its strong relations with relevance logics. After the first meeting we met many times, and each time gave me an opportunity to appreciate more and more Mike's abilities, as well as his personality. Needless to say, in everything connected with relevance logics, Mike has always been for me the first person to consult.

Mike's contribution is mainly devoted to the semi-relevant system **RM**, of which he is one of the parents. So was my own contribution to *his* volume. The two papers are in a way complimentary to each other, since Mike's paper contains interesting information on **RM** which mine lacks (and vice versa). This includes technical issues, like Urquhart's results about the complexity of **RM**, as well as a fascinating account of the history of **RM**. Mike points out that the short story I tell about the birth of **RM** in my paper on it is wrong. (The motivation I described there is something that Mike actually discovered only after **RM** had been invented.) The truth is that I do not feel *too* bad about it. What I wrote was not really meant to be faithful to the actual events that had led to **RM**; rather, it was a rational reconstruction of how **RM** could (or even should) have been born. But it is always very interesting and telling to learn how discoveries are really reached. Anyhow, together our papers contain almost everything that someone interested in **RM** might like to know (as well as my own contributions to the study of this system).

A very important point for me in Mike's paper is his description and discussion of the pragmatist approach he now has to logic. This approach is in complete contrast to that of Anderson and Belnap in Anderson and Belnap (1975) (a book to which Mike has contributed a lot), where the goal has been to find the "one true logic". It would be very interesting for me to know whether this was Mike's approach from the start, and if not—when and why he turned to it. As for me, I have never hidden the fact (noted also in connection with Sara Negri's contribution), that I am a classical logician, who freely applies classical logic in his investigations of non-classical logics. My motivation in studying the latter have always been exactly what Mike describes as his in his paper, and I am very glad to see that we now share this approach.

Before I end, I would like to make two small clarifications.

1. It is true that I had not been the first to use structures similar to hypersequents (as noted in Sect. 6.1 of Mike's paper). Nevertheless, it has been my *independent* creation (including the name "hypersequent"). Moreover: I was the first to turn hypersequents into a general tool for different families of logics.

2. Unlike what is said in footnote 9, "theory" in the sense I use in my papers is not *my* word. It is a rather usual terminology. (See, e.g., the definition of "theory" in Hodges (2001).)

14.7 Fitting

Mel's paper connects two topics to which I have devoted a great part of my research. One is of course the subject of bilattices. The other is the general, rather difficult, foundational question, which has always interested me (and was the subject of some of my own papers): "What is a logic?". What I particularly like about this paper is that this time it uses the first, rather technical, subject in order to shed new light on the second.

Bilattices is a subject I have first heard about in a lecture of Mel about 30 years ago. That lecture made me very interested in the idea, and so I devoted some time to learn it. Again I did it from Mel's papers. Finally, I reached the point of being able to contribute to this subject myself (and even to solve a problem left open by Mel himself about the structure of interlaced bilattices). It is very satisfying to see that in this paper, in turn, Mel is using some results and ideas concerning bilattices due to Ofer Arieli and me (like logical bilattices and bifilters), in order to investigate fine distinctions between logical concepts, and an interesting new class of logics (the strict/tolerant ones). This is how global scientific research should be progressing!

As for the question "What is a logic?", I said above that Mel's paper is connected with it, even though Mel himself explicitly writes at the beginning of his paper that he is not going to directly deal with this question, but only with its companion, "When are logics the same?". However, I do not see this as just a companion question, but as an essential part of the general question. Thus if we take as distinct two logics which are identical in their language and consequence relations, but differ on the metalevel (as suggested, with many examples, involving general constructions, in Mel's paper), then necessarily the metalevel consequence should be included as an essential component in any answer to the question "What is a logic?". I find this idea as very interesting and quite appealing—even though accepting it means that what I have presented in my papers as the definition of a "propositional logic" is wrong, and should be taken only as the definition of the *first level* of a logic.

14.8 Gabbay

Among the contributors to this volume, my acquaintance with Dov is by far the longest: I first met him when I was still a Ph.D. student. Even though he is only seven years older than me, he was then already a Professor of Mathematics, and one of the world's leading logicians. Since that first meeting. Dov has been having great influence on my work. For example, it was from him that I learned for the first time about the notion of a consequence relation (that from that point became central for my own work), and its crucial importance for logic. Dov has also been the one who made me interested in big philosophical questions concerning logic, like: "What is a logic?" or "What is an implication?". Both of us have been devoting a lot of work over the years to answer such questions. Therefore there is little wonder that Dov's contribution to this volume again deals with a problem of this sort: the problem of what is a negation. In more precise words, the question is: when are we justified in seeing as a negation a given unary connective of some logic? Working a lot on paraconsistent logics, this question has always been particularly important for my research, and (exactly like Dov) I gave several answers to it over the years. (One of my papers about this subject was published in Gabbay and Wansing (1999), a volume of which Dov was one of the editors, as well as one of the contributors.) However, Dov's answer(s) in this volume is different from any of those that had been given by me.

What I would like to understand better about Dov's characterization of negation is first of all its relations with the definition of negation which is given in Avron et al. (2018). After Question 1.6 Dov writes: "If we write too many axioms we may get only classical negation, and even that is not guaranteed because maybe we do not know how the negation axioms are supposed to interact with other connectives". These words leave it unclear whether if $\neg$ is a negation in a logic $\mathbf{L}$ then $\mathbf{L}$ should be contained in classical logic (or, more accurately, be coherent with it, according to the definition in our book). If so, negation in Dov's sense is negation in our sense (but not necessarily vice versa). But does this *follow* from Dov's definitions? For example, what about the negation of connexive logics? It is not a negation according to our definition. Is it a negation according to Dov's definition, and/or his intuition?

Several other comments which I have had about Dov's contribution are now quoted in its final version, so I shall not dwell on them here. Let me just note that I find Dov's suggestion in this paper very interesting, and calling for extensive further research. Accordingly, I perfectly understand, why despite its length and rich content, Dov is still classifying his contribution as a "position paper".

14.9 Mares

I enjoyed reading all the papers in this volume, and I am very grateful to all the contributors. Still, I have been particularly delighted to read Ed Mares' contribution to it. The reason: it is the very first work known to me that refers to, and even tries to apply, the main content of my Ph.D. thesis. In that thesis I developed a new approach to relevance logics that was based on certain semantic ideas (which I believed then, and still do, to be rather intuitive and appealing), and on the use of languages that have just purely relevant connectives, without any extensional one. The main part of the thesis was published in three parts in the JSL and NDFOL. Later I tried to promote their content and to explain their ideas in another paper ("Whither Relevance

Logic?"). However, unlike many of my other papers, those that were devoted to the logic **RMI**, which was introduced and investigated in my thesis, where almost totally ignored. At a certain point I gave up the hope of getting the logical community's attention to this logic. As a result, I have never published the chapter on its first-order extension that was included in my thesis. Well, it is difficult to describe what a nice surprise it is to finally see, thirty-six years after my thesis was submitted, a paper that is based on it, continues the work done in it, and applies it to the philosophy and practice of science—which was exactly the area in which I expected it to be applied. Unfortunately, I myself do not have sufficient knowledge in this area to do it myself, and being in a computer science department has necessarily dictated devoting my time to the study of other topics. Anyhow, Ed's applications look very promising, and of course I welcome any additions or changes in my original framework (like those made in this paper of Ed) that he or others might find useful. I myself, in turn, am now strongly motivated by this paper to publish at last my old work on adding quantifiers to **RMI**. As noted by Ed (and according to footnote 8 also by one of the referees of his paper), this step is very important for a really successful application of **RMI**. At present there is nothing about it in the literature, but I intend to change this situation already this year.

14.10 Mundici

I have many things to be grateful to Danielle for. The most important of them is that he is the person who has made me an official member of the fuzzy logic community (in which I have now several other great friends). This happened about twenty years ago, when he invited me to the first (out of two great ones I participated in) workshop on the subject in Garnagno, on the shore of the beautiful Garda lake.

My main contribution to the research on fuzzy logics was the introduction of the framework of hypersequents, which is now the main tool used in the proof theory of fuzzy logics (Metcalfe et al. 2009). In fact, one of the two first logics for which I have developed proof calculi based on hypersequents was Gödel logic (more accurately: Gödel-Dummett logic), which is known as one of the three basic fuzzy logics (the other two being product logic and Łukasiewicz logic). I have returned from time to time to investigate Gödel logic also after this first contribution. In contrast, I personally contributed practically nothing to the research on the oldest and arguably the most important fuzzy logic: Łukasiewicz logic. The reason was not because I thought that this logic is worthless. (On the contrary: it is the most natural logic within which one may be able to develop naive set theory without falling into contradictions.) It was because the methods I am accustomed to use are discrete in nature, and not so suitable for dealing with Łukasiewicz logic. This is one particular reason for me to admire Daniele's successful work on Łukasiewicz logic in general, and his contribution to this volume in particular. This contribution contains the most convincing and interesting (at least for me) application of Łukasiewicz logic I have ever seen. I am also very impressed by his success to provide semantics for this logic that validates a compactness theorem and a *strong* soundness and completeness theorem for the standard axiomatization of this logic. This semantics comes from the continuous world of analysis (including the use of derivatives)—and this is a completely new idea for me. The only small reservation I have about this paper is that I would have liked it to contain some abstract general definition of an "implication connective", according to which the connective $\rightarrow$ of Łukasiewicz logic is an implication. (Such a definition would necessarily be different from my own notion of semi-implication, which is mentioned in Danielle's paper.)

14.11 Negri

Sara is of course one of the world's greatest experts on structural proof theory. Accordingly, there have always been strong connections between her research and mine, since both of us have devoted a great part of our work to the proof theory of non-classical logics. However, this contribution of Sara is particularly interesting for me, because of its relevance to two other mathematical areas which are close to my heart. One is Euclidean geometry. The other is the development of frameworks for mathematical reasoning and for the mechanization of mathematics that go beyond first-order logic, but are not based on all the ontological commitments of full second-order logic. As mentioned in Sara's paper, the logic I find particularly suitable for this task is ancestral logic, either in classical version or in an intuitionistic one. (I myself am a classical logician, and I freely use classical logic in my study of non-classical logics. Still, here too I see the benefits of using also a non-classical logic. In this case it is intuitionistic logic which is the most obvious choice.) Another feature of my research in this area is that I have restricted myself to absolutely finite systems, and among them especially to those that are predicatively justified. Sara too is investigating both a classical version and an intuitionistic version of the systems she developed in her contribution to this volume. However, in contrast to me, she is investigating here the use of *infinitary* proof systems. Nevertheless, her systems are restricted in a way that may make them predicatively acceptable. Indeed, proof systems similarly restricted have been successfully used by Schutte's school of proof theory for investigating predicative finite systems (like **PA**, as well as much stronger ones). I see great potential in Sara's results to serve as a promising starting point to do the same with theories that are based on what are called in her paper coherent or geometric axioms. I hope that in future continuation of this research an attention will be given to the realization of this potential. One direction in which I have particular interest is what Sara justly calls my favorite logic: Ancestral logic. Another one that I am fond of is formalizations of Euclidean geometry that include Archimedes Axiom.

14.12 Odintsov, Skurt, and Wansing

The paper of Sergey, Daniel, and Heinrich is a part of a research program of Heinrich that is devoted to the family of connexive logics. This family is not dealt with in the book of Ofer, Anna, and me on paraconsistent logics. Still, the connexive logics investigated in this paper are implicitly shown in it to belong to what we have called in our book *effective* paraconsistent logics. They are proved in it to be decidable; analytic proof systems are provided for them; and they are endowed with useful types of semantics (both Kripke-style semantics and algebraic semantics). Moreover, these logics are based on logical bilattices. These are structures that have been introduced by Ofer and me, and *are* studied in our book. Accordingly, the question whether a second edition of our book should contain a chapter on connexive logics depends on the question to what extent the unary connective $\sim$ of these logics may be taken as a *negation*. Unlike in the past, I tend now to think that it does.

The question "What is negation?" has been central for both Heinrich and me for a long time. (This is even a name of a book with a contribution of mine that Heinrich has edited.) In the past I thought that a connective $\neg$ of a logic $\mathbf{L}$ cannot be taken as negation in case there is a formula φ such that both φ and $\neg\varphi$ are logically valid in $\mathbf{L}$. However, as is pointed out in this paper, this is what happens even in some of the logics which I have introduced and study (together with Ofer Arieli), in case connectives of non-classical nature are present. Accordingly, in our book we define $\neg$ as a negation of $\mathbf{L}$ if $\mathbf{L}$ has a connective $\diamond$, which is either a disjunction for $\mathbf{L}$, or a conjunction for it, or a semi-implication for it, so that the $\{\neg, \diamond\}$-fragment of $\mathbf{L}$ is contained in its classical counterpart. This means that whether $\neg$ is a negation in $\mathbf{L}$ depends on its relations with other connectives of the language of $\mathbf{L}$. I believe that the authors of this paper would agree at least with the last thesis. Thus the four principles that a connexive logic should respect that are given at the beginning of the paper are valid for *classical negation* in case $\supset$ is interpreted as the classical *biconditional*. Hence their acceptance or rejection depends on the question "What is implication?" even more than it depends on the question "What is negation?". These fundamental questions are not directly discussed in this paper. Nevertheless, its content is a significant contribution to the study of them.

References

Anderson, A. R., & Belnap, N. D. (1975). *Entailment: The Logic of Relevance and Necessity* (Vol. I). Princeton: Princeton University Press.

Avron, A., Arieli, O., & Zamansky, A. (2018). *Theory of Effective Propositional Paraconsistent Logics* (Vol. 75), Studies in Logic. College Publications.

Avron, A., & Konikowska, B. (2012). Finite-valued logics for information processing. *Fundamenta Informaticae*, *114*, 1–30.

Avron, A., & Zohar, Y. (2019). Rexpansions of nondeterministic matrices and their applications in nonclassical logics. *The Review of Symbolic Logic*, *12*(1), 173–200.

Gabbay, D., & Wansing, H. (Eds.). (1999). *What is Negation?* (Vol. 13), Applied logic series. Kluwer Academic Publishers.
Hodges, W. (2001). Elementary predicate logic. In D. M. Gabbay & F. Guenthner (Eds.), *Handbook of Philosophical Logic* (2nd ed., Vol. 1, pp. 1–129). Kluwer Academic Publishers.
Metcalfe, G., Olivetti, N., & Gabbay, D. (2009). *Proof Theory for Fuzzy Logics*. Springer.

List of Publication

Arnon Avron, Ph.D.

Books

1. A. Avron, **Gödel's Theorems and The Problem of the Foundations of Mathematics** (written in Hebrew), Broadcasted University Series, Ministry of Defense Publications, Israel, 1998
2. A. Avron, **Introduction to Discrete Mathematics** (written in Hebrew), Tel-Aviv University, Israel, 2001
3. A. Avron, O. Arieli, and A. Zamansky, **Theory of Effective Propositional Paraconsitent Logics**, Studies in Logic (Mathematical Logic and Foundations) 75, College Publications, 2018.

Articles

1. A. Avron, *Relevant Entailment - Semantics and formal systems*, Journal of Symbolic Logic 49, 334–342 (1984).
2. A. Avron, *On Modal Systems having arithmetical interpretations*, Journal of Symbolic Logic 49, 935–942 (1984).
3. A. Avron, *On Purely Relevant Logics*, Notre Dame Journal of Formal Logic 27, 180–194 (1986).
4. A. Avron, *On an Implication Connective of RM*, Notre Dame Journal of Formal Logic 27, 201–209 (1986).
5. A. Avron, *Theorems on Strong Constructibility with a Compass alone*, Journal of Geometry 30, 28–35 (1987).
6. A. Avron, *A Constructive Analysis of RM*, Journal of symbolic Logic 52, 939–951 (1987).
7. A. Avron, *The Semantics and Proof Theory of Linear Logic*, Theoretical Computer Science 57, 161–184 (1988).
8. A. Avron, *Construction Problems - Why and How?*, International Journal of Mathematical Education in Science and Technology 20, 273–287 (1989).

O. Arieli and A. Zamansky (eds.), *Arnon Avron on Semantics and Proof Theory of Non-Classical Logics*, Outstanding Contributions to Logic 21,
https://doi.org/10.1007/978-3-030-71258-7

9. A. Avron, F. Honsell, I.A. Mason , *An Overview of the Logical Framework*, in **Current Trends in Hardware Verification and Automated Theorem Proving** (G. Birtwistle, S.A. Subrahamyam, eds.), 323–340, Springer-Verlag, 1989.
10. A. Avron, *Gentzenizing Schroeder-Heister's Natural Extension of Natural Deduction*, Notre Dame Journal of Formal Logic 31, 127–135 (1990).
11. A. Avron, *On Strict Strong Constructibility with a Compass Alone*, Journal of Geometry 38, 12–15 (1990).
12. A. Avron, *Relevance and Paraconsistency - A New Approach*, Journal of Symbolic Logic 55, 707–732 (1990).
13. A. Avron, *Relevance and Paraconsistency - A New Approach. Part II: the Formal systems* , Notre Dame Journal of Formal Logic 31, 169–202 (1990).
14. A. Avron, *Relevance and Paraconsistency - A New Approach. Part III: Cut-Free Gentzen-type Systems*, Notre-Dame Journal of Formal Logic 32, 147–160 (1991).
15. A. Avron, *Simple Consequence Relations*, Information and Computation 92, 105–139 (1991).
16. A. Avron, *Natural 3-valued Logics—Characterization and Proof Theory*, Journal of Symbolic Logic 56, 276–294 (1991).
17. A. Avron, *Hypersequents, Logical Consequence and Intermediate Logics for Concurrency*, Annals of Mathematics and Artificial Intelligence 4, 225–248 (1991).
18. A. Avron, *A Note on Provability, Truth and Existence*, The Journal of Philosophical Logic 20, 403–409 (1991).
19. A. Avron, J. Hirshfeld, *On First Order Database Query Languages*, in **Proceedings of the 6th annual IEEE symposium on Logic in Computer Science (LICS91)**, IEEE Press, 226–231 (1991).
20. A. Avron , *Axiomatic Systems, Deduction and Implication*, Journal of Logic and Computation 2, 51–98 (1992).
21. A. Avron, *Whither Relevance logic?*, Journal of Philosophical Logic 21, 243–281 (1992).
22. A. Avron, F.A. Honsell, I.A. Mason, R. Pollack, *Using Typed Lambda Calculus to Implement Formal Systems on a Machine* , The Journal of Automated Reasoning 9, 309–354 (1992).
23. A. Avron, *Gentzen-Type Systems, Resolution and Tableaux*, Journal of Automated Reasoning 10, 265–281 (1993).
24. A. Avron, *An Exercise in An Interactive Geometrical research*, Annals of Mathematics and Artificial Intelligence 9, 239–252 (1993).
25. A. Avron , *What is a Logical System?*, in **What is a Logical System** (D. Gabbay, ed.), Studies in Logic and Computation, 4, 217–238, Oxford University Press (1994).
26. A. Avron, J. Hirshfeld, *Queries Evaluation, Relative Safety and Domain Independence in First Order Database Languages with Functions*, Methods of Logic in Computer Science, 1, 261–277 (1994).
27. A. Avron , *Some Properties of Linear Logic Proved by Semantic Methods*, Journal of Logic and Computation 4, 929–938 (1994).

28. A. Avron, N. Sasson, *Stability, Sequentiality and Demand Driven Evaluation in Dataflow*, Formal Aspects of Computing 6, 620–642 (1994).
29. O. Arieli , A. Avron, *Logical Bilattices and Inconsistent Data*, in **Proceedings of the 9th IEEE annual symposium on Logic in Computer Science (LICS94)**, IEEE Press, 468–476 (1994).
30. A. Avron, *A Note on the Structure of Bilattices*, Journal of Mathematical Structures in Computer Science 5, 431–438 (1995).
31. A. Avron, *The Method of Hypersequents in Proof Theory of Propositional Non-Classical Logics*, In: **Logic: Foundations to Applications** (W. Hodges, M. Hyland, C. Steinhorn and J. Truss eds.) 1–32, Oxford Science Publications (1996).
32. O. Arieli, A. Avron, *Reasoning with Logical Bilattices*, Journal of Logic, Language and Information 5, 25–63 (1996).
33. A. Avron, *The Structure of Interlaced Bilattices*, Journal of Mathematical Structures in Computer Science 6, 287–299 (1996).
34. O. Arieli, A. Avron , *Four-Valued Diagnoses for Stratified Knowledge-Bases*, in **Proceedings of the 1996 Annual Conference of the European Association for Computer Science Logic (CSL96) - Selected Papers** (D.Van-dalen, M.Benzem, eds.), LNCS 1258, 1–17, Springer Verlag (1997).
35. A. Avron, *Multiplicative Conjunction as an Extensional Conjunction*, Logic Journal of the IGPL 5, 181–208 (1997).
36. A. Avron, *Formulas for which Contraction is Admissible*, Logic Journal of the IGPL 6, 43–48 (1998).
37. A. Avron, F. Honsell, M. Miculan, C. Paravano, *Encoding Modal Logics in Logical Frameworks*, Studia Logica 60, 161–208 (1998).
38. O. Arieli, A. Avron, *The Logical Role of the Four-Valued Bilattice*, in **Proceedings of the 13th IEEE annual symposium on Logic in Computer Science (LICS98)**, IEEE Press, 118–126 (1998).
39. O. Arieli, A. Avron, *The Value of the Four Values*, Artificial Intelligence 102, 97–141 (1998).
40. A. Avron, *Multiplicative Conjunction and an Algebraic Meaning of Contraction and Weakening*, Journal of Symbolic Logic 63, 831–859 (1998).
41. A. Avron, *Two Types of Multiple-conclusion Systems*, Logic Journal of the IGPL 6, 695–717 (1998).
42. A. Avron, *On the Proof Theory of Natural Many-valued Logics* , in Collegium Logicum (Annals of the Kurt Gödel Society) 3, 51–59 (1999).
43. O. Arieli, A. Avron, *A Model-theoretic Approach for Recovering Consistent Data from Inconsistent Knowledge Bases*, Journal of Automated Reasoning 22, 263–309 (1999).
44. A. Avron, *On the Expressive Power of Three-valued and Four-valued Languages*, Journal of Logic and Computation 9, 977–994, (1999).
45. A. Avron, *Negation: Two points of view*, in **What is Negation?** (D. Gabbay and H. Wansing, eds.), Applied Logic Series 13, 3–22, Kluwer Academic Publishers (1999).

46. O Arieli, A Avron, *Nonmonotonic and Paraconsistent Reasoning: From Basic Entailments to Plausible Relations*, in **Proceedings of the 5th European Conf. on Symbolic and Quantitative Approaches to Reasoning with Uncertainty (ECSQARU'99)**, LNCS 1638, 11–22, Springer (1999).
47. O. Arieli, A. Avron , *Bilattices and Paraconsistency*, in **Frontiers in Paraconsistent Logic** (D. Batens, C. Mortensen, G. Priest, and J. Van-Bengedem, eds.), Studies in Logic and Computation 8, 11–28, Research Studies Press (2000).
48. O. Arieli, A. Avron, *General Patterns for Nonmonotonic Reasoning: From Basic Entailments to Plausible Relations*, Logic Journal of the IGPL 8, 119–148 (2000).
49. A. Avron, *Implicational F-Structures and Implicational Relevance Logics*, Journal of Symbolic Logic 65, 788–802 (2000).
50. A. Avron , *A Tableaux System for Gödel-Dummett Logic Based on a Hypersequential Calculus*, in **Proceedings of the 9th International Conference on Analytic Tableaux and Related Methods (TABLEAUX'2000)** (R. Dyckhoff, Ed.), LNAI 1847, 98–111, Springer Verlag (2000).
51. A. Avron, and B. Konikowska, *Decomposition Proof Systems for Gödel Logics*, Studia Logica 69, 197–219 (2001).
52. A. Avron, I. Lev, *Canonical Propositional Gentzen-Type Systems*, in **Proceedings of the 1st International Joint Conference on Automated Reasoning (IJCAR 2001)** (R. Goré, A Leitsch, T. Nipkow, eds.), LNAI 2083, 529–544, Springer Verlag (2001).
53. A. Avron, I. Lev, *A Formula-Preferential Base for Paraconsistent and Plausible Reasoning Systems*, in **Proceedings of the Workshop on Inconsistency in Data and Knowledge (KRR-4)**, 60–70, International Joint Conference on Artificial Intelligence (IJCAI) (2001).
54. A. Avron , *On Negation, Completeness and Consistency*, in **Handbook of Philosophical Logic** (D. Gabbay and F. Guenthner, eds.) Vol. 9, 287–319, Kluwer Academic Publishers (2002).
55. A. Avron , *Classical Gentzen-type Methods in Propositional Many-Valued Logics*, in **Beyond Two: Theory and Applications of Multiple-Valued Logic** (M. Fitting and E. Orlowska, eds.), Studies in Fuzziness and Soft Computing 114, 117–155, Physica Verlag (2003).
56. A. Avron, *Transitive Closure and the Mechanization of Mathematics*, in **Thirty Five Years of Automating Mathematics** (F. Kamareddine, ed.), 149–171, Kluwer Academic Publishers (2003).
57. A. Avron, *Tableaux with Four Signs as a Unified Framework*, in **Proceedings of TABLEAUX 2003** (M.C. Mayer and F. Pirri, eds.), 4–16, LNAI 2796, Springer (2003).
58. A. Avron, *Formalizing Set Theory as It Is Actually Used*, in **Proceedings of Mathematical Knowledge Management (MKM 2004)** (A. Asperti, G. Banecerek, and A. Trybulec, eds.), 32–43, LNCS 3119, Springer (2004).
59. A. Avron, *Safety Signatures for First-order Languages and Their Applications*, in **First-Order Logic Revisited** (Hendricks et all,, eds.), 37–58, Logos Verlag Berlin (2004).

60. A. Avron, *Combining Classical Logic, Paraconsistency and Relevance*, Journal of Applied Logic 3, 133–160 (2005).
61. A. Avron, I. Lev, *Non-Deterministic Multiple-valued Structures*, Journal of Logic and Computation 15, 241–261 (2005).
62. A. Avron, *Non-deterministic Matrices and Modular Semantics of Rules*, in **Logica Universalis** (J.-Y. Beziau, ed.), 149–167, Birkhäuser Verlag (2005).
63. A. Avron, *Non-deterministic Semantics for Paraconsistent* **C**-*systems*, **Proceedings of 8th ECSQARU (Euro. Conf. Symb. and Quant. Approaches to Reasoning on Uncertainty)** (L. Godo, ed.), 625–637, LNAI 3571, Springer (2005).
64. A. Avron, *A Nondeterministic View on Nonclassical Negations*, Studia Logica 80, 159–194 (2005).
65. A. Avron, and B. Konikowska, *Multi-valued Calculi for Logics Based on Non-determinism*, Logic Journal of the IGPL 13, 365–387 (2005).
66. A. Avron, *Logical Non-determinism as a Tool for Logical Modularity: An Introduction*, in **We Will Show Them: Essays in Honor of Dov Gabbay**, Vol. 1 (S. Artemov, H. Barringer, A. S. d'Avila Garcez, L. C. Lamb, and J. Woods, eds.), 105–124, College Publications (2005).
67. A. Zamansky, A. Avron, *Cut-Elimination and Quantification in Canonical Systems*, Studia Logica 82, 157–176 (2006).
68. A. Avron, *From Constructibility and Absoluteness to Computability and Domain Independence*, in **Logical Approaches to Computational Barriers** (proceedings of the second conference on Computability in Europe CiE 2006, edited by A. Beckmann, U. Berger, B. Lowe and J. V. Tucker), 11–20, LNCS 3988, Springer (2006).
69. A. Zamansky and A. Avron, *Non-deterministic Semantics for First-order Paraconsistent Logics*, in **Proceedings of the 10th International Conference on Knowledge Representation (KR'06)**, 431–440 (2006).
70. A. Zamansky and A. Avron, *Canonical Gentzen-type calculi with (n,k)-ary quantifiers*, in **Proceedings of the Third International Joint Conference in Automated Reasoning (IJCAR06)** (Furbach and Shankar eds.), 251–265, LNAI 4130, Springer (2006).
71. A. Avron and A. Zamansky, *Canonical calculi with (n,k)-ary quantifiers*, in **Proceedings of Logical Methods in Computer Science (LFCS 2007)** (Artemov S. and Nerode A. eds.), 26–40, LNCS 4514, Springer (2007).
72. A. Avron, *Non-deterministic Semantics for Logics with a Consistency Operator*, International Journal of Approximate Reasoning 45, 271–287 (2007).
73. A. Avron, J. Ben-Naim, and B. Konikowska, *Cut-free Ordinary Sequent Calculi for Logics Having Finite-Valued Semantics*, Logica Universalis 1, 41–69 (2007).
74. A. Avron, *Non-deterministic Semantics for Families of Paraconsistent Logics*, in **Handbook of Paraconsistency** (J.-Y. Beziau, W. Carnielli, and D. M. Gabbay, eds.), 285–320, Studies in Logic 9, College Publications (2007).
75. A. Avron and A. Zamansky, *Many-valued Non-deterministic Semantics for First-order Logics of Formal (In)consistency*, in **Algebraic and Proof-theoretic**

Aspects of Non-classical Logics, (S. Aguzzoli, A. Ciabattoni, B. Gerla, C. Manara, and V. Marra, eds.), 1–24, LNAI 4460, Springer (2007).

76. A. Zamansky and A. Avron , *Effective Non-deterministic Semantics for First-order LFIs*, Multiple-Valued Logic and Soft Computing 13, 335–372 (2007).
77. A. Avron, *A Framework for Formalizing Set Theories Based on the Use of Static Set Terms*, in **Pillars of Computer Science**, (A. Avron, N. Dershowitz, and A. Rabinovich, eds.), 87–106, LNCS 4800, Springer (2008).
78. A. Avron, *Constructibility and Decidability versus Domain Independence and Absoluteness* , Theoretical Computer Science 394, 144–158 (2008).
79. A. Avron and A. Zamansky, *Canonical Systems with (n,k)-ary Quantifiers*, Logical Methods in Computer Science 4, 1–23 (2008).
80. A. Avron, and B. Konikowska, *Rough Sets and 3-valued Logics*, Studia Logica 90, 69–92 (2008).
81. A. Avron and A. Zamansky, *A Triple Correspondence in Canonical Calculi: Strong Cut-Elimination, Coherence, and Non-deterministic Semantics*, in **Computer Science — Theory and Applications: Proceedings of the Third International Computer Science Symposium in Russia (CSR 2008)**, (E. A. Hirsch, A. A. Razborov, A. L. Semenov, A. Slissenko, eds.), LNCS 5010, 52–64, Springer (2008).
82. A. Avron, *5-valued Non-deterministic Semantics for the Basic Paraconsistent Logic* **mCi**, Studies in Logic, Grammar and Rhetoric 14, 127–136 (2008).
83. A. Avron, J. Ben-Naim, and B. Konikowska, *Processing Information from a Set of Sources*, in **Towards Mathematical Philosophy**, (D. Makinson, J. Malinowski, and H. Wansing, eds.), Trends in Logic 28, 165–186, Springer (2009).
84. A. Avron, and B. Konikowska, *Proof Systems for Reasoning about Computation Errors*, Studia Logica 91, 273–293 (2009).
85. A. Avron, *Multi-valued Semantics: Why and How*, Studia Logica 92, 163–182 (2009).
86. A. Avron, A. Ciabattoni, and A. Zamansky,, *Canonical Calculi: Invertibility, Axiom expansion and (Non)-determinism*, in **Computer Science — Theory and Applications: Proceedings of 4th International Computer Science Symposium in Russia (CSR 2009)**, (A. Frid, A. Morozov, A. Rybalchenko, K. W. Wagner, eds.), LNCS 5675, 26–37, Springer (2009).
87. A. Avron and A. Zamansky, *Canonical Signed Calculi, Non-deterministic Matrices and Cut-elimination*, in **Proc. of the International Symposium on Logical Foundations of Computer Science (LFCS09)**, (Sergei Artemov and Anil Nerode eds.), LNCS 5407, 31–46, Springer (2009).
88. A. Avron, *Modular Semantics for Some Basic Logics of Formal Inconsistency*, in **The Many Sides of Logic** (W. A. Carnielli, M. E. Coniglio, I. M. L. D' Ottaviano, eds.), Studies in Logic 21, 15–26, College Publications (2009).
89. A. Avron and O. Lahav, *Canonical Constructive Systems* , In **Proc. of the 18th international Conference on Automated Reasoning with Analytic Tableaux and Related Methods (TABLEAUX 2009)**, (M. Giese, A. Waaler, eds.), LNAI 5607, 62–76, Springer (2009).

90. A. Avron, O. Arieli, and A. Zamansky, *On Strong Maximality of Paraconsistent Finite-Valued Logics*, in **Proceedings of the 25th IEEE Symposium on Logic in Computer Science (LICS 2010)**, 304–313, IEEE Press (2010).
91. A. Avron, *A New Approach to Predicative Set Theory*, in **Ways of Proof Theory** (R. Schindler, ed.), 31–63, onto series in mathematical logic, onto verlag (2010).
92. A. Avron, *Tonk - A Full Mathematical Solution*, in **Hues of Philosophy: Essays in Memory of Ruth Manor** (A. Biletzki, ed.), 17–42, College Publications (2010).
93. A. Avron and O. Lahav, *Strict Canonical Constructive Systems*, in **Fields of Logic and Computation: Essays Dedicated to Yuri Gurevich on the Occasion of His 70th Birthday**, (A. Blass, N. Dershowitz, W. Reisig, eds.), 75–94, LNCS 6300, Springer (2010).
94. A. Avron and O. Lahav, *On Constructive Connectives and Systems*, Logic Methods in Computer Science 5, 1–20 (2010).
95. A. Avron, *A Simple Proof of Completeness and Cut-elimination for Propositional Gödel Logic*, Journal of Logic and Computation 21, 813–821 (2011).
96. O. Arieli, A. Avron, and A. Zamansky, *Maximal and Premaximal Paraconsistency in the Framework of Three-Valued Semantics*, Studia Logica 97, 31–60 (2011).
97. A. Avron and A. Zamansky, , *Non-deterministic Semantics for Logical Systems*, in **Handbook of Philosophical Logic** (D. M. Gabbay, and F. Guenthner, eds.), Vol. 16, 227–304, Springer (2011).
98. A. Avron, O. Lahav, *Kripke Semantics for Basic Sequent Systems*, in **Proceedings of the 20th international Conference on Automated Reasoning with Analytic Tableaux and Related Methods (TABLEAUX 2011)**, (K. Brünler, G. Metcalfe, eds.), LNCS 6973, 43–57, Springer (2011).
99. O. Arieli, A. Avron, and A. Zamansky, *Ideal Paraconsistent Logics*, Studia Logica, 99, 31–60 (2011).
100. A. Avron, O. Lahav, *A Multiple-Conclusion Calculus for First-Order Gödel Logic*, in **Computer Science — Theory and Applications: Proceedings of the Third International Computer Science Symposium in Russia (CSR 2011)**, (A. Kulikov, N. Vereshchagin, eds.), LNCS 6651, 456–479 (2011).
101. O. Arieli, A. Avron and A. Zamansky, *What is an ideal logic for reasoning with inconsistency?*, in **Proceedings of the 22nd International Joint Conference on Artificial Intelligence (IJCAI 2011)**, AAAI Press, 706–711 (2011).
102. A. Avron, B. Konikowska, and A. Zamansky, *Analytic Calculi for Logics of Formal Inconsistency*, In: **Logic without Frontiers**, (J. Y. Beziau and M. E. Coniglio, eds.), 265–275, College Publications (2011).
103. A. Zamansky and A. Avron, *Canonical Signed Calculi with Multi-ary Quantifiers*, Annals of Pure and Applied Logic 163, 951–960 (2012).
104. A. Avron, and B. Konikowska, *Finite-valued Logics for Information Processing*, Fundamenta Informaticae 114, 1–30 (2012).
105. A. Avron, B. Konikowska, and A. Zamansky, *Modular Construction of Cut-free Sequent Calculi for Paraconsistent Logics*, in **Proceedings of the 27th Annual IEEE Symposium on Logic in Computer Science (LICS'12)**, 85–94 (2012).

106. O. Lahav and A. Avron, *A Semantic Proof of Strong Cut-admissibility for First-Order Gödel Logic*, Journal of Logic and Computation 23, 59–86 (2013).
107. A. Avron, B. Konikowska, and A. Zamansky, *Cut-free Sequent Calculi for C-systems With Generalized Finite-valued Semantics*, Journal of Logic and Computation 23, 517–540 (2013).
108. O. Lahav, and A. Avron, *A Unified Semantic Framework for Fully-structural Propositional Sequent Systems*, Transactions on Computational Logic 14 (4), article 27, 2013.
109. A. Avron, *What is Relevance Logic?*, Annals of Pure and Applied Logic 165, 26–48 (2014).
110. A. Avron, *The Classical Constraint on Relevance*, Logica Universalis 8, 1–15 (2014).
111. A. Avron, *Paraconsistency, Paracompleteness, Gentzen Systems, and Trivalent Semantics*, Journal of Applied Non-Classical Logic 24, 12–34 (2014).
112. A. Avron, B. Konikowska, and A. Zamansky, *Efficient Reasoning with Inconsistent Information Using C-systems*, Information Sciences 296, 219–236 (2015).
113. A. Avron, *Semi-implication: A Chapter in Universal Logic*, In: **The Road to Universal Logic**, I, (A. Koslow and A. Buchsbaum, eds.), 59–72, Studies in Universal Logic, Birkhäuser (2015).
114. O. Lahav, and A. Avron, *A Cut-Free Calculus for Second-Order Gödel Logic*, Fuzzy Sets and Systems 276, 1–30 (2015).
115. O. Arieli and A. Avron, *Three-Valued Paraconsistent Propositional Logics*, In: **New Directions in Paraconsistent Logic** (J.-Y.Beziau, M.Chakraborty and S.Dutta, eds.), 91–129, Springer India (2015).
116. A. Avron and L. Cohen, *Formalizing Scientifically Applicable Mathematics in a Definitional Framework*, Journal of Formalized Reasoning 9, 53–70 (2016).
117. A. Avron, *Paraconsistent fuzzy logic preserving non-falsity*, Fuzzy Sets and Systems 292, 75–84 (2016).
118. A. Avron, *RM and Its Nice Properties*, In: **J. Michael Dunn on Information Based Logics** (K. Bimbo, ed.), 15–43, Outstanding Contributions to Logic 8, Springer (2016).
119. A. Avron, *A Logical Framework for Developing and Mechanizing Set Theories*, in **Proceedings of the International Joint Conference on Automated Reasoning (IJCAR 2016)**, (N. Olivetti, A. Tiwari eds.), LNCS 9706, 3–8, Springer, Cham (2016).
120. A. Avron, and A. Zamansky, *A Paraconsistent View on* **B** *and* **S5**, in **Advances in Modal Logic 11** (L. Beklemishev, S. Demri and A. Máté, eds.), College Publications, 21–37 (2016).
121. A. Avron and N. Dershowitz, *Cayley's Formula: A Page from* The Book, American Mathematical Monthly 123 (7), 699–700 (2016).
122. A. Avron, *Cut-elimination in* **RM** *Proved Semantically*, IFCoLog Journal of Logics and their Applications 4, 605–621 (2017).
123. A. Avron and J.-Y. Beziau, *Self-extensional Three-valued Paraconsistent Logics Have No Implication*, Logic Journal of the IGPL 25, 183–194 (2017).

124. A. Avron, *Self-extensional Three-Valued Paraconsistent Logics*, Logica Universalis 11, 297–315 (2017).
125. O. Arieli and A. Avron, *Four-Valued Paradefinite Logics*, Studia Logica 105 (6), 1087–1122 (2017).
126. A. Avron and A. Zamansky, *Paraconsistency, Self-extensionality, Modality*, Logic Journal of the IGPL, https://doi.org/10.1093/jigpal/jzy064 (2018).
127. A. Avron and L. Cohen, *A Minimal Computational Theory of a Minimal Computational Universe*, in **Proc. of the International Symposium on Logical Foundations of Computer Science (LFCS 2018)** (S. Artemov and A. Nerode eds.), LNCS 10703, 37–54, Springer (2018).
128. A. Avron and L. Cohen, *Applicable Mathematics in a Minimal Computational Theory of Sets*, Logical Method in Computer Science 14 (4), DOI: 10.23638/LMCS-14(4:1)2018 (2018).
129. A. Avron, *Quasi-canonical Systems and Their Semantics*, Synthese, https://doi.org/10.1007/s11229-018-02045-0 (2018).
130. A. Avron, S. Lev, and N, Levy, *Safety, Absoluteness, and Computability*, in **Proceedings of 27th EACSL Annual Conference on Computer Science Logic (CSL 2018)** (D. Ghica, A. Jung, eds.), Leibniz International Proceedings in Informatics (LIPIcs) 119, 8:1–8:17, Schloss Dagstuhl–Leibniz-Zentrum fuer Informatik, Dagstuhl, Germany (2018).
131. A. Avron and O. Lahav, *A Simple Cut-Free Hypersequential System for* **S5**, Advances in Modal Logic 12, 29–42, College Publications (2018).
132. A. Avron, *Paraconsistency and The Need for Infinite Semantics*, Soft Computing 23, 2167–2175 (2019).
133. L. Cohen , and A. Avron, *The Middle Ground – Ancestral Logic*, Synthese 196, 2671–2693 (2019).
134. A. Avron and Y. Zohar, *Rexpansions of Non-deterministic Matrices and Their Applications in Non-classical Logics*, Review of Symbolic Logic 12, 173–200 (2019).
135. B. Konikowska and A. Avron, *Reasoning about Covering-based Rough Sets Using Three Truth Values*, Journal of Applied Logics - IfColog Journal of Logics and their Applications 6 (2), 359–378 (2019).
136. Y. Dvir and A. Avron, *First-Order Quasi-canonical Proof Systems*, in **Automated Reasoning with Analytic Tableaux and Related Methods (TABLEAUX 2019)** (S. Cerrito and A. Popescu, eds.) LNCS 11714, Springer, 77–93 (2019).
137. A. Avron, *Why Predicative Sets*, In **Fields of Logic and Computation III**, LNCS 12180, Springer, 30–45 (2020).
138. A. Avron, *The Normal and Self-extensional Extension of Dunn-Belnap Logic*, Logica Universalis (2020), DOI: 10.1007/s11787-020-00254-1.
139. A. Avron, *Weyl Reexamined: "Das Kontinuum" 100 Years Later*, The Bulletin of Symbolic Logic (2020), 1–57, doi:10.1017/bsl.2020.23.
140. A. Avron, *The Problematic Nature of Gödel's Disjunctions and Lucas-Penrose's Theses*, to appear in a special issue of Semiotic Studies.

The manufacturer's authorised representative in the EU is Springer Nature Customer Service Centre GmbH, Europaplatz 3, 69115 Heidelberg, Germany. If you have any concerns regarding our products, please contact ProductSafety@springernature.com

Printed and bound by CPI Group (UK) Ltd, Croydon, CR0 4YY
28/09/2026
02224090-0020